AF598338

# ADAPTIVE STRUCTURES

# ADAPTIVE STRUCTURES

## Dynamics and Control


**ROBERT L. CLARK, Ph.D.**
Duke University
Durham, NC

**WILLIAM R. SAUNDERS, Ph.D.**
Virginia Polytechnic Institute and State University
Blacksburg, VA

**GARY P. GIBBS, Ph.D.**
NASA–Langley Research Center
Hampton, VA

A Wiley-Interscience Publication

**JOHN WILEY & SONS, INC.**

New York • Chichester • Weinheim • Brisbane • Singapore • Toronto

Designations used by companies to distinguish their products are often claimed as trademarks. In all instances where John Wiley & Sons, Inc., is aware of a claim, the product names appear in initial capital or all capital letters. Readers, however, should contact the appropriate companies for more complete information regarding trademarks and registration.

This text is printed on acid-free paper. ♾



This publication is designed to provide accurate and authoritative information in regard to the subject matter covered. It is sold with the understanding that the publisher is not engaged in rendering legal, accounting, or other professional services. If legal advice or other expert assistance is required, the services of a competent professional person should be sought.

***Library of Congress Cataloging-in-Publication Data***

Clark, Robert L., 1964–
Adaptive structures : dynamics and control / Robert L. Clark, William Saunders, Gary P. Gibbs.
p. cm.
Includes index.
ISBN 0-471-12262-9 (cloth : alk. paper)
1. Structural control (Engineering). 2. Smart structures.
I. Saunders, William, Dr. II. Gibbs, Gary P. III. Title.
TA654.9.C54 1997
624.1′7–dc21 97-20308

Printed in the United States of America

10 9 8 7 6 5 4 3 2

# CONTENTS

# PREFACE

Adaptation is typically associated with a modification, change, or adjustment that might occur over some period of time. In biology, adaptation is considered to be *a characteristic of an organism that makes it better able to live in its environment.*[*] The concept of an adaptive structure is thus biologically inspired and is motivated by the need to modify a structure's response to various stimuli (disturbances) and environmental conditions (design uncertainties) to meet some predetermined response characteristics. Control system design forms a critical path in the development of an adaptive structure and essentially classifies the level of adaptation possible. For example, active control systems, in feedforward or feedback configurations, are often used to minimize the response of structures subjected to external disturbances. In the event that a fixed-gain control law is implemented, the resulting system is defined to be a *constrained* adaptive system. The structure has been given the ability to modify itself within its environment, but it must do so within a constrained set of control inputs. When the parameters used to implement the structural control system are adjusted as a function of time, external disturbances, and/or environmental conditions, then the adaptability of the system more closely resembles the biological systems that inspire us to model engineering systems after natural systems.

The evolution of adaptive structures technology, and its current knowledge base, was made possible by the explosive growth in the microprocessor industry over the past decade. It is now feasible to implement multivariable, digital control laws with the bandwidths required for structural control. Digital parameters are more readily adjusted than their analog equivalents, leading to a real capabil-

[*]K. Arms and P. S. Gamp. *Biology*, CBS College Publishing, New York, 1982, p. 11.

ity to provide adaptation in an on-line environment. The other critical element for adaptive structure design has been the development of material technologies, primarily electromechanical materials. New design philosophies have spawned innovative blends of electromechanical sensors, actuators, and material systems in a diverse variety of applications, including automotive systems, computer hardware systems, industrial machine design and tooling, aircraft systems, and civil engineering structural systems. This textbook develops fundamental relationships describing the interaction of electromechanical transduction devices and host structure materials under the direction of microprocessor-based control algorithms.

During the past decade, research on adaptive structures has accelerated rapidly, leading to a vast information base that consists of engineering realities, expectations, and dreams for truly intelligent structural systems. More recently, important progress has been made toward realization of practical engineering structures equipped to perform as adaptive structures. This book focuses on the realities of practical adaptive structure designs, placing a much lesser emphasis on the expectations and dreams for tomorrow's intelligent systems. Throughout the book, we have tried to include discussions on implementation aspects that we feel are of interest to engineers doing hands-on work with these contemporary structures. The emphasis of the text is on the integration of control strategies, novel sensing methods (in physical coordinates or transformed coordinates), and appropriate actuator configurations for adaptive structures. The selection of specific active material components is application sensitive, and thus, constitutive relations presented in this text have been arranged to show clearly both the exclusivity and exchangeability of the chosen material. To be consistent with the research community and industrial demand, an emphasis has been placed on the use of piezoelectric materials in the examples presented in the text. However, the basic design philosophy presented is applicable to a variety of actuator and sensor devices.

Our goal is to use this textbook to reach engineers from a broad range of disciplines who may envision important roles for adaptive structures in their specific field of study. We have entered a new era in engineering, where multidisciplinary skills are absolutely critical in order to explore new frontiers. Because of this, the text has been written primarily for graduate students and experienced engineers who wish to develop fundamental skills required for design of adaptive structures. Since it is impossible to exhaustively cover all disciplines required for adaptive structure design and synthesis in a single manuscript, this book is written with the assumption that the reader has successfully completed introductory courses in vibration theory, classical controls, matrix operations, linear differential and difference equations, and boundary value problems. It would be most helpful if the reader has been exposed to linear system theory (state-space modeling and control system design, including optimal control), and adaptive signal processing theory, as well as having some background in solid mechanics and advanced structural dynamics. These subject areas are studied here in the context of adaptive structures; however, previous experience in these areas of study will facilitate the use of the text as a blended theoretical and practical guide to building adaptive structures.

Simulation methods are essential to adaptive structure design and are exploited throughout this book. Due to the overwhelming use of MATLAB software throughout the adaptive structures community and during our own careers, we have adopted the software package for the demonstration problems. There is sufficient information in the following chapters and supplemental MATLAB software programs to ensure that a serious student will be able, after finishing the material, to design, at the very least, simple adaptive structures. More experienced readers, we hope, will be able to augment their capabilities with many of the techniques presented here. A computer disk is included on the back cover with the necessary script files to repeat example problems included in the text and to develop models of simple adaptive structures. We hesitated slightly with this approach, as the reader must have purchased MATLAB software to reap the benefits of the supplemental programs; however, we are of the biased opinion that all serious students and practitioners of adaptive structures either already own MATLAB software or should purchase it.

One of the primary contributions of this text rests in its presentation of adaptive feedforward control within the framework of classical and modern control theories. Adaptive structure design and implementation are driven by the physics of each desired application, and thus the best controller is always that which provides the desired performance with the least complexity and lowest cost. Adaptive signal processing methods have proven to be viable candidates for certain types of structural control, especially in regards to complexity and system stability. We have labored over the unification of adaptive feedforward control and feedback control architectures, as it is our belief that adaptive structure design is facilitated by a clear understanding of both control system design approaches.

Another contribution of this text results from the emphasis placed on the transformation between physical models, modal models, and wave models for structural dynamic analysis and control. For adaptive structure design, one must understand the dynamics of the structure and identify the most efficient method of sensing and actuating the desired structural response. The size, placement, and choice of active materials for sensing and actuation depend largely upon the boundary conditions of the adaptive structure, which often dictate the type of model chosen. By understanding the advantages of different modeling approaches, it may be possible to minimize the number of components required to achieve the performance goals for adaptive structures. Most of the existing literature presents the modal and wave models as separate concepts. Throughout this text, we have illustrated the physics that link the two modeling methods. Whenever possible, functional relationships and qualitative descriptions of the crossover from physical coordinates to modal or wave coordinate systems are provided.

A third contribution of the text lies in its presentation of spatial signal processing techniques for novel methods of sensing and actuating adaptive structures. Methods of blending spatial and temporal signal processing of structural response measurements are detailed for developing frequency-shaped control cost functions. A primary motivation for this approach is to incorporate certain dynamic input-output relationships that may be difficult to measure directly. The design of spatially convolving sensors and actuators is also shown to relieve some of the computational burden normally placed upon the digital controller implementation.

To familiarize the reader with the text in advance, a summary of the chapters is presented. In Chapter 1, the reader is introduced to the concept of adaptive structures, and to the definitions adopted by the authors for this text. In Chapter 2, the fundamentals of structural dynamics for distributed systems are discussed. Solution methods leading to modal expansions and traveling wave representations are detailed, with a number of examples provided for simple structures. There is also a strong emphasis on mathematical descriptions of structural damping in this chapter. Selected topics in linear system theory are reviewed in Chapter 3 to familiarize the reader with concepts such as controllability, observability, stability, and other control theory terminology necessary for the later chapters. In Chapter 4, we review a number of fundamental concepts in digital signal processing theory—concepts that are important to the development of adaptive feedforward control and understanding spatial signal processing of structural measurements. Chapter 5 is devoted to methods of modeling transduction devices typically used for adaptive structure designs. We stress an understanding of the coupled dynamics between active materials and host structure materials through detailed development of a piezostructure's dynamic behavior. The coupled dynamic descriptions are critical for accurate representations of adaptive structures in the presence of control. In Chapter 6, methods of using actuators and sensors for spatial signal processing are discussed, along with procedures for combining spatial and temporal signal processing for real-time sampling of selective system output variables. Chapter 7 provides an overview of classical controls confined to the perspective of controlling reverberant systems; structural plants are considered as a subset. The purpose of Chapter 8 is to describe a number of different control system architectures and algorithms that are available to the design engineer, including feedforward, adaptive feedforward, and feedback for multivariable systems. The reader is encouraged to approach the adaptive structure design process in two stages: selection of the appropriate control system architecture, followed by selection of appropriate adaptive feedforward and/or feedback control system design and synthesis procedure. Chapter 9 presents numerous adaptive structure applications, using simulated and experimental results to lead the reader through the design process.

## ACKNOWLEDGMENTS

The authors would like to acknowledge the numerous researchers in industrial and government laboratories and at academic institutions who have contributed to the field of adaptive structures and whose names do not appear in the bibliography included. The authors made every attempt to credit those individuals upon whose work the presented material is based. We would also like to thank our respective academic advisors, Professor Chris Fuller and Professor Harry Robertshaw of Virginia Polytechnic Institute and State University, for providing the initial foundation for our research efforts. Additionally, we would like to acknowledge the benefits of daily technical dialogues with close colleagues, including Dr. Kenneth Frampton, Russell Groves, Clark Smith, and Dr. Jeffrey Vipperman of Duke University, Daniel Cole and Dr. William Baumann of Virginia Polytechnic Institute and State University, David Cox and Dr. Richard Silcox of the NASA–Langley Research Center, Dr. Shawn Burke of the Center for Applied Photonics Research at Boston University, and Dr. Dennis Bernstein of the University of Michigan for their insightful comments and suggestions during the development and review of this manuscript. The authors are also indebted to the Air Force Office of Scientific Research, the NASA–Langley Research Center, the National Science Foundation, and the Office of Naval Research for funding much of the research from which this book was formed.

Finally, we would like to thank our families (Dana, Trey, and Amelia Clark, Ben, Laura, and Emily Saunders) for their support and acceptance of our dedication throughout this project, for without it, we could not have completed this task.

Robert L. Clark
William R. Saunders
Gary P. Gibbs

# 1

# OVERVIEW OF ADAPTIVE STRUCTURES

## 1.1 MOTIVATION FOR ADAPTIVE STRUCTURES

In biology, adaptation is considered to be *a characteristic of an organism that makes it better able to live in its environment* (Arms and Gamp, 1982). Consider for example the chameleon. This small lizard is capable of changing the color of its skin to blend with the color of the surroundings. Thus, camouflage to the chameleon is essential to survival and is a characteristic that can be modified according to changes in the environment. This is an example of *short-term adaptation*, since the relative timescale over which the organism adapts is small in comparison to that of its evolution. As an example of *long-term adaptation*, consider the long bone of a bird's wing. The wing has little or no weight to carry, yet it must withstand significant bending moments while in flight. In birds capable of sustained flight, such as the albatross, the bone takes the form of a thin, cylindrical shell. Thus, the load bearing member has evolved over a significant timescale to the optimal mechanical structure for the application (Thompson, 1942).

With engineering structures, we are more concerned with short-term adaptation than with long-term adaptation, since the structure is not living and thus cannot synthesize mass from energy, redistribute its mass, or reproduce a structure better suited to the environment. It is the absence of these lifelike characteristics that modifies our perception of long-term adaptation for structures. For engineering structures, incremental scientific and technological advances provide the forum for long-term adaptation, since the design and performance of systems and structures are slowly improved as a function of these advances. One must simply compare the biplanes of World War I to the Stealth Bomber, made possible by aeroelastic tailoring of composites and high-performance jet engines, to visualize this evolutionary process. The engineering science of adaptive structures has only recently begun. At this juncture, we may view this exciting technology as an opportunity for ultimate long-term adaptation, or we can examine the impact that adaptive structures will have on the ability of engineering systems to exhibit short-term adaptation. Our focus is on the characterization of short-term adaptation strategies. It is shown that these strategies are designed to provide material systems and structures with the capability for short-term adaptation in the presence of changing environmental conditions, component failures, or external disturbances.

The study of adaptive structure technology is important because of its demonstrated potential to outperform conventional structures in a number of different applications. First, the presence of active control typically provides vibration suppression at significantly lower frequencies, as compared to available passive methods. Second, the presence of adaptive control algorithms provides critical short-term adaptation mechanisms in response to changes in a variety of system parameters. Third, dual-use strategies can often be implemented to combine any number of performance objectives including improved structural performance (lower vibration, reduced acoustic emission, etc.), self-health monitoring, and damage mitigation using a common adaptive structure platform. The use of adaptive structures for such improved performance and reliability has only begun. This book is intended to encourage the study of adaptive structure systems and design methods with a critical focus on the advantages and disadvantages that accompany different applications of interest.

## 1.2 BACKGROUND AND HISTORY

One of the earliest studies of adaptive structures was completed by Swigert and Forward (1981). They conducted a theoretical and experimental study that involved an "electronic damper" implementing a system of electromechanical transducers made from lead zirconium titanate (PZT) as elements of electronic feedback loops to control the mechanical vibration of an end-supported mast. The structure consisted of a hollow fiberglass cylinder called an omni-antenna mast, and the concept was to apply electronic damping with surface-mounted piezoelectric sensors and actuators. A schematic diagram of the omni-antenna is presented in Figure 1.1. The outputs of the sensors were amplified and appropriately shifted in time to provide control inputs for actuators positioned symmetrically on the cylindrical surface. This initial study opened the doors for interest in strain actuation and sensing of structures.

Bailey and Hubbard (1985) developed the first adaptive structure using polyvinylidene fluoride (PVDF) as an active damper for structural vibration control of a cantilever beam. By implementing both constant gain and constant amplitude controllers, they experimentally demonstrated that the PVDF actuator could significantly increase the measured loss factor (i.e., the system damping) when the structure was subjected to some initial displacement at its end.

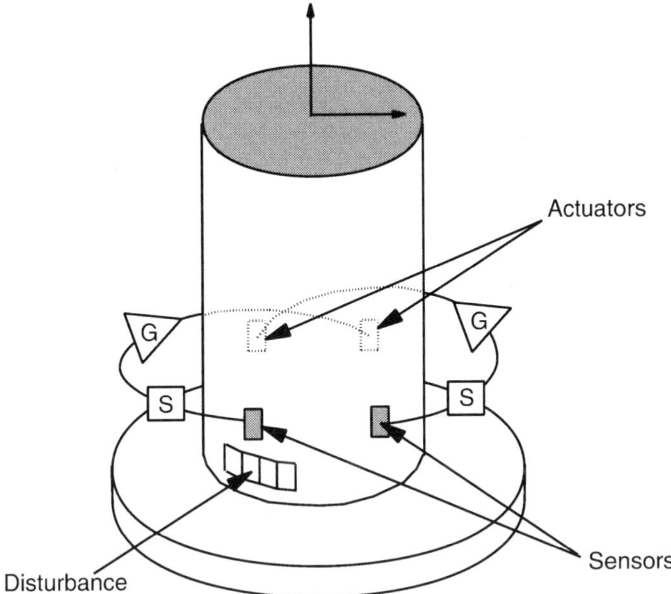


**Figure 1.1:** Artist's rendering of an early adaptive structure, the omni-mast adapted from Swigert and Forward (1981).

Thus, the experiment performed by Bailey and Hubbard was similar in many ways to the one performed earlier by Swigert and Forward (1981), the major difference being the use of PVDF actuation as opposed to PZT actuation and the application of energy-based control system design (Lyapanov's 2nd method).

By the late 1980s, research papers began to appear on any imaginable blend of active materials, host structure materials, and control algorithms. Fuller (1988) demonstrated the use of vibrational inputs and adaptive signal processing to control the sound radiation from vibrating structures. Rogers et al. (1989) incorporated nitinol, a shape memory alloy (SMA), as a strain actuated device for structural control. Their demonstration of different applications for the shape memory material coincided with experimentation using a number of "intelligent" materials for structural sensing and actuation, including fiber optics, electrorheological fluids, electrostrictive, and magnetostrictive materials. The field of adaptive structures was beginning to emerge, although standard terminology was still very much in question. Crawley and de Luis (1987) are generally credited with creating the term "intelligent structures" in their paper titled: *Use of Piezoelectric Actuators as Elements of Intelligent Structures*. Wada et al. (1990) published a paper that defined a number of terms associated with smart, adaptive, and intelligent structures. The vocabulary defined in their presentation is reviewed in the following; however, the terminology adopted by the authors of this text differs slightly and will be illuminated in the following sections.

**TABLE 1.1 Adaptive Structure Terminology**

- Adaptive structures
- Intelligent structures
- Smart structures
- Active structures
- Smart sensors
- Smart actuators
- Intelligent material systems and structures

Table 1.1 lists a number of descriptive terms for adaptive structures, as outlined by Wada et al. (1990). The partial or total integration of transduction devices with host structures provided the framework for categorizing the different configurations. As illustrated in Figure 1.2, the various types of structural entities were defined as a subset of other fields of research. "Adaptive structures" were defined to be those structures that have actuators distributed throughout (the authors of this text choose to call such structures "actuated structures"). The term "sensory structure" was used to describe those structures configured with distributed sensors, and the integration of sensory and adaptive structures with a closed-loop control system was termed a "controlled structure." Wada et al. defined structures with embedded components serving some function in the load bearing properties of the system as "active structures."

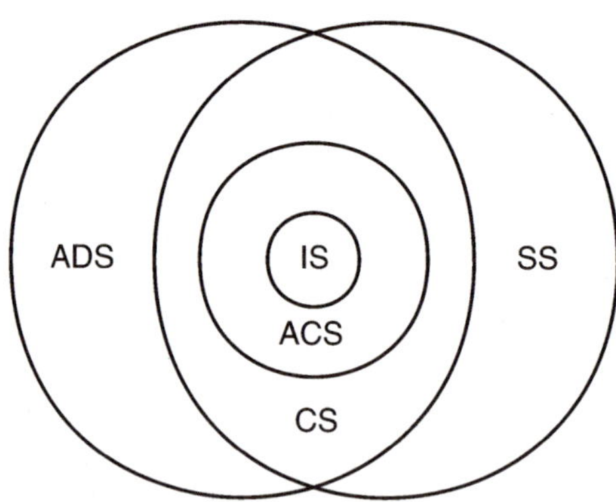


**Figure 1.2:** Diagram of intelligent structures as detailed by Wada et al. (1990).

Finally, the term "intelligent structure" was reserved for "those which incorporate actuators and sensors that are highly integrated into the structure and have structural functionality, as well as highly integrated control logic, signal conditioning and power amplification electronics." This list of descriptive titles is by no means exhaustive but demonstrates the ever-increasing volume of vocabulary used to describe the merging of active materials, controls, and structures in composite systems. In this book, we will refer to "adaptive structures" as defined in the following section; however, the reader should keep in mind that subtle differences in interpretation may result with the use of different titles.

Adaptive structures were created largely by combining technological advances in microelectronics, active materials, and selected disciplines of adaptive control theory. Other engineering fields have also played significant roles, the most important contributions coming from the fields of digital signal processing and composite materials. It is beyond the scope of this text to concentrate on all of the contributing subject areas, and our focus remains on the integration of active materials and control system architectures with host structures. Active materials in this book are designated as those materials that are capable of converting electrical, thermal, or some other form of energy to mechanical energy and/or vice versa. Some examples of active materials pertinent to adaptive structures are SMAs, electrostrictive materials, magnetostrictive materials, and piezoelectric materials. By far, piezoelectric materials have been most commonly used in the development of adaptive structures and must be recognized as the leading material associated with the evolution of adaptive structures.

As detailed by Collins et al. (1990), the acceptance of active materials as critical elements in adaptive structure designs is relatively new but investigation of their potential began during the Cold War era. The piezoceramic material most commonly utilized for actuator design, PZT, was discovered by Jaffe et al. (1954). Before this discovery, Von Hippel et al. (1946) and Wul and Vereshchagin (1945) independently discovered barium titanate ($BaTiO^3$). Since the Curie temperature of this ceramic material was only 120°C, excitement over the discovery was limited (Collins et al., 1990). In contrast, not only does PZT exhibit better piezoelectric properties than $BaTiO^3$, but it has a higher Curie temperature: 340°C (Jaffe et al., 1971). A second popular form of piezoelectric materials, the piezoelectric polymers, were emphasized by Japanese research and development. In 1969, Kawai discovered the piezoelectric effect in polarized films such as PVDF, and although other films have been studied, PVDF has dominated the market in research and development applications since its discovery. It differs from the ceramic PZT material since it is 30 times more compliant and 4 times less dense. Thus, it showed little authority over the dynamics of stiff structures, constructed from steel or aluminum, and has been employed mostly for sensing functions in adaptive structure applications.

Development of adaptive structures technology has occurred, in some areas, without significant interaction with the established disciplines of adaptive control theory. Thus, a distinction may exist for certain readers between the use of the term "adaptive" in adaptive structures and adaptive control theory. However, a careful review of both technical areas shows that there are substantial similarities in purpose and implementations of control methods. The relatively new existence of adaptive structures, compared to adaptive control theory, is partly responsible for the subtle confusion in how the two fields overlap. Therefore, before formal definitions of adaptive structures are presented, we wish to briefly summarize some of the historical aspects of adaptive control theory that have influenced the presentation of adaptive structures in this book.

The motivation behind the development of adaptive control theories can be traced to a desire to generate controller designs that were capable of performing over a broad range of operating conditions. By the early 1960s, engineers were beginning to address critical issues that surround the use of adaptation in closed-loop control systems. According to Whitaker (1963), an adaptive control system was defined to be one that "by measurable adjustments of controllable parameters or by measurable well-defined changes in dynamic characteristics, can adapt to a changed environment, a changed character of input signals, or a changed system or component characteristic in such a manner that the specified desired performance is maintained." The emphasis is on change; environmental changes, changing disturbance input signals, and changing system dynamics. When the control system can successfully adapt to such changes, then it can legitimately be called an adaptive control system.

For structures, change and uncertainties often require elements of adaptation to be present in control implementations. Many times, there is not sufficient a priori knowledge of the plant dynamics for robust closed-loop design. The complexities of structural plants, with a multitude of welded joints, nonlinear transmission paths, and time-varying static loads, can introduce enough uncertainty to the controlled plant dynamics to make high-performance goals unreachable and closed-loop instability a likely result. In addressing these real world obstacles, adaptive control theorists have promoted a large number of topologically different adaptive control architectures. For adaptive structures, we are concerned with three of the most popular adaptive control design approaches: adaptive signal processing methods for control, model reference adaptive control (MRAC), and self-tuning regulators (STR). Although each approach is based on somewhat different theoretical foundations, all three methods have a common purpose: *to eliminate the effect of variations in disturbance signature and plant dynamics on the controlled system performance.*

Adaptive signal processing can be categorized as an adaptive control approach because of its ability to self-adjust controller parameters in response to real-time system response measurements (Narendra and Monopoli, 1980). Most adaptive signal processing algorithms for structural control applications operate in the feedforward path. One of the earliest demonstrations of adaptive signal processing for structural control was by Fuller (1988), where he used an acoustic cost function in conjunction with structural inputs to control farfield acoustic pressures. A large number of research efforts followed that style of adaptive feedforward control system implementation, and due to its relative robustness and ease of implementation, the control approach has been integrated into many practical control systems, such as the Lord Active Isolation Control System, developed by the Lord Corporation for the Cessna Aircraft Company. Specific issues associated with adaptive feedback control of structures were initially summarized by Balas and Johnson (1980). Two popular feedback design methods, MRAC and STR, were studied extensively during the past 30 years but mostly for nonstructural plants (Landau, 1979; Narendra and Monopoli, 1980; Astrom and Wittenmark, 1989).

To conclude this background discussion, we wish to summarize the present stage of development for adaptive structures. Their brief history, less than 20 years, has been dominated by continuing efforts to refine the blends of compact, integrated active material elements and active control algorithms, for use in any number of applications. Transducers constructed from a wide variety of active materials have already been implemented in schemes for persistent disturbance rejection, active damping, and as local or distributed strain or strain-rate sensors. Actuators and sensors have been shaped, optimally positioned, and digitally filtered to control or monitor the response of selected structural modes, structural waves, or appropriate frequency-shaped combinations of the structural modes and/or waves. Many important application areas have been identified, some of which are demonstrated in the later chapters of this book. We make the observation that adaptive structure designs, either in constrained adaptive or purely adaptive architectures, have been proven to be a credible engineering alternative for many different forms of application physics. The most obvious omission in adaptive structure technology, to date, is found in the prediction of power requirements and economic feasibility for specific applications. In later chapters, we provide the theory that is necessary to simplify engineering calculations of these two important features of adaptive structures.

## 1.3 WHAT IS AN ADAPTIVE STRUCTURE?

In this book, an *adaptive structure* is defined to be a structure configured with distributed actuators and sensors and directed by a controller capable of modifying the dynamic response of the structure in the presence of time-varying environmental and operational conditions. The structural dynamic modification process has essentially only three mechanisms: gain adjustment, zero placement, or pole placement. This is consistent with well-established system control concepts, so that a principal objective of this book is to define the structural system in a manner consistent with the control literature and examine the controlled system response. We find that there are two fundamental forms of adaptive structures: *constrained adaptive* or *purely adaptive*. A constrained adaptive structure operates under the direction of a fixed-gain control law and is constrained to performance limits that lie in the range space of the controlled system. A purely adaptive structure operates in the presence of a control law whose parameters are modified as a function of the system output measurements or a fixed control law that operates on real-time estimates of the plant parameters.

Many applications for adaptive structures exhibit superior performance when the control law is *model insensitive*, meaning that the controlled response will be insensitive to changes in system parameters over the range of operation. The term model insensitive is used, as opposed to model independent, because all control systems, either purely adaptive or constrained-adaptive, display some degree of dependence upon modeled dynamics, that is, some model of the system. When the dynamic system parameters are uncertain, the control signals are inevitably influenced by corresponding deviations between the expected and measured outputs. This model dependence, which we examine in detail, is a fundamental characteristic of any controlled system and is primarily a result of implementing controllers on real adaptive structures whose infinite-dimensional dynamics cannot be completely described, or even accurately measured, over the entire operational bandwidth. Therefore, the goal is to design controllers with sufficient insensitivity to either changing or poorly defined structural parameters (i.e., *robustness*).

Based upon the previous definition of an adaptive structure and model insensitive control, one recognizes that there are different levels of complexity that can be designed into adaptive structures. The adaptation properties depend upon the complexity of the control system and ultimately dictate the number and type of components required to fabricate the structure. In the later chapters, we concentrate on architectural styles and design methods that can generate varying levels of sophistication in the adaptive structure. In the next section, we introduce the basic building blocks that are used to describe the performance of adaptive structures.

## 1.4 GENERAL ARCHITECTURE OF AN ADAPTIVE STRUCTURE

The basic components that are essential to the operation of adaptive structures are actuators, sensors, and a controller. It may be argued that a controller is not an essential element for certain classes of adaptive structures, which we refer to as *passive-adaptive structures*. One example of a passive-adaptive structure is a composite panel with embedded SMA fibers. When the fibers receive a thermal input, the resulting change in their length alters the structural stiffness. This open-loop process effectively adapts the structure's performance without a controller; however, the ability to precisely maintain the desired adapted response does not exist for passive-adaptive structures. For this reason, we include the controller as an essential element in the adaptive structure's architecture.

### 1.4.1 Transduction Devices

A large number of different transduction devices are available for sensing and actuation of structures. A representative list is shown in Table 1.2. The most

**TABLE 1.2 Transduction Devices for Adaptive Structures**

- Accelerometer (sensor)
- Electrodynamic shaker (actuator)
- Electrostrictive material (actuator)
- Ferroelectric material (actuator)
- Magnetostrictive material (actuator)
- Optical fibers (sensor)
- Piezoelectric material (actuator or sensor)
- Shape memory alloy (actuator or sensor)
- Strain gauge (sensor)

common application of each material is listed in parentheses; however, electromechanical materials can typically be used as sensors or actuators, and sometimes as both simultaneously. Chapter 5 presents technical background and design approaches for selected transduction devices.

Once appropriate materials have been selected for actuation and sensing, they must be integrated into the structural design with the electrical wiring and buses necessary for communication with the controller. In a composite structure, all components are an integral part of the structure and must be part of the structural design. However, for a number of structures, the actuators and sensors can be surface mounted. For steel and aluminum structures, this is the only practical alternative. Schematic diagrams of the two types of structures, composite and isotropic, are presented in Figure 1.3.

Each of the two configurations has advantages and disadvantages. As illustrated in Figure 1.3*a*, the embedded devices in the composite structure are shielded by the laminates; however, if one of the devices fails during operation or fabrication, the entire composite section must be replaced. On the other hand, the surface-mounted devices on the isotropic structure are readily accessible in the event of failure, but they are more susceptible to damage from external forces and electrical disturbances. However, in both cases, you may destroy the structure when replacing the transducers. The surface-mounted devices offer an additional benefit for control of flexural waves, since greater levels of structural excitation result as the actuator is displaced from the neutral axis of the structure.

### 1.4.2 Control System Architectures

Whether the actuators and sensors are surface mounted on an isotropic structure or embedded within the matrix of a composite structure, the general architecture is not complete without a controller. The emphasis of this text is the controlled structural plant, where the controller must always be selected such that the structure meets the desired performance requirements. This results in various forms of adaptive control architectures. For example, for vibration or structural acoustic suppression, a number of different disturbance inputs are possible. The statistics

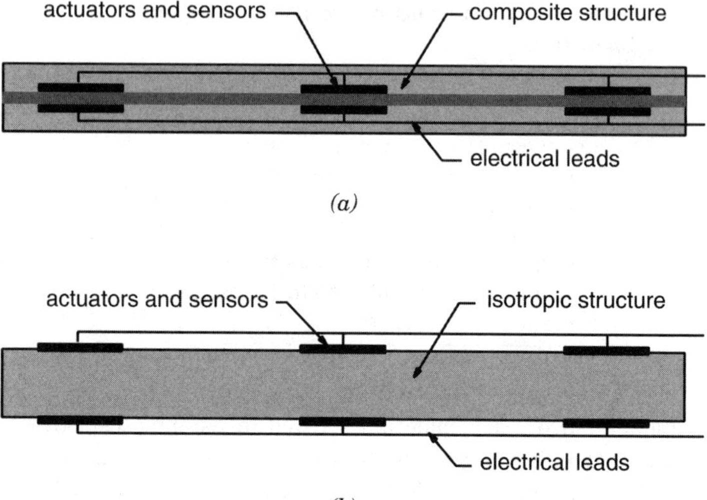


**Figure 1.3:** (*a*) Schematic diagram of composite adaptive structure. (*b*) Schematic diagram of isotropic adaptive structure.

of the structural disturbance(s) influence the type of control: adaptive feedforward, feedback, or hybrid blends of adaptive feedforward and feedback. The types of input disturbances and some general control alternatives are shown in Table 1.3.

The input disturbance(s) to the system may be some combination of the various types shown in the table, and hence the resulting control system can logically be any combination of the two basic types of control architectures. Several levels of adaptation can be built into the structure, depending on the selection of the control process.

Fixed-gain feedback control, coupled with a fully instrumented structure constitutes the *constrained-adaptive structure* introduced earlier. A block diagram of a constrained-adaptive structure is presented in Figure 1.4. The open-loop structure is represented by the transfer function, designated $\mathbf{P}(z)$, with disturbance input $\mathbf{u}_d(z)$, and with control input $\mathbf{u}_c(z)$, which is produced by

**TABLE 1.3 Basic Control Architectures as a Function of Input Disturbances**

| Disturbance | Adaptive Feedforward | Feedback | Hybrid |
|---|---|---|---|
| Harmonic | X | | X |
| Multifrequency | X | | X |
| Broadband | X | X | X |
| Impulsive | | X | X |

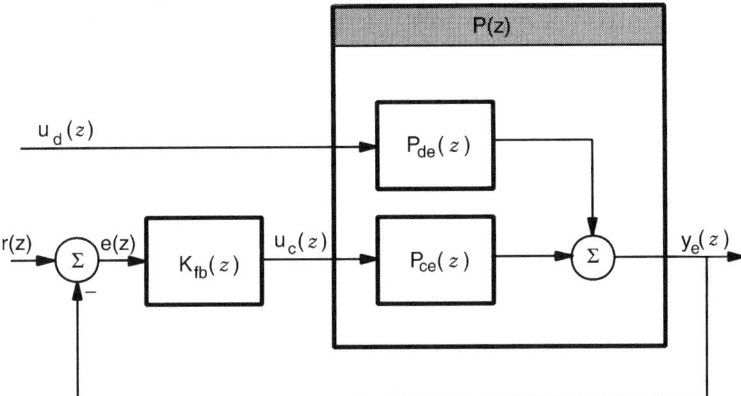


**Figure 1.4:** Block diagram of adaptive structure with active control.

a compensator $\mathbf{K}_{fb}(z)$ with inputs from a reference $\mathbf{r}(z)$ and the output $\mathbf{y}_e(z)$. The constrained-adaptive structure is designed to be closed-loop stable while it maintains its performance objectives over a specified operating bandwidth. Since the feedback gain matrix is fixed over a prescribed bandwidth, the controlled structural response cannot be maintained in the presence of large uncertainties or time-varying behavior of the structural parameters. Thus, in a constrained sense, the structure can adapt its response to a disturbance input, but if the gain matrix of the controller is fixed, then the performance envelope of the response is also fixed. The constrained-adaptive structure is thus the lowest level of an adaptive structure. Applications for a constrained-adaptive structure include persistent disturbance rejection on well-known structural dynamics or low-fidelity, active damping of resonant response.

A higher level of adaptation is associated with adaptive feedforward control since the input to the structure evolves with time as a function of changing environmental or operational conditions. A steepest descent algorithm, such as the least-mean-squares (LMS) algorithm, is typically employed to adapt the coefficients of a finite impulse response (FIR) filter necessary to generate the control input. For structures, a filtered reference must be generated to compensate for the propagation time in the structure. The filtered-$x$ LMS algorithm is then used in conjunction with the measured (i.e., modeled) control-to-error transfer function implicit in the filtered reference. A block diagram of the general architecture is illustrated in Figure 1.5. The open-loop structure is represented by the transfer function, designated $\mathbf{P}(z)$, with disturbance input $\mathbf{u}_d(z)$, and with control input $\mathbf{u}_c(z)$, which is produced by a compensator $\mathbf{K}_{ff}(z)$ with inputs from a reference signal correlated with the disturbance $\mathbf{u}_d(z)$. The compensator filter coefficients are adjusted by an adaptive algorithm based upon a cost functional incorporating a measure of the error $\mathbf{y}_e(z)$. An alternative adaptive algorithm to the

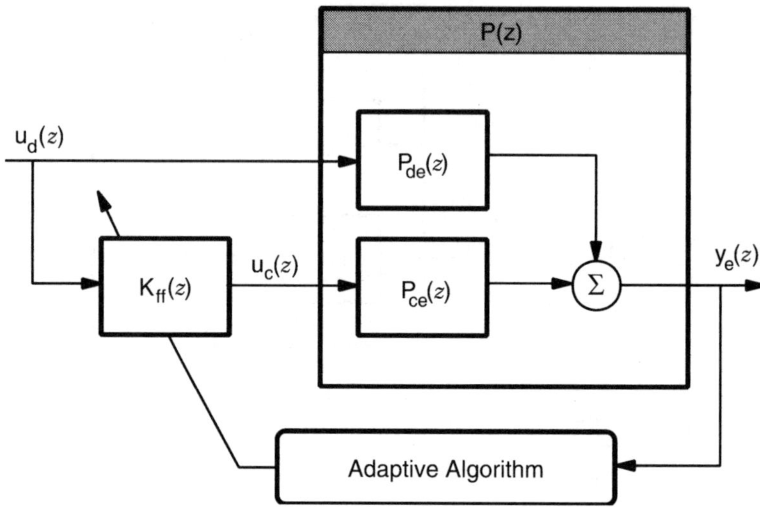


**Figure 1.5:** Block diagram of adaptive structure with adaptive feedforward control.

filtered-$x$ LMS algorithm is the time-averaged gradient (TAG) descent algorithm. A TAG descent approach utilizes a statistical estimate of the mean-squared error to determine the gradient, and thus, does not require a measure of the control-to-error transfer function. This approach is truly model insensitive but typically takes more time to converge to the optimal filter. Since feedforward control only operates on the zeros of the system, it is ideal for persistent disturbance rejection of single frequency, multifrequency, or broadband structural disturbances for which there exists an uncontrollable reference signal correlated with the disturbance. An *uncontrollable reference* is defined to be a signal that is unaffected by any control input to the structural system. Fortunately, in the case of persistent disturbances, an uncontrollable reference signal is frequently available. For example, structure-borne transformer noise occurs at harmonics of 60 Hz, and thus a reference is always available. In addition, the rotational speed of any form of rotating machinery, such as AC or DC motors, turbines, compressors, or gears, can be measured using shaft tachometers or encoders to provide uncontrollable reference signals for control of vibration resulting from disturbances correlated with such rotational speeds and harmonics.

Quite often, the input disturbance is some combination of persistent and impulsive disturbances. In fact, during the adaptation of the FIR filter in feedforward control, the time-varying control input during the convergence of the filter presents an impulsive disturbance to the structure due to the discrete changes in the filter coefficients used to generate the control input. In such cases, a combination of adaptive feedforward control and feedback control is warranted, and the combination is defined as *hybrid control*. Hybrid control is defined in this text to be the combination of adaptive feedforward control and feedback control in a single control system architecture. A block diagram of a hybrid controller

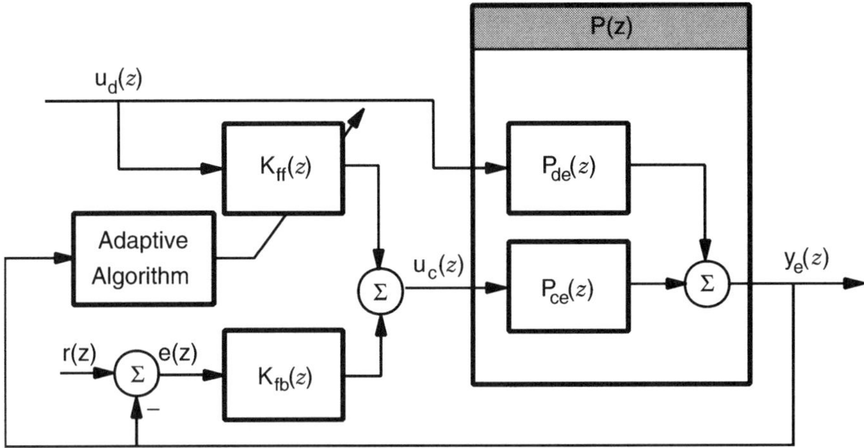


**Figure 1.6:** Block diagram of adaptive structure with hybrid control.

is illustrated in Figure 1.6. As illustrated, the dynamic system (the structure) is modified through feedback control, which could involve static or dynamic compensators, and simultaneously, persistent disturbance rejection is achieved with adaptive feedforward control. Figure 1.6 represents the integration of Figures 1.4 and 1.5, and the symbols used in the diagram are consistent with the previous description for each respective figure. Use of hybrid control architectures can provide significant performance advantages, and these advantages are discussed in detail in Chapter 8. The hybrid adaptive structure represents the most complex adaptive structure detailed in this work, but also affords the most flexibility and performance advantages.

Thus, we have presented three basic levels of adaptive structures, levels that revolve around the control architecture implemented to modify the system response. Various sublevels can also be defined, depending upon the order of the adaptive filter required, the order of the multi-input, multi-output (MIMO) system, or the specific feedback and feedforward algorithms selected for control implementation. The advantages and disadvantages of the available control architectures are conveyed to the reader in the following chapters through theory and examples based upon current applications.

## 1.5 APPLICATIONS FOR ADAPTIVE STRUCTURES

All structural models are inaccurate in their representation of the actual structural entities we wish them to represent. As we show in the following chapters, the uncertainty can be dealt with in a number of ways using adaptive structures. Consider, for example, the dynamics of an airfoil of a commercial or military aircraft. The structural response of this system depends upon certain environmental parameters (i.e., ambient temperature, humidity, pressure, density, etc.) of the fluid, flight speed, and payload, as well as upon transient excitations due to gusts of wind. The control law for such systems, subject to time-varying conditions, must also be a function of these parameters. One of the most important choices made by the adaptive structure design engineer is whether or not to develop a constrained-adaptive or purely adaptive structure for the problem identified. A fixed control law, even if valid over a wide range of expected conditions for the nominally designed plant, is subject to robustness problems if the actual plant is significantly different from the design model.

Another important application for short-term adaptive structures is in the control of sound radiation from vibrating structures. Because certain structural modes are more efficient sound radiators than others, an adaptive structure approach may be used so that the coupled fluid-structure response is dominated by acoustically inefficient structural modes. Other potential applications for adaptive structures are shape control of structural members (e.g., pointing antennas and astronomical telescopes), position control for robotic applications, and seismic and vibration control of large-scale civil engineering structures. In each of these examples, some form of control law must be exercised to adapt the structural response. Whether or not the control algorithm is capable of modifying its actions as a function of changes in environmental conditions, external disturbances, or failures in structural components depends on the choices made by the system designer: *you*. In the following chapters, we present some of the tools required to build both constrained-adaptive and purely adaptive structures (with lots of room for future improvements of course).

## 1.6 SUMMARY

As indicated by the title, one of the primary purposes of this text is to provide the reader with alternative methods of modeling, measuring, or estimating the dynamics of structures for the design of adaptive structures. The adaptive structure design process typically begins with some performance objective that necessitates a basic understanding of the dynamics of the structural system. In this book, we concentrate on dynamics represented by either physical, traveling wave, or modal variables, with insight into the relationship between the realizations and potential advantages and disadvantages of each. A comprehensive presentation of structural control law designs is presented as well. Our emphasis is on the use of a unified control theory that encompasses both feedforward and feedback control operated independently or in combined "hybrid" configurations. *Adaptive* or *model insensitive* algorithms, capable of adjusting to slowly time-varying parameters caused by the environment, or failures in structural components such as actuators or sensors and various input disturbances, are emphasized. To distinguish this work from previous dynamics and controls textbooks, alternative methods of modeling strain actuated and sensing devices are summarized with a focus on the relationship between the *transduction dynamics* and the structural dynamics. Simulated and experimental results are presented for different physical problems to demonstrate the transition from concepts to implementation of *adaptive structures*.

## BIBLIOGRAPHY


Arms, K. and P. S. Gamp, 1982. *Biology*, CBS College Publishing, New York, p. 11.

Astrom, K. J. and B. Wittenmark, 1989. *Adaptive Control*, Addison-Wesley.

Bailey, T. and J. E. Hubbard, Jr., 1985. "Distributed Piezoelectric-Polymer Active Vibration Control of a Cantilever Beam," *Journal of Guidance and Control*, **8**(5), 605–611.

Balas, M. J. and R. C. Johnson, Jr., 1979. "Adaptive Control of Distributed Parameter Systems: The Ultimate Reduced-order Problem," Proceedings of the IEEE Conference on Decision and Control, Ft. Lauderdale, FL, Dec. 1979, 1013–1017.

Collins, S. A., D. W. Miller, and A. H. von Flotow, 1990. "Sensors for Structural Control Applications Using Piezoelectric Polymer Film," *SERC #12-90*, MIT Space Engineering Research Center, Massachusetts Institute of Technology, Cambrige, MA.

Crawley, E. F., 1993. "Intelligent Structures: A Technology Overview and Assessment," *Proceedings of the 1993 SDM Conference*, Vol. 6, pp. 1–16.

Crawley, E. F. and J. de Luis, 1987. "Use of Piezoelectric Actuators as Elements of Intelligent Structures," *AIAA Journal*, **25**(10), 1373–1385.

Forward, R. L., 1981. "Electronic Damping of Orthogonal Bending Modes in a Cylindrical Mast-Experiment," *Journal of Spacecraft and Rockets*, **18**(1), 11–17.

Fuller, C. R., 1988. "Analysis of Active Control of Sound Radiation from Elastic Plates by Force Inputs," *Proceedings of Inter-Noise '88*, Avignon, France, Vol. 2, 1061–1064.

Jaffe, B., R. S. Roth, and S. Marzullo, 1954. "Piezoelectric Properties of Lead Zirconate-Lead Titanate Solid-Solution Ceramics," *Journal of Applied Physics*, **25**(6), 809–810.

Jaffe, B., W. R. Cook, Jr., and H. Jaffe, 1971. *Piezoelectric Ceramics*, Volume 3 of *Nonmetallic Solids*, edited by J. P. Roberts and P. Popper, Academic Press, London, p. 317.

Kawai, H., 1969. "The Piezoelectricity of Polyvinylidene Fluoride," *Japan Journal of Applied Physics*, **8**(7), 975–976.

Landau, Y. D., 1979. *Adaptive Control, The Model Reference Approach*, Marcel Dekker, New York.

Narendra, K. S. and R. V. Monopoli, 1980. *Applications of Adaptive Control*, Academic Press, New York.

Rogers, C. A., C. Liang, and D. K. Barker, 1989. "Dynamic Control Concepts Using Shape Memory Alloy Reinforced Plates," In *Smart Materials, Structures and Mathematical Issues*, edited by C. A. Rogers, Technomic Press, Lancaster, PA.

Swigert, C. J. and R. L. Forward, 1981. "Electronic Damping of Orthogonal Bending Modes in a Cylindrical Mast-Theory," *Journal of Spacecraft and Rockets*, **18**(1), 5–10.

Thompson, D. W., 1942. *On Growth and Form*, Cambridge University Press, Cambridge, England.

Von Hippel, A., R. G. Breckenridge, F. G. Chesley, and L. Tisza, 1946. "High Dielectric Constant Ceramics," *Ind. Eng. Chem.*, **38**(11), 1097–1109.

Wada, B. K., J. L. Fanson, and E. F. Crawley, 1990. "Adaptive Structures," *Journal of Intelligent Material Systems and Structures*, **1**, 157–174.

Warkentin, D. J., E. F. Crawley, and S. D. Senturia, 1992. "The Feasibility of Embedded Electronics for Intelligent Structures," *Journal of Intelligent Material Systems and Structures*, **3**, 462–482.

Whitaker, H. P., 1963. "Adaptive Control Systems," *Air, Space, and Instruments, Draper Anniversary Volume*, McGraw-Hill, New York, pp. 207–234.

Wul, B. M. and L. F. Vereshchagin, 1945. "Dependence of the Dielectric Constant of Barium Titanate Upon the Pressure," *C. r. (Sov.)*, **48**(9), 634–636.

# 2

# A REVIEW OF STRUCTURAL DYNAMICS

## 2.1 INTRODUCTION

Adaptive structures usually rely on some form of system modeling, at one or more stages in the design, construction, or implementation process. From a design perspective, a model of the structure is essential for analysis of control-structure performance and interaction. Thus, adaptive structure designs often begin with analytical dynamics, formulating the equations of motion and associated boundary conditions of the structure in the language of mathematics. While developing the equations of motion, one can choose from a number of solution methodologies, based upon the modal approach, upon a wave approach, or upon a variety of numerical techniques such as finite element analysis. For each approach, one must make a number of physical assumptions about the structure in order to compute the response to transient or stationary inputs. For example, one must decide on a damping model that is physically satisfying for the application, as well as mathematically feasible within the solution methods selected by the designer. Additional assumptions must be made about the relative magnitudes of the structural displacements and the number of modes, or finite elements, to be included in the structural model.

The following presentation of selected topics in analytical dynamics should be helpful for students who are interested in formulating simple models of structures that capture the essence of more complex multivariable systems encountered in practice. Dynamics of continuous, as opposed to discrete, systems is the focus of this chapter, since the emphasis in this text is placed upon the integration of distributed sensors and actuators for the control of vibration and structure-borne sound radiation. From a practical perspective, methods in analytical dynamics are more amenable to developing the equations of motion for continuous systems than Newton's approach. Had it been possible, the authors would have included a physical structure with the textbook; however, since this is not possible, the next best thing is a model that can be used to simulate the dynamic response of a physical system. Adaptive structure design is based upon the integration of signal processing, adaptive signal processing, transduction device dynamics, and various control approaches with respect to the structure. Simple models of continuous structures yielding the mass, stiffness, and damping matrix of the system can be used to build simulations that convey many of the basic principles of adaptive structures. For more complex structures, the mass, stiffness, and damping matrix might be obtained through finite element analysis or experimental system identification. It is important to realize that, once the model of the structure is obtained and placed in this form, the design procedures are generic.

## 2.2 PRINCIPLES OF DYNAMICS

The equations of motion for distributed parameter systems are typically derived from analytical mechanics. Although Newton's laws certainly hold for distributed parameter systems, one typically does not use that approach to derive the equations of motion, since it is a vector approach and interacting forces between system components must be computed. In contrast, analytical mechanics provides a method of considering the system as an entity, operating with scalars as opposed to vectors and eliminating the need to compute interacting forces. One typically formulates the problem in a set of generalized coordinates with generalized forces and uses the calculus of variations to obtain the equations of motion and all possible geometric and natural boundary conditions simultaneously. The power of analytical mechanics is conveyed by examples throughout this text.

### 2.2.1 Hamilton's Principle

Hamilton's principle is based upon the concepts of virtual work and is derived from D'Alembert's principle as outlined by Meirovitch (1967). Consider, for example, a system with $N$ particles. From D'Alembert's principle,

$$\sum_{n=1}^{N} (m_n \ddot{\mathbf{r}}_n - \mathbf{F}_n) \cdot \delta \mathbf{r}_n = 0 \tag{2.1}$$

where $\mathbf{r}_n$ is the position vector of the $n$th particle (i.e., the generalized coordinate for that particle), $\mathbf{F}_n$ is the vector force applied to the $n$th particle, $m_n$ is the mass of the $n$th particle, and $\delta \mathbf{r}_n$ is the virtual displacement of the $n$th particle.

Virtual displacements are defined as infinitesimal, hypothetical changes in the coordinate system from the particle's actual path. Since virtual displacements are hypothetical or "imagined" displacements, there is no time change associated with them. Each virtual displacement is subject to the system constraints. For example, if one were to describe the dynamics of a bead traveling on a thin wire, the bead could not be virtually displaced off the wire. The virtual work is defined in terms of the virtual displacement as follows:

$$\delta W = \sum_{n=1}^{N} \mathbf{F}_n \cdot \delta \mathbf{r}_n \tag{2.2}$$

The symbol $\delta$ is used to describe the character of the variations, and the rules of elementary calculus apply to this operator. Hence, the first term in equation (2.1) can be expressed as

$$\ddot{\mathbf{r}}_n \cdot \delta \mathbf{r}_n = \frac{d}{dt}(\dot{\mathbf{r}}_n \cdot \delta \mathbf{r}_n) - \delta\left(\frac{1}{2}\dot{\mathbf{r}}_n \cdot \dot{\mathbf{r}}_n\right) \tag{2.3}$$

Substituting equations (2.2) and (2.3) into equation (2.1), one obtains

$$\sum_{n=1}^{N}\left[m_n\left(\frac{d}{dt}(\dot{\mathbf{r}}_n \cdot \delta \mathbf{r}_n) - \delta\left(\frac{1}{2}\dot{\mathbf{r}}_n \cdot \dot{\mathbf{r}}_n\right)\right)\right] - \delta W = 0. \tag{2.4}$$

Now, the total kinetic energy of the $N$ particles is defined as

$$T = \sum_{n=1}^{N} \tfrac{1}{2} m_n \dot{\mathbf{r}}_n \cdot \dot{\mathbf{r}}_n \tag{2.5}$$

Thus, we can rewrite equation (2.4) as follows:

$$\delta T + \delta W = \sum_{n=1}^{N} m_n \frac{d}{dt}(\dot{\mathbf{r}}_n \cdot \delta \mathbf{r}_n) \tag{2.6}$$

The instantaneous response of the dynamic system is thus described in $N$ generalized coordinates as a function of time. A "varied" path is chosen such that the virtual displacement is zero (i.e., $\delta \mathbf{r}_n = 0$) at the time limits $t_1$ and $t_2$. To derive the generalized Hamilton's principle, multiply equation (2.6) by $dt$ and integrate between the limits $t_1$ and $t_2$:

$$\begin{aligned}\int_{t_1}^{t_2} (\delta T + \delta W)dt &= \int_{t_1}^{t_2} \sum_{n=1}^{N} m_n \frac{d}{dt} (\dot{\mathbf{r}}_n \cdot \delta\mathbf{r})dt \\ &= \sum_{n=1}^{N} \int_{t_1}^{t_2} m_n \frac{d}{dt} (\dot{\mathbf{r}}_n \cdot \delta\mathbf{r})dt \\ &= \sum_{n=1}^{N} m_n(\dot{\mathbf{r}}_n \cdot \delta\mathbf{r})\Big|_{t_1}^{t_2} \\ &= 0 \end{aligned} \tag{2.7}$$

The virtual work can be divided into a portion due to conservative forces and a portion due to nonconservative forces:

$$\delta W = \delta W_c + \delta W_{nc} \tag{2.8}$$

In a conservative field, the virtual work is simply the negative of the change in the virtual potential energy:

$$\delta W_c = -\delta V \tag{2.9}$$

A conservative force is one that performs no work around a closed dynamical path. For example, consider the mass-spring system illustrated in Figure 2.1. If the spring is given some initial displacement, $X_0$, in the absence of any damping the mass will simply oscillate about the static equilibrium position, $X_{\mathrm{eq}}$, with a

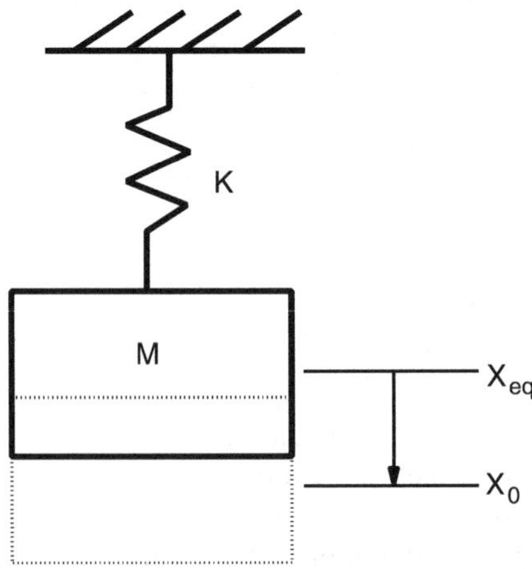


**Figure 2.1:** Schematic of simple mass-spring system.

cyclic change between kinetic and potential energy. However, each time the mass returns to the original point of initial displacement, the total work performed by the conservative spring force is zero. In contrast, if a damper is introduced into the system, then energy is continuously dissipated. Thus, the force due to a damper is nonconservative and leads to nonconservative work. In general, nonconservative forces can increase, decrease, or maintain the total energy within a system. External disturbances are typically those nonconservative forces that increase the total energy of the system. External forces used to decrease the total energy of the system are considered stabilizing and provide a mechanism for controlling the structural response. In adaptive structures, we exploit electromechanical materials to generate nonconservative forces necessary to remove energy from the system or to modify the primary input impedance of the structure.

For Hamilton's principle, as opposed to the generalized Hamilton's principle, the nonconservative work is set equal to zero such that

$$\int_{t_1}^{t_2} (\delta T - \delta V)dt = 0 \tag{2.10}$$

The Lagrangian is defined as

$$L = T - V \tag{2.11}$$

Substituting equation (2.11) into equation (2.10) and assuming that the constraints on the system are functions of the coordinates and time, one obtains Hamilton's principle:

$$\delta \int_{t_1}^{t_2} L\ dt = 0 \tag{2.12}$$

This principle states that, for any particle acted upon by conservative forces, the particle moves such that the difference between the time average of the kinetic and potential energies is stationary (Forray, 1968). Thus, Hamilton's principle is a formulation of the problems of dynamics, and the condition that yields the integral stationary leads to the equations of motion of the system and associated boundary conditions (Meirovitch, 1967).

**Example 2.1: Tapered Beam in Tension** Hamilton's principle is applied to derive the equation of extensional motion (one-dimensional) for the tapered beam illustrated in Figure 2.2. Uniform deformation is assumed over each cross section of the beam. The strain energy for a beam in extension can be expressed as (Meirovitch, 1967)

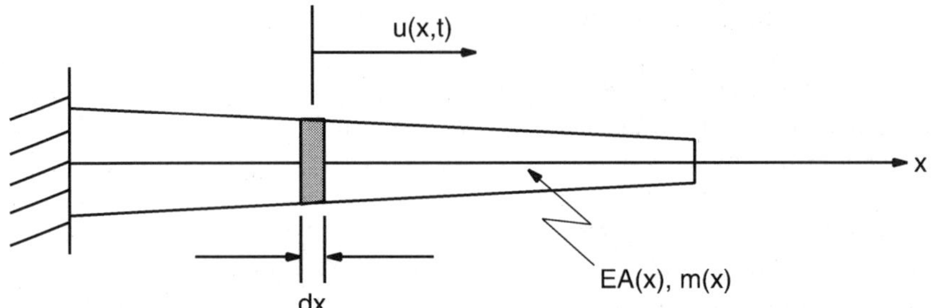


**Figure 2.2:** Schematic diagram of tapered beam in extension.

$$V(t) = \frac{1}{2} \int_0^L EA(x) \left[ \frac{\partial u(x,t)}{\partial x} \right]^2 dx \tag{2.13}$$

where $EA(x)$ is the product of the Young's modulus and the cross-sectional area as a function of $x$. Now, the kinetic energy of the beam can be expressed as follows:

$$T(t) = \frac{1}{2} \int_0^L m(x) \left[ \frac{\partial u(x,t)}{\partial t} \right]^2 dx \tag{2.14}$$

where $m(x)$ is the mass per unit length. Applying Hamilton's principle,

$$\delta \int_{t_1}^{t_2} \left( \frac{1}{2} \int_0^L m(x) \left[ \frac{\partial u}{\partial t} \right]^2 dx - \frac{1}{2} \int_0^L EA(x) \left[ \frac{\partial u}{\partial x} \right]^2 dx \right) dt = 0 \tag{2.15}$$

Since the integrations with respect to $t$ and $x$ are interchangeable, the integrals can be evaluated one by one. Taking the time integral of the kinetic energy term,

$$\frac{1}{2} \int_{t_1}^{t_2} m(x) \delta \left( \frac{\partial u}{\partial t} \right)^2 dt = \int_{t_1}^{t_2} m(x) \frac{\partial u}{\partial t} \delta \left( \frac{\partial u}{\partial t} \right) dt \tag{2.16}$$

Now, the differential operators and the virtual operator commute, so

$$\int_{t_1}^{t_2} m(x)\,\frac{\partial u}{\partial t}\,\delta\left(\frac{\partial u}{\partial t}\right)\,dt$$

$$= \int_{t_1}^{t_2} m(x)\,\frac{\partial u}{\partial t}\,\frac{\partial}{\partial t}\,(\delta u)\,dt \tag{2.17}$$

$$= m(x)\,\frac{\partial u}{\partial t}\,\delta u\big|_{t_1}^{t_2} - \int_{t_1}^{t_2} m(x)\,\frac{\partial^2 u}{\partial t^2}\,\delta u\,dt \tag{2.18}$$

$$= -\int_{t_1}^{t_2} m(x)\frac{\partial^2 u}{\partial t^2}\,\delta u\,dt \tag{2.19}$$

Similarly, we can integrate the strain energy term by parts:

$$\frac{1}{2}\,\delta\int_0^L EA(x)\left[\frac{\partial u}{\partial x}\right]^2 dx$$

$$= \int_0^L EA(x)\,\frac{\partial u}{\partial x}\,\delta\left(\frac{\partial u}{\partial x}\right)\,dx \tag{2.20}$$

$$= \int_0^L EA(x)\,\frac{\partial u}{\partial x}\,\frac{\partial}{\partial x}\,(\delta u)\,dx \tag{2.21}$$

$$= EA(x)\,\frac{\partial u}{\partial x}\,\delta u\big|_0^L - \int_0^L \frac{\partial}{\partial x}\left(EA(x)\,\frac{\partial u}{\partial x}\right)\,\delta u\,dx \tag{2.22}$$

Substituting equations (2.19) and (2.22) into equation (2.15), one obtains:

$$\int_{t_1}^{t_2}\left\{\int_0^L\left[\frac{\partial}{\partial x}\left(EA(x)\,\frac{\partial u}{\partial x}\right) - m(x)\frac{\partial^2 u}{\partial t^2}\right]\delta u\,dx - EA(x)\,\frac{\partial u}{\partial x}\,\delta u\big|_0^L\right\}dt = 0 \tag{2.23}$$

Equation (2.23) must hold for those $\delta u$ that are identically zero at $x = 0$ and $x = L$, but $\delta u$ is otherwise arbitrary over the domain. Thus, the integral is rendered stationary when

$$\frac{\partial}{\partial x}\left(EA(x)\frac{\partial u(x,t)}{\partial x}\right) - m(x)\,\frac{\partial^2 u(x,t)}{\partial t^2} = 0 \tag{2.24}$$

and

$$EA(x) \frac{\partial u(x,t)}{\partial x} \delta u(x,t)|_0^L = 0 \tag{2.25}$$

which is the differential equation of motion and corresponding geometric and natural boundary conditions of the structure.

**Example 2.2: Uniform Beam in Transverse Vibration** The generalized Hamilton's principle is applied to derive the equation of transverse motion for a nonuniform beam subjected to a distributed load $f(x,t)$, as illustrated in Figure 2.3*a*. A free-body diagram of a differential element of the beam is illustrated in

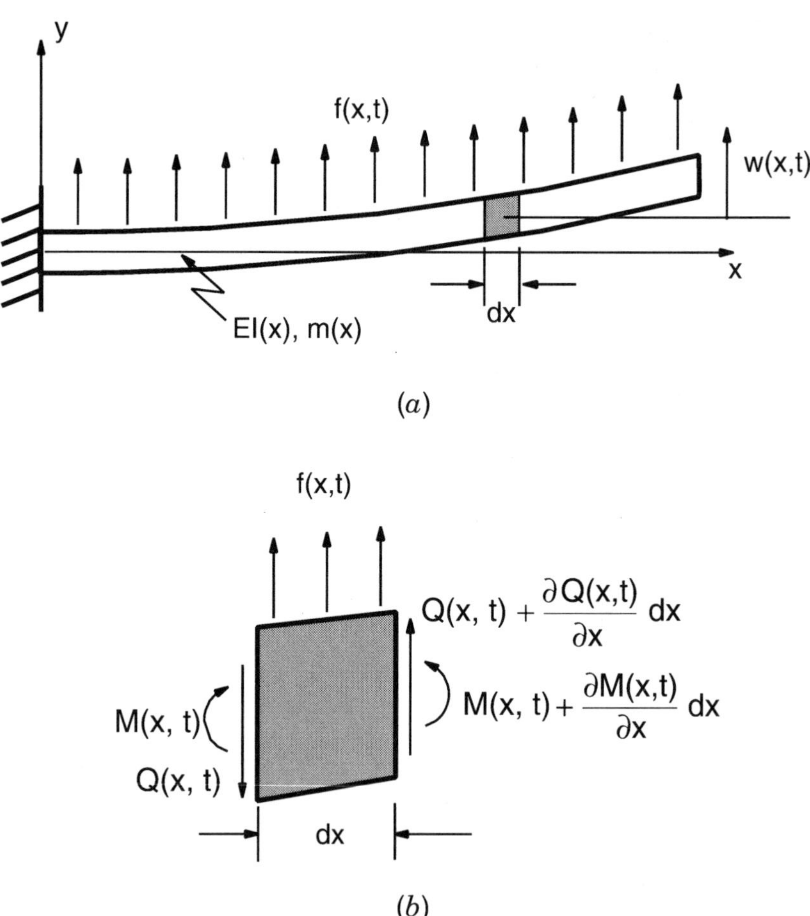


**Figure 2.3:** (*a*) Schematic diagram of beam in transverse vibration. (*b*) Free-body diagram of differential element.

Figure 2.3*b*. For the purpose of this example, we assume that the shear deformation of the beam is small with respect to the bending deformation and that the rotatory inertia is negligible in comparison to that due to translation. The total virtual work can be attributed to the conservative work due to the potential energy of the structure and the nonconservative work due to the distributed load. Based upon the previous assumptions, the potential energy of the beam can be expressed as follows (Meirovitch, 1967):

$$V(t) = \frac{1}{2} \int_0^L M(x,t) \frac{\partial^2 w(x,t)}{\partial x^2} \, dx \tag{2.26}$$

where

$$M(x,t) = EI(x) \frac{\partial^2 w(x,t)}{\partial x^2} \tag{2.27}$$

and $EI(x)$ is the product of the Young's modulus and the area moment of inertia as a function of $x$. The virtual work due to the nonconservative forces can be computed as follows:

$$\delta W_{nc}(t) = \int_0^L f(x,t)\delta w(x,t) dx \tag{2.28}$$

Now, the kinetic energy of the beam due to translation can be expressed as follows:

$$T(t) = \frac{1}{2} \int_0^L m(x) \left[ \frac{\partial w(x,t)}{\partial t} \right]^2 dx \tag{2.29}$$

where $m(x)$ is the mass per unit length. Applying the extended Hamilton's principle equation (2.7),

$$\int_{t_1}^{t_2} \left\{ \delta \left( \frac{1}{2} \int_0^L m(x) \left[ \frac{\partial w}{\partial t} \right]^2 \right) dx - \delta \left( \frac{1}{2} \int_0^L EI(x) \left[ \frac{\partial^2 w}{\partial x^2} \right]^2 \right) dx \right.$$
$$\left. + \int_0^L f(x,t)\delta w \, dx \right\} dt = 0. \tag{2.30}$$

Proceeding as in Example 2.1, we integrate the kinetic energy term by parts with respect to time:

$$\frac{1}{2}\int_{t_1}^{t_2} m(x)\delta\left(\frac{\partial w}{\partial t}\right)^2 dt = -\int_{t_1}^{t_2} m(x)\,\frac{\partial^2 w}{\partial t^2}\,\delta w\, dt \tag{2.31}$$

Integration of the potential energy term by parts with respect to the spatial variable yields

$$\frac{1}{2}\int_0^L EI(x)\delta\left(\frac{\partial^2 w}{\partial x^2}\right)^2 dx = EI(x)\,\frac{\partial^2 w}{\partial x^2}\,\delta\left(\frac{\partial w}{\partial x}\right)\Big|_0^L - \int_0^L \frac{\partial}{\partial x}\left(EI(x)\,\frac{\partial^2 w}{\partial x^2}\right)\frac{\partial}{\partial x}(\delta w)dx \tag{2.32}$$

Integrating the term on the right-hand side of equation (2.32) by parts again:

$$\int_0^L \frac{\partial}{\partial x}\left(EI(x)\frac{\partial^2 w}{\partial x^2}\right)\frac{\partial}{\partial x}(\delta w)dt = \frac{\partial}{\partial x}\left(EI(x)\frac{\partial^2 w}{\partial x^2}\right)\delta w\Big|_0^L - \int_0^L \frac{\partial^2}{\partial x^2}\left(EI(x)\,\frac{\partial^2 w}{\partial x^2}\right)\delta w\, dx \tag{2.33}$$

Substituting equation (2.33) into equation (2.32) and equations (2.31) and (2.32) into equation (2.30) and collecting terms, one obtains

$$\int_{t_1}^{t_2}\left\{\int_0^L\left[-m(x)\,\frac{\partial^2 w}{\partial t^2} - \frac{\partial^2}{\partial x^2}\left(EI(x)\,\frac{\partial^2 w}{\partial x^2}\right) + f(x,t)\right]\delta w\, dx - EI(x)\,\frac{\partial^2 w}{\partial x^2}\,\delta\left(\frac{\partial w}{\partial x}\right)\Big|_0^L + \frac{\partial}{\partial x}\left(EI(x)\,\frac{\partial^2 w}{\partial x^2}\right)\delta w\,\Big|_0^L\right\} dt \tag{2.34}$$

Thus, the equations of motion and corresponding boundary conditions are obtained from the solution that renders the integral stationary:

$$m(x)\,\frac{\partial^2 w}{\partial t^2} + \frac{\partial^2}{\partial x^2}\left(EI(x)\,\frac{\partial^2 w}{\partial x^2}\right) = f(x,t) \tag{2.35}$$

subject to

$$EI(x)\frac{\partial^2 w}{\partial x^2}\,\delta\left(\frac{\partial w}{\partial x}\right)\Big|_0^L = 0 \tag{2.36}$$

and

$$\frac{\partial}{\partial x}\left(EI(x)\,\frac{\partial^2 w}{\partial x^2}\right)\,\delta w\,\Big|_0^L = 0 \tag{2.37}$$

Equations (2.35)–(2.37) yield a boundary-value problem, as do equations (2.24)–(2.25). Thus, the variational mechanics approach yields the equation of motion for the system as well as all possible geometric and natural boundary conditions. For example, in equation (2.37), two potential boundary conditions exist. Either the displacement (geometric boundary condition) or the shear force (natural boundary condition) of the structure must be identically zero at the boundaries. The boundary conditions imposed on the structure ensure that the solution to the equation of motion is unique. Methods of solving the system of equations are presented in subsequent sections.

### 2.2.2 Lagrange's Equations

The generalized Hamilton's principle can be used to develop Lagrange's equations. Consider an $n$-dimensional system that can be described in $n$ generalized coordinates. Generalized coordinates are defined to be the minimum number of independent coordinates necessary to describe the system (Meirovitch, 1967). For a system of $p$ particles, each with displacement vectors $\mathbf{r}_1$, $\mathbf{r}_2$, ..., $\mathbf{r}_p$, the number $n$ of independent coordinates depends upon the number, $c$, of constraint equations:

$$n = 3p - c \tag{2.38}$$

For example, consider the planar motion of a pendulum, as illustrated in Figure 2.4. The motion of the pendulum can be described in three Cartesian coordinates, $x$, $y$, and $z$. However it is subject to the following two constraint equations:

$$x^2 + y^2 = L$$
$$z = 0$$

Thus, there is only one generalized coordinate, $\theta$, for the mass concentrated at the end of the rigid, massless rod. One can express the position vector of each particle in terms of the generalized coordinates:

$$\mathbf{r}_i = \mathbf{r}_i(q_1, q_2, \ldots, q_n, t), \qquad i = 1, 2, \ldots, p \tag{2.39}$$

Since the kinetic energy of a system of $p$ particles can be expressed as follows:

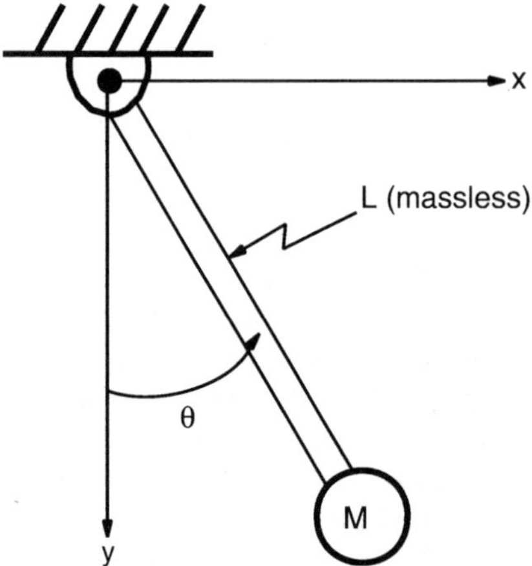


**Figure 2.4:** Schematic diagram of a simple pendulum.

$$T = \frac{1}{2} \sum_{i=1}^{p} m_i \dot{\mathbf{r}}_i \cdot \dot{\mathbf{r}}_i \tag{2.40}$$

where

$$\dot{\mathbf{r}}_i = \sum_{r=1}^{n} \frac{\partial \mathbf{r}_i}{\partial q_r} \dot{q}_r + \frac{\partial \mathbf{r}_i}{\partial t} \tag{2.41}$$

upon substituting equation (2.41) into equation (2.40), one obtains

$$T = \frac{1}{2} \sum_{i=1}^{p} m_i \left\{ \sum_{r=1}^{n} \sum_{s=1}^{n} \frac{\partial \mathbf{r}_i}{\partial q_r} \cdot \frac{\partial \mathbf{r}_i}{\partial q_s} \dot{q}_r \dot{q}_s + 2 \frac{\partial \mathbf{r}_i}{\partial t} \cdot \sum_{r=1}^{n} \frac{\partial \mathbf{r}_i}{\partial q_r} \dot{q}_r + \frac{\partial \mathbf{r}_i}{\partial t} \cdot \frac{\partial \mathbf{r}_i}{\partial t} \right\} dt \tag{2.42}$$

The kinetic energy is clearly a function of the generalized displacements, generalized velocities, and time. To compute $\delta T$, one proceeds as follows:

$$\delta T = \sum_{r=1}^{n} \frac{\partial T}{\partial q_r} \delta q_r + \sum_{r=1}^{n} \frac{\partial T}{\partial \dot{q}_r} \delta \dot{q}_r \tag{2.43}$$

where

$$\delta \dot{q}_r = \frac{d}{dt} \, \delta q_r \tag{2.44}$$

The virtual work can be computed from the product of the forces acting on each of the $p$ particles and the virtual displacement of each particle:

$$\delta W = \sum_{k=1}^{p} \mathbf{F}_k \cdot \delta \mathbf{r}_k \tag{2.45}$$

where

$$\delta \mathbf{r}_k = \sum_{r=1}^{n} \frac{\partial \mathbf{r}_k}{\partial q_r} \, \delta q_r \tag{2.46}$$

Substituting equation (2.46) into equation (2.45), one obtains

$$\delta W = \sum_{k=1}^{p} \sum_{r=1}^{n} \mathbf{F}_k \cdot \frac{\partial \mathbf{r}_k}{\partial q_r} \, \delta q_r \tag{2.47}$$

$$= \sum_{r=1}^{n} Q_r \, \delta q_r \tag{2.48}$$

where

$$Q_r = \sum_{k=1}^{p} \mathbf{F}_k \cdot \frac{\partial \mathbf{r}_k}{\partial q_r} \tag{2.49}$$

The generalized forces (in generalized coordinates) are designated by $Q_r$. One should note that, to completely control the system response, one must in general be able to arbitrarily specify the generalized forces, which leads to the concept of controllability. (Controllability is discussed in Chapter 3.) Substituting equations (2.43) and (2.48) into the generalized Hamilton's principle and integrating the kinetic energy term by parts (Meirovitch, 1967), one obtains

$$\int_{t_1}^{t_2} (\delta T + \delta W) dt = - \sum_{r=1}^{n} \int_{t_1}^{t_2} \left[ \frac{d}{dt} \left( \frac{\partial T}{\partial \dot{q}_r} \right) - \frac{\partial T}{\partial q_r} - Q_r \right] \delta q_r \, dt = 0 \tag{2.50}$$

Since the virtual displacements are arbitrary and independent of one another, one can choose to set all but one equal to zero. Stepping through this process one generalized coordinate at a time yields $n$ independent equations of motion; more specifically, Lagrange's equations of motion:

$$\frac{d}{dt}\left(\frac{\partial T}{\partial \dot{q}_r}\right) - \frac{\partial T}{\partial q_r} - Q_r = 0, \qquad r = 1, 2, \ldots, n \tag{2.51}$$

The generalized forces are in general composed of both conservative and nonconservative forces:

$$Q_r = Q_{rc} + Q_{rnc} \tag{2.52}$$

The conservative forces result in conservative work, which is described in terms of the potential energy $V$ of the system:

$$\delta W_c = -\delta V = \sum_{r=1}^{n} Q_{rc}\,\delta q_r \tag{2.53}$$

Now,

$$\delta V = \sum_{r=1}^{n} \frac{\partial V}{\partial q_r}\,\delta q_r \tag{2.54}$$

Thus, the generalized conservative forces are simply equal to the partial derivative of the potential energy with respect to each generalized coordinate:

$$Q_{rc} = -\frac{\partial V}{\partial q_r} \tag{2.55}$$

Substituting equation (2.55) into equation (2.51), one obtains Lagrange's equations in terms of the kinetic energy, potential energy, and the nonconservative forces:

$$\frac{d}{dt}\left(\frac{\partial T}{\partial \dot{q}_r}\right) - \frac{\partial T}{\partial q_r} + \frac{\partial V}{\partial q_r} = Q_{rnc}, \qquad r = 1, 2, \ldots, n \tag{2.56}$$

For a conservative system, the term $Q_{rnc} = 0$. In the control of adaptive structures, disturbance inputs typically arise through nonconservative forces, and to control the system response, one must generate additional nonconservative forces to "remove energy" from the system or modify the input impedance of the disturbance, for example. This important concept is explored in Chapter 9.

### 2.2.3 The Eigenvalue Problem

Before proceeding to the section on solution methodologies for the boundary value problem, the derivation of the eigenvalue problem from the boundary value problem is presented. As mentioned earlier, calculation of the eigenvalues for a structural system is an important step in many aspects of adaptive structure design. Generation of state space matrices, certain forms of spatial-temporal windows, test-analysis correlation methods, and a number of other design techniques all rely on knowledge of the system eigenvalues. The procedure begins with the typical assumption that the solution of the boundary value problem is separable in space and time:

$$w(x,t) = W(x)\eta(t) \tag{2.57}$$

Consider, for example, the partial differential equation describing the response of the beam under transverse vibration. Substituting equation (2.57) into equation (2.35) and separating variables, one obtains

$$\frac{1}{m(x)W(x)}\frac{d^2}{dx^2}\left(EI(x)\frac{d^2W(x)}{dx^2}\right) = -\frac{1}{\eta(t)}\frac{d^2\eta(t)}{dt^2} \tag{2.58}$$

Notice that the left-hand side of equation (2.58) is strictly a function of $x$, while the right-hand side is strictly a function of $t$. Since the variables are independent, both sides must be constant for the existence of a solution (Meirovitch, 1967). The constant typically chosen is $\omega^2$:

$$\frac{1}{m(x)W(x)}\frac{d^2}{dx^2}\left(EI(x)\frac{d^2W(x)}{dx^2}\right) = -\frac{1}{\eta(t)}\frac{d^2\eta(t)}{dt^2} = \omega^2 \tag{2.59}$$

One thus obtains two ordinary differential equations:

$$\frac{d^2}{dx^2}\left[EI(x)\frac{d^2W(x)}{dx^2}\right] - \omega^2 m(x)W(x) = 0 \tag{2.60}$$

$$\frac{d^2\eta(t)}{dt^2} + \omega^2\eta(t) = 0 \tag{2.61}$$

The nontrivial solution of equation (2.60), satisfying the homogeneous boundary conditions for distinct values of $\omega^2$, is termed the eigenvalue problem. The eigenvalues are designated by $\omega^2$, and the associated eigenfunctions are the nontrivial solutions $W(x)$, corresponding to each solution $\omega^2$.

**Example 2.3: The Eigenvalue Problem for a Simply Supported Beam** Consider the uniform property beam with hinged supports at each end presented

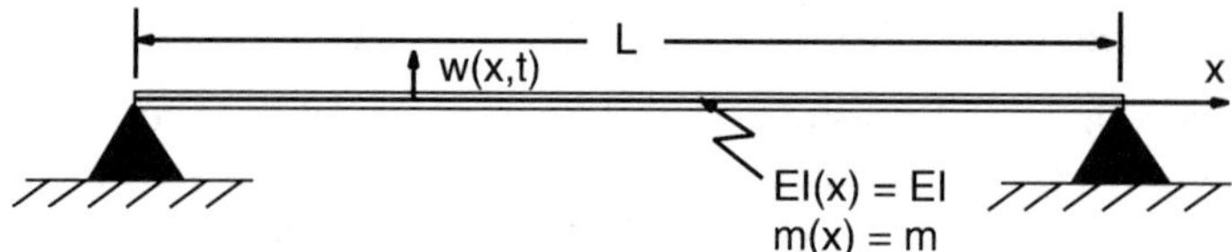


**Figure 2.5:** Schematic diagram of a simply supported beam.

in Figure 2.5, often termed a simply supported beam. The eigenvalue problem for this structure has previously been derived in equations (2.60) and (2.61). The boundary conditions can be obtained as follows, since the deflection and bending moment must be zero at each end of the structure:

$$W(x)|_{x=0,L}\eta(t) = 0 \tag{2.62}$$

$$EI(x)\,\frac{d^2W(x)}{dx^2}|_{x=0,L}\eta(t) = 0 \tag{2.63}$$

Since the solution to the eigenvalue problem was assumed separable in space and time, the boundary conditions become

$$W(x)|_{x=0,L} = 0 \tag{2.64}$$

$$EI(x)\,\frac{d^2W(x)}{dx^2}|_{x=0,L} = 0 \tag{2.65}$$

The spatial portion of the differential equation of motion, shown previously in equation (2.60), can be expressed as

$$\frac{d^4W(x)}{dx^4} - \gamma^4 W(x) = 0 \tag{2.66}$$

where

$$\gamma^4 = \frac{\omega^2 m}{EI} \tag{2.67}$$

The general solution to the differential equation is (Boyce and DiPrima, 1986)

$$W(x) = A_1 \sin(\gamma x) + A_2 \cos(\gamma x) + A_3 \sinh(\gamma x) + A_4 \cosh(\gamma x) \tag{2.68}$$

Applying the boundary conditions, one obtains

$$W(0) = 0 = A_2 + A_4 \tag{2.69}$$

$$\frac{d^2W(0)}{dx^2} = 0 = \gamma^2(-A_2 + A_4) \tag{2.70}$$

Thus, $A_2 = A_4 = 0$.

Now,

$$W(L) = 0 = A_1 \sin(\gamma L) + A_3 \sinh(\gamma L) \tag{2.71}$$

$$\frac{d^2W(L)}{dx^2} = 0 = \gamma^2(-A_1 \sin(\gamma L) + A_3 \sinh(\gamma L)) \tag{2.72}$$

Thus, $A_3 = 0$ and the characteristic equation is

$$\sin(\gamma_r L) = 0 \tag{2.73}$$

Solution of this equation for discrete values of $\gamma_r$ produces a corresponding set of eigenvalues of the system:

$$\omega_r^2 = \frac{EI}{m}\left(\frac{r\pi}{L}\right)^4, \qquad r = 1, 2, \ldots \tag{2.74}$$

and corresponding eigenfunctions:

$$W_r(x) = A_r \sin(\gamma_r x) \tag{2.75}$$

The constants $A_r$ are arbitrary, but as demonstrated in Section 2.2.4, it is often convenient to normalize them with respect to the mass.

Now, consider an alternative solution to the differential equation of motion, often termed the traveling wave solution:

$$W(x) = A_1 \exp(-jkx) + A_2 \exp(jkx) + A_3 \exp(kx) + A_4 \exp(-kx) \tag{2.76}$$

Applying the boundary conditions at $x = 0$, one obtains

$$W(0) = 0 = A_1 + A_2 + A_3 + A_4 \tag{2.77}$$

$$\frac{d^2W(0)}{dx^2} = 0 = k^2[-A_1 - A_2 + A_3 + A_4] \tag{2.78}$$

Applying the boundary conditions at $x = L$, one obtains

$$W(L) = 0 = A_1 \exp(-jkL) + A_2 \exp(jkL) + A_3 \exp(kL) + A_4 \exp(-kL) \tag{2.79}$$

$$\frac{d^2W(L)}{dx^2} = 0 = k^2[-A_1 \exp(-jkL) - A_2 \exp(jkL) + A_3 \exp(kL) + A_4 \exp(-kL)] \tag{2.80}$$

After manipulation, the characteristic equation is

$$\frac{\exp(jkL) - \exp(-jkL)}{2i} = 0 \tag{2.81}$$

The eigenvalues are the same as shown previously in equation (2.73), and the corresponding eigenfunctions are given as follows:

$$W_r(x) = A_r \frac{\exp(jk_rL) - \exp(-jk_rL)}{2i} \tag{2.82}$$

where

$$k_r = \sqrt[4]{\frac{m}{EI}}\sqrt{\omega_r} \tag{2.83}$$

Thus, as one might expect, the eigenvalues and eigenfunctions can be obtained from the alternative form of the assumed solution. The concept of traveling wave analysis and the implications of the alternative solution form are discussed in detail later in this chapter.

### 2.2.4 General Form of the Eigenvalue Problem, Orthogonality, and the Expansion Theorem

Consider a continuous system described by the following partial differential equation:

$$\mathcal{L}[w(\mathbf{x}, t)] + \mathcal{M}\left[\frac{\partial^2 w(\mathbf{x}, t)}{\partial t^2}\right] = f(\mathbf{x}, t) \tag{2.84}$$

over the domain $D$ with spatial variables defined by $\mathbf{x}$. $\mathcal{L}$ is defined to be a linear homogeneous differential operator of order $2p$, and $\mathcal{M}$ is defined to be a linear homogeneous differential operator of order $2q$, where $q \leq p$. The corresponding eigenvalue problem can be expressed as follows:

$$\mathcal{L}[W(\mathbf{x})] = \omega^2 \mathcal{M}[W(\mathbf{x})] \tag{2.85}$$

There are $p$ corresponding boundary conditions associated with the partial differential equation, which can be represented as follows:

$$\mathcal{B}_i[W(\mathbf{x})] = \omega^2 \mathcal{C}_i[W(\mathbf{x})], \qquad i = 1, 2, \ldots, p \tag{2.86}$$

where $\mathcal{B}_i$ and $\mathcal{C}_i$ are linear homogeneous differential operators of order less than or equal to $2p - 1$.

For the eigenvalue problem presented in Example 2.3, the differential operators are defined as follows:

$$\mathcal{L} = \frac{d^2}{dx^2}\left(EI\,\frac{d^2}{dx^2}\right) \tag{2.87}$$

$$\mathcal{M} = m \tag{2.88}$$

and the differential operators for the boundary conditions are

$$\mathcal{B}_1 = 1 \tag{2.89}$$

$$\mathcal{C}_1 = 0 \tag{2.90}$$

$$\mathcal{B}_2 = EI\,\frac{d^2}{dx^2} \tag{2.91}$$

$$\mathcal{C}_2 = 0 \tag{2.92}$$

Thus, for the example provided, the eigenvalue does not appear in the boundary conditions. For this special case, there are three types of functions that can be described, as outlined by Meirovitch (1967): admissible functions, comparison functions, and eigenfunctions. *Admissible functions* are any set of functions that satisfy the geometric boundary conditions of the eigenvalue problem and are differentiable $p$ times. *Comparison functions* are any set of functions that satisfy both the geometric and natural boundary conditions and are $2p$ times differentiable. The final set of functions are the *eigenfunctions*, and these special functions satisfy all boundary conditions and the differential equation of the eigenvalue problem. For the purpose of developing the general principle of orthogonality, we define the self-adjoint eigenvalue problem. For any two arbitrary comparison functions $w_m$ and $w_n$, the eigenvalue problem is self-adjoint if

$$\int_D w_m \mathcal{L}[w_n]dD = \int_D w_n \mathcal{L}[w_m]dD \tag{2.93}$$

and

$$\int_D w_m \mathcal{M}[w_n] dD = \int_D w_n \mathcal{M}[w_m] dD \tag{2.94}$$

Now, let $\omega_m^2$ and $\omega_n^2$ be two distinct eigenvalues with corresponding eigenfunctions $w_m$ and $w_n$ resulting from the solution of the self-adjoint eigenvalue problem. Then one can readily show that

$$\int_D w_m \mathcal{M}[w_n] dD = 0, \qquad \text{for } \omega_m^2 \neq \omega_n^2 \tag{2.95}$$

Equation (2.95) is known as the generalized condition of orthogonality, and from the statement of the eigenvalue problem, it follows that

$$\int_D w_m \mathcal{L}[w_n] dD = 0, \qquad \text{for } \omega_m^2 \neq \omega_n^2 \tag{2.96}$$

For simplicity, the eigenfunctions are typically normalized with respect to $\mathcal{M}$ such that

$$\int_D w_m \mathcal{M}[w_n] dD = \delta_{mn} \tag{2.97}$$

where $\delta_{mn}$ is the Kronecker delta. Thus, equation (2.96) can be expressed as

$$\int_D w_m \mathcal{L}[w_n] dD = \delta_{mn} \omega_m^2 \tag{2.98}$$

We have now developed expressions for a family of orthonormal eigenfunctions (modes) based upon equation (2.97), and this set of modes forms a basis that spans the solution space of the eigenvalue problem. Thus, the response of the system can be computed at any arbitrary point in the domain as a linear combination (superposition) of these basis functions:

$$W(\mathbf{x}) = \sum_{m=1}^{\infty} a_m w_m(\mathbf{x}) \tag{2.99}$$

where the principle of orthogonality can be used to determine the coefficients:

$$a_m = \int_D W(\mathbf{x})\mathcal{M}[w_m(\mathbf{x})]dD, \qquad m = 1, 2, \ldots \tag{2.100}$$

Equations (2.99) and (2.100) are the mathematical statement of the expansion theorem for continuous systems, which is one of the most widely used tools in the field of vibration analysis.

In summary, the eigenvalue problem solution is a method for determining a set of orthogonal functions that are generalized coordinates spanning the solution space of the system. Expressing the continuous system response in terms of a set of orthonormal eigenfunctions leads to the concept of modal analysis. While conceptually this method of analysis is ideal for modeling system response, the eigenfunctions of adaptive structures often cannot be determined in practice, due to inhomogeneous properties of the structure, local changes in stiffness and mass due to the embedded or surface-mounted active materials, complexity of the structure, and other such factors. (In addition, it is impractical to construct a model with an infinite number of modes!) However, eigenfunctions of simple systems often serve as comparison functions for more complex systems and can be used to construct "approximate" modal solutions based upon techniques such as the Rayleigh-Ritz method, assumed-modes method, or Galerkin's method. In the next section, modal analysis solutions for simple systems are discussed, followed by a brief summary of approximate methods and finite element methods that are applicable for more complex dynamic systems.

## 2.3 SOLUTION METHODOLOGIES

Before discussing the various methods, exact and approximate, of solving the boundary value problem, an example is provided to motivate the section. Consider a beam undergoing transverse vibration, rigidly fixed at one end and free at the other (i.e., cantilevered). The eigenfunctions for this structure can be derived from the equations of motion and, as outlined by Blevins (1984), are given as follows:

$$\phi_r(x) = \cosh\left(\frac{\lambda_r x}{L}\right) - \cos\left(\frac{\lambda_r x}{L}\right) - \frac{\sinh(\lambda_r) - \sin(\lambda_r)}{\cosh(\lambda_r) + \cos(\lambda_r)}\left(\sinh\left(\frac{\lambda_r x}{L}\right) - \sin\left(\frac{\lambda_r x}{L}\right)\right) \tag{2.101}$$

where $\phi_r(x)$ is the eigenfunction and $\lambda_r$ is the corresponding eigenvalue. The eigenfunctions and eigenvalues can be used to construct the response in modal coordinates based upon modal analysis or Galerkin's method, as discussed in the subsequent section, since the eigenfunctions are simply a subset of comparison functions (i.e., functions that satisfy both the geometric and natural boundary conditions). Unless the structure is exceedingly simple, finding the eigenfunctions is most likely not possible. Thus, alternative methods of approximating the solution have been proposed, based upon a number of techniques, such as the Rayleigh-Ritz method and finite element analysis to name only a couple.

For the purpose of this discussion, the assumed-modes method is employed to approximate the structural response in terms of a finite number of admissible functions (i.e., functions that satisfy the geometric boundary conditions). The geometric boundary conditions of a cantilevered beam are such that the displacement and slope of the beam are zero at the fixed boundary. Hence, the following set of functions satisfy these boundary conditions and are thus admissible functions:

$$\psi_r(x) = \frac{x}{L}\sin\left(\frac{r\pi x}{2L}\right) \tag{2.102}$$

The functions should not be "blindly" selected. One should consider the response at the opposite end of the beam as well, even though there is no geometric boundary condition imposed. By letting the argument of the sine function vary from 0 to $r\pi/2$, the admissible functions yield significant response at $x = L$, which is desirable for the transverse vibration response.

The frequency response of the structure for a colocated point force disturbance and displacement sensor are presented in Figure 2.6, using the eigenfunc-

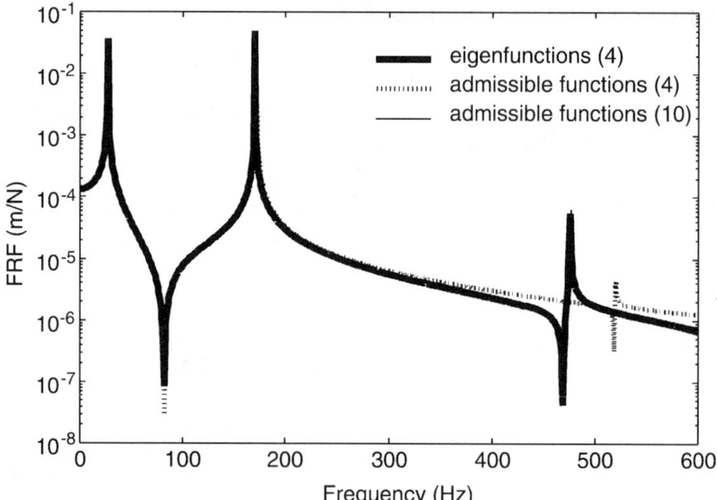


**Figure 2.6:** A comparison between the frequency response function (FRF) obtained from the modal model and that obtained by an assumed modes approach for a cantilevered beam.

tions to generate the response in one case and the admissible functions in the other two cases. The details of the model are included in the MATLAB files **galkbeam.m** and **assmbeam.m**, which have been used to generate the state-space model. The MATLAB file **plotmthd.m** can be executed to recreate a plot of the frequency response function (FRF) if desired. As illustrated in Figure 2.6, when using four admissible functions to compute the FRF, the response of the system at the first two resonant frequencies is reasonably accurate; however, the response begins to deviate more significantly near the third resonant frequency. As the number of admissible functions used in the structural model is increased, the estimated response more accurately models the "true" modal response. As discussed by Meirovitch (1980), the eigenvalues of the approximate model approach the eigenvalues of the modal model monotonically from above. As indicated in this example, one needs between two and three times the number of admissible functions to represent the system response. Thus, the price paid for using an approximate method is in the number of admissible functions required to generate an accurate estimate of the system response over a desired frequency range.

### 2.3.1 Modal Analysis

Consider a continuous system represented in operator notation, as outlined in the previous section:

$$\mathcal{L}[w(\mathbf{x},t)] + \mathcal{C}\left[\frac{\partial w(\mathbf{x},t)}{\partial t}\right] + \mathcal{M}\left[\frac{\partial^2 w(\mathbf{x},t)}{\partial t^2}\right] = f(\mathbf{x},t) \tag{2.103}$$

defined over some domain $D$ with spatial coordinates $\mathbf{x}$. The force is considered to be some arbitrary input defined by some spatial aperture of application and time-dependent characteristic. The differential operators are consistent with the definitions presented in the previous section, with the exception of $\mathcal{C}$, which is defined as a linear combination of the mass and stiffness operators, $\mathcal{M}$ and $\mathcal{L}$ respectively:

$$\mathcal{C} = c_1\mathcal{L} + c_2\mathcal{M} \tag{2.104}$$

This special operator is commonly used to incorporate *proportional* damping. While in practice, this might not be the most realistic model for damping in a continuous structure, as discussed in a subsequent section, it provides a method of including damping in the model amenable to modal analysis.

The boundary conditions are assumed to be independent of the eigenvalues such that

$$\mathcal{B}_i[w(\mathbf{x},t)] = 0, \qquad i = 1, 2, \ldots, p \tag{2.105}$$

where $\mathcal{B}_i$ was defined as a linear homogeneous differential operator of order less than or equal to $2p - 1$.

Ignoring the proportional damping and solving the eigenvalue problem yields the eigenvalues and eigenfunctions of the continuous system. From the expansion theorem, the system response at any arbitrary coordinate in the domain can be expressed as a linear combination of the eigenfunctions:

$$w(\mathbf{x}, t) = \sum_{m=1}^{\infty} w_m(\mathbf{x})\eta_m(t) \tag{2.106}$$

where $\eta_m(t)$ are the generalized coordinates of the system. The eigenfunctions are normalized as outlined in the previous section such that

$$\int_D w_m(\mathbf{x})m(x)w_n(\mathbf{x})dD(\mathbf{x}) = \delta_{mn} \tag{2.107}$$

and

$$\int_D w_m(\mathbf{x})\mathcal{L}[w_n(\mathbf{x})]dD(\mathbf{x}) = \delta_{mn}\omega_m^2 \tag{2.108}$$

Since proportional damping is assumed,

$$\int_D w_m(\mathbf{x})\mathcal{C}[w_n(\mathbf{x})]dD(\mathbf{x}) = \delta_{mn}2\zeta_m\omega_m \tag{2.109}$$

where

$$\zeta_m = \frac{1}{2\omega_m}(c_1\omega_m^2 + c_2) \tag{2.110}$$

Substituting equation (2.106) into equation (2.103), one obtains:

$$\mathcal{L}\left[\sum_{m=1}^{\infty} w_m(\mathbf{x})\eta_m(t)\right] + \mathcal{C}\left[\frac{\partial}{\partial t}\sum_{m=1}^{\infty} w_m(\mathbf{x})\eta_m(t)\right] + \mathcal{M}\left[\frac{\partial^2}{\partial t^2}\sum_{m=1}^{\infty} w_m(\mathbf{x})\eta_m(t)\right] = f(\mathbf{x}, t) \tag{2.111}$$

Taking advantage of the fact that the eigenfunctions are orthonormal, we choose to multiply equation (2.111) by $w_n(\mathbf{x})$ and integrate over the domain:

$$\sum_{m=1}^{\infty} \eta_m(t)\delta_{mn}\omega_m^2 + \sum_{m=1}^{\infty} \dot{\eta}_m(t)\delta_{mn}2\zeta_m\omega_m + \sum_{m=1}^{\infty} \ddot{\eta}_m(t)\delta_{mn} = Q_n(t) \qquad (2.112)$$

where

$$Q_n(t) = \int_D w_n(\mathbf{x}) f(\mathbf{x}, t) dD(\mathbf{x}) \qquad (2.113)$$

is defined as the $n$th generalized force. Due to orthogonality, the summation in equation (2.112) only holds when $m = n$. Thus, one obtains an infinite set of uncoupled ordinary differential equations:

$$\ddot{\eta}_m(t) + 2\zeta_m\omega_m\dot{\eta}_m(t) + \omega_m^2\eta_m(t) = Q_m(t), \qquad m = 1, 2, \ldots \qquad (2.114)$$

The system is now described by an infinite set of second-order differential equations that can be solved independently. The solution to each second-order differential equation can be solved by Laplace transforms, or we can define state variables and reduce each second-order equation into two first-order differential equations. The latter approach is used in this book as the emphasis is placed upon control of the system response, and modern control theory is based upon the state variable representation. While in practice an infinite number of modes are required to predict the system response, the higher order modes can be truncated if the frequency range is limited, which is almost always the case in the control of adaptive structures, due to finite bandwidths of sensors, actuators, instrumentation, amplifiers, and limitations on the sample rate if digital signal processors are used. However, as demonstrated by Clark (1995), if colocated transducers are used in the control, one must take care to include sufficient modes for convergence of the transfer function zeros. Convergence of the zeros assures that the low-frequency compliance of the structural transfer function has been accurately modeled.

### 2.3.2 Approximate Solutions

Even the simplest adaptive structure typically exhibits nonuniform system properties due to embedded or surface-mounted active materials required for sensing and actuation. The resulting continuous system yields an eigenvalue problem that is, in general, impossible to solve. However, one is still confronted with the need to predict the resonant characteristics of the system and the response to various input disturbances, as well as the need to control inputs, if the objective is to eventually predict and control the system response. One is thus confronted with the challenge of developing an approximate modal model. A number of alternative methods exist for constructing this model:

1. Rayleigh-Ritz method (a variational approach)
2. Assumed-modes method (a variational approach)
3. Galerkin's method (a differential approach)
4. Colocation method (a differential approach)
5. Finite element method

***Rayleigh-Ritz Method*** The Rayleigh-Ritz method is a variational approach for deriving the equations of motion for a system with self-adjoint differential operators. It can be shown that the solution to the eigenvalue problem also renders the Rayleigh quotient stationary, where Rayleigh's quotient can be expressed as follows:

$$\omega^2 = R(w(\mathbf{x})) = \frac{\int_D w(\mathbf{x})\mathcal{L}[w(\mathbf{x})]dD}{\int_D w(\mathbf{x})\mathcal{M}[w(\mathbf{x})]dD} \tag{2.115}$$

for boundary conditions that are not dependent upon the eigenvalues. In the Rayleigh-Ritz method, one assumes a solution based upon a finite number of admissible functions:

$$w(\mathbf{x}) = \sum_{r=1}^{N} \phi_r(\mathbf{x})a_r \tag{2.116}$$

where $N$ linearly independent admissible functions $\phi_r(\mathbf{x})$, over the domain $D$, are used to expand the solution in terms of the undetermined coefficients $a_r$. Based upon this assumed solution, Rayleigh's quotient must be rendered stationary. As one includes more independent admissible functions in the expansion, the computed eigenvalues from the truncated model approach the actual eigenvalues of the system monotonically from above (Meirovitch, 1980). In essence, one obtains better approximations of the structural response in the solution space by "curve-fitting" the "true" modes with a finite set of admissible functions.

***Assumed-Modes Method*** An alternative to the Rayleigh-Ritz method is the assumed-modes method, which is also considered a variational approach since it is based upon Lagrange's equations. Again, a solution is assumed based upon a set of admissible functions as in equation (2.116); however, in this case, the undetermined coefficients are the time-dependent generalized coordinates $\eta'_r(t)$:

$$w(\mathbf{x}, t) = \sum_{r=1}^{N} \phi_r(\mathbf{x})\eta'_r(t) \tag{2.117}$$

where $\eta'_r(t)$ is used as opposed to $\eta_r(t)$ to distinguish the response in "discrete" modal coordinates from that obtained in the "true" (or infinite) modal coordinates, respectively. The assumed solution is substituted into Lagrange's equations (2.56), reducing the system of equations to discrete form. In general, both the Rayleigh-Ritz method and the assumed-modes method lead to $N$ equations of motion:

$$\sum_{s=1}^{N} (m_{rs}\ddot{\eta}'_s(t) + k_{rs}\eta'_s(t)) = 0, \qquad r = 1, 2, \ldots, N \tag{2.118}$$

The eigenvalue problem of the discretized system is readily derived from equation (2.118) with a discrete set $N$ of eigenvalues and eigenvectors. Thus, the approximate mode shapes can be constructed from the discrete eigenvectors since the $r$th element of the eigenvector serves to weight the $r$th admissible function used to expand the solution over the domain.

**Example 2.4: A Simply Supported Beam with a Surface Mounted Active Material** Consider the simply supported beam presented in Example 2.3. However, in this case, two piezoceramic elements are bonded symmetrically to the structure about its neutral axis, as illustrated in Figure 2.7. We assume that the elements are perfectly bonded and that the glue layer is infinitesimal. At this point, the focus is placed on the effect that the stiffness and mass of the elements will have on the dynamic response of the structure. The assumed-modes method is used to derive the equations of motion for the structure. We proceed by developing expressions for the potential energy and kinetic energy, respectively:

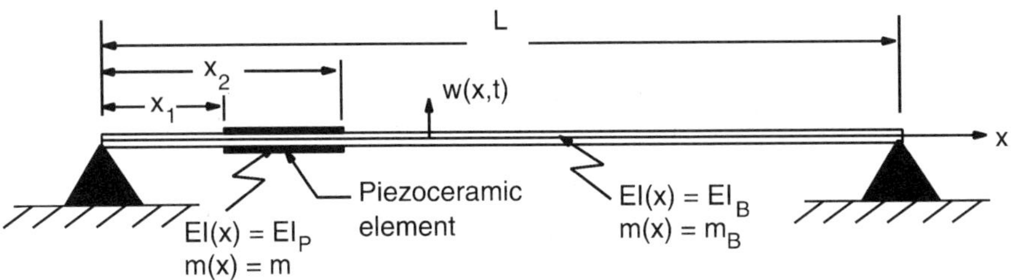


**Figure 2.7:** Schematic diagram of a simply supported beam configured with a distributed piezoelectric device.

$$V(t) = \frac{1}{2}\int_0^L EI_b \left(\frac{\partial^2 w(x,t)}{\partial x^2}\right)^2 dx$$
$$+ \frac{1}{2}\int_0^L EI_p \left(\frac{\partial^2 w(x,t)}{\partial x^2}\right)^2 [u(x-x_1) - u(x-x_2)]dx, \quad (2.119)$$

and

$$T(t) = \frac{1}{2}\int_0^L m_b \left(\frac{\partial w(x,t)}{\partial t}\right)^2 dx$$
$$+ \frac{1}{2}\int_0^L m_p \left(\frac{\partial w(x,t)}{\partial t}\right)^2 [u(x-x_1) - u(x-x_2)]dx \quad (2.120)$$

where $EI_b$ is the stiffness per unit length of the beam, $EI_p$ is the effective stiffness per unit length of the combined piezoceramic elements, $m_b$ is the mass per unit length of the beam, $m_p$ is the mass per unit length of the combined piezoceramic elements, and $u(x)$ is the spatial step function used to indicate the area of application of the piezoceramic element.

Now, we expand the solution in terms of a finite set of comparison functions:

$$w(x,t) = \sum_{r=1}^{N} \phi_r(x)\eta'_r(t) \quad (2.121)$$

where

$$\phi_r(x) = \sin\left(\frac{r\pi x}{L}\right) \quad (2.122)$$

which are the eigenfunctions of the uniform simply supported beam. Substituting equation (2.121) into equations (2.119) and (2.120), one obtains

$$V(t) = \frac{1}{2}\sum_{r=1}^{N}\sum_{s=1}^{N} \eta'_r(t)\eta'_s(t) \int_0^L \left\{EI_b \frac{d^2\phi_r(x)}{dx^2}\frac{d^2\phi_s(x)}{dx^2}\right.$$
$$\left. + EI_p \frac{d^2\phi_r(x)}{dx^2}\frac{d^2\phi_s(x)}{dx^2} [u(x-x_1) - u(x-x_2)]\right\} dx \quad (2.123)$$
$$= \frac{1}{2}\sum_{r=1}^{N}\sum_{s=1}^{N} k_{rs}\eta'_r(t)\eta'_s(t) \quad (2.124)$$

and

$$T(t) = \tfrac{1}{2} \textstyle\sum_{r=1}^{N} \sum_{s=1}^{N} \dot{\eta}'_r(t)\dot{\eta}'_s(t) \int_0^L \{m_b\phi_r(x)\phi_s(x)$$

$$+ m_p\phi_r(x)\phi_s(x)[u(x - x_1) - u(x - x_2)]\} \, dx \tag{2.125}$$

$$= \tfrac{1}{2} \textstyle\sum_{r=1}^{N} \sum_{s=1}^{N} m_{rs}\dot{\eta}'_r(t)\dot{\eta}'_s(t) \tag{2.126}$$

where

$$k_{rs} = \int_0^L EI_b \frac{d^2\phi_r(x)}{dx^2} \frac{d^2\phi_s(x)}{dx^2} dx + \int_{x_1}^{x_2} EI_p \frac{d^2\phi_r(x)}{dx^2} \frac{d^2\phi_s(x)}{dx^2} dx \tag{2.127}$$

and

$$m_{rs} = \int_0^L m_b\phi_r(x)\phi_s(x)dx + \int_{x_1}^{x_2} m_p\phi_r(x)\phi_s(x)dx \tag{2.128}$$

From Lagrange's equations:

$$\frac{d}{dt}\left(\frac{\partial T(t)}{\partial \dot{\eta}'_s(t)}\right) - \frac{\partial T(t)}{\partial \eta'_s(t)} + \frac{\partial V}{\partial \eta'_s(t)} = Q'_r(t)$$

Taking derivatives of equations (2.124) and (2.126) with respect to the generalized coordinates, and substituting into Lagrange's equations, one obtains

$$\sum_{s=1}^{N} m_{rs}\ddot{\eta}'_s(t) + \sum_{s=1}^{N} k_{rs}\eta'_s(t) = Q'_r(t), \qquad r = 1, 2, \ldots, N \tag{2.129}$$

where the generalized forces are derived from the virtual work as follows:

$$\delta W = \int_0^L f(x,t)\delta w(x,t)dx \tag{2.130}$$

$$= \sum_{r=1}^{N} \int_0^L f(x,t)\phi_r(x)\delta\eta'_r(t)dx \tag{2.131}$$

$$= \sum_{r=1}^{N} \int_0^L f(x,t)\phi_r(x)dx\ \delta\eta'_r(t) \tag{2.132}$$

$$= \sum_{r=1}^{N} Q'_r(t)\delta\eta'_r(t) \tag{2.133}$$

where

$$Q'_r(t) = \int_0^L f(x,t)\phi_r(x)dx \tag{2.134}$$

In matrix notation:

$$\mathbf{M}\ddot{\eta}'(t) + \mathbf{K}\eta'(t) = \mathbf{Q}'(t) \tag{2.135}$$

where **M** and **K** are $N \times N$ matrices, and $\eta'(t)$ and $\mathbf{Q}'(t)$ are $N \times 1$ column vectors. The elements of the matrices can be computed as follows for $r = s$:

$$k_{rr} = \left(\frac{r\pi}{L}\right)^4 \left\{\frac{EI_bL}{2} + EI_p\left(\frac{x_2 - x_1}{2}\right) + \frac{EI_pL}{4\pi r}\left[\sin\left(\frac{2\pi r x_1}{L}\right) - \sin\left(\frac{2\pi r x_2}{L}\right)\right]\right\} \tag{2.136}$$

$$m_{rr} = \frac{m_bL}{2} + m_p\left(\frac{x_2 - x_1}{2}\right) + \frac{m_pL}{4\pi r}\left[\sin\left(\frac{2\pi r x_1}{L}\right) - \sin\left(\frac{2\pi r x_2}{L}\right)\right] \tag{2.137}$$

For $r \neq s$:

$$k_{rs} = \frac{EI_pL}{\pi}\left(\frac{rs\pi^2}{L^2}\right)^2\left[\frac{r\cos(r\pi x/L)\sin(s\pi x/L)}{(s^2-r^2)}\right.$$
$$\left.+\frac{s\cos(s\pi x/L)\sin(r\pi x/L)}{(r^2-s^2)}\right]_{x_1}^{x_2} \tag{2.138}$$

$$m_{rs} = \frac{m_pL}{\pi}\left[\frac{r\cos(r\pi x/L)\sin(s\pi x/L)}{(s^2-r^2)}+\frac{s\cos(s\pi x/L)\sin(r\pi x/L)}{(r^2-s^2)}\right]_{x_1}^{x_2} \tag{2.139}$$

Specific values of the mass and stiffness coefficients can be computed upon defining the dimensions and material properties of the piezostructure. If the system response must be computed repetitively, then the system of equations can be transformed to discrete modal coordinates. In general, the effect that active elements have on the structural response depends upon their relative location on the structure with respect to nodal lines of modes and their relative distributed mass and stiffness properties with respect to that of the structure.

***Galerkin's Method*** Galerkin's method is more restrictive in the class of functions chosen (i.e., comparison functions as opposed to admissible functions) with respect to the previous two methods; however, Galerkin's method is not based upon variational principles. The eigenvalue problem is derived directly from the differential equation and corresponding boundary conditions, and thus, the stiffness operator $\mathcal{L}$ need not be self-adjoint. A solution is assumed such that

$$w(\mathbf{x}, t) = \sum_{r=1}^{N} \psi_r(\mathbf{x})\eta'_r(t) \tag{2.140}$$

where $N$ linearly independent comparison functions $\psi_r(\mathbf{x})$ are used to expand the solution over the domain $D$ in generalized coordinates $\eta'_r(t)$. Substituting equation (2.140) into equation (2.103), multiplying by $\psi_s(\mathbf{x})$, and integrating over the domain, one obtains

$$\sum_{r=1}^{N} m_{sr}\ddot{\eta}'_r + \sum_{r=1}^{N} c_{sr}\dot{\eta}'_r + \sum_{r=1}^{N} k_{sr}\eta'_r = Q'_s(t), \qquad s = 1, 2, \ldots, N \tag{2.141}$$

where

$$m_{sr} = \int_D \psi_s(\mathbf{x})\mathcal{M}[\psi_r(\mathbf{x})]dD(\mathbf{x}), \qquad r, s = 1, 2, \ldots, N \tag{2.142}$$

$$c_{sr} = \int_D \psi_s(\mathbf{x})\mathcal{C}[\psi_r(\mathbf{x})]dD(\mathbf{x}), \qquad r, s = 1, 2, \ldots, N \tag{2.143}$$

$$k_{sr} = \int_D \psi_s(\mathbf{x})\mathcal{L}[\psi_r(\mathbf{x})]dD(\mathbf{x}), \qquad r, s = 1, 2, \ldots, N \tag{2.144}$$

The damping and stiffness coefficients are in general nonsymmetric, and $Q_s'(t)$ are the generalized forces:

$$Q_s'(t) = \int_D \psi_s(\mathbf{x})f(\mathbf{x}, t)dD(\mathbf{x}), \qquad s = 1, 2, \ldots, N \tag{2.145}$$

In the event that $\mathcal{L}$ is a self-adjoint differential operator (the stiffness coefficients are symmetric), and in the absence of any nonconservative forces, the equations of motion obtained by Galerkin's method are identical to those obtained by the Rayleigh-Ritz method or the assumed-modes method as in equation (2.118), provided that comparison functions are used in the expansion. For the special case of proportional damping and a self-adjoint differential operator $\mathcal{L}$, the stiffness and damping coefficients are symmetric and the discretized system can be represented in matrix form:

$$\mathbf{M}\ddot{\eta}'(t) + \mathbf{C}\dot{\eta}'(t) + \mathbf{K}\eta'(t) = \mathbf{Q}'(t) \tag{2.146}$$

where $\mathbf{M}$ is an $N \times N$ matrix of the coefficients $m_{sr}$, $\mathbf{C}$ is an $N \times N$ matrix of the coefficients $c_{sr}$, $\mathbf{K}$ is an $N \times N$ matrix of the coefficients $k_{sr}$, $\mathbf{Q}'(t)$ is an $N \times 1$ vector of generalized forces, $Q_s'(t)$, and $\boldsymbol{\eta}'(t)$ is an $N \times 1$ vector of generalized coordinates $\eta_r'(t)$.

**Example 2.5: A Simply Supported Beam with Nonuniform Properties** Consider the simply supported beam presented in Example 2.3. However, in this case, the mass per unit length and stiffness are given as

$$m(x) = m\left(1 - \frac{x}{2L}\right) \tag{2.147}$$

$$EI(x) = EI\left(1 - \frac{x}{2L}\right) \tag{2.148}$$

The eigenfunctions of the uniform property beam satisfy both the geometric and natural boundary conditions for the given case, and thus, form a set of comparison functions that can be used to formulate the solution in terms of a finite expansion:

$$w(x,t) = \sum_{r=1}^{N} \psi_r(x)\eta'_r(t) \tag{2.149}$$

where

$$\psi_r(x) = \sin\left(\frac{r\pi x}{L}\right) \tag{2.150}$$

Substituting equation (2.149) into equation (2.35) derived earlier, one gets

$$\sum_{r=1}^{N} \left[\frac{d^2}{dx^2}\left(EI(x)\,\frac{d^2\psi_r(x)}{dx^2}\right)\eta'_r(t) + m(x)\psi_r(x)\ddot{\eta}'_r(t)\right] = f(x,t) \tag{2.151}$$

Multiplying equation (2.151) by $\psi_s(x)$ and integrating over the domain, one obtains

$$\sum_{r=1}^{N} (k_{sr}\eta'_r(t) + m_{sr}\ddot{\eta}'_r(t)) = Q'_s(t) \tag{2.152}$$

or, in matrix notation,

$$\mathbf{M}\ddot{\eta}'(t) + \mathbf{K}\eta'(t) = \mathbf{Q}'(t) \tag{2.153}$$

where

$$k_{sr} = EI\int_0^L \psi_s(x)\,\frac{d^2}{dx^2}\left[\left(1 - \frac{x}{2L}\right)\frac{d^2\psi_r(x)}{dx^2}\right] dx \tag{2.154}$$

$$m_{sr} = m\int_0^L \psi_s(x)\left(1 - \frac{x}{2L}\right)\psi_r(x)dx \tag{2.155}$$

and

$$Q'_s(t) = \int_0^L \psi_s(x)f(x,t)dx \tag{2.156}$$

For $s = r$:

$$k_{ss} = EI\,\frac{3L}{8}\left(\frac{s\pi}{L}\right)^4 \tag{2.157}$$

$$m_{ss} = m\,\frac{3L}{8} \tag{2.158}$$

and for $s \neq r$

$$k_{sr} = -\frac{EI(sr)^2}{L}\left(\frac{\pi}{L}\right)^4\left[\frac{L^2 sr}{(s^2 - r^2)^2\pi^2}\,(\cos(s\pi)\cos(r\pi) - 1)\right] \tag{2.159}$$

$$m_{sr} = -\frac{m}{L}\left[\frac{L^2 sr}{(s^2 - r^2)^2\pi^2}\,(\cos(s\pi)\cos(r\pi) - 1)\right] \tag{2.160}$$

Upon choosing the number of terms to include in the expansion $N$, the mass and stiffness matrices can be constructed. Notice from the previous set of equations that both the mass matrix and stiffness matrix are symmetric. This is due to the fact that the differential operators,

$$\mathcal{L} = \frac{d^2}{dx^2}\left[EI\left(1 - \frac{x}{2L}\right)\frac{d^2}{dx^2}\right] \tag{2.161}$$

$$\mathcal{M} = m\left(1 - \frac{x}{2L}\right) \tag{2.162}$$

are self-adjoint.

***Colocation Method*** Another alternative to the previous formulations is the colocation method. In this method, one assumes a solution based upon some linear combination of functions such that

$$w(\mathbf{x}) = \sum_{m=1}^{N} w_m(\mathbf{x})a_m \tag{2.163}$$

but the functions $w_m(\mathbf{x})$ need not satisfy the differential equation or the boundary conditions in general. The discussion here is limited to those functions that satisfy the geometric and natural boundary conditions (i.e., comparison functions). One proceeds by discretizing the system spatially into $N$ discrete coordinates, $\mathbf{x}_n$ for $n = 1, 2, \ldots, N$. The objective is to guarantee that the differential equation is satisfied at each spatial coordinate, $\mathbf{x}_n$. For the eigenvalue problem presented in equation (2.85), letting $W(\mathbf{x}) = w(\mathbf{x}_n)$, one obtains $N$ independent equations for $N$ unknowns:

$$\mathcal{L}[w(\mathbf{x}_n)] = \Omega^2 \mathcal{M}[w(\mathbf{x}_n)], \qquad n = 1, 2, \ldots, N \tag{2.164}$$

where $\Omega^2$ is an estimate of the eigenvalue. Now, upon substituting equation (2.163) into equation (2.164):

$$\sum_{m=1}^{N} (\mathcal{L}[w_m(\mathbf{x}_n)] - \Omega^2 \mathcal{M}[w_m(\mathbf{x}_n)])a_m = 0, \qquad n = 1, 2, \ldots, N \tag{2.165}$$

Letting

$$k_{nm} = \mathcal{L}[w_m(\mathbf{x}_n)] \tag{2.166}$$

$$m_{nm} = \mathcal{M}[w_m(\mathbf{x}_n)] \tag{2.167}$$

one defines stiffness coefficients and mass coefficients, respectively, which are not symmetric, even if the operators $\mathcal{L}$ and $\mathcal{M}$ are self-adjoint. However, if the differential operators are self-adjoint, the eigenvalues of the discrete system of equations will be real, even though the mass and stiffness matrices are not symmetric. In this case, we solve for expansion coefficients $a_m$, which guarantee that the solution satisfies the differential equation at each position $x_n$. For a more detailed discussion on the solution of eigenvalue problems lacking symmetry, the reader is referred to *Analytical Methods in Vibrations* (Meirovitch, 1967).

***Finite Element Method*** If the geometry of the adaptive structure is complex, and is thus not amenable to the previously described discretization techniques, the finite element method offers an alternative approach to formulating the system of equations describing the structural response. Like the Rayleigh-Ritz method, the finite element method incorporates the use of admissible functions to construct the structural response. However, the admissible functions used in the finite element method are confined to "local" regions on the structure (i.e., zero at all other locations). Thus, the generalized displacements in the finite element method are actually the displacements at the nodal points of the structure (i.e., the desired physical coordinates), which is more tractable for output feedback control than modal control. For more information on dynamic modeling of structures using the finite element method the reader is refered to the texts by Reddy (1984), Segerland (1984), Rao (1998), Zienkiewicz and Taylor (1989), and Bathe (1995), to name a few.

***Summary*** Even in the design of simple adaptive structures, one is confronted with the dilemma of how to model the structural response. In all cases, the model inevitably results from some discretization process, and the method chosen depends upon the general complexity of the geometry of the structure, discontinuities, and the type of information desired from the model. Regardless of the method chosen, due to the discretization process, the model is always less than ideal. In general, the greater the number of degrees of freedom used in the structural model, the greater the accuracy in predicting the actual response. If the model is based upon a discretization with a finite number of expansion functions, the best rule of thumb is to increase the number of expansion functions used until the resonant frequencies in the bandwidth of operation converge. Additionally, for colocated FRFs, convergence of the zeros indicates that the low-frequency compliance of the structure has been accurately modeled.

### 2.3.3 Wave Solutions

An alternative approach for solving linear partial differential equations is based upon the method of assumed solutions, which for structures is more generally described as the wave approach (Hildebrand, 1976). In contrast to the modal approach, where the solution is expanded in terms of a set of basis functions (typically the eigenfunctions of the homogeneous system), the wave approach is based upon an assumed solution of a finite set of traveling waves. For lightly damped structures of finite dimensions (i.e., with discontinuities resulting from boundary conditions), the modal approach is typically chosen, since the system is characterized by an infinite number of standing waves, each associated with a resonant frequency (related to the natural frequency) of the system. The wave approach lends itself well for modeling systems in which there are minimal reflections at the boundaries (i.e., systems of infinite length or more practically, those that exhibit significant structural damping). However, one should recognize that any modeling approach is simply based upon a mathematical technique for solving the partial differential equation, and thus each method will yield the same characteristic system response. The objective is to determine which method of modeling is most appropriate (or easiest) for analysis and design of any particular structural system.

To illustrate the characteristic form of the assumed solution for traveling waves, consider the aforementioned partial differential equation of motion presented in equation (2.85). If there exists a set of solutions such that

$$W(x) = A_1 \exp(kx) \tag{2.168}$$

where $k$ is in general complex and is called the wavenumber, then the solutions are termed wave solutions. Typically, the wavenumber $k$ takes on one of four characteristic forms for undamped systems:

$$k = \pm\alpha, \ \pm j\alpha \tag{2.169}$$

where $\alpha$ is a finite real quantity. Assuming a positive time convention (i.e.,

$e^{j\omega t}$), $+\alpha$ corresponds to an evanescent (nearfield) wave that decays in the negative $x$ direction, $-\alpha$ corresponds to an evanescent (nearfield) wave that decays in the positive $x$ direction, $+j\alpha$ corresponds to a traveling wave in the negative $x$ direction, and $-j\alpha$ corresponds to a traveling wave in the positive $x$ direction. The traveling waves continue in their respective directions with constant amplitude. To provide an example, consider the system response characterized by a single positive traveling wave:

$$w(x, t) = A \exp(-jkx) \exp(j\omega t) \tag{2.170}$$

This function is a subset of functions that have the general form

$$w(x, t) = f(ct - x) \tag{2.171}$$

and that are characteristic of the wave approach as outlined by Hildebrand (1976). A particular point of constant phase on this traveling wave is seen to travel in the positive $x$ direction with a particular speed $c$, called the phase speed (or wave speed). In equation (2.171), the phase speed is

$$c = \frac{\omega}{k} \tag{2.172}$$

Two basic solution methodologies exist for resolving both the decaying waves and traveling waves of a partial differential equation such as that presented in equation (2.85). The first method is based upon assuming a solution of the form shown in equation (2.170). The assumed solution is substituted into equation (2.85) to determine the characteristic equation. This characteristic equation dictates all possible values of the wavenumber $k$ for which a solution exists. The group of candidate solutions is then summed as part of an expansion to obtain the general solution, and the boundary conditions are applied to determine the unknown wave amplitudes (i.e., $A$ in equation (2.170)) necessary to construct the system response.

In the second approach, the equations of motion (2.85) and boundary conditions are transformed into the wavenumber space using either the Laplace transform or Fourier transform. (A discussion of each transform method is presented in Appendix A.) Upon applying the appropriate boundary conditions, the inverse Laplace transform or inverse Fourier transform, respectively, can be used to determine the corresponding solution. Frequently, the inverse transform will contain poles and corresponding residues that yield the solution set. For applications in which either method can be applied, the poles from the second method will correspond to the possible wavenumbers determined from the characteristic equation of the first method.

An example is provided to illustrate the two solution methodologies on a semi-infinite beam.

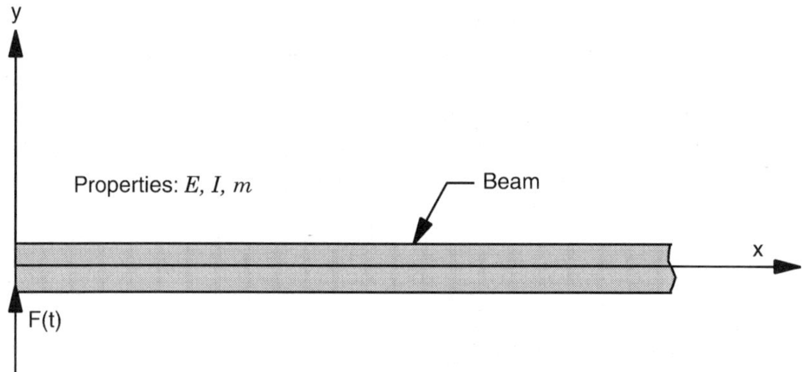


**Figure 2.8:** Schematic diagram of a semi-infinite beam.

**Example 2.6: Wave Response of a Semi-infinite Beam** Consider a uniform semi-infinite beam presented in Figure 2.8. The differential operators are as previously presented for the simply supported beam in equations (2.87) and (2.88). In this case, the homogeneous equation of motion is

$$EI\,\frac{\partial^4 w(x,t)}{\partial x^4} + m\,\frac{\partial^2 w(x,t)}{\partial t^2} = 0 \tag{2.173}$$

In the first portion of this example, we assume a wave solution. The beam is subjected to a point force at the free end, which for this part of the example will be handled in the boundary condition. The solution is assumed separable in space and time, which produces the spatial differential equation of motion shown previously in equation (2.66), where the wavenumber $k$ is equal to $\gamma$, as defined previously for the modal solution. The assumed response of the beam is comprised of four waves: two traveling waves and two decaying (nearfield) waves, respectively:

$$W(x) = A_1 \exp(-jkx) + A_2 \exp(jkx) + A_3 \exp(kx) + A_4 \exp(-kx) \tag{2.174}$$

where $A_1$ is the unknown positive traveling wave coefficient, $A_2$ is the unknown negative traveling wave coefficient, $A_3$ is the unknown negative decaying nearfield wave coefficient, $A_4$ is the unknown positive decaying nearfield wave coefficient, and $k$ is the flexural wavenumber defined previously in equation (2.83). Since the beam is semi-infinite, it is impossible for waves to arrive from infinity (Sommerfeld radiation condition), and thus wave coefficients $A_2$ and $A_3$ must be identically zero. Wave amplitudes $A_1$ and $A_4$ are found by applying the boundary conditions at $x = 0$. The *natural boundary conditions* at the free end ($x = 0$) require that the moment be zero and that the shear force be equal to the excitation force:

$$EI\,\frac{d^2W(0)}{dx^2} = 0 = A_4 - A_1 \tag{2.175}$$

$$EI\,\frac{d^3W(0)}{dx^3} = F = EIk^3(-A_4 - jA_1) \tag{2.176}$$

Combining equations (2.175) and (2.176) and substituting back into the assumed response presented in equation (2.174) produces the following relation:

$$W(x) = \frac{jF}{2EIk}(\exp(-jkx) + \exp(-kx)) \tag{2.177}$$

Thus, the response of the beam is characterized by a positive traveling wave and a positive decaying wave, since no reflections occur at the infinite boundary, as one might expect. The complete response for the system is found from equation (2.57) where harmonic time dependence is included.

Now, we solve the same problem by the second method, using integral transforms. We begin by taking the Laplace transform (in spatial coordinates) of the spatial portion of the equation of motion as follows:

$$EI\,[s^4\hat{W}(s) - s^3\hat{W}(0) - s^2\hat{W}'(0) - s\hat{W}''(0) - \hat{W}'''(0)] - m\omega^2\hat{W}(s)$$
$$= \int_0^\infty \exp(-sx)\hat{F}(s)\delta(x)dx \tag{2.178}$$

Upon including the boundary conditions, the transformed equation of motion becomes:

$$\hat{W}(s)(s^4 - k^4) = \frac{2\hat{F}(s)}{EI} + s^3\hat{W}(0) + s^2\hat{W}'(0) \tag{2.179}$$

Now, the inverse Laplace transform is utilized to obtain the forced response as follows:

$$W(x) = \frac{2F}{EI}\left[\left(\frac{1}{4k^3} + \frac{w(0)}{4} + \frac{w'(0)}{4k}\right)\exp(kx) + \right.$$

$$+\left(-\frac{1}{4k^3}+\frac{w(0)}{4}-\frac{w'(0)}{4k}\right)\exp(-kx)$$

$$+\left(-\frac{1}{4k^3}+\frac{w(0)}{4}+\frac{w'(0)}{4jk}\right)\exp(jkx)$$

$$+\left(\frac{1}{4k^3}+\frac{w(0)}{4}-\frac{w'(0)}{4jk}\right)\exp(-jkx)\Bigg] \tag{2.180}$$

We now invoke the infinite boundary condition (waves cannot arrive from infinity), which in this case requires that the terms in front of $\exp(kx)$ and $\exp(jkx)$ must be identically zero, producing the following relations:

$$\left(\frac{1}{4k^3}+\frac{w(0)}{4}+\frac{w'(0)}{4k}\right)=0 \tag{2.181}$$

$$\left(-\frac{1}{4k^3}+\frac{w(0)}{4}+\frac{w'(0)}{4jk}\right)=0 \tag{2.182}$$

Solving the above equations simultaneously, and substituting into the response relation presented in equation (2.180), one arrives at the following final solution:

$$W(x)=\frac{jF}{2EIk}(\exp(-jkx)+\exp(-kx)) \tag{2.183}$$

which is consistent with the result found previously. It is again emphasized that the solution is expressed in terms of a *continuous* wavenumber $k$, in contrast to the solutions found by solving the eigenvalue problem where the forced response is expressed in terms of an expansion over an infinite number of discrete waves.

### 2.3.4 Wave Models Versus Modal Models

In the previous sections, both the modal approach and the wave approach have been presented as methods of solving linear partial differential equations. It is important to recognize applications for which each solution methodology is best utilized. Regardless of the boundary conditions, the wave approach or the modal approach can be applied; however, solutions of the form

$$W(x)=A_1\sin(\gamma x)+A_2\cos(\gamma x)+A_3\sinh(\gamma x)+A_4\cosh(\gamma x)$$

are typically associated with finite boundary conditions, while solutions of the form

$$W(x) = A_1 \exp(-jkx) + A_2 \exp(jkx) + A_3 \exp(kx) + A_4 \exp(-kx)$$

are typically associated with some form of infinite boundary condition. However, either general form of the solution can be applied to the *homogeneous* equation of motion to determine the eigenfunctions and eigenvalues, as illustrated in Example 2.3. This is due to the fact that the sine and cosine function can be expressed in terms of complex exponentials and the hyperbolic sine and hyperbolic cosine can be expressed in terms of real exponentials. However, the utility of the wave approach is realized in solving for the forced response of systems in which there is a small number of discontinuities (including the boundaries). Thus, the wave model is typically employed when the structure is described as semi-infinite or infinite, as has been illustrated in Example 2.6. The wave model can be applied to finite systems with a limited number of discrete discontinuities quite easily, as demonstrated in Example 2.7 below. The primary advantage of solving for the forced response in terms of the wave model is that the system can be described in terms of a finite set of waves characterized by *continuous* wavenumbers, as opposed to an expansion of an infinite set of *discrete* eigenfunctions and corresponding eigenvalues (wavenumbers) as in the modal model.

**Example 2.7: Wave Solution of a Forced-Free Beam** In this section, a forced-free beam is examined using the wave model described in the previous section. A schematic diagram of the structure is presented in Figure 2.9. The equation of motion for this system has previously been presented in equation (2.173), which is separated in space and time, producing the spatial equation (2.66), and the boundary conditions for the forced end are identical to those presented for the semi-infinite beam in equations (2.176) and (2.175). The natural boundary conditions at the free end ($x = L$) require the shear force and bending moment to be zero, which produces the following set of equations corresponding to both sets of boundary conditions:

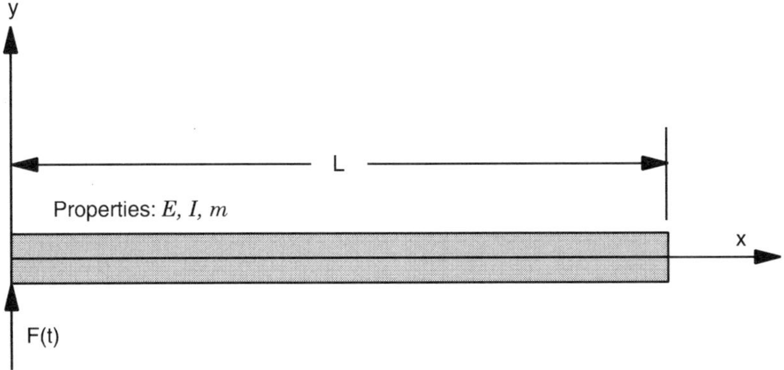


**Figure 2.9:** Schematic diagram of forced-free beam.

$$EI\,\frac{d^2W(0)}{dx^2} = 0 \tag{2.184}$$

$$EI\,\frac{d^3W(0)}{dx^3} = F \tag{2.185}$$

$$EI\,\frac{d^2W(L)}{dx^2} = 0 \tag{2.186}$$

$$EI\,\frac{d^3W(L)}{dx^3} = 0 \tag{2.187}$$

The assumed solution for the forced response of the semi-infinite beam, as presented in equation (2.174), will again be substituted into each of the previous equations, producing the following set of relations for the unknown wave amplitudes $A_1$, $A_2$, $A_3$, and $A_4$:

$$\begin{bmatrix} j & -j & 1 & -1 \\ -1 & -1 & 1 & 1 \\ -e^{-jkL} & -e^{jkL} & e^{kL} & e^{-kL} \\ je^{-jkL} & -je^{jkL} & e^{kL} & -e^{-kL} \end{bmatrix} \begin{pmatrix} A_1 \\ A_2 \\ A_3 \\ A_4 \end{pmatrix} = \begin{pmatrix} \dfrac{F}{EIk^3} \\ 0 \\ 0 \\ 0 \end{pmatrix} \tag{2.188}$$

This $4 \times 4$ system of equations can be solved in closed form (Mathematica was used for this example), producing the following relations for the unknown coefficients:

$$\begin{pmatrix} A_1 \\ A_2 \\ A_3 \\ A_4 \end{pmatrix} = \begin{pmatrix} \dfrac{(1+j)\exp(jkL)(-j\exp(jkL)+(1+j)\exp(kL)-\exp((2+j)kL))F}{(1+\exp(j2kL)-4\exp((1+j)kL)+\exp(2kL)+\exp((2+2j)kL))2EIk^3} \\ \dfrac{(1+j)(1-(1+j)\exp((1+j)kL)+j\exp(2kL))F}{(1+\exp(j2kL)-4\exp((1+j)kL)+\exp(2kL)+\exp((2+2j)kL))2EIk^3} \\ \dfrac{(1+j)(1-j\exp(2jkL)+(j-1)\exp((1+j)kL))F}{(1+\exp(j2kL)-4\exp((1+j)kL)+\exp(2kL)+\exp((2+2j)kL))2EIk^3} \\ \dfrac{(1+j)\exp(kL)((j-1)\exp(jkL)-j\exp(kL)+\exp((1+2j)kL))F}{(1+\exp(j2kL)4\exp((1+j)kL)+\exp(2kL)+\exp((2+2j)kL))2EIk^3} \end{pmatrix} \tag{2.189}$$

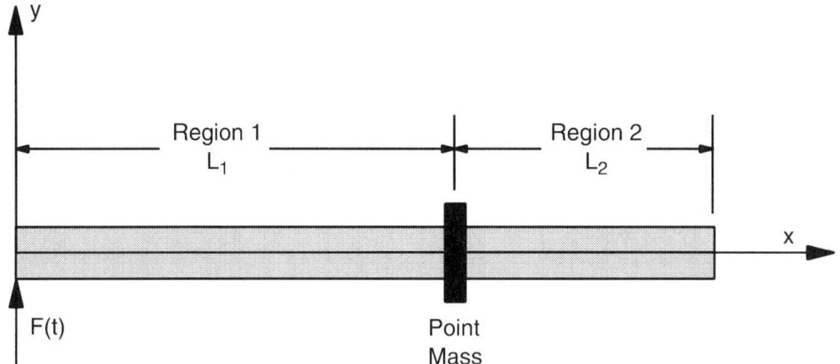


**Figure 2.10:** Schematic diagram of beam with multiple discontinuities.

The values for each coefficient can be substituted into the assumed response expressed in equation (2.174) to determine the net forced response of the forced-free beam. Note the denominator of each wave is the same and has roots that correspond to the transfer function poles or modal approach eigenvalues. It should be noted that the wave model yields a more compact solution than the equivalent modal model, which would include a modal expansion (infinite weighted summation) of the discrete eigenvalues and corresponding eigenvectors. The wave approach can be expanded to include multiple discontinuities when necessary. For example, each discontinuity in the system, whether an input force or boundary condition, can be used to break the structure into finite regions. The partial differential equation is then applied to each respective section with appropriate matching boundary conditions between sections of the structure. Consider the forced-free beam with a lumped mass in the middle, as illustrated in Figure 2.10. The beam can be subdivided into two discrete sections, and upon applying boundary conditions, a total of eight equations and eight unknowns results. In general, for one-dimensional structures, the number of independent equations can be computed in advance as a function of the number of discontinuities:

$$o = 4(d - 1) \tag{2.190}$$

where $o$ is the number of equations and unknowns, and $d$ is the number of discontinuities. For multiple inputs, the order of the system of equations can become somewhat overwhelming, in which case the modal model might be more appropriate, specifically for finite-dimensional structures. Either solution method is valid, and the method chosen should be determined based on ease of computation and accuracy of the solution.

Another consideration in deciding which model to use for a given system can be the variables to be observed. The modal approach will provide an estimate of the kinetic or potential energy of the structural response, while the wave approach provides a compact representation of the total energy flow in a structure (or region of a structure). Consider, for example, a thin uniform beam, as previously described by equation (2.173), whose total response can be described by four waves.

$$w(x,t) = (A_1 \exp(-jkx) + A_2 \exp(jkx) + A_3 \exp(kx) + A_4 \exp(-kx)) \exp(j\omega t) \tag{2.191}$$

where the wave amplitudes were described previously in this section. The power flow due to the flexural motion is found by the following relation (Cremer et al., 1988):

$$P_f(\omega) = \tfrac{1}{2} \operatorname{Re} (F_w v_w^* + M\dot{\theta}^*) \tag{2.192}$$

where $F_w$ is the shear force, $v_w$ is the out-of-plane velocity, $M$ is the flexural moment, and $\dot{\theta}$ is the rotational velocity given by

$$v_w(x,t) = \frac{\partial w}{\partial t} = j\omega(A_1 \exp(-jkx) + A_2 \exp(jkx) + A_3 \exp(kx) + A_4 \exp(-kx)) \exp(j\omega t) \tag{2.193}$$

$$F_w(x,t) = EI \frac{\partial^3 w}{\partial x^3} = EIk^3 \, (jA_1 \exp(-jkx) - jA_2 \exp(jkx) + A_3 \exp(kx) - A_4 \exp(-kx)) \exp(j\omega t) \tag{2.194}$$

$$\dot{\theta}(x,t) = \frac{\partial v_w}{\partial x} = j\omega k(-jA_1 \exp(-jkx) + jA_2 \exp(jkx) + A_3 \exp(kx) - A_4 \exp(-kx)) \exp(j\omega t) \tag{2.195}$$

$$M(x,t) = -EI \frac{\partial^2 w}{\partial x^2} = -EIk^2(-A_1 \exp(-jkx) - A_2 \exp(jkx) + A_3 \exp(kx) + A_4 \exp(-kx)) \exp(j\omega t) \tag{2.196}$$

and the superscript * denotes the complex conjugate. When equations (2.193)–(2.196) are further reduced to the form of equation (2.192), the resulting flexural power flow is

$$P_f(\omega) = EIk^3\omega(|A_1|^2 - |A_2|^2 - 2 \operatorname{Imag}(A_3 A_4{}^*)) \tag{2.197}$$

It is interesting to note that the power flow consists of three primary components. The first part is associated with the positive traveling wave, and as indicated, the total energy is positive and proportional to the magnitude of the wave $A_1$ squared. Similarly, the second part is associated with the negative traveling wave, and the total energy is negative and proportional to the magnitude of the wave $A_2$ squared. In passing, one recognizes that two traveling waves of equal magnitude will result in zero power flow through the structure in absence of any nearfield interaction. Thus, the objective in active control may be, for example, to observe both traveling waves and generate the necessary contribution to the traveling waves to minimize the power flow.

The final contribution to the total energy is proportional to the imaginary part of the complex product of the two nearfields (i.e., the nearfield interaction). Thus, if the nearfields are not totally in-phase or out-of-phase, there will exist a net flow of energy as shown in equation (2.197). If the nearfields *are* totally in-phase or out-of-phase, there will be no power flow associated with nearfield interaction. For a large number of systems, the energy associated with the nearfield is insignificant in comparison to that associated with the traveling waves. Thus, it can be seen that the total energy flow in a beam section can be described by the waves that "carry" the energy. If there exists a method of observing each wave in the structure independently with some form of transduction device, then the total energy can be observed and controlled with a relatively simple control system design. This concept is discussed in detail in Chapter 6.

## 2.4 DAMPING MODELS

In the previous sections, we have discussed methods of modeling a structure's dynamic response characteristics. Analytical approaches have been developed using displacement coordinates for distributed parameter realizations, modal amplitudes, elastic wave amplitudes, and displacements of finite element discretizations. For all of those modeling approaches, we have mentioned only briefly how to account for *damping* effects exhibited by all real-life structures. Damping is defined to be the irreversible loss of energy from the structure that occurs through a variety of physical mechanisms. In the absence of damping, the cyclic exchange of energy between potential and kinetic energy, referred to in Section 2.2.1 for a mass-spring system, takes place without any work being done by the structure. Using the notation of Section 2.2.2, the nonconservative damping force performs work according to

$$\mathbf{F}_{nc} \cdot \dot{\mathbf{r}} = \frac{d}{dt}\,(\mathrm{T} + \mathrm{V}) \tag{2.198}$$

which is equal to the change in the total system energy. Clearly, if we can generate methods to introduce added nonconservative forces, in addition to those created by natural forms of structural damping, we can remove energy from the structural system. This is one major focus of adaptive structure design.

Structural damping is often associated with a conversion of mechanical energy to heat. If we imagine a control volume whose boundaries coincide with the structural boundaries, it is easy to visualize energy transfer across the control volume surface. There are many important forms of energy transfer across such an imaginary control volume surface, in addition to the radiation of heat. Adaptive structures are often designed to exploit mechanisms that have the ability to dissipate energy in the structural system, that is, to transfer structural disturbance energy away from the structure. In the following sections, we motivate a perspective on damping mechanisms that facilitates adaptive structures design and review the mathematics that describe some of the more important damping mechanisms.

For adaptive structures, knowledge of damping mechanisms is extremely important for several reasons. First, one of the most important applications of adaptive structures is the suppression of vibration response. Therefore, it is essential to be able to predict the effects that structural damping have on a structure's passive performance. It is equally important to be able to quantify the actively controlled structure's performance using the same metrics established for quantifying passive damping mechanisms. This capability will ensure that the nonconservative forces introduced by a properly designed adaptive control system can provide cost-effective reductions to the structure's nominal vibration response, acoustic radiation, and so on. Once the designer has an accurate damping model, the work required from (and on) the adaptive structure can be identified and used to specify appropriate power sources and compensators required to achieve the desired performance. Because of the importance of damping force calculations for assessing adaptive structure design strategies and performance, the following sections attempt to summarize a large amount of information. It is our hope that the summary will be supplemented with readings from the appropriately referenced material.

Development of models for structural damping has been a popular research topic since the 1950s. Early analytical treatments of various forms of structural damping were published by Ruzicka (1959) and Lazan (1968). More recently, Torvik (1980) edited a number of articles on structural damping mechanisms and Nashif et al. (1985) presented passive damping design procedures that can be used to solve industrial noise and vibration problems. For our purpose, we wish to define the impact that damping models have on the ultimate objective of accurately modeling the dynamic response of different adaptive structures. Table 2.1 summarizes certain design objectives that are explicitly linked to the structural damping model(s) selected by the design engineer.

Discussions on damping in structures are usually approached from three distinct perspectives (at least). In Section 2.3.1, the concept of proportional damping has been introduced. Although not stated explicitly there, proportional damping allows one to decouple the equation of motion for multivariable systems using normal mode vectors (Meirovitch, 1967; Klosterman, 1971). This

**TABLE 2.1 Design Objectives Requiring Accurate Damping Models**

1. Define peak amplitudes for open-loop and closed-loop forced response
2. Predict rise times, settling times, etc., for transient structural response
3. Identify open-loop adaptive structure poles and zeros for accurate compensator design
4. Quantify energy dissipation for passive versus active damping

approach to damping, easily extended to matrix equations, is an important one in the development of state-variable models for structural response and is discussed further in Chapter 3. The second popular approach to damping introduces damping models physically motivated by the experimental characterization of solid material properties. These models are discussed in the following section. Finally, all undergraduate vibration textbooks present viscous and hysteretic damping models based on a single-degree-of-freedom (SDOF) system, as shown in Figure 2.11. For viscous damping, the dashpot shown in Figure 2.11*a* introduces a damping force equal to $c\dot{w}(t)$, while the complex spring in Figure 2.11*b* is associated with a damping force of $(k\eta/\omega)\dot{w}(t)$. Although these models are simple, they lead to useful expressions for deducing damping values from structural response measurements. At the conclusion of this section, we summarize the various types of damping mechanisms that can occur in open-loop and closed-loop adaptive structures.

### 2.4.1 Inherent Damping of Structures

The existence of damping in structures is obvious to anyone who has observed structural vibration amplitudes decay over some finite time period after the source of excitation is removed. A second important result of structural dynamics, developed long ago, is that mathematical models of structures as conservative systems lead to a prediction of infinite response when the structures are forced at their natural frequencies $\omega_n$. The traditional analysis of a mass-damper-spring system can be used to illustrate these two important characteristics, which are qualitatively representative of all structures, adaptive or not. We begin with the impulse response function for the viscously damped system shown in Figure 2.11. As discussed by Meirovitch (1986) and Inman (1989), the SDOF response to an impulsive load with zero initial conditions is

$$w(t) = \left\{ \begin{array}{ll} \dfrac{\exp(-\zeta\omega_r t)\sin\omega_d t}{m\omega_d} & t > 0 \\ 0, & t < 0 \end{array} \right\} \tag{2.199}$$

where $\omega_d = (1 - \zeta^2)^{1/2}\omega_n$ is called the damped natural frequency. The damper,

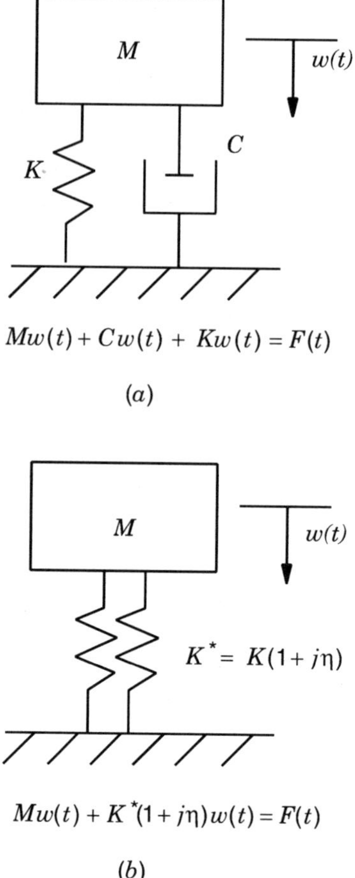


**Figure 2.11:** Basic models of damping. (*a*) Viscous damping model. (*b*) Hysteretic damping model.

with a damping ratio $\zeta = c/2\sqrt{km}$, thus imposes *asymptotic stability* on the mass-spring system. Loosely, it may be stated that an asymptotically stable system response $w(t)$ approaches the zero state as $t$ increases to infinity (Chen, 1984). This property is clearly indicated by the exponential term in equation (2.199).

The magnitude of the steady-state response of the mass-spring-damper system to a harmonic input of $F_o \sin \omega_d t$ (again for zero initial conditions) is written in the frequency domain as

$$\left| \frac{Wk}{F_o} \right| = \frac{1}{\sqrt{[1 - (\omega/\omega_n)^2]^2 + [2\zeta(\omega/\omega_n)]^2}} \tag{2.200}$$

This expression for steady-state response indicates that the amplitude will be finite even when the excitation frequency coincides with a natural frequency $\omega_n$ of the system. This matches observed behavior for real structures and is referred to as *bounded-input, bounded-output stable* (BIBO stable). A detailed discussion of the relationship between damping and system stability is presented in Chapter 7. Notice that for the undamped form of the system differential equation, that is, for $\zeta = 0$, the predicted response is infinite at $\omega = \omega_n$. This result is clearly not possible for real structures, although structures with very little damping do have the ability to reach dangerously large response amplitudes when excited at resonant frequencies. In Chapters 7 and 9, it is shown that a primary function of adaptive structures is to increase the effective damping ratio(s) in order to limit the peak response amplitudes.

The transient and steady-state expressions above illustrate the stabilizing influence that damping has on a simple mass-spring system. Although engineers never have the opportunity to work with such basic systems, it is possible to extend many concepts from simple systems to multi-degree-of-freedom (MDOF) structural models. At present, we are concerned with the need to represent the energy dissipation observed for real structures, which implies multivariable systems in most cases, and there are numerous physical mechanisms that have been postulated for such systems. All of these mechanisms, ranging from the common critical damping ratio $\zeta$ to more advanced material damping models, introduce dissipation of mechanical vibrational energy. As mentioned previously, most of the dissipation occurs as mechanical energy is converted into heat, leaving an imaginary control volume that encloses the structural system. However, the structural system may include both *external damping* and *internal damping* mechanisms. External damping processes are more closely associated with energy transfer away from the structural system, although the final stage of such processes may still be conversion to heat. It is important to make a distinction between external and internal damping processes because internal damping is associated with the internal state of the structural material itself and cannot easily be altered without some level of materials engineering. Even then, real-time modifications of internal material damping are mostly beyond the present capabilities of adaptive structures technologies. One exception is the use of rheological fluids that may be indicative of future trends in this field.

### 2.4.2 Material Damping Models

For many applications, it is necessary to evaluate the importance of the externally generated damping forces compared to the damping forces that result from the *material damping*, where material damping refers to the internal energy dissipation within the micro- and macrostructure of the material. In this section, material damping models are examined to demonstrate the functional dependences that structural response has on energy dissipated within the host structure. The introduction of external damping treatments, significantly more important to adaptive structure design and performance, is considered in the following section.

***Phenomenological Modeling Approaches*** Mathematical descriptions of material damping were originally generated by modifying the basic equations of elasticity to account for dissipative forces acting in addition to elastic forces (Meyer, 1874; Boltzmann, 1876). It is interesting to note that even early researchers were inclined to suggest viscous friction forces to account for damping, as noted by Cremer (1988). For example, Meyer developed a modified Hooke's law, which can be written as

$$\sigma = D\left(\epsilon + c\,\frac{d\epsilon}{dt}\right) \tag{2.201}$$

where the modulus term $D$ applies to any form of wave in the solid medium, and $c$ is a scalar, which implies that the damping has a linear dependence on frequency. When a periodic strain of the form $\epsilon = \epsilon_o \sin(\omega t)$ is substituted into equation (2.201), the resulting stress is no longer in phase with strain, as seen in the following expression:

$$\sigma_o = D\epsilon_o\sqrt{1+\omega^2c^2}\,\sin(\omega t + \tan^{-1}(\omega c)) \tag{2.202}$$

It may have already occurred to the reader that the existence of viscous forces in homogenous solid materials is not immediately apparent. Boltzmann was the first to suggest that a more appropriate relation between stress and strain might be obtained. He reasoned that the stress in a material at time $t$ actually depends on the strain at time $t$ in addition to some functional relationship of the strain history $\epsilon(t - \Delta t)$. This parametric approach results in a convolution integral, equivalent to a moving average model, which is written

$$\sigma(t) = D\epsilon(t) - \int_0^\infty \epsilon(t-\Delta t)\mathcal{R}(\Delta t)d(\Delta t) \tag{2.203}$$

If a relaxation function, corresponding to the decay times of molecular oscillations, is chosen of the form

$$\mathcal{R}(\Delta t) = \frac{C}{\tau}\exp -\frac{\Delta t}{\tau} \tag{2.204}$$

one again finds a phase shift between cyclic stress and strain as follows:

$$\sigma(t) = \left( D - \frac{C}{\omega^2\tau^2 + 1} \right) \epsilon_o \cos \omega t - \left( C \frac{\omega\tau}{\omega^2\tau^2 + 1} \right) \epsilon_o \sin \omega t \qquad (2.205)$$

This phase difference between stress and strain is a fundamental effect of material damping mechanisms. The phase difference is indicative of energy dissipation, as heat, in the material. The form of equation (2.205) shows that the amount of energy dissipation depends on the damping proportionality constant $C$, the relaxation time $\tau$ of the molecular structure, and the frequency $\omega$ of the excitation.

Today, the science of rheology is associated with the study of deformation and flow of matter. Rheologists use molecular level physics to develop expressions similar to equations (2.201) and (2.205). These types of models are referred to as *phenomenological models* and derive from state equations based on thermodynamics of irreversible processes. This approach is most helpful for high-polymer materials such as polyvinylchloride, rubbers, and different combinations of plastics. *For adaptive structures, these advanced phenomenological models are used mostly for viscoelastic materials employed as passive or active constrained layers on the main structure.* The important concept for the reader is that one can build models of the appropriate energy state equation from experimental measurements of a specific material.

Now, two distinct forms of advanced models are presented, forms that are simplifications to a generalized linear polynomial equation for stress and strain given by

$$\sigma_o = \epsilon_o \left\{ \frac{\sum_{n=0}^{M} B_n(j\omega)^n}{\sum_{n=0}^{N} A_n(j\omega)^n} \right\} \qquad (2.206)$$

This autoregressive moving average (ARMA) model has the familiar form of an FRF, which can be very helpful for control system design. However, the difficulty lies in curve fitting the values of the coefficients. One simplification to the general case shown in equation 2.206 is the *standard linear model*, which includes first-order dynamics for the stress-strain relationship:

$$\sigma + A \frac{d\sigma}{dt} = D\epsilon + BD \frac{d\epsilon}{dt} \qquad (2.207)$$

Using this model, we can write the transient variation of stress to an initial strain $\epsilon = \epsilon_o$ (Nashif et al., 1985) as

$$\sigma(t) = D\epsilon_o(1 - \exp(-t/\tau)) \tag{2.208}$$

where $\tau$ is a time constant associated with stress relaxation, generated by the molecular relaxation times discussed earlier. The steady-state form of the standard linear damping model derives from substituting $\sigma = \sigma_o \exp(j\omega t)$ and $\epsilon = \epsilon_o \exp(j\omega t)$ into equation (2.207):

$$\sigma_o = D\epsilon_o \left\{ \frac{1 + j\omega B}{1 + j\omega A} \right\} \tag{2.209}$$

The previous two equations are familiar forms of first-order systems and are simplifications of the general form of equation (2.206). Unfortunately, experience has shown that it is usually necessary to include several higher order terms to obtain sufficient agreement between experimental measurements of material damping and the phenomenological models. This leads to the *generalized standard linear model* given by

$$\sigma + \sum_{n=1}^{\infty} A_n \frac{d^n \sigma}{dt^n} = D\epsilon + D \sum_{n=1}^{\infty} B_n \frac{d^n \epsilon}{dt^n} \tag{2.210}$$

It is easily seen that this is of the same form as equation (2.206), and the remaining discussion is in regard to how many terms must be included in the model. This can only be determined by processing the experimental data for specific material specimens, but a general rule of thumb can be obtained from the work of Golla and Hughes (1985), McTavish (1988), and Slater (1991). Based upon the generalized standard linear model, a phenomenological model using the Laplace variable $s = j\omega$ has been proposed by Golla, Hughes, and McTavish (GHM). This GHM model is based on a summation of second-order curve fits and can be written in the form

$$\frac{\sigma(s)}{\epsilon(s)} = D^* \left( 1 + \sum_{n=1}^{N} \tilde{A}_n \frac{s^2 + 2\tilde{\zeta}_n \tilde{\omega}_n s}{s^2 + 2\tilde{\zeta}_n \tilde{\omega}_n s + \tilde{\omega}_n^2} \right) \tag{2.211}$$

Slater showed good results fitting the GHM model to a viscoelastic strut constructed with Scotchdamp SJ-2015X Type 112 for $2 \leq n \leq 4$ terms. Other researchers have also indicated that good results can be obtained with low-order curvefits. Finally, one additional advantage of the GHM damping model is that the Laplace domain representation is better suited for transient response analyses than the other models discussed in this section.

***Complex Modulus Representations*** For many of the subsequent discussions on the relationship between adaptive structure performance and structural damping, it is helpful to review the concept of a *complex modulus*. Recall that the independent variables in the models above have been either time or frequency. In most cases, the time variables are converted to frequency variables through an assumption of harmonic relationships ($\exp(j\omega t)$) between stress and strain. This is advantageous for system identification purposes because experimental data is usually frequency domain data. For this reason, we want to emphasize the complex modulus representation of the phenomenological models.

The coefficients $A_n$ and $B_n$ in equation (2.206) are generally complex, so the stress-strain relationship can be written as

$$\sigma_o = (D_r + jD_i)\epsilon_o \tag{2.212}$$

where $D_r$ and $D_i$ are real numbers that are functions of frequency (and temperature for most materials). It is very important to keep in mind that this is a frequency domain representation for harmonic response, implying that transform methods must be used to generate equivalent time-domain representations. A more compact form of the complex modulus is given by

$$\sigma_o = D^*(1 + j\eta)\epsilon_o \tag{2.213}$$

where $D^* = D(1 + j\eta)$ is the complex stiffness and $\eta$ is called the *loss factor* of the material system. The concept of loss factors is associated with a hysteresis loop, as shown in Figure 2.12. For SDOF systems, it is common to assume that $\eta$ is a constant. The effects of making this assumption need to be clarified with regard to structural modeling. First, it has been shown that noncausal behavior, that is,

$$w(t) \neq 0, \qquad t < 0 \tag{2.214}$$

is predicted when the transient response of the system is recovered using the inverse Fourier transform (Crandall, 1970; Nashif et al., 1985). Clearly, this cannot physically occur and is related to the mathematics imposed on the problem. One can avoid the noncausal behavior in the time response by retaining the frequency dependence in $\eta(\omega)$ and $D^*(\omega)$, which agrees with the curvefit data of equation (2.206). The second implication of choosing a constant loss factor can be observed by examining the expression for energy dissipated *at resonance* by a structure modeled with complex stiffness. The steady-state response of a system to a force $F_o \sin \omega t$ can be written

$$w(t) = \frac{F_o}{\sqrt{(k - m\omega^2)^2 + (k\eta)^2}} \sin\left(\omega t - \arctan\left(\frac{k\eta}{k - m\omega^2}\right)\right) \tag{2.215}$$

(Notice that the phase shift in the response is dependent on $\eta$, as mentioned

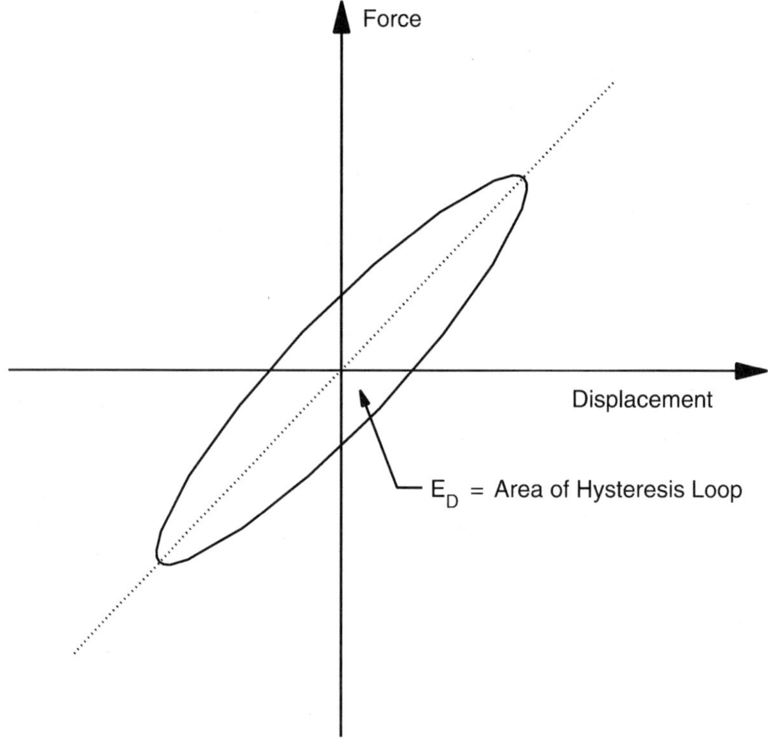


**Figure 2.12:** Schematic diagram of hysteresis loop.

previously.) The energy dissipated per cycle is found by integrating the product of the force and the velocity over one period:

$$E_d = \int_0^{2\pi/\omega} F\dot{w}\, dt \tag{2.216}$$

Referring to the magnitude of equation (2.215) as $W$, and substituting expressions for the force and velocity into the energy integral, the energy dissipated by hysteretic damping is found to be

$$E_d = \pi k \eta W^2 \tag{2.217}$$

When viscous damping is used (see equation (2.200)) the energy dissipation is

$$E_d = \pi c \omega W^2 \tag{2.218}$$

**TABLE 2.2 External and Enhanced Damping Mechanisms**

| Type of Damping | Description | Category |
|---|---|---|
| Radiation damping | Conversion of mechanical energy acoustic energy | External |
| Coulomb friction damping | Relative motion between two contacting surfaces | External |
| Fluid entrapment (air pumps) | Lossy fluid enclosure adjacent to structural members | External |
| Adhesive surface treatments | Energy loss due to shear deformation | Enhanced |
| Electronic shunts | Conversion of mechanical energy to resistive heating | Enhanced |
| Active damping | Conversion of mechanical energy to resistive heating | Enhanced |

Comparing the two energy expressions shows that the dissipation at resonance is independent of frequency for an assumption of constant loss factor. For real structures, this is generally not a problem if the designer uses several different values of $\eta$ across different bandwidths. Notice that the viscous damping model, also referred to as Kelvin-Voight damping, predicts a linear dependence on frequency. This frequency dependence tends to overestimate the amount of energy dissipated at higher frequencies for real structures. Thus, these simple models are useful but care must be exercised when using them for higher order models. The use of multiple damping constants, either $\zeta$ or $\eta$, for MDOF models is addressed when state-variable realizations of decoupled systems are presented in Chapter 3.

### 2.4.3 External Damping and Enhanced Damping of Structures

The vibrational energy of a structure is naturally dissipated through a variety of physical mechanisms that we refer to as *external damping*. In addition, some of these external damping processes are created by engineered systems, designed specifically for the purpose of removing those energies. We refer to the engineered damping mechanisms as *enhanced damping*. Enhanced damping refers to the tailoring of adaptive structures, using active or passive/active architectures, for increased rates of mechanical energy dissipation compared to the open-loop structure. Some examples of natural and engineered damping processes are given in Table 2.2.

## 2.5 SUMMARY

This chapter was devoted to a review of analytical dynamics and solution methodologies used in the modeling and analysis of distributed parameter systems. Hamilton's principle was derived and used to obtain Lagrange's equations. Hamilton's principle was also used to develop boundary value problems, and upon assuming a separable solution in space and time, the eigenvalue problem was developed. Solution methodologies based upon modal models and wave-domain models were developed and compared, and various models of damping were presented.

## BIBLIOGRAPHY

Bathe, J.-K., 1995. *Finite Element Procedures*, Prentice-Hall, New York.

Boltzmann, L., 1876. "Ann Physik," *Eng. Bd.*, **7,** 624–654.

Boyce, W. E. and R. C. DiPrima, 1977. *Elementary Differential Equations and Boundary Value Problems*, Wiley, New York.

Chen, C., 1984. *Linear System Theory and Design*, Holt, Rinehart and Winston, New York. pp. 403–404.

Clark, R. L., 1995. "Accounting for Out-of-Bandwidth Modes in the Assumed Modes Approach: Implications on Colocated Output Feedback Control," Accepted by *Journal of Dynamic Systems, Measurement, and Control.*

Crandall, S. H., 1970. "The Role of Damping in Vibration Theory," *Journal of Sound and Vibration*, **11**(1), 3–18.

Cremer, L., M. Heckl, and E. E. Ungar, 1988. *Structure-Borne Sound*, 2nd ed., Springer-Verlag, Berlin.

Forray, M. J., 1968. *Variational Calculus in Science and Engineering*, McGraw-Hill, New York.

Golla, D. F. and P. C Hughes, 1985. "Dynamics of Viscoelastic Structures—A Time-Domain, Finite Element Formulation," *Journal of Applied Mechanics*, **52** (Dec.), 897–906.

Hildebrand, F. B., 1976. *Advanced Calculus for Applications*, Prentice Hall, Englewood Cliffs, NJ, p. 397.

Inman, D. J., 1989. *VIBRATION with Control, Measurement, and Stability*, Prentice Hall, Englewood Cliffs, NJ, p. 10.

Klosterman, A. L., 1971. *On the Experimental Determination and Use of Model Representations of Dynamic Characteristics*, University Microfilms, Ann Arbor, MI.

Lazan, B. J., 1968. Damping of Materials and Members in Structural Mechanics, Pergamon Press, Oxford.

McTavish, D. J., 1988. "The Mini-Oscillator Technique: A Finite Element Method for the Modeling of Linear Viscoelastic Structures," University of Toronto Institute for Aerospace Studies, Report Number 323, Toronto, Ontario, March.

Meirovitch, L., 1967. *Analytical Methods in Vibrations*, Macmillan, New York.

Meirovitch, L., 1980. *Computational Methods in Structural Dynamics*, Sijthoff & Noordhoff, Rockville, MD.

Meirovitch, L., 1986. *Elements of Vibration Analysis*, McGraw-Hill, New York.

Meyer, O. E., 1874. "J. Reine U.," *Angewandte Math.*, **78,** 130–135.

Nashif, A.D., D. I. Jones, and J. P. Henderson, 1985. *Vibration Damping*, Wiley, New York.

Rao, S. S., 1989. *The Finite Element Method in Engineering*, Pergamon Press, New York.

Reddy, J. N., 1984. *Energy and Variational Methods in Applied Mechanics: With an Introduction to the Finite-Element Method*, Wiley, New York.

Ruzicka, J. E., Ed., 1959. *Structural Damping*, A colloquium on structural damping sponsored by ASME Shock and Vibration Committee, New York.

Segerland, L. J., 1984. *Applied Finite Element Analysis*, Wiley, New York.

Slater, J. C., 1991. "Modeling of Non-Proportional Frequency Dependent Viscoelastic Damping in Space Structures," M.S. Thesis, State University of New York at Buffalo, Oct.

Torvik, P. J., Ed., 1980. *Damping Applications for Vibration Control*, ASME Winter Annual Meeting, Chicago, IL, Nov., pp. 123–132.

Zienkiewicz, O. and R. L. Taylor, 1989. *The Finite Element Method: Basic Formulation and Linear Problems*, McGraw-Hill, New York.

# 3

# LINEAR SYSTEMS AND SIGNALS

## 3.1 INTRODUCTION

In the preceding chapter, methods of deriving the equations of motion for dynamic systems have been presented, which is the first step in the design and analysis of adaptive structures. Upon developing the equations of motion, one must cast the equations in a form that is tractable for computer simulations and control system design. Since the emphasis in this text is placed upon linear adaptive structures, concepts from linear systems and signals are reviewed. It should be clear to the reader that once the problem is cast in the standard linear system format, the relative importance of the spatial compensation that can be achieved through transducer selection and placement is embedded in the mathematics of the model. From a systems approach, one is concerned with the mapping of the inputs to the outputs, which is an important element in the design and synthesis of an adaptive structure. However, the *selection* of appropriate inputs and outputs for desired spatial compensation is an important step in the design of adaptive structures that is somewhat "forgotten" at the systems level. The transducer deposition and aperture that constitute this spatial compensation are the emphasis of Chapter 6.

A simple test can be performed to determine if a system is linear or nonlinear. For example, consider the general expression for a continuous system presented in Chapter 2:

$$\mathcal{L}[w(\mathbf{x},\mathbf{t})] + C\left[\frac{\partial w(\mathbf{x},t)}{\partial t}\right] + \mathcal{M}\left[\frac{\partial^2 w(\mathbf{x},t)}{\partial t^2}\right] = f(\mathbf{x},t) \tag{3.1}$$

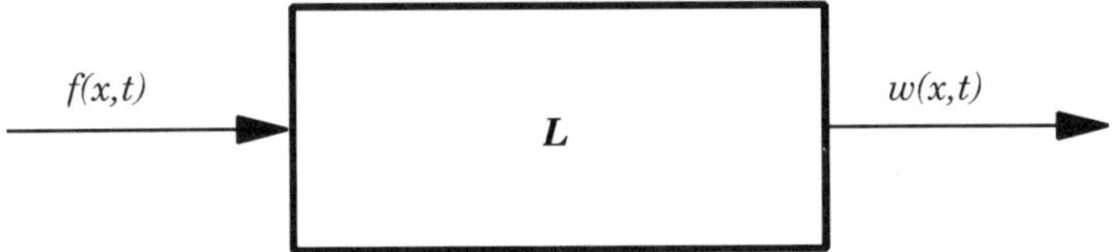


**Figure 3.1:** Schematic of linear system.

The system can be represented in block diagram format as illustrated in Figure 3.1, where

$$L = \mathcal{L} + \mathcal{C}\left(\frac{\partial}{\partial t}\right) + \mathcal{M}\left(\frac{\partial^2}{\partial t^2}\right) \tag{3.2}$$

One can express the system response to two unique inputs as follows:

$$f_1(\mathbf{x}, t) = L[w_1(\mathbf{x}, t)] \tag{3.3}$$

$$f_2(\mathbf{x}, t) = L[w_2(\mathbf{x}, t)] \tag{3.4}$$

Now, let the system input be a linear combination of $f_1(\mathbf{x}, t)$ and $f_2(\mathbf{x}, t)$ such that:

$$f_3(\mathbf{x}, t) = a_1 f_1(\mathbf{x}, t) + a_2 f_2(\mathbf{x}, t) \tag{3.5}$$

If

$$f_3(\mathbf{x}, t) = L[w_3(\mathbf{x}, t)] = a_1 L[w_1(\mathbf{x}, t)] + a_2 L[w_2(\mathbf{x}, t)] \tag{3.6}$$

then the system is said to be linear.

In reality, all systems are nonlinear; however, engineers tend to linearize the system by assuming small displacements or restricting the operating range to some specific linear region. The mathematical foundations for the design and analysis of linear systems are well developed, and it is suggested in this book that the initial design and analysis of adaptive structures can usually be restricted to a linearized system. Nonlinearities must eventually be addressed, but the analysis of linear adaptive structures will serve to guide the basic design process and aid in physical understanding.

## 3.2 TERMINOLOGY

A number of outstanding texts devoted to the topic of linear systems theory exist and should be studied for more detailed coverage of the material (Kailath,

1980; Chen, 1984). There is a well-established terminology in the field of linear systems theory that is reviewed here for background. In general, the variable $t$ is considered to be a real, scalar-valued variable that typically has units of time, but is not restricted to units of time. The variable $\mathbf{x}(t)$ is used to denote a time-varying vector in $n$-dimensional euclidean space:

$$\mathbf{x}(t) \in R^n \tag{3.7}$$

The $n$-column vector $\mathbf{x}(t)$ is defined in terms of its components with respect to the basis $R_n$ as follows:

$$\mathbf{x}(t) = \begin{bmatrix} x_1(t) \\ x_2(t) \\ x_3(t) \\ \vdots \\ x_n(t) \end{bmatrix} \tag{3.8}$$

and is typically termed the state vector. The state vector represents a minimum set of variables required to completely describe the internal dynamics of the system.

The time-varying input vector $\mathbf{u}(t)$ is represented in the $m$-dimensional euclidean space:

$$\mathbf{u}(t) \in R^m \tag{3.9}$$

and can be expressed in vector notation as follows:

$$\mathbf{u}(t) = \begin{bmatrix} u_1(t) \\ u_2(t) \\ u_3(t) \\ \vdots \\ u_m(t) \end{bmatrix} \tag{3.10}$$

The state vector and input vector are used to describe the dynamic response of the system at any chosen output. The forced, linear, time-invariant (LTI) ordinary vector differential equation of order $n$ for a system is expressed as follows:

$$\dot{\mathbf{x}}(t) = \mathbf{A}\mathbf{x}(t) + \mathbf{B}\mathbf{u}(t), \tag{3.11}$$

where $\dot{\mathbf{x}}(t)$ is the time derivative of the time-varying state vector. The matrix **A** is often termed the system matrix and is of dimension $n \times n$ with real, constant coefficients for the current discussion. The input gain matrix $\mathbf{B}$ is an $n \times m$ time-invariant matrix with real constant elements.

## 3.3 AN OVERVIEW OF LTI SYSTEMS

### 3.3.1 Solution of the Homogeneous LTI Vector Differential Equations

The solution of differential equations requires the definition of boundary conditions that when refering to time-dependent equations, are often termed initial conditions. The initial conditions provide a point of origination for the system. For example, if $\mathbf{x}(t_0)$ and $\mathbf{u}(t_0)$ are known, then the derivative of the state vector can be obtained:

$$\dot{\mathbf{x}}(t_0) = \mathbf{A}\mathbf{x}(t_0) + \mathbf{B}\mathbf{u}(t_0) \tag{3.12}$$

The time derivative of the state vector provides a measure of the gradient with respect to time that implies direction. A Taylor series expansion can then be used to deduce the response after some infinitesimal increment in time.

If $\mathbf{u}(t) = 0$ for all $t$, then the homogeneous differential equation is obtained:

$$\dot{\mathbf{x}}(t) = \mathbf{A}\mathbf{x}(t) \tag{3.13}$$

Assuming that the initial conditions of the system are defined by $\mathbf{x}(0)$, the solution of the unforced vector differential equation can be obtained such that the initial conditions and the homogeneous differential equation are satisfied. We assume a solution

$$\phi(t) = \exp(\mathbf{A}t)\mathbf{x}(0) \tag{3.14}$$

where the matrix exponential is defined by (Kailath, 1980)

$$\exp(\mathbf{A}t) \doteq \sum_{n=0}^{\infty} \frac{1}{n!} \mathbf{A}^n t^n \tag{3.15}$$

Substituting equation (3.14) into equation (3.13), one obtains

$$\frac{d}{dt} \phi(t) = \mathbf{A}\phi(t) \tag{3.16}$$

Taking the derivative of the assumed solution yields

$$\frac{d}{dt}\,\phi(t) = \mathbf{A}\exp(\mathbf{A}t)\mathbf{x}(0) = \mathbf{A}\phi(t) \tag{3.17}$$

which is the desired result. One can also show that the solution is unique.

### 3.3.2 Evaluating the Matrix Exponential and Geometric Interpretation

There are several techniques used in evaluating the matrix exponential, $\exp(\mathbf{A}t)$, as detailed by Chen (1984), one of which is based upon numerical methods and the use of a digital computer. However, in the event that distinct eigenvalues of the system matrix $\mathbf{A}$ can be obtained and if the corresponding eigenvectors are real, a similarity transform can be used to compute the matrix exponential. This limiting case is presented to provide the reader with a geometric interpretation of the matrix exponential. Under these assumptions, it can be shown that

$$\exp(\mathbf{A}t) = \mathbf{P}\exp(\mathbf{\Lambda}t)\mathbf{P}^{-1} \tag{3.18}$$

where $\mathbf{\Lambda}$ is a diagonal matrix of eigenvalues of $\mathbf{A}$ and $\mathbf{P}$ is a nonsingular matrix of the corresponding eigenvectors:

$$\mathbf{\Lambda} = \mathbf{P}^{-1}\mathbf{A}\mathbf{P} \tag{3.19}$$

where the $j$th column $\mathbf{p}_j$ of the matrix $\mathbf{P}$ is the corresponding eigenvector of the $j$th eigenvalue $\lambda_j$ of the matrix $\mathbf{\Lambda}$. For a geometric interpretation, consider providing the system with an initial condition such that $\mathbf{x}(0) = \mathbf{p}_j$ (i.e., uniquely equal to the $j$th eigenvector). The system response for all time $t$ can be expressed as follows:

$$\mathbf{x}(t) = \exp(\mathbf{A}t)\mathbf{p}_j = \exp(\lambda_j t)\mathbf{p}_j \tag{3.20}$$

Thus, the system response is constrained to follow the geometric path defined by the eigenvector for all time. For asymptotic stability, the states return to the origin along the vector $\mathbf{p}_j$ over some timescale defined by the eigenvalue $\lambda_j$.

### 3.3.3 Solution of the Forced LTI Vector Differential Equations

The forced vector differential equation has been presented in equation (3.11) and is repeated here for reference:

$$\dot{\mathbf{x}}(t) = \mathbf{A}\mathbf{x}(t) + \mathbf{B}\mathbf{u}(t) \tag{3.21}$$

Assuming that an initial condition is prescribed as well, $\mathbf{x}(t_0) = \mathbf{x}_{t_0}$, and that

$\mathbf{Bu}(\tau)$ is bounded for $\tau \in [t_0, t]$, then the solution of the forced response can be obtained from convolution:

$$\phi(t) = \exp(\mathbf{A}(t - t_0)) \left[ \mathbf{x}_{t_0} + \int_{t_0}^{t} \exp(-\mathbf{A}(\tau - t_0))\mathbf{Bu}(\tau)d\tau \right] \tag{3.22}$$

It can be shown that the solution is unique (Kailath, 1980; Chen, 1984).

### 3.3.4 LTI System Stability

Referring to equation (3.11), we are to determine the stability of the system of vector differential equations by assuming that the input vector is zero $\mathbf{u}(t) = \mathbf{0}$ for all $t \geq 0$ and constrain our analysis to the system with prescribed initial conditions:

$$\dot{\mathbf{x}}(t) = \mathbf{Ax}(t), \qquad \mathbf{x}(0) = \mathbf{x}_0 \tag{3.23}$$

If the system is dissipative, then the initial energy stored within the system will be dissipated as a function of time such that all of the states tend to zero. Formally, the system is asymptotically stable if

$$\lim_{t \to \infty} \|\mathbf{x}(t)\| = 0, \qquad \forall\ \mathbf{x}_0 \in R_n \tag{3.24}$$

where $\|\cdot\|$ is any norm of the vector $\mathbf{x}(t)$ . Since $\mathbf{x}_0$ is some arbitrary constant vector, one can show that a system is asymptotically stable if and only if

$$\lim_{t \to \infty} \| \exp(\mathbf{A}t)\| = 0 \tag{3.25}$$

Assuming that the matrix $\mathbf{A}$ has distinct eigenvalues, a sufficient condition for asymptotic stability is

$$\lim_{t \to \infty} \| \exp(\mathbf{\Lambda}t)\| = 0 \tag{3.26}$$

For this condition to hold, the real part of each eigenvalue must be less than zero.

***Theorem 1*** Given the system $\dot{\mathbf{x}}(t) = \mathbf{Ax}(t)$ with $\mathbf{x}(0) = \mathbf{x}_0$. Assume that the eigenvalues of $\mathbf{A}$, $\lambda_1, \lambda_2, \ldots, \lambda_n$, are all distinct. The system is asymptotically stable if and only if all of the real parts of the eigenvalues are negative.

The theorem holds for repeated eigenvalues as well (Kwakernaak and Sivan, 1972). If any one of the eigenvalues is positive, then the system grows with-

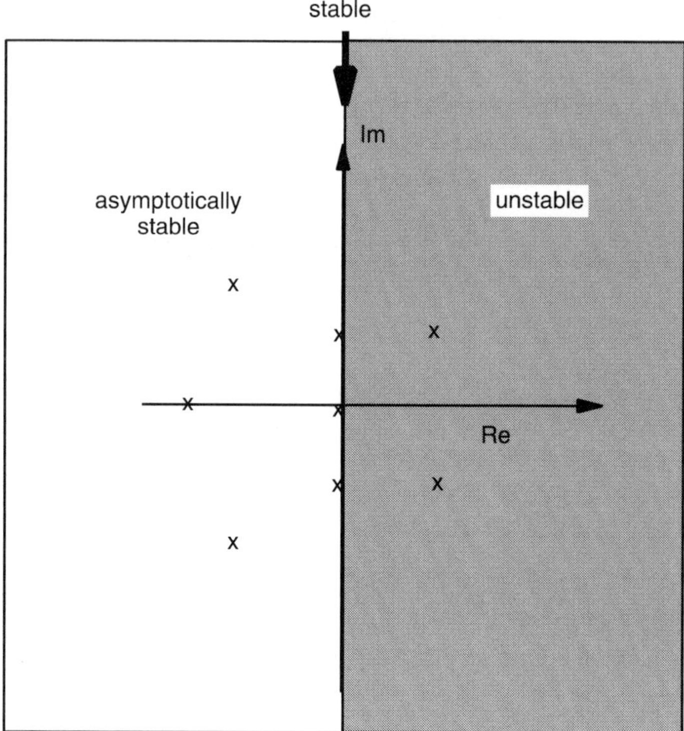


**Figure 3.2:** Stability regions in the complex $s$-plane.

out bound as $t \rightarrow \infty$, which defines an unstable system. Consider the schematic diagram presented in Figure 3.2 of possible eigenvalues in the complex plane. Systems with all left-half plane (left of the $j\omega$ axis where $\lambda_i = \sigma_i + j\omega_i$) eigenvalues are asymptotically stable. Systems with any one or more right-half plane eigenvalues are unstable. The limiting case occurs when eigenvalues are found on the imaginary ($j\omega$) axis. For this case, the system is considered stable, but not asymptotically stable. Regardless of whether there is one or more eigenvalues on the $j\omega$ axis, the system is stable. Repeated eigenvalues on the $j\omega$ axis result in an unstable system.

## 3.4 STATE-VARIABLE FEEDBACK

Some typical goals in the development of adaptive structures for active control of sound and vibration are to modify the existing open-loop system response, increase the input impedance of the system to exogenous inputs, modify the transient response behavior to increase the level of dissipation, or modify the structural response so as to minimize sound radiation (structural acoustic control). A number of different alternative control approaches, based upon feed-forward, adaptive feedforward, and feedback control, are presented in Chapter 7 and Chapter 8. However, with respect to feedback control for adaptive structures, much can be learned from state-variable feedback since this formal approach provides the designer with a physical limit on the performance that can be achieved under an ideal feedback control system case (full information and no uncertainty). Typical objectives might be to stabilize an otherwise unstable system, achieve asymptotic stability in a system that is simply stable, increase the dissipation in a lightly damped system, or decrease the transient response time.

To implement state-variable feedback control, one assumes that the input $\mathbf{u}(t)$ can be made proportional to the states $\mathbf{x}(t)$ through some $m \times n$ constant state-feedback gain matrix $\mathbf{G}$, which must be determined:

$$\mathbf{u}(t) = \mathbf{G}\mathbf{x}(t) \tag{3.27}$$

Substituting equation (3.27) into equation (3.11), one obtains the following vector differential equation:

$$\dot{\mathbf{x}}(t) = \mathbf{A}\mathbf{x}(t) - \mathbf{B}\mathbf{G}\mathbf{x}(t) \tag{3.28}$$

which can be factored to obtain

$$\dot{\mathbf{x}}(t) = (\mathbf{A} - \mathbf{B}\mathbf{G})\mathbf{x}(t) \tag{3.29}$$

The matrix $\mathbf{A} + \mathbf{B}\mathbf{G}$ is often refered to as the closed-loop system matrix. The closed-loop system initial condition response can be expressed as follows:

$$\mathbf{x}(t) = \exp((\mathbf{A} - \mathbf{B}\mathbf{G})t)\mathbf{x}_0 \tag{3.30}$$

Hence the closed-loop system response is controlled by the eigenvalues of the closed-loop system matrix $\mathbf{A}-\mathbf{B}\mathbf{G}$. The objective is to determine the appropriate state-feedback/gain matrix such that the eigenvalues of the closed-loop system matrix can be specified as desired. However, this objective can only be achieved if the pair $[\mathbf{A}, \mathbf{B}]$ is a controllable pair.

***Definition 1*** Given an arbitrary $n \times n$ matrix $\mathbf{A}$ and an arbitrary $n \times m$ matrix $\mathbf{B}$, $[\mathbf{A}, \mathbf{B}]$ is said to be a controllable pair if and only if the rank of the $n \times (nm)$ matrix $\mathbf{C}_o$ defined by

$$\mathbf{C}_o = [\mathbf{B}|\mathbf{A}\mathbf{B}|\mathbf{A}^2\mathbf{B}| \;\cdots\; |\mathbf{A}^{n-1}\mathbf{B}] \tag{3.31}$$

is equal to $n$.

An alternative method of determining if a system is controllable is to verify that the following Gramian matrix has full rank for any $t > 0$:

$$\mathcal{C}_c(t) \equiv \int_0^t \exp(\mathbf{A}\tau)\mathbf{B}\mathbf{B}^{\mathrm{T}} \exp(\mathbf{A}^{\mathrm{T}}\tau)d\tau \tag{3.32}$$

For a stable system, one can evaluate the integral at time $t = \infty$, which yields the controllability Gramian:

$$\mathcal{C}_c \equiv \int_0^\infty \exp(\mathbf{A}\tau)\mathbf{B}\mathbf{B}^{\mathrm{T}} \exp(\mathbf{A}^{\mathrm{T}}\tau)d\tau \tag{3.33}$$

One can readily obtain the controllability Gramian from the solution of the following Lyapunov equation:

$$\mathbf{A}\mathcal{C}_c + \mathcal{C}_c\mathbf{A}^{\mathrm{T}} = -\mathbf{B}\mathbf{B}^{\mathrm{T}} \tag{3.34}$$

***Theorem 2*** Consider a system with $n \times n$ system matrix $\mathbf{A}$ and $n \times m$ input gain matrix $\mathbf{B}$, and where the pair $[\mathbf{A}, \mathbf{B}]$ is controllable. An $m \times n$ state feedback gain matrix $\mathbf{G}$ can be found that allows for an arbitrary assignment of the eigenvalues of the closed-loop system matrix $(\mathbf{A}+\mathbf{B}\mathbf{G})$, where the complex eigenvalues are assigned as complex conjugate pairs.

From this theorem, one observes that, if the system is controllable and full knowledge of the system states is accessible, the dynamics of the closed-loop system can be assigned arbitrarily (Kwakernaak and Sivan, 1972). Thus, state-feedback control is often used to predict the upper bound on performance that can be achieved, and is an important tool in the design of feedback control systems for adaptive structures.

## 3.5 LTI SYSTEM REPRESENTATIONS: STATE-VARIABLE AND INPUT-OUTPUT FORMS

If one builds a system model from differential equations, a choice between state-variable form or a polynomial transfer function (input-output) form is typically made. Since either method can be used to describe and predict the response of an LTI system, it is useful to understand the relationships between the two forms. However, more often than not, practical implementation of a control system involves a system identification process that requires the synthesis of a system model from a set of measured data, including either the time-dependent impulse responses through each input-output path or a set of frequency response functions that describe the dynamics of each input-output path. A basic understanding between the different forms of system representation allows the designer to move easily from one form to another and gain greater insight into the system identification process and model reduction approaches typically employed.

### 3.5.1 Single-Input, Single-Output (SISO), LTI Systems

The SISO, LTI system can be described in terms of a scalar differential equation that maps the input $u(t)$ of the system to the output $y(t)$:

$$\begin{aligned} y^n(t) + a_{n-1}y^{n-1}(t) + \cdots + a_1 y^1(t) + a_0 y(t) \\ = b_m u^m(t) + b_{m-1}u^{m-1}(t) + \cdots + b_1 u^1(t) + b_0 u(t) \end{aligned} \tag{3.35}$$

where

$$y^k(t) \doteq \frac{d^k y(t)}{dt^k} \tag{3.36}$$

$$u^k(t) \doteq \frac{d^k u(t)}{dt^k} \tag{3.37}$$

and $a_i$ and $b_i$ are real constant scalar coefficients. A schematic diagram of the system is depicted in Figure 3.3. All real systems are constrained such that the highest order of the differential operator applied to the input is less than the highest order of the differential operator applied to the output: $m \leq n$. The order of the system is measured by the order of the highest differential operator applied to the output: $n$. The system transfer function can be obtained from the differential equation by applying the Laplace transform with zero initial conditions. For greater detail on this integral transform, refer to Appendix A. The Laplace transform of equation (3.35) is

$$H(s) \doteq \frac{y(s)}{u(s)} \doteq \frac{N(s)}{D(s)} \doteq \frac{b_m s^m + b_{m-1}s^{m-1} + \cdots + b_1 s^1 + b_0}{s^n + a_{n-1}s^{n-1} + \cdots + a_1 s + a_0} \tag{3.38}$$

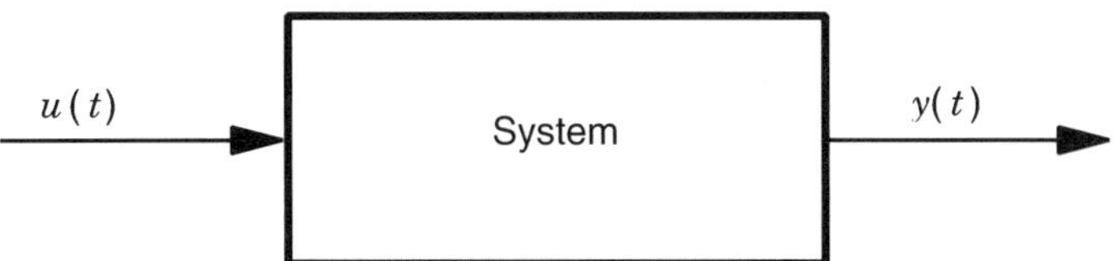


**Figure 3.3:** Schematic diagram of SISO system.

where $N(s)$ is the numerator polynomial and $D(s)$ is the denominator polynomial of the system transfer function $H(s)$ . The roots of the numerator polynomial are termed the *zeros* of the transfer function:

$$N(s) = \prod_{i=1}^{m} (s - z_i) \tag{3.39}$$

and the roots of the denominator polynomial are termed the *poles* of the transfer function:

$$D(s) = \prod_{i=1}^{n} (s - p_i) \tag{3.40}$$

In the event that $m \le n - 1$, and if the poles of $H(s)$ are distinct, then one can express the transfer function in pole-residue form:

$$H(s) = \sum_{i=1}^{n} \frac{r_i}{s - p_i} \tag{3.41}$$

As indicated in Appendix A, the residues of the system are quite important in the evaluation of the inverse Laplace transform, which leads to the time-dependent solution of the differential equation.

Let us now consider the state-variable form of the SISO system:

$$\dot{\mathbf{x}}(t) = \mathbf{A}\mathbf{x}(t) + \mathbf{b}u(t) \tag{3.42}$$

$$y(t) = \mathbf{c}\mathbf{x}(t) + du(t) \tag{3.43}$$

where $\mathbf{A}$ is an $n \times n$ matrix of real constants, $\mathbf{b}$ is an $n \times 1$ column vector of real constants, $\mathbf{c}$ is a $1 \times n$ row vector of real constants, and $d$ is a real scalar constant. Taking the Laplace transform of equation (3.43) with zero initial conditions yields

$$s\mathbf{x}(s) = \mathbf{A}\mathbf{x}(s) + \mathbf{b}u(s) \tag{3.44}$$

$$y(s) = \mathbf{c}\mathbf{x}(s) + du(s) \tag{3.45}$$

Solving for $x(s)$ and substituting into the output equation yields the following relationship:

$$y(s) = \left[\mathbf{c}[s\mathbf{I} - \mathbf{A}]^{-1}\mathbf{b} + d\right]\ u(s) \tag{3.46}$$

Dividing equation (3.46) by $u(s)$ yields

$$\frac{y(s)}{u(s)} = H(s) = \left[\mathbf{c}[s\mathbf{I} - \mathbf{A}]^{-1}\mathbf{b} + d\right] \tag{3.47}$$

Using basic properties of matrices, one can express the system transfer function as follows:

$$H(s) = \frac{\mathbf{c}\,\mathrm{adj}(s\mathbf{I} - \mathbf{A})\mathbf{b} + d\det(s\mathbf{I} - \mathbf{A})}{\det(s\mathbf{I} - \mathbf{A})} \tag{3.48}$$

From equation (3.48) one can obtain expressions for the numerator polynomial $N(s)$ and the denominator polynomial $D(s)$:

$$N(s) = \mathbf{c}\,\mathrm{adj}(s\mathbf{I} - \mathbf{A})\mathbf{b} + d\det(s\mathbf{I} - \mathbf{A}) \tag{3.49}$$

$$D(s) = \det(s\mathbf{I} - \mathbf{A}) \tag{3.50}$$

The only case in which the order of the numerator polynomial is equal to that of the denominator polynomial is when the feedthrough term $d$ is nonzero. As discussed in Chapter 5, this feedthrough term is important in the field of adaptive structures when implementing a piezoelectric transduction device concurrently as a sensor and an actuator, and it is nonzero when the electrical dynamics of the transducer are present in the output. Additionally, for a dynamic system, $d$ will be nonzero when the output variable is proportional to acceleration, since a feedthrough term associated with the forcing input is required to predict the output.

In general, a state-variable representation of a system is nonunique in that there are many different forms that can be used to represent the input-output system. This can be used to the designer's advantage by choosing states that are readily observed with sensors. One can perform a similarity transform on the system matrix since there are an infinite number of matrices that are similar to $\mathbf{A}$. Through a change of variables alternative states can be defined that yield the same input-output relationship. One can choose from a number of different canonical forms for state-variable representations. The more common forms are the *controllable form* and the *observable form*. Consider, for example, an input-output system described by

$$H(s) = \frac{b_{n-1}s^{n-1} + \cdots + b_1 s + b_0}{s^n + a_{n-1}s^{n-1} + \cdots + a_1 s + a_0} \tag{3.51}$$

The controllable representation is such that

$$\mathbf{A} = \begin{bmatrix} 0 & 1 & 0 & \cdots & 0 \\ 0 & 0 & 1 & \cdots & 0 \\ \vdots & \vdots & \vdots & \vdots & \vdots \\ -a_0 & -a_1 & -a_2 & \cdots & -a_{n-1} \end{bmatrix} \tag{3.52}$$

$$\mathbf{b} = \begin{bmatrix} 0 \\ 0 \\ \vdots \\ 1 \end{bmatrix} \tag{3.53}$$

$$\mathbf{c} = [b_0 \quad b_1 \quad \cdots \quad b_{n-1}] \tag{3.54}$$

$$d = 0 \tag{3.55}$$

If the system has feedthrough dynamics, the observable form is used. Given a system input-output function

$$H(s) = \frac{b_n s^n + b_{n-1}s^{n-1} + \cdots + b_1 s + b_0}{s^n + a_{n-1}s^{n-1} + \cdots + a_1 s + a_0} \tag{3.56}$$

the observable representation is such that

$$\mathbf{A} = \begin{bmatrix} 0 & 1 & 0 & \cdots & 0 \\ 0 & 0 & 1 & \cdots & 0 \\ \vdots & \vdots & \vdots & \vdots & \vdots \\ -a_0 & -a_1 & -a_2 & \cdots & -a_{n-1} \end{bmatrix} \tag{3.57}$$

$$\mathbf{b} = \begin{bmatrix} b_{11} \\ b_{21} \\ \vdots \\ b_{n1} \end{bmatrix} \tag{3.58}$$

$$\mathbf{c} = [1 \quad 0 \quad 0 \quad \cdots \quad 0] \tag{3.59}$$

$$d = b_n \tag{3.60}$$

where

$$b_{01} = b_n \tag{3.61}$$

$$\begin{aligned} b_{11} &= b_{n-1} - b_{01}a_{n-1} \\ b_{21} &= b_{n-1} - b_{01}a_{n-2} - b_{11}a_{n-1} \\ &\vdots \\ b_{n1} &= b_0 - \sum_{j=0}^{n-1} a_i b_{i1} \end{aligned} \tag{3.62}$$

The distinct difference between the controllable and observable forms is in the representations of the **b** and **c** vectors.

### 3.5.2 Multiple-Input, Multiple-Output (MIMO), LTI Systems

The MIMO, LTI system can be described in terms of a vector differential equation that maps the $r$ inputs $\mathbf{u}(t)$ of the system to the $m$ outputs $\mathbf{y}(t)$ . The form is analogous to equation (3.43):

$$\dot{\mathbf{x}}(t) = \mathbf{A}\mathbf{x}(t) + \mathbf{B}\mathbf{u}(t) \tag{3.63}$$

$$\mathbf{y}(t) = \mathbf{C}\mathbf{x}(t) + \mathbf{D}\mathbf{u}(t) \tag{3.64}$$

Upon applying the Laplace transform and performing minor substitutions, the system of equations can be expressed as follows:

$$\mathbf{y}(s) = [\mathbf{C}(s\mathbf{I} - \mathbf{A})^{-1}\mathbf{B} + \mathbf{D}]\mathbf{u}(s) \tag{3.65}$$

The $m \times n$ matrix transfer function $\mathbf{H}(s)$ is defined as the matrix that premultiplies the input vector $\mathbf{u}(s)$:

$$\mathbf{H}(s) = \mathbf{C}(s\mathbf{I} - \mathbf{A})^{-1}\mathbf{B} + \mathbf{D} \tag{3.66}$$

Again, one can reduce the above expression to the following:

$$\mathbf{H}(s) = \frac{\mathbf{C}\ \mathrm{adj}(s\mathbf{I} - \mathbf{A})\mathbf{B} + \mathbf{D}\det(s\mathbf{I} - \mathbf{A})}{\det(s\mathbf{I} - \mathbf{A})} \tag{3.67}$$

where the denominator polynomial forms an expression of the characteristic equation

$$D(s) \doteq \det(s\mathbf{I} - \mathbf{A}) \tag{3.68}$$

and the numerator is a polynomial matrix defining the zeros of the $r \times m$ system:

$$\mathbf{N}(s) \doteq \mathbf{C}\,\mathrm{adj}(s\mathbf{I} - \mathbf{A})\mathbf{B} + \mathbf{D}\det(s\mathbf{I} - \mathbf{A}) \tag{3.69}$$

Hence, for the MIMO system, any desired SISO transfer function can be obtained by considering the appropriate entry of the transfer matrix $\mathbf{H}(s)$ . The transfer matrix $\mathbf{H}(s)$ is termed a *strictly proper* rational matrix if it has more poles than zeros, which corresponds to $\mathbf{D} = 0$. However, if $\mathbf{D}$ is nonzero, the order of each polynomial in $\mathbf{N}(s)$ will equal that of $D(s)$ , and the matrix is termed *proper* with an equal number of poles and zeros.

While the derivation of a transfer matrix from state equations is relatively simple, the reverse transformation is nonunique. One can readily show through a change of basis in the state-space that an infinite number of $n$th-order realizations of $\mathbf{H}(s)$ exist. However, in the modeling and design of control systems, if output feedback control is to be employed, one is often interested in a minimum (irreducible) realization of the system.

***Definition 2*** For all possible realizations of $\mathbf{H}(s)$ , the state-space representation $\{\mathbf{A}, \mathbf{B}, \mathbf{C}, \mathbf{D}\}$ is said to be a minimum realization if the corresponding state space has the smallest possible dimension.

Thus, a minimum realization of a system can be used to predict the response between the defined inputs and outputs of the system; however, information about the internal structure of the system is lost. For analytical work, using a minimum realization offers little advantage unless one is faced with a system of high-order and needs to find the minimum realization to reduce the order of the system. For experimental work, which typically involves system identification, the transfer matrix obtained (typically from frequency response data) is always a minimum realization since nothing can be determined about the internal structure that is absent from the input-output data measured. In general, a transfer matrix $\mathbf{H}(s)$ and its minimum realization simply describe the subsystem that is both completely controllable and completely observable. In fact, the technique used to find a minimum realization involves decomposing a realization of $\mathbf{H}(s)$ into a subsystem that is completely observable and completely controllable (Brogan, 1985).

***Definition 3*** Given an arbitrary $n \times n$ matrix $\mathbf{A}$ and an arbitrary $m \times n$ matrix $\mathbf{C}$, $[\mathbf{C}, \mathbf{A}]$ is said to be an observable pair if and only if the rank of the $n \times (nm)$ matrix $\mathbf{O}_o$, defined by

$$\mathbf{O}_o = [\mathbf{C}^{\mathrm{T}} | \mathbf{A}^{\mathrm{T}}\mathbf{C}^{\mathrm{T}} | (\mathbf{A}^{\mathrm{T}})^2\mathbf{C}^{\mathrm{T}} | \cdots | (\mathbf{A}^{\mathrm{T}})^{n-1}\mathbf{C}^{\mathrm{T}}] \tag{3.70}$$

is equal to $n$.

Alternatively, for a stable system, one can determine if the system is state observable by determining if the observability Gramian, $O_o$, has full rank:

$$O_o \equiv \int_0^\infty \exp(\mathbf{A}^{\mathrm{T}}\tau)\mathbf{C}^{\mathrm{T}}\mathbf{C}\exp(\mathbf{A}\tau)d\tau \tag{3.71}$$

One can readily obtain the observability Gramian from the solution of the following Lyapunov equation:

$$\mathbf{A}^{\mathrm{T}}O_o + O_o\mathbf{A} = -\mathbf{C}^{\mathrm{T}}\mathbf{C} \tag{3.72}$$

For the sake of terminology, a transfer matrix realization can be expressed in terms of the quadruple of matrices {**A**, **B**, **C**, **D**}:

$$\mathbf{H}(s) = \left[\begin{array}{c|c} \mathbf{A} & \mathbf{B} \\ \hline \mathbf{C} & \mathbf{D} \end{array}\right] \tag{3.73}$$

This notation is frequently used to represent MIMO system realizations. The number of poles in the transfer matrix $\mathbf{H}(s)$ is known as the *McMillan degree* of $\mathbf{H}(s)$ , which is equal to the dimension of the state vector in the minimum realization (Green and Limebeer, 1995). For a given system realization, the pair $(\mathbf{A}, \mathbf{B})$ is said to be *stabilizable* if every unstable eigenspace is controllable, and the pair $(\mathbf{A}, \mathbf{C})$ is said to be *detectable* if every unstable eigenspace is observable. Thus, one must be able to observe and control all unstable poles of a given realization to achieve stability. This is a less severe requirement than that of controllability and observability, precluding arbitrary pole placement but allowing for the development of a closed-loop stable system based upon a chosen array of sensors and actuators.

### 3.5.3 Building a MIMO Model of a Structure

Thus far, we have discussed the representation of LTI systems in very general terms. However, the objective is to develop realizations of structural models in state-variable form to evaluate various control strategies for adaptive structures. With this thrust in mind, consider the MIMO representation of an LTI system:

$$\begin{aligned}\dot{\mathbf{x}}(t) &= \mathbf{A}\mathbf{x}(t) + \mathbf{B}\mathbf{u}(t)\\ \mathbf{y}(t) &= \mathbf{C}\mathbf{x}(t) + \mathbf{D}\mathbf{u}(t)\end{aligned}$$

Assume that the states are defined in terms of the generalized coordinates of the structure such that

$$\mathbf{x}^{\mathrm{T}} = [\mathbf{r}_{N\times 1}^{\mathrm{T}} \quad \dot{\mathbf{r}}_{N\times 1}^{\mathrm{T}}] \tag{3.74}$$

where $\mathbf{r}(t)$ and $\dot{\mathbf{r}}(t)$ are the generalized displacements and generalized velocities, respectively, of the structural model. Then from a Ritz model or assumed-modes approach, one can express $\mathbf{A}$ in terms of the structural properties:

$$\mathbf{A} = \begin{bmatrix} \mathbf{0}_{N\times N} & \mathbf{I}_{N\times N} \\ -[\mathbf{M}]^{-1}_{N\times N}[\mathbf{K}]_{N\times N} & -[\mathbf{M}]^{-1}_{N\times N}[\mathbf{D}]_{N\times N} \end{bmatrix}, \tag{3.75}$$

where one assumes that there are $N$ expansion functions used in the model, and $[\mathbf{M}]$ is the mass matrix, $[\mathbf{K}]$ is the stiffness matrix, and $[\mathbf{D}]$ is a damping matrix. In the event that proportional damping is chosen, and the eigenfunctions (normalized with respect to the mass) are used in the expansion, $\mathbf{A}$ can be expressed as follows as previously outlined by Meirovitch (1990):

$$\mathbf{A} = \begin{bmatrix} \mathbf{0}_{N\times N} & \mathbf{I}_{N\times N} \\ -\mathbf{\Lambda}^2_{N\times N} & -2\xi\mathbf{\Lambda}_{N\times N} \end{bmatrix} \tag{3.76}$$

where $\mathbf{\Lambda}$ is a diagonal matrix of the natural frequencies, and $\xi$ is the proportional damping ratio (assumed equal for all natural frequencies).

The input gain matrix can be expressed as follows for $M$ inputs to the system:

$$\mathbf{B} = \begin{bmatrix} \mathbf{0}_{N\times M} \\ [\mathbf{M}]^{-1}_{N\times N}\mathbf{F}_{N\times M} \end{bmatrix} \tag{3.77}$$

where $[\mathbf{M}]^{-1}\mathbf{F}$ is a matrix describing the influence of each of the $M$ forcing inputs in the $N$ generalized coordinates. This "participation" of the force in generalized coordinates is, in general, defined by the type of actuator used (i.e., point-force, induced-strain, etc.), its location, and spatial aperture, as discussed in Chapter 5. The input signal $\mathbf{u}(t)$ is expressed in terms of units appropriate to the type of input chosen.

Recall that the displacement response for a structure can be expressed in terms of the expansion function:

$$w(\mathbf{x}, t) = \sum_{n=1}^{N} \phi_n(\mathbf{x}) r_n(t)$$

where $\phi_n(\mathbf{x})$ represents the participation of each generalized coordinate $r_n(t)$ in the structural response at position $\mathbf{x}$. The observed output can be expressed in terms of any linear combination of the states or their derivatives. For example, if one were to observe the displacement at $P$ points on the structure, then for our chosen states,

$$\mathbf{C} = [\boldsymbol{\Phi}_{P\times N} \quad \mathbf{0}_{P\times N}] \tag{3.78}$$

where $\boldsymbol{\Phi}$ represents the participation of each of the $N$ generalized coordinates in the response at the $P$ locations on the structure. However, the influence of the generalized velocities do not influence the displacement response. If one were to observe velocities at the same points as opposed to displacements, then the submatrices in the previous expression would be "switched" to define $\mathbf{C}$. As was the case for the inputs, this participation matrix $\boldsymbol{\Phi}$ is affected by the type, position, and aperture of the transducer, as expanded upon in Chapter 5. If displacements or velocites are observed, then the feedthrough dynamics are zero such that

$$\mathbf{D} = [\mathbf{0}]_{P\times M} \tag{3.79}$$

For accelerations, the $\mathbf{D}$ matrix is nonzero.

## 3.6 SIGNAL AND SYSTEM NORMS

We introduce the definition of signal and system norms in this section to provide a foundation for the chapters devoted to control that follow. Our motivation for computing norms is to provide a consistent method of characterizing the *size* of a signal or in the case of a system, the size of an output signal produced by a particular input signal. In either case, the general objective is to develop a consistent means of evaluating the performance of a control system based upon a consistent measure of the size of the error signals, control signals, and so on. One should take care to distinguish between the norm of a signal and the norm of a system; the difference is reviewed in this section.

### 3.6.1 The $L_2$ Norm of a Signal

The $L_2$ norm of a signal is defined as follows:

$$\|u\|_2 \doteq \left( \int_0^\infty u(t)^2 \, dt \right)^{1/2} \tag{3.80}$$

This norm is representative of signals with finite total energy and details the integral root mean square (RMS) power over time. Applying Parseval's theorem, the frequency domain equivalent of the $L_2$ norm can be expressed as follows:

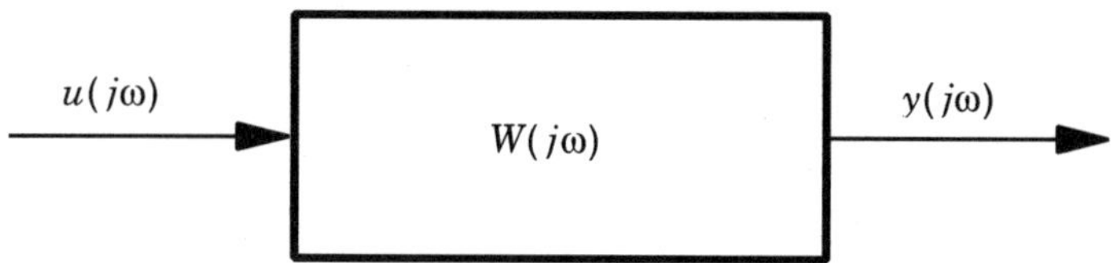


**Figure 3.4:** Schematic diagram of a filtered signal.

$$\|u\|_2 = \left( \frac{1}{2\pi} \int_{-\infty}^{\infty} |U(j\omega)|^2 \, d\omega \right)^{1/2} \tag{3.81}$$

Consider now a filtered signal $y(\omega)$, as illustrated in Figure 3.4. The power spectral density of the filtered signal can be expressed as follows:

$$S_{yy}(\omega) = S_{uu}(\omega)|W(j\omega)|^2 \tag{3.82}$$

where $W(j\omega)$ is the frequency response function of the linear system and $S_{uu}(\omega)$ is the autospectrum of the input (Bendat and Piersol, 1986).

The $L_2$ norm of the resulting output signal can thus be expressed as follows:

$$\|y\|_2 = \left( \frac{1}{2\pi} \int_{-\infty}^{\infty} S_{uu}(\omega)|W(j\omega)|^2 \, d\omega \right)^{1/2} \tag{3.83}$$

The filter $W(j\omega)$ serves to weight the power spectrum, increasing the contribution where the magnitude of the filter is large and decreasing the contribution where the magnitude of the filter is small. The design and selection of frequency weighted filters for design and synthesis of control systems is integral to the closed-loop performance of the system and is the method by which the "physics" of the problem is incorporated into the design, as demonstrated in Chapter 9.

### 3.6.2 The $L_\infty$ Norm of a Signal

The $L_\infty$ norm of a signal is defined as follows:

$$\|u\|_\infty \doteq \sup_{t \geq 0} |u(t)| \tag{3.84}$$

The $L_\infty$ norm of a signal provides an alternative method of characterizing whether a signal is small or large. However, one should observe that transients will play a significant role in the $L_\infty$ norm by definition. Thus, spurious large signals produce a large $L_\infty$ norm. However, if a signal must be maintained below some threshold for all time, the $L_\infty$ norm provides the necessary measure.

### 3.6.3 The $\mathcal{H}_2$ Norm of a System

***The SISO, LTI System*** Consider the expression for the $L_2$ norm of a weighted signal as in equation (3.83). If the input power spectrum is set to unity ($S_{uu}(\omega) = 1$), then one obtains an expression for the $\mathcal{H}_2$ norm of the filter:

$$\|W\|_2 \doteq \left( \frac{1}{2\pi} \int_{-\infty}^{\infty} |W(j\omega)|^2 \, d\omega \right)^{1/2} \tag{3.85}$$

Thus, the $\mathcal{H}_2$ norm of a filter represents the RMS response of the filter output when driven by white noise input (Boyd and Baratt, 1991). Additionally, by Parseval's theorem,

$$\|W\|_2 = \|w\|_2 = \left( \int_0^{\infty} w(t)^2 \, dt \right)^{1/2} \tag{3.86}$$

where $\|w\|_2$ is the $L_2$ norm of the impulse response $w(t)$ of the LTI filter. Thus, the $\mathcal{H}_2$ norm of a system can be obtained from the $L_2$ norm of the signal defined by the system's impulse response. It is important to recognize the difference between the norm of a signal and the norm of a system.

***The MIMO, LTI System*** Let $\mathbf{H}(j\omega)$ represent a stable system with an input to the system defined by the power spectral density matrix of $\mathbf{u}$, $\mathbf{S}_{uu}(j\omega)$ . The RMS norm of the system response with respect to the input can be expressed as follows:

$$\|\mathbf{H}\|_{RMS} = \left( \mathrm{Tr} \, \frac{1}{2\pi} \int_{-\infty}^{\infty} \mathbf{H}(j\omega)\mathbf{S}_{uu}(j\omega)\mathbf{H}^*(j\omega) d\omega \right)^{1/2} \tag{3.87}$$

Now, if the power spectral density matrix is assumed to be a white process (i.e., $\mathbf{S}_{uu}(j\omega) \approx 1$), then the $\mathcal{H}_2$ norm of the system can be expressed as follows:

$$\|\mathbf{H}\|_2 \doteq \left( \mathrm{Tr} \, \frac{1}{2\pi} \int_{-\infty}^{\infty} \mathbf{H}(j\omega)\mathbf{H}^*(j\omega) d\omega \right)^{1/2} \tag{3.88}$$

The physical interpretation is that the $\mathcal{H}_2$ norm is the RMS output when the inputs are driven concurrently by independent, stochastic input signals. The time-domain equivalent of the norm is obtained from Parseval's theorem:

$$\|\mathbf{H}\|_2 \doteq \left( \mathrm{Tr} \int_0^{\infty} \mathbf{h}(t)\mathbf{h}^{\mathrm{T}}(t)\, dt \right)^{1/2} \tag{3.89}$$

where $\mathbf{h}(t)$ is the impulse response of the system $\mathbf{H}(j\omega)$.

It turns out that in the analysis of MIMO systems, the *singular values* of a system transfer matrix play an important role. For a SISO system, the singular values correspond to the magnitude of the Bode plot, and for a MIMO system, they provide analogous information about the system response. The singular values can also be used to compute the $\mathcal{H}_2$ norm of the system $\mathbf{H}(j\omega)$:

$$\|\mathbf{H}\|_2 \doteq \left( \mathrm{Tr} \frac{1}{2\pi} \int_{-\infty}^{\infty} \sum_{j=1}^{n} \sigma_j(\mathbf{H}(j\omega))^2\, d\omega \right)^{1/2} \tag{3.90}$$

where $n$ is the minimum of the number of inputs or outputs. Thus, the $\mathcal{H}_2$ norm is obtained from the sum of the squares of all of the singular values of the system transfer matrix $\mathbf{H}$. Stating in words the expression given in equation (3.90): the $\mathcal{H}_2$ norm of a system transfer matrix is proportional to the square root of the area under the curve defined by the summation of the squared singular values of the system transfer matrix. To provide an example, the MATLAB script file **svd_ex.m** can be executed to display Figure 3.5. As indicated in Figure 3.5, the sum of the squared singular values as a function of frequency for the system developed in the script file is presented. Since the system is a two-input, two-output system, there are two singular values at each frequency. The square root of the area under this curve defines the $\mathcal{H}_2$ norm.

Practically, the $\mathcal{H}_2$ norm can be computed from the observability Gramian or the controllablity Gramian. For a system described in state-variable form such that $\mathbf{G}(s) = \mathbf{C}(s\mathbf{I} - \mathbf{A})^{-1}\mathbf{B}$, the $\mathcal{H}_2$ norm can be computed as follows:

$$\|\mathbf{G}(s)\|_2 = \sqrt{\mathrm{Tr}(\mathbf{B}^{\mathrm{T}} \mathcal{O}_o \mathbf{B})} \tag{3.91}$$

or

$$\|\mathbf{G}(s)\|_2 = \sqrt{\mathrm{Tr}(\mathbf{C} \mathcal{C}_c \mathbf{C}^{\mathrm{T}})} \tag{3.92}$$

where $\mathcal{O}_o$ and $\mathcal{C}_c$ are the observability and controllability Gramians, respectively.

### 3.6.4 The $\mathcal{H}_\infty$ Norm of a System

***The SISO System*** The $\mathcal{H}_\infty$ norm of a system transfer function is defined as follows:

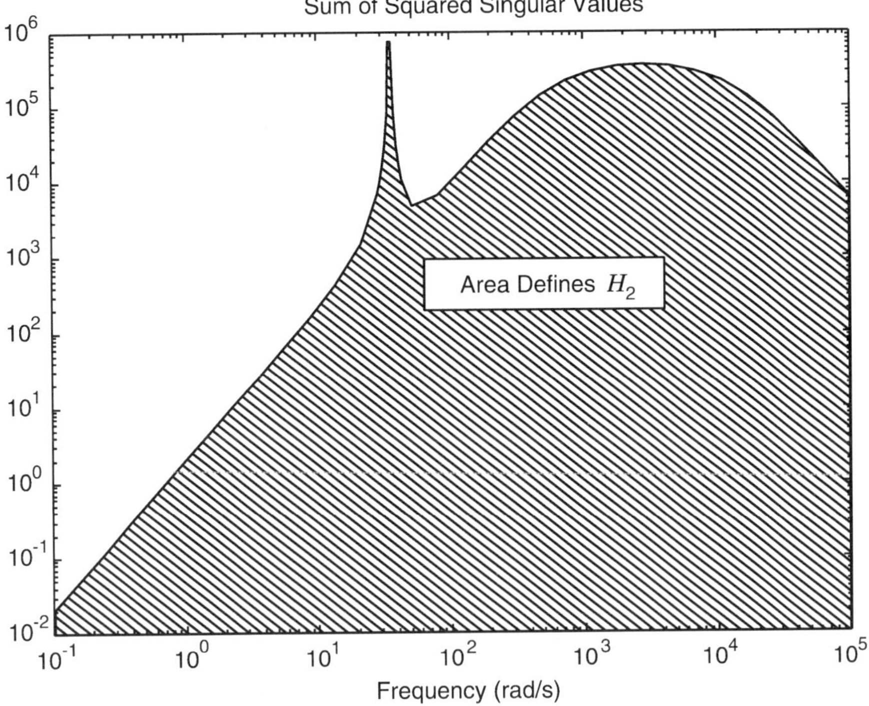


**Figure 3.5:** Sum of squared singular values as a function of frequency for a fourth-order system.

$$\|H\|_\infty = \|H\|_{\text{RMS}_{\text{gain}}} \doteq \sup_{\|u\|_{\text{RMS}} \neq 0} \frac{\|Hu\|_{\text{RMS}}}{\|u\|_{\text{RMS}}} \tag{3.93}$$

However, the RMS gain of $H$ is simply the $L_2$ gain, and the infinity norm can be expressed as follows:

$$\|H\|_\infty = \sup_{\|u\|_2 \neq 0} \frac{\|Hu\|_2}{\|u\|_2} \tag{3.94}$$

***The MIMO System*** The $\mathcal{H}_\infty$ norm of a MIMO system transfer matrix is representative of the RMS gain of a system and is defined as follows:

$$\|\mathbf{H}\|_\infty \doteq \sup_\omega \ \sigma_{\max}(\mathbf{H}(j\omega)) \tag{3.95}$$

Again, the script file (**svd_ex.m**) used to generate the singular values used in the

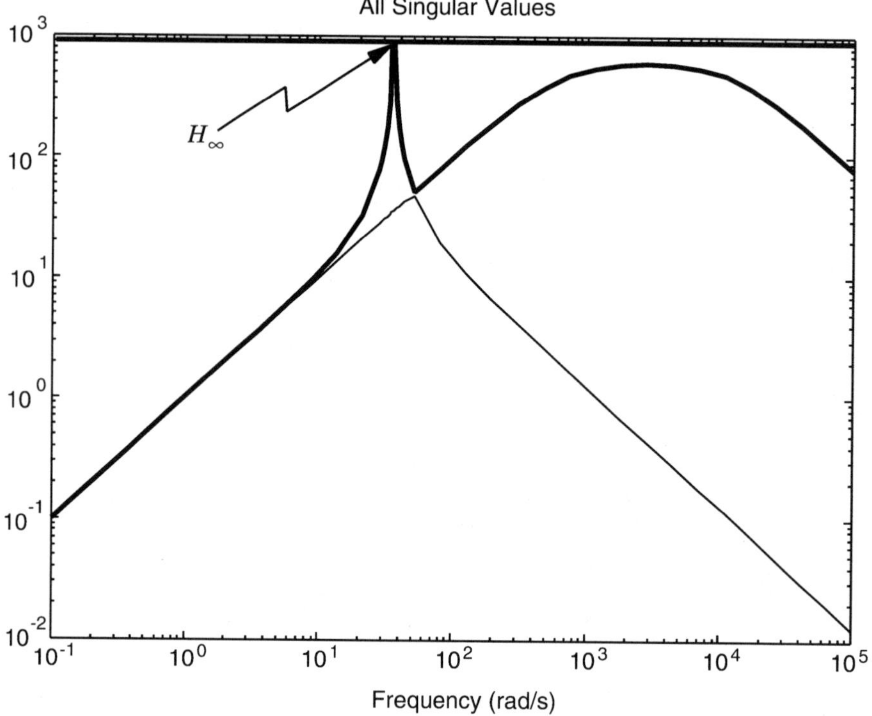


**Figure 3.6:** Plot of all singular values of a fourth-order system.

computation of the $\mathcal{H}_2$ norm was used to generate Figure 3.6. As illustrated, only the maximum singular values are important in the consideration of the $\mathcal{H}_\infty$ norm. The remaining singular values have no impact. The $\mathcal{H}_\infty$ norm thus readily emphasizes *peaks* that might occur in the system transfer matrix. Detailed use of the norm for control system design is discussed in subsequent chapters.

The $\mathcal{H}_\infty$ norm can be computed through numerical iteration to determine the smallest value of $\gamma$ such that the Hamiltonian matrix has no eigenvalues on the imaginary axis. The Hamiltonian matrix is defined as follows for a state-space realization of a system:

$$\mathbf{H} = \begin{bmatrix} \mathbf{A} + \mathbf{B}\mathbf{M}^{-1}\mathbf{D}^{\mathrm{T}}\mathbf{C} & \mathbf{B}\mathbf{M}^{-1}\mathbf{B}^{\mathrm{T}} \\ -\mathbf{C}^{\mathrm{T}}(\mathbf{I} + \mathbf{D}\mathbf{M}^{-1}\mathbf{D}^{\mathrm{T}})\mathbf{C} & -(\mathbf{A} + \mathbf{B}\mathbf{M}^{-1}\mathbf{D}^{\mathrm{T}}\mathbf{C})^{\mathrm{T}} \end{bmatrix} \tag{3.96}$$

where $\mathbf{M} = \gamma^2\mathbf{I} - \mathbf{D}^{\mathrm{T}}\mathbf{D}$ (Zhou et al., 1996). To iterate, one chooses a large value of $\gamma$ to begin and then continues the iterative process until some specified level of convergence is reached (i.e., the imaginary eigenvalues result).

## 3.7 THE LINEAR QUADRATIC REGULATOR (LQR) PROBLEM

The LQR problem is important in the analysis and design of feedback control systems, including those considered for adaptive structures, because it provides a means of evaluating the optimal control that can be achieved. The optimal control is constrained by a given array of actuators and some predefined state-variable peformance weighting matrix and control effort weighting matrix. This optimization problem involves finding the appropriate state-feedback controller that minimizes the following cost functional:

$$J = \int_{t_0}^{t_f} (\mathbf{x}^{\mathrm{T}}(t)\mathbf{Q}(t)\mathbf{x}(t) + \mathbf{u}^{\mathrm{T}}(t)\mathbf{R}(t)\mathbf{u}(t))\,dt + \mathbf{x}^{\mathrm{T}}(t_f)\mathbf{M}\mathbf{x}(t_f) \tag{3.97}$$

where $\mathbf{M}$ and $\mathbf{Q}(t)$ are symmetric positive semidefinite marices, and $\mathbf{R}(t)$ is a symmetric positive definite matrix. The first term in the integral serves to penalize the states and defines the desired control system performance through an appropriate choice of coefficients in the state weighting matrix, $\mathbf{Q}(t)$ . The second term in the integral serves to penalize the control effort, and $\mathbf{R}(t)$ serves as a "knob" that can be adjusted to guard against unrealistic control input signals. The final term in the cost function defines the penalty placed upon the final value of the states. The weighting matrix $\mathbf{M}$ provides a means of conveying the importance of the final state $\mathbf{x}(t_f)$. The cost function is subject to the dynamic constraint of the system:

$$\dot{\mathbf{x}}(t) = \mathbf{A}(t)\mathbf{x}(t) + \mathbf{B}(t)\mathbf{u}(t), \quad \mathbf{x}(t_0) = \mathbf{x}_0 \tag{3.98}$$

The optimal control is obtained through full state-feedback with a control law defined as follows:

$$\mathbf{u}(t) = -\mathbf{G}(t)\mathbf{x}(t) \tag{3.99}$$

where the gain matrix is given by

$$\mathbf{G}(t) = \mathbf{R}^{-1}(t)\mathbf{B}^{\mathrm{T}}(t)\mathbf{K}(t) \tag{3.100}$$

and $\mathbf{K}(t)$ is obtained from the solution of the matrix Riccati differential equation:

$$\frac{d}{dt}\mathbf{K}(t) = -\mathbf{K}(t)\mathbf{A}(t) - \mathbf{A}^{\mathrm{T}}(t)\mathbf{K}(t) - \mathbf{Q}(t) + \mathbf{K}(t)\mathbf{B}(t)\mathbf{R}^{-1}(t)\mathbf{B}^{\mathrm{T}}(t)\mathbf{K}(t) \tag{3.101}$$

The solution of the Riccati differential equation is subject to the boundary condition at the final time:

$$\mathbf{K}(t_f) = \mathbf{M} \tag{3.102}$$

The optimal controller is designed for the current response of the system based upon the future desired system response through the application of the boundary condition at time $t_f$.

While the development thus far has been for a time-varying system, the solution for the LTI problem is rendered by letting the optimization time-interval $t_f$ approach infinity (i.e., $t_f \rightarrow \infty$). Consider the MIMO, LTI control system design, for which one typically desires to minimize the outputs associated with some system. The following quadratic cost functional can be formed to express the desired objective:

$$J = \int_0^\infty (\mathbf{y}^{\mathrm{T}}(t)\mathbf{y}(t) + \mathbf{u}^{\mathrm{T}}(t)\mathbf{R}\mathbf{u}(t))\, dt \tag{3.103}$$

Letting $\mathbf{Q} = \mathbf{C}^{\mathrm{T}}\mathbf{C}$, one can express equation (3.103) as follows:

$$J = \int_0^\infty (\mathbf{x}^{\mathrm{T}}(t)\mathbf{Q}\mathbf{x}(t) + \mathbf{u}^{\mathrm{T}}(t)\mathbf{R}\mathbf{u}(t))\, dt \tag{3.104}$$

subject to the LTI dynamic constraint:

$$\dot{\mathbf{x}}(t) = \mathbf{A}\mathbf{x}(t) + \mathbf{B}\mathbf{u}(t) \tag{3.105}$$

Given the matrix Riccati differential equation (3.101), subject to the boundary condition $\mathbf{K}(T) = 0$, with $\mathbf{Q} = \mathbf{C}^{\mathrm{T}}\mathbf{C}$, if one makes the assumptions that the pair $[\mathbf{A}, \mathbf{B}]$ is controllable and the pair $[\mathbf{A}, \mathbf{C}]$ is observable, then the unique optimal state-feedback control is

$$\mathbf{u}(t) = -\mathbf{G}\mathbf{x}(t) \tag{3.106}$$

where the constant coefficient gain matrix is

$$\mathbf{G} = \mathbf{R}^{-1}\mathbf{B}\mathbf{K} \tag{3.107}$$

and $\mathbf{K}$ is a symmetic, positive definite, constant coefficent matrix obtained from the solution of the algebraic Riccati equation (ARE):

$$\mathbf{0} = -\mathbf{K}\mathbf{A} - \mathbf{A}^{\mathrm{T}}\mathbf{K} - \mathbf{Q} + \mathbf{K}\mathbf{B}\mathbf{R}^{-1}\mathbf{B}^{\mathrm{T}}\mathbf{K} \tag{3.108}$$

The assumption of controllability and observability guarantees a unique solution for the optimal contol law that provides the desired closed-loop, asymptotic stability.

One can further relax the conditions placed on the system, namely that the pair [**A**, **B**] is stabilizable and the pair [**A**, **C**] is detectable, to produce a unique positive semidefinite matrix **K** that is the solution of the ARE, and an optimal control defined as in equation (3.106).

To provide an example, an LQR controller has been designed for a uniform property plate modeled with pinned boundary conditions. The dimensions and material properties of the plate are defined in the script file **plate.m**. For the given example, the objective is to minimize the response of the structure at the location of the disturbance force using two auxiliary, colocated point force actuators and point-velocity sensors. The script file **lqr_demo.m** is used to generate the optimal control law, and the results are presented in Figure 3.7. As indicated, a significant level of attenuation is obtained near the resonant frequencies of the structural modes for a performance weighting, $Q = 1$, for all states and a control effort penalty of $R = 10^{-16}$ for the two control inputs. The basic commands used in the script file to design the control system are part of the MATLAB Control System Toolbox.

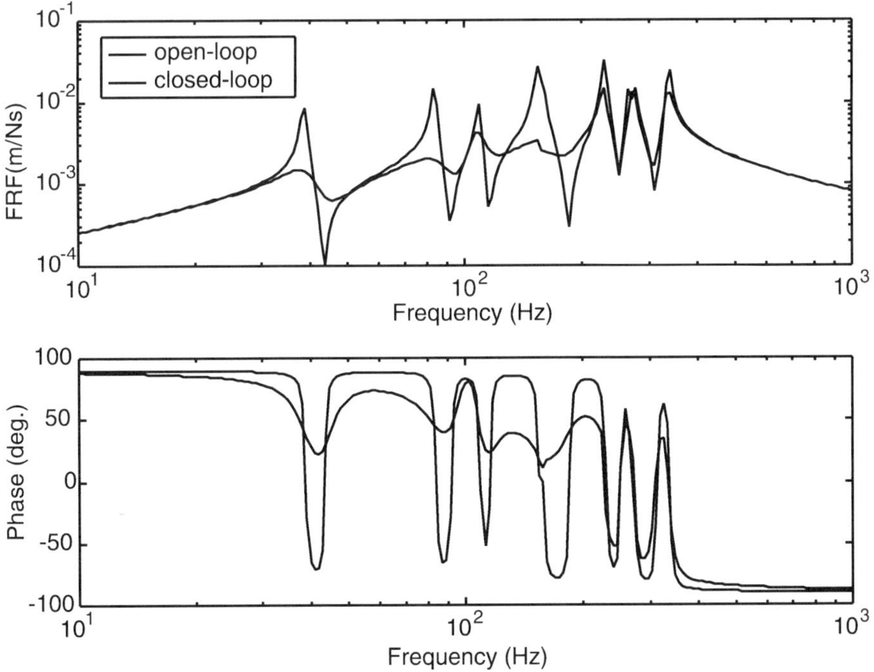


**Figure 3.7:** Comparison of open-loop and closed-loop (LQR) frequency responses of a plate configured with a colocated transducer pair.

## 3.8 THE LINEAR QUADRATIC GAUSSIAN (LQG) PROBLEM

The LQG problem is devoted to the solution of the output feeback control problem where system information is limited to some linear combination of the internal states. One assumes that these measured outputs are composed of the desired system response and noise. The discussion is limited to the LTI problem, and the first step is to develop a method of estimating the internal states of the system. A Kalman-Bucy filter is used for stochastic estimation. Consider the estimation problem for a stochastic system:

$$\dot{\mathbf{x}}(t) = \mathbf{A}\mathbf{x}(t) + \mathbf{B}\mathbf{u}(t) + \mathbf{B}_1\mathbf{w}(t) \tag{3.109}$$

and

$$\mathbf{y}(t) = \mathbf{C}\mathbf{x}(t) + \mathbf{v}(t) \tag{3.110}$$

where $\mathbf{w}(t)$ is the process noise and $\mathbf{v}(t)$ is the measurement noise. One assumes that the noise sources are uncorrelated, zero-mean, Gaussian, white-noise random vectors with correlation matrices defined as follows:

$$E\{\mathbf{v}(t)\mathbf{v}'(t)\} = \mathbf{V}\,\delta(t-\tau) \tag{3.111}$$

and

$$E\{\mathbf{w}(t)\mathbf{w}'(t)\} = \mathbf{W}\,\delta(t-\tau) \tag{3.112}$$

Defining the estimator error in terms of the true and estimated state vector, one obtains

$$\tilde{\mathbf{x}}(t) \doteq \mathbf{x}(t) - \hat{\mathbf{x}}(t) \tag{3.113}$$

where $\tilde{\mathbf{x}}(t)$ is the estimator error, and $\hat{\mathbf{x}}(t)$ is the state estimate. One can express the estimator dynamics as follows where a feedback gain matrix $\mathbf{L}$ is used to correct the estimate:

$$\dot{\hat{\mathbf{x}}}(t) = \mathbf{A}\hat{\mathbf{x}}(t) + \mathbf{B}\mathbf{u}(t) + \mathbf{L}(\mathbf{y}(t) - \mathbf{C}\hat{\mathbf{x}}(t)) \tag{3.114}$$

As indicated, the disturbance presented to the estimator system takes the form of the difference between the actual and estimated output. Subtracting equation (3.114) from equation (3.109), one obtains

$$\dot{\tilde{\mathbf{x}}}(t) = (\mathbf{A} - \mathbf{L}\mathbf{C})\tilde{\mathbf{x}}(t) + \mathbf{B}_1\mathbf{w}(t) - \mathbf{L}\mathbf{v}(t) \tag{3.115}$$

Since the linear combination of two white processes is white, one can express equation (3.115) as follows:

$$\dot{\tilde{\mathbf{x}}}(t) = (\mathbf{A} - \mathbf{LC})\tilde{\mathbf{x}}(t) + \boldsymbol{\psi}(t) \tag{3.116}$$

Due to the dualism between stochastic and deterministic systems (Dorato et al., 1995), one can show that the optimal gain $\mathbf{L}$ is expressed as follows:

$$\mathbf{L} = \mathbf{PC}^{\mathrm{T}}\mathbf{V} \tag{3.117}$$

where $\mathbf{P}$ is obtained from the filter algebraic Riccati equation (FARE):

$$0 = \mathbf{AP} + \mathbf{PA}^{\mathrm{T}} + \Xi - \mathbf{PC}^{\mathrm{T}}\mathbf{V}^{-1}\mathbf{CP} \tag{3.118}$$

The existence of the solution requires that the pair $[\mathbf{A}^{\mathrm{T}}, \mathbf{C}^{\mathrm{T}}]$ be stabilizable, and the pair $[\mathbf{A}^{\mathrm{T}}, \mathbf{M}^{\mathrm{T}}]$ be detectable, where $\Xi = \mathbf{MM}^{\mathrm{T}}$, and $\mathbf{V}$ is positive definite ($\Xi$ and $\mathbf{V}$ are analogous to $\mathbf{Q}$ and $\mathbf{R}$ in the LQR problem).

Armed with a means of estimating the states, we now proceed with the solution of the LQG problem. The objective is to find a compensator that minimizes the performance measure:

$$J = \lim_{t \to \infty} E\{\mathbf{x}^{\mathrm{T}}(t)\mathbf{Qx}(t) + \mathbf{u}^{\mathrm{T}}(t)\mathbf{Ru}(t)\} \tag{3.119}$$

using measurements of $\mathbf{u}(t)$ and $\mathbf{y}(t) = \mathbf{Cx}(t) + \mathbf{v}(t)$ and the stochastic system with white noise processes $\mathbf{v}(t)$ and $\mathbf{w}(t)$ . The solution of this problem hinges upon the separation principle, which is discussed in a number of texts (Kwakernaak and Sivan, 1972; Kailath, 1980; Chen, 1984; Dorato et al., 1995). The separation principle states that the optimal LQG problem may be solved by independently solving the optimal regulator problem and the optimal estimation problem. Hence, the solution requires that one solve the ARE twice: once for the regulator problem and once for the filter problem. The resulting dynamic compensator will be of the same order as that of the original dynamic system under consideration.

As indicated in Figure 3.8, one has two alternative forms of implementation. In Figure 3.8*a*, the estimator-controller form is presented whereby the Kalman-Bucy filter used as the estimator is implemented separately from the control law. The advantage of this form is that the estimator is always a stable realization; however, the disadvantage is that one must obtain a measure of the control input generated and pass it to the estimator. In the alternative compensator form presented in Figure 3.8*b*, the control law and estimator are combined into a single filter. The disadvantage of this form is that there is no guarantee that the resulting compensator is stable. As discussed in Chapter 8, the compensator form is directly analogous to the resulting compensator obtained from the design of an $\mathcal{H}_2$ controller when the input and output weightings of the two-port model are selected to represent the processes representative of the LQG problem.

To provide an example, a script file (**lqg_demo.m**) has been written to implement an LQG controller on a plate with pinned boundary conditions. Locations

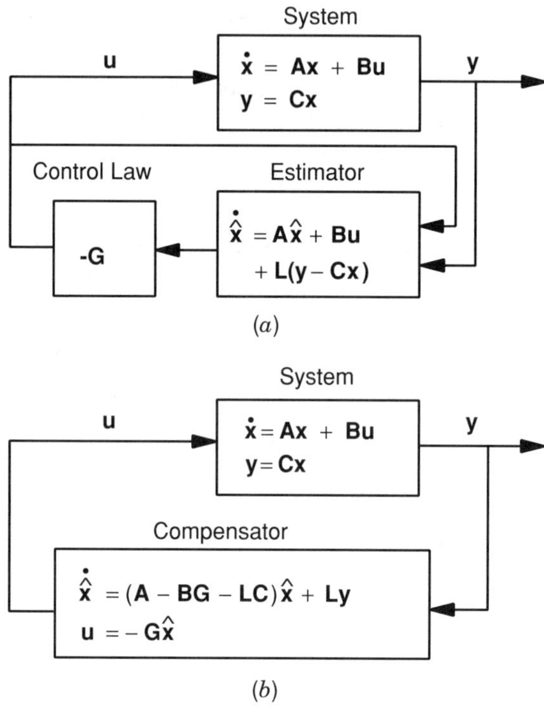


**Figure 3.8:** Alternative forms of implementation for an LQG control system design. (*a*) Estimator-controller form. (*b*) Compensator form.

of control inputs, sensors, and disturbance forces, as well as material properties and dimensions, are detailed in the script file **plate.m**. The system is the same as that used in the demonstration of the previous section; however, in this case, one assumes that the internal states must be estimated to implement the control system. As indicated from the results presented in Figure 3.9, the performance of the closed-loop system parallels that of the LQR controller for the chosen levels of process noise, measurement noise, control effort penalty, and error penalty. The basic commands used in the script file to design this controller are part of the MATLAB Control System Toolbox.

## 3.9 SUMMARY

In this chapter, we reviewed basic elements of linear systems and signals as needed for communicating the concepts presented in modeling and control of adaptive structures. System realizations were described, including controllability, observability, stabilizability, and detectability. Transfer function descriptions were reviewed, including concepts of poles, zeros, and their relationship to sys-

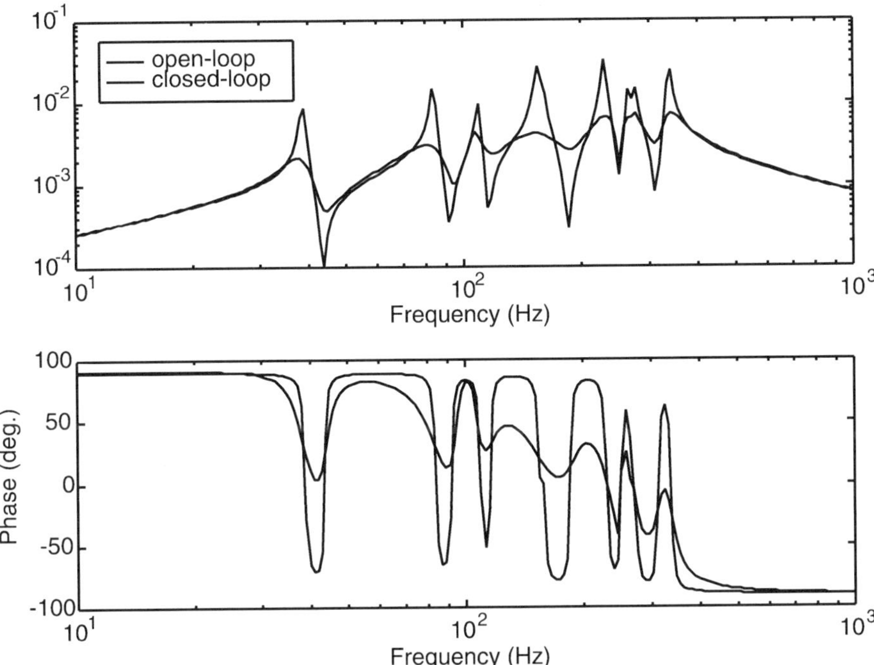


**Figure 3.9:** Comparison of open-loop and closed-loop (LQG) frequency responses of a plate configured with a colocated transducer pair.

tem realizations. State-feedback control was reviewed, and a method of casting the structural dynamic problem in state-variable form was presented. Norms of signals and systems were presented to provide a foundation for the development of cost functions for optimal control. Additionally, the LQR and LQG control problem were reviewed.

## BIBLIOGRAPHY

Bendat, J. S. and A. G. Piersol, 1986. *Random Data*, Wiley, New York.

Boyd, S. and C. Baratt, 1991. *Linear Controller Design: Limits of Performance*, Prentice Hall, Englewood Cliffs, NJ.

Brogan, W. L., 1985. *Modern Control Theory*, Prentice Hall, Englewood Cliffs, NJ.

Chen, C.-T., 1984. *Linear Systems Theory and Design*, Harcourt Brace College Publishers, New York.

Dorato, P., C. Abdallah, and V. Cerone, 1995. *Linear-Quadratic Control*, Prentice Hall, Englewood Cliffs, NJ.

Green, M., and D. J. N. Limebeer, 1995. *Linear Robust Control*, Prentice Hall, Englewood Cliffs, NJ.

Kailath, T., 1980. *Linear Systems*, Prentice Hall, Englewood Cliffs, NJ.

Kwakernaak, H. and R. Sivan, 1972. *Linear Optimal Control Systems*, Wiley-Interscience, New York.

Meirovitch, L., 1990. *Dynamics and Control of Structures*, Wiley, New York.

Zhou, K., J. C. Doyle, and K. Glover, 1996. *Robust and Optimal Control*, Prentice-Hall, Upper Saddle River, NJ.

# 4

# SIGNAL PROCESSING AND DIGITAL FILTERS

## 4.1 INTRODUCTION

Digital signal processing is applicable to any arbitrary sequence; however, it is most commonly used to process continuous-time signals that are sampled in discrete time to construct the sequence. Signal processing is critical to communication systems, radar, sonar, acoustics, image processing and a host of other applications. In terms of adaptive structures, signal processing is an essential element for both the control and system identification process. Digital filters are often employed to model the dynamic response of structures to applied control inputs, and adaptive signal processing is used to adjust the coefficients of adaptive filters required to generate the control input to the dynamic system. However, the application of adaptive signal processing to the control of dynamic systems is complicated by the fact that the structural response is characterized by waveforms that propagate in both space and time. Thus, for adaptive structures, spatial and temporal signal processing is necessary to control the system response.

Consider, for example, a one-dimensional flexural standing wave along the axis of the beam illustrated in Figure 4.1. For now, we consider the response of a single mode and assume that the wave was launched at some initial time $t_0$. If a local observer (a sensor) is placed on the structure at the spatial coordinate where the wave is launched, the structural motion can be recorded as a function of time as the equal and opposite waves propagate back and forth between the boundaries. In the case of an underdamped system, a decaying sinusoidal wave results, as illustrated in Figure 4.1, with a period derived from the damped natural frequency of oscillation. However, if the observer is moved far enough

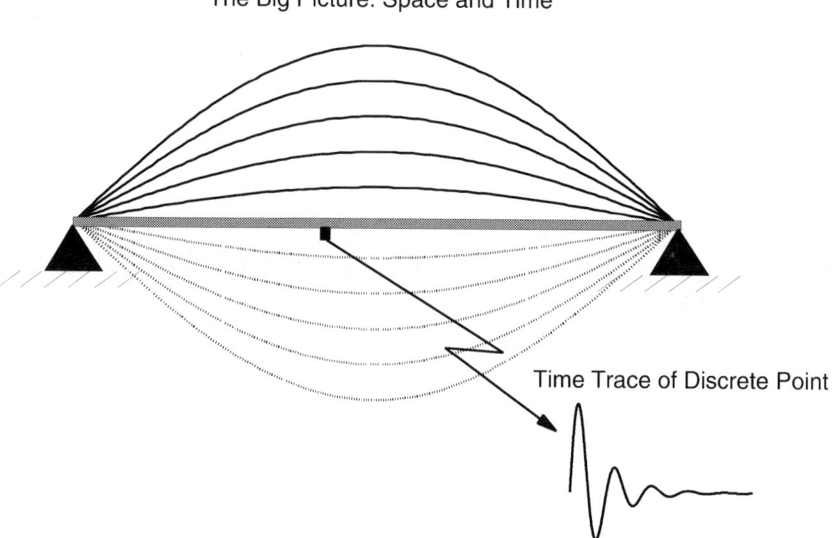


**Figure 4.1:** Illustration of spatial and temporal response of a dynamic system.

from the structure to capture the "big picture," then the wave propagation is seen as a function of space and time.

An understanding of the space and time relationship of the structural response is critical to the successful implementation of control approaches. In general, wave propagation in dispersive mediums (i.e., where the wave speed is a function of frequency) requires the most signal processing for successful implementation of the controller over a wide bandwidth of frequencies. Thus, signal processing, adaptive signal processing, and digital filters are essential elements in the design and implementation of adaptive structures. Although only the basics are presented in this chapter, references are provided to give the reader a more detailed understanding of the material.

## 4.2 BASIC TYPES OF SIGNALS

In the control of adaptive structures, we are concerned primarily with three distinct types of signals: periodic, transient and stationary random. Before proceeding to a discussion of pertinent signal processing terminology, it is appropriate to discuss the distinguishing characteristics of each type of signal.

### 4.2.1 Multifrequency (Periodic) Signals

The analysis of multifrequency signals is important with respect to persistent disturbance rejection. For example, the vibration of mechanical systems is typically associated with some periodic motion of the machinery and multiples of that fundamental frequency. Adaptive feedforward control is ideally suited for the control of such disturbances. A multifrequency signal can, in general, be defined in terms of a sum over all frequencies such that:

$$x(t) = \sum_{k=1}^{\infty} X_k \sin(\omega_k t + \phi_k) \tag{4.1}$$

where $X_k$ is the magnitude, $\omega_k$ is the frequency and $\phi_k$ is the phase of the $k$th frequency component. If the ratio of $\omega_r/\omega_s$ (where $\omega_s$ is the sample period) is a rational number for all frequency components present, then the signal has a finite period. However, if the ratio of $\omega_r/\omega_s$ yields an irrational number, then the period of the signal is infinite. For example, the sound and vibration in the fuselage of a turbo-prop aircraft will be due to periodic signals from each engine, but will in general have an infinite period since the engines are not synchronized.

An example of a multifrequency signal that is periodic, along with one that is not periodic, is presented in Figure 4.2. As illustrated, the signal $x_2(t)$ does not have a finite period since the ratio of the two frequencies is an irrational number. This lack of periodicity brings about an interesting point with respect

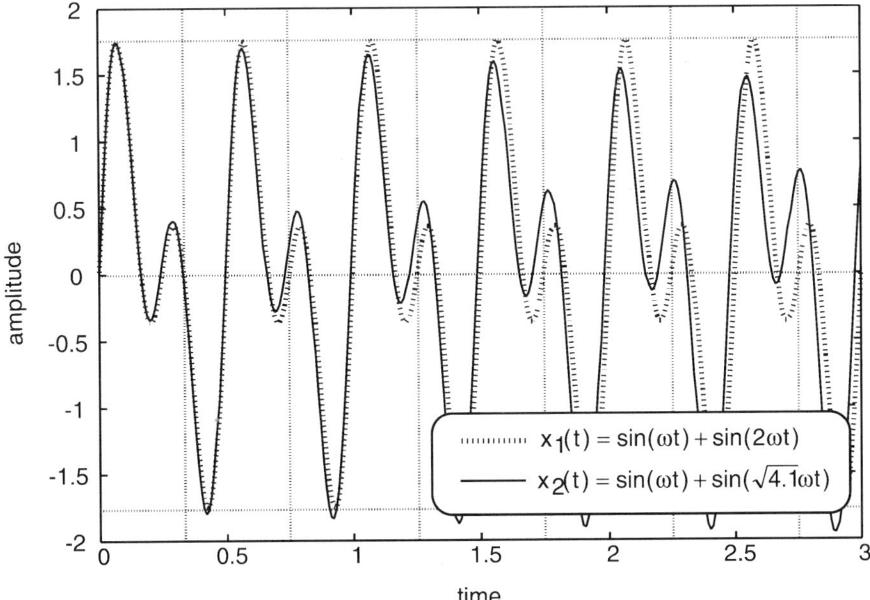


**Figure 4.2:** A multifrequency signal that is periodic and one that is not.

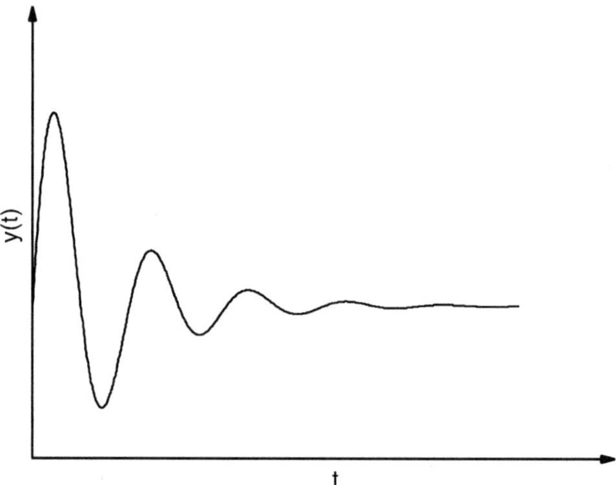


**Figure 4.3:** A transient signal.

to digital signal processing and sampled data. Unless synchronous sampling is employed for a periodic signal, the resulting sampled data sequence will be nonperiodic, which is one of the reasons that the concept of windowing the data set is important in digital signal processing.

### 4.2.2 Transient Signals

Transient signals are another form of nonperiodic signals, since the signal typically decays as a function of time or in general has some finite duration, as illustrated in Figure 4.3. The damped response of a structural system to an impulsive input is an example of a transient signal. In the control of adaptive structures, one typically employs some form of feedback control to increase the damping in the system and thus minimize the transient response to impulsive disturbances. In addition, the system stability can typically be inferred from the transient response characteristics and is thus important in the analysis and design of adaptive structures.

### 4.2.3 Random Signals

Random signals are also important in the control of adaptive structures, since structural disturbances often vary randomly as a function of time. For example, consider the turbulent boundary layer that forms over the surface of the fuselage of a subsonic aircraft. Sound radiation within the fuselage is a function of disturbances generated in this turbulent boundary layer. Since random signals are nonperiodic, they are typically classified in terms of statistical parameters such as the mean value and variance. Random signals can generally be broken into

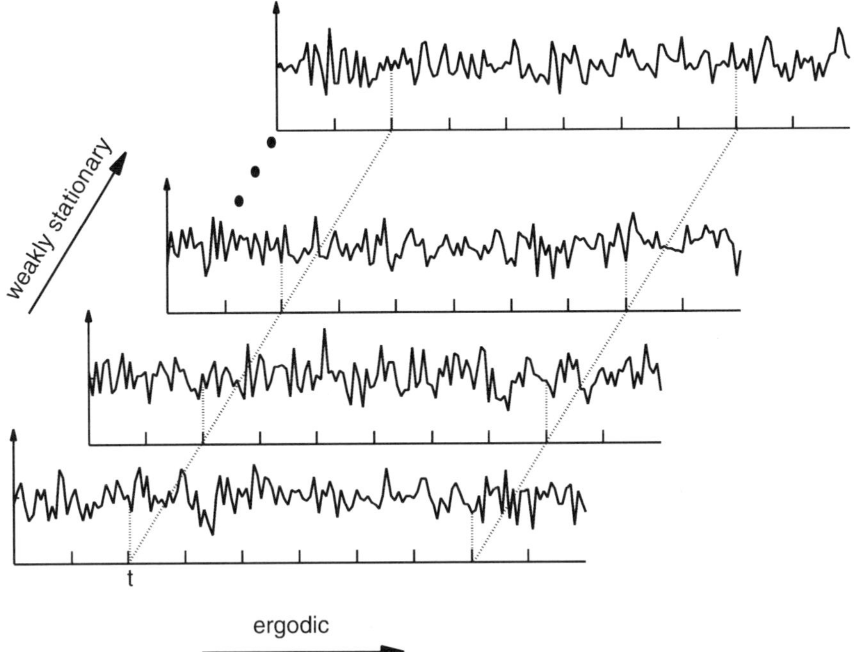


**Figure 4.4:** Illustration of ergodic random signal and weakly stationary random signal (after Bendat and Piersol, 1986).

two classifications: weakly stationary and ergodic. As illustrated in Figure 4.4 (after Bendat and Piersol, 1986), an ergodic signal is one in which the statistical properties of a given sequence (for example, over $x_1(t)$) are dependent only on $\tau$, the time lag, but not $t$, the time. A weakly stationary signal is one in which the statistical properties averaged over the ensemble (i.e., all $M$ sequences) at some fixed $t$ and $\tau$ of each sequence are equivalent. For a stationary ergodic process:

$$\mu_x = E[x(t)] = \lim_{T \to \infty} \frac{1}{T} \int_{-T/2}^{T/2} x(t)dt, \tag{4.2}$$

and

$$\overline{x^2(t)} = E[x^2(t)] = \lim_{T \to \infty} \frac{1}{T} \int_{-T/2}^{T/2} x^2(t)dt, \tag{4.3}$$

where $\mu_x$ is the mean value and $\overline{x^2(t)}$ is the mean-square value of $x(t)$. The concept of weakly stationary stochastic signals are paramount to the field of signal processing.

## 4.3 A QUICK LOOK AT SIGNAL PROCESSING

As evident in the literature, signal processing is a significant field of study that cannot be covered in detail within this text. However, enough information is provided in this section for the reader to understand the terminology and apply the basics of signal processing to the different types of signals discussed in the previous section. For the reader interested in greater detail, a number of references are available (Bendat and Piersol, 1986; Oppenheim and Schafer, 1989).

### 4.3.1 Correlation Functions

The cross-correlation function between two functions $x(t)$ and $y(t)$ is defined as

$$r_{xy}(\tau) = E[x(t)y(t+\tau)], \tag{4.4}$$

where $\tau$ is some time lag of the signal relative to $t$. Replacing $y(t + \tau)$ by $x(t + \tau)$, one obtains the autocorrelation function

$$r_{xx}(\tau) = E[x(t)x(t+\tau)] \tag{4.5}$$

In the event that $\tau \equiv 0$,

$$r_{xx}(0) = E[x^2(t)], \tag{4.6}$$

which is the mean-square value as presented in equation (4.3) and also the variance of a signal with zero mean value. The cross-correlation and autocorrelation for a stationary ergodic process can be expressed by averaging over time, respectively, as follows:

$$r_{xy}(\tau) = \lim_{T\to\infty} \frac{1}{T}\int_{-T/2}^{T/2} x(t)y(t+\tau)dt \tag{4.7}$$

and

$$r_{xx}(\tau) = \lim_{T\to\infty} \frac{1}{T}\int_{-T/2}^{T/2} x(t)x(t+\tau)dt \tag{4.8}$$

One should recognize that, for periodic signals (i.e., signals with a finite period), the cross-correlation and autocorrelation can be expressed as follows:

$$r_{xy}(\tau) = \frac{1}{T} \int_{-T/2}^{T/2} x(t)y(t+\tau)dt \tag{4.9}$$

and

$$r_{xx}(\tau) = \frac{1}{T} \int_{-T/2}^{T/2} x(t)x(t+\tau)dt \tag{4.10}$$

where $T$ now represents the period of the signal. In this case, one could integrate over infinity (an infinite number of periods) and divide by the corresponding time period, producing the same result.

For real-valued sequences $\{x(k)\}$, the autocorrelation function is symmetric. (In this book, we deal with real-valued sequences exclusively.) Hence,

$$r_{xx}(\tau) = r_{xx}(-\tau) \tag{4.11}$$

and the autocorrelation function is thus observed to be an even function. One can also show that

$$r_{xy}(\tau) = r_{yx}(-\tau) \tag{4.12}$$

$$r_{yx}(\tau) = r_{xy}(-\tau) \tag{4.13}$$

These properties are useful in the development of optimal filter theory, as demonstrated in a later section.

### 4.3.2 Spectral Density Functions

The traditional method of defining the spectral density functions is in terms of the Fourier transform of the auto-correlation and crosscorrelation functions. Thus, the autospectral density function of the sequence $\{x(k)\}$ is defined as

$$S_{xx}(\omega) = \int_{-\infty}^{\infty} r_{xx}(\tau) \exp(-j\omega\tau)d\tau \tag{4.14}$$

where $S_{xx}(\omega)$ is the autospectral density function. Similarly, the cross-spectral density function can be defined in terms of two sequences $\{x(k)\}$ and $\{y(k)\}$:

$$S_{xy}(\omega) = \int_{-\infty}^{\infty} r_{xy}(\tau) \exp(-j\omega\tau)d\tau \tag{4.15}$$

Reciprocally, the inverse Fourier transform can be employed to obtain the autocorrelation function and the cross-correlation function from the autospectral density function and cross-spectral density function, respectively, as follows:

$$r_{xx}(\tau) = \frac{1}{2\pi} \int_{-\infty}^{\infty} S_{xx}(\omega) \exp(j\omega\tau) d\omega \tag{4.16}$$

$$r_{xy}(\tau) = \frac{1}{2\pi} \int_{-\infty}^{\infty} S_{xy}(\omega) \exp(j\omega\tau) d\omega \tag{4.17}$$

Since the autocorrelation function is an even function, one can readily demonstrate that the autospectral density function is an even function:

$$S_{xx}(\omega) = S_{xx}(-\omega) \tag{4.18}$$

For two stationary random stochastic processes, $\{x(k)\}$ and $\{y(k)\}$, sampled over a finite period $T$, one can show (Bendat and Piersol, 1986):

$$S_{xy}(\omega) = \lim_{T \to \infty} \frac{1}{T} \mathrm{E}[X_k^*(\omega) Y_k(\omega)], \tag{4.19}$$

where the asterisk is used to indicate the complex conjugate of $X_k^*$, and the Fourier transform of the sequences $x_k(t)$, and $y_k(t)$:

$$X_k(\omega) = \int_0^T x_k(t) \exp(-j\omega t) dt \tag{4.20}$$

$$Y_k(\omega) = \int_0^T y_k(t) \exp(-j\omega t) dt \tag{4.21}$$

Letting $Y_k(\omega) = X_k(\omega)$ in equation (4.19) one obtains a similar expression for the autospectral density function:

$$S_{xx}(\omega) = \lim_{T \to \infty} \frac{1}{T} \mathrm{E}[|X_k(\omega)|^2] \tag{4.22}$$

The autospectral density function and the cross-spectral density function can thus be obtained from the Fourier transform of an ensemble of sequences with the length of the sequences approaching infinity. Even the fastest digital signal processor (DSP) takes an infinite amount of time to find the expected value as $T \to \infty$. Thus, an ensemble of sequences with finite record length must be used in practice. Before discussing the practical implementation of the discrete Fourier transform (DFT) presented in Appendix A, the effect of windowing a function is discussed.

### 4.3.3 Fourier Transform of a Windowed Function

In general, one deals with ensembles of finite-length data records in the analysis of dynamic systems (Harris, 1978). An example of a truncated signal is illustrated in Figure 4.5. A sample interval of duration $T$ is considered for the cosine wave with circular frequency $\omega_0$. The truncation of the data for $t < -T/2$ and $t > T/2$ is equivalent to multiplying the function by the boxcar window also illustrated in Figure 4.5. Thus,

$$x_b(t) = x(t)w_b(t) \tag{4.23}$$

where $x_b(t)$ is the response weighted by the boxcar function $w_b(t)$. The Fourier integral transform of equation (4.23) can be expressed as follows:

$$X_b(\omega) = \int_{-\infty}^{\infty} x(t)w_b(t)\exp(-j\omega t)dt \tag{4.24}$$

where

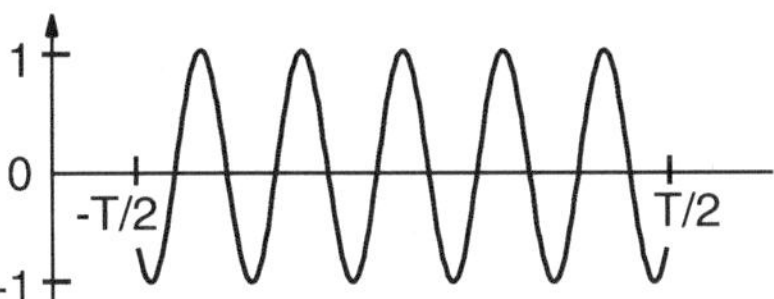


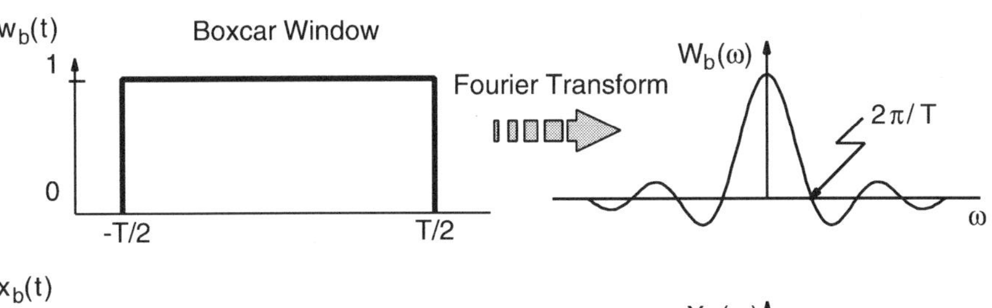


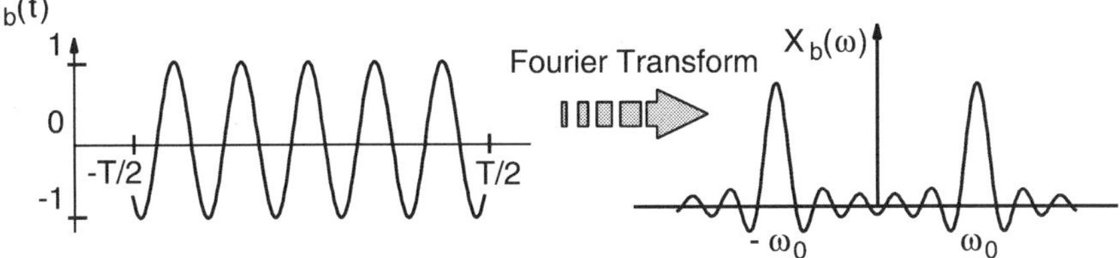


**Figure 4.5:** Application of a boxcar window.

$$x(t) = \frac{1}{2\pi} \int_{-\infty}^{\infty} X(\omega) \exp(j\omega t) d\omega \tag{4.25}$$

and

$$w_b(t) = \frac{1}{2\pi} \int_{-\infty}^{\infty} W_b(\omega) \exp(j\omega t) d\omega \tag{4.26}$$

Substituting equations (4.25) and (4.26) into equation (4.24), one can show that

$$X_b(\omega) = \frac{1}{2\pi} \int_{-\infty}^{\infty} X(\omega_d) W(\omega - \omega_d) d\omega_d, \tag{4.27}$$

where $\omega_d$ is a dummy variable of integration. One should recognize that the multiplication of two time-domain signals results in a convolution of the Fourier transforms of the signals in the frequency domain, as shown in Figure 4.5. The boxcar window illustrated in Figure 4.5 can be expressed as follows:

$$w_b(t) = u\left(t - \left(\frac{-T}{2}\right)\right) - u\left(t - \frac{T}{2}\right) \tag{4.28}$$

where $u(t)$ is the unit step function. The Fourier transform of this window can be represented as follows:

$$W_b(\omega) = T \, \frac{\sin(\omega T/2)}{\omega T/2} \tag{4.29}$$

and is illustrated in Figure 4.5. Taking the Fourier transform of the windowed cosine signal, one obtains:

$$X_b(\omega) = \frac{T}{2} \left( \frac{\sin(\omega - \omega_0)T/2}{(\omega - \omega_0)T/2} + \frac{\sin(\omega + \omega_0)T/2}{(\omega + \omega_0)T/2} \right) \tag{4.30}$$

and the function is plotted against frequency in Figure 4.5. Since the cosine wave is not periodic within the boxcar window, the spectral energy is distributed over the entire frequency range. The length of the window $T$ dictates the spectral resolution of the signal, and increasing the length of the window with respect to the number of cycles of the cosine wave can reduce the spillover of energy or "leakage" into the other frequency components.

However, the purpose of considering a finite-length data record was to reduce

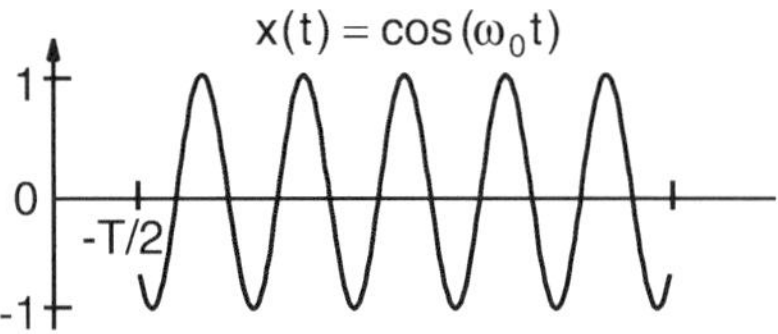


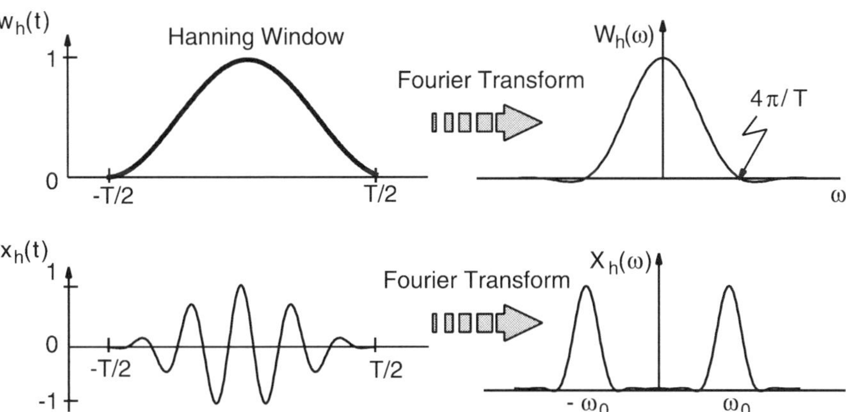


**Figure 4.6:** Application of a Hanning window.

the burden in terms of calculation time when implementing the DFT; therefore, an alternative to increasing the length of the data record to reduce the effects of leakage must be considered. One can choose alternative windows that have a more gradual weighting distribution over the length of the record, such as the Hanning window. The Hanning window is illustrated in Figure 4.6 and can be expressed as follows:

$$w_h(t) = \cos^2\left(\frac{\pi t}{T}\right), \qquad -\frac{T}{2} \leq t \leq \frac{T}{2} \tag{4.31}$$

The Fourier transform of the Hanning window is also illustrated in Figure 4.6 and can be expressed as

$$W_h(\omega) = \frac{T}{2}\,\frac{\sin(\omega T/2)}{(\omega T/2)[1 - (1/\pi^2)(\omega T/2)^2]} \tag{4.32}$$

It is obvious that less leakage will result with the application of the Hanning window; however, less resolution of the frequency of interest results since the bandwidth of the window in the frequency domain is wider (twice that of the boxcar). The cosine signal weighted by the Hanning window and the Fourier transform of the weighted signal are also illustrated in Figure 4.6. As expected, the leakage is suppressed. A number of different windows have been studied in signal processing applications, and one typically has the choice of multiple windows on spectrum analyzers. For an in-depth discussion of windows, the reader is referred to the paper by Harris (1978).

### 4.3.4 Processing Finite-Length Data Records

To compute estimates of the autospectral density function and the cross-spectral density function, we must first compute the DFT of an ensemble of data sequences. Thus, the first step in the processing of finite length data records is to obtain the DFT of each ensemble. The DFT of a finite-length data sequence can be expressed as follows:

$$X_N(\exp(j\omega T)) = \sum_{n=0}^{N-1} x_N(nT)\exp(-j\omega nT) \tag{4.33}$$

where the sequence is obtained by applying an $N$-point filter to the data sequence as discussed in Appendix A. For convenience, we choose to define

$$T = \frac{2\pi}{\omega_s} \tag{4.34}$$

where $\omega_s$ is the circular sampling frequency used in acquiring the sequence. Letting

$$\frac{\omega}{\omega_s} = \frac{k}{N} \tag{4.35}$$

we can express equation (4.33) as follows:

$$X_N\left(\exp\left(\frac{jnk2\pi}{N}\right)\right) = \sum_{n=0}^{N-1} x_N(nT)\exp\left(\frac{-jnk2\pi}{N}\right), \qquad 0 \le k \le N-1 \tag{4.36}$$

The inverse of the DFT presented in equation (4.36) can be expressed as follows:

$$x_N(nT) = \frac{1}{N}\sum_{k=0}^{N-1} X_N\left(\exp\left(\frac{jnk2\pi}{N}\right)\right)\exp\left(\frac{jnk2\pi}{N}\right), \qquad 0 \le n \le N-1 \tag{4.37}$$

It is assumed that sequence $x_N(nT)$ is periodic over $N$, and therefore that $X_N(\exp(jnk2\pi/N))$ is periodic over $N$ as well. In practice, the DFT is rarely used. Most spectrum analyzers and software packages employ the fast Fourier transform (FFT), which is roughly two orders of magnitude faster (less instructions) in the computation of the transform.

Once the DFT of each data record in the ensemble is computed, the autospectral density function and cross-spectral density function can be estimated, respectively, as follows:

$$\hat{S}_{xx}\left(\exp\left(\frac{jnk2\pi}{N}\right)\right)=\frac{1}{M}\sum_{r=1}^{M}\frac{1}{T}\left|X_r\left(\exp\left(\frac{jnk2\pi}{N}\right)\right)\right|^2 \qquad k=0,1,\ldots,N-1 \tag{4.38}$$

and

$$\hat{S}_{xy}\left(\exp\left(\frac{jnk2\pi}{N}\right)\right)=\frac{1}{M}\sum_{r=1}^{M}\frac{1}{T}X_r^*\left(\exp\left(\frac{jnk2\pi}{N}\right)\right)\cdot Y_r\left(\exp\left(\frac{jnk2\pi}{N}\right)\right), \qquad k=0,1,\ldots,N-1 \tag{4.39}$$

where $M$ is the number of data records in the ensemble, and $X_r$ and $Y_r$ are DFTs of the $r$th data records of $x_N(nT)$ and $y_N(nT)$, respectively. Notice that the autospectral density function is purely real, while the cross-spectral density function is in general complex. The autospectral density function and the cross-spectral density function can be used to estimate the frequency response function (FRF) of a system as follows ($H_1$ estimate):

$$\hat{H}_{xy}(\omega)=\frac{\hat{S}_{xy}}{\hat{S}_{xx}} \tag{4.40}$$

This estimate assumes noise on the output and thus is more accurate under these conditions. An alternate estimate, $H_2$, which assumes noise on the input, is as follows:

$$\hat{H}_{xy}(\omega)=\frac{\hat{S}_{yy}}{\hat{S}_{xy}} \tag{4.41}$$

Further discussion of the effects of noise on these estimates can be found in Bendat and Piersol (1986). In addition, the autospectral density function of the input and output can be used to estimate the magnitude of the FRF:

$$|\hat{H}_{xy}|^2 = \frac{\hat{S}_{yy}}{\hat{S}_{xx}} \tag{4.42}$$

As discussed by Bendat and Piersol (1986), measurement noise is often present either on the input, output, or both and must be considered when data is obtained from transduction devices and associated instrumentation. However, equations (4.41) and (4.42) serve to demonstrate the advantages afforded with frequency domain analysis in terms of extracting the frequency response characteristics of a system from typical auto and cross-spectrum analysis. The fundamental drawback of frequency domain analysis in the design of feedforward control systems is that the resulting design is typically not physically realizable in a broadband sense, due to causality and stability requirements.

## 4.4 DIGITAL FILTERS

Digital filters perform an important function in the design and control of adaptive structures. For example, in developing a prototype adaptive structure, one typically begins with a dynamic model of the structure, as outlined in Chapter 2, and formulates the equations of motion in terms of state-variables for simulations, as discussed in Chapter 3. Upon determining the appropriate locations for actuators and sensors based upon the model, the prototype structure is constructed in hardware. It is highly unlikely that the dynamic response of the physical system will exactly match that of the initial analytical or numerical model. Thus, before finalizing the control system design, a system identification is typically performed on the prototype structure to model the dynamic characteristics of the real system. For complex systems that are not readily modeled by analytical or numerical methods, the design process typically begins here. In any event, the model is constructed from sampled data sequences obtained by exciting the structure with a known input sequence (weakly stationary random) $\{u_m(n)\}$ at the $m$th input and measuring the response of the system $\{y_l(n)\}$ at the $l$th output. A schematic diagram of the general multi-input, multi-output (MIMO) system is presented in Figure 4.7*a*.

Each transfer function path (i.e., from the $m$th input to the $l$th output) can be thought of as a mapping of the input sequence $\{u_m(n)\}$ into an output sequence $\{y_l(n)\}$, as illustrated in Figure 4.7*b*. The mathematical description of this discrete-time system can be expressed as follows (the discrete delta function):

$$y_l(n) = \mathrm{L}_{lm}[u_m(n)] \tag{4.43}$$

where the linear operator $\mathrm{L}_{lm}$ defines the transformation. Any sequence $\{u_m(n)\}$ can be expressed in terms of the unit sample sequence:

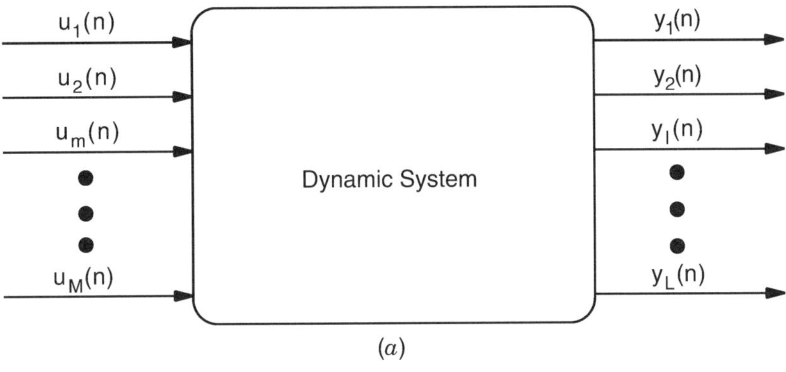


(*a*)

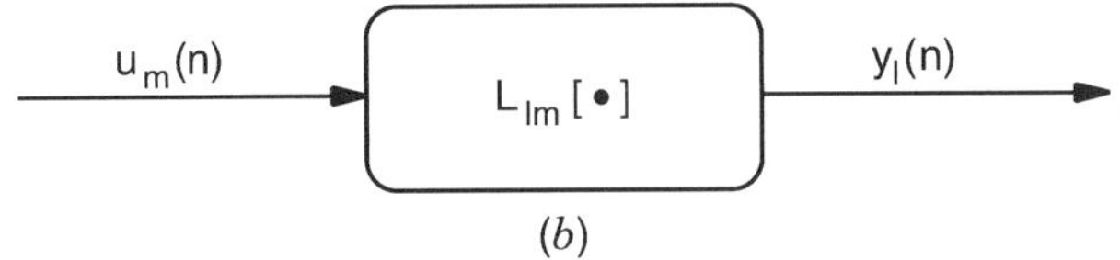


(*b*)

**Figure 4.7:** Schematic diagram of a general dynamic system and the specific discrete-time interpretation. (*a*) General MIMO dynamic system. (*b*) Discrete-time system transformation.

$$\{\delta(n)\} = \begin{cases} 0, & n \neq 0 \\ 1, & n = 0 \end{cases} \tag{4.44}$$

such that

$$u_m(n) = \sum_{k=-\infty}^{\infty} u_m(k)\delta(n-k) \tag{4.45}$$

Substituting equation (4.45) into equation (4.43), one obtains

$$y_l(n) = \mathrm{L}_{lm}\left[\sum_{k=-\infty}^{\infty} u_m(k)\delta(n-k)\right] \tag{4.46}$$

Thus, from the principle of superposition for linear systems, one recognizes that

$$y_l(n) = \sum_{k=-\infty}^{\infty} u_m(k)\mathrm{L}_{lm}\left[\delta(n-k)\right] \tag{4.47}$$

In the event that the differential operator is time-invariant, one can define the discrete-time system response in terms of the impulse response:

$$y_l(n) = \sum_{k=-\infty}^{\infty} h_{lm}(n-k)u_m(k) \tag{4.48}$$

where

$$h_{lm}(n-k) \equiv \mathrm{L}_{lm}[\delta(n-k)] \tag{4.49}$$

Equation (4.48) is recognized as the convolution sum for the linear time-invariant (LTI) system. Now, taking the $z$-transform (see Appendix A) of equation (4.48), one obtains

$$Y_l(z) = \sum_{k=-\infty}^{\infty} u_m(k) \sum_{n=-\infty}^{\infty} h_{lm}(n-k)z^{-n} \tag{4.50}$$

$$= \sum_{k=-\infty}^{\infty} u_m(k)z^{-k} \sum_{n=-\infty}^{\infty} h_{lm}(n)z^{-n} \tag{4.51}$$

$$= U_m(z)H_{lm}(z) \tag{4.52}$$

where $H_{lm}(z)$ is recognized as the $z$-transfer function. Hence, the discrete impulse response of the system can be obtained directly from the inverse $z$-transform of the $z$-transfer function of the system, and likewise, the $z$-transfer function can be obtained from the $z$-transform of the discrete impulse response of the system. Obviously, if we let $z = \exp(j\omega T)$, an expression for the DFT is obtained. Due to this mapping from the $z$-plane to the frequency domain, the FRF plays an important role in the design of discrete-time filters for modeling systems. However, before discussing alternative design methods, we proceed by outlining the basic operation of digital filters for LTI systems.

In digital signal processing, the autoregressive moving average (ARMA) model is often used to describe a sequence in terms of its previous inputs and outputs:

$$y_l(n) = \sum_{j=0}^{J} a_j u_m(n-j) + \sum_{k=1}^{K} b_k y_l(n-k) \tag{4.53}$$

where $a_0, a_2, \ldots, a_J$ are constants known as the moving average (MA) parameters and $b_1, b_2, \ldots, b_K$ are constants known as the autoregressive (AR) parameters. Taking the $z$-transform of equation (4.53):

$$Y_l(z) = \sum_{j=0}^{J} a_j \sum_{n=-\infty}^{\infty} u_m(n-j)z^{-n} + \sum_{k=1}^{K} b_k \sum_{n=-\infty}^{\infty} y_l(n-k)z^{-n}$$

$$= U_m(z)\sum_{j=0}^{J} a_j z^{-j} + Y_l(z)\sum_{k=1}^{K} b_k z^{-k} \tag{4.54}$$

Solving in terms of $Y_l(z)$:

$$Y_l(z) = U_m(z)\,\frac{\sum_{j=0}^{J} a_j z^{-j}}{1 - \sum_{k=1}^{K} b_k z^{-k}} \tag{4.55}$$

Thus, one can obtain an expression for the $z$-transfer function in terms of the coefficients of the ARMA model:

$$H_{lm}(z) = \frac{Y_l(z)}{U_m(z)} = \frac{\sum_{j=0}^{J} a_j z^{-j}}{1 - \sum_{k=1}^{K} b_k z^{-k}} \tag{4.56}$$

To obtain the discrete impulse response of the system, we must find the inverse $z$-transform as outlined in Appendix A. The $z$-transfer function of a system can be expressed as follows:

$$H_{lm}(z) = \frac{c_l \operatorname{adj}[zI - \Phi]\gamma_m}{|zI - \Phi|} + d_{lm} \tag{4.57}$$

where $c_l$ is the $l$th row of the $L \times 2n$ observation matrix $C$, $\gamma_m$ is the $m$th column of the $2n \times M$ input matrix $\Gamma$, and $d_{lm}$ is the corresponding entry of the $L \times M$ feedthrough matrix $D$. Comparing equations (4.57) and (4.56), one can equate the numerator and denominator, respectively, to obtain:

$$c_l \operatorname{adj}[zI - \Phi]\gamma_m + d_{lm}|zI - \Phi| = \sum_{j=0}^{J} a_j z^{-j} \tag{4.58}$$

and

$$|zI - \Phi| = 1 - \sum_{k=1}^{K} b_k z^{-k} \tag{4.59}$$

Hence, the numerator of the transfer function obtained from the ARMA model (the MA parameters) describes the zeros of the system, and the denominator (the AR parameters) describes the poles of the system. As discussed in Appendix A, the poles of the transfer function must reside within the unit circle in the $z$-plane to assure stability.

One of the challenges presented in digital signal processing is to find the coefficients $a_j$ and $b_k$ such that the output sequence $\{y_l(n)\}$ can be computed from a known input sequence $\{u_m(n)\}$. A number of methods exist for designing digital filters:

- Impulse invariance (infinite impulse response (IIR) filter)
- Bilinear transform (IIR filter)
- Analog prototypes (IIR filter)
- Optimization: Parks-McClellan algorithm (finite impulse response (FIR) filter)
- Windowing (FIR filter)
- Inverse filter design (IIR or FIR filter)

The previously outlined design approaches are addressed in detail in Chapter 7 of *Discrete-Time Signal Processing* by Oppenheim and Schafer (1989). In the modeling of dynamic systems for the design of adaptive structures, we focus primarily on the inverse filter design approach, since it is based upon a design optimization that requires that the filter response match *both* the magnitude and phase response in the frequency domain. A MATLAB subroutine **invfreqz** can be used to design digital filters from FRFs; however, before providing an example of such filter design approaches, we explore the two basic types of filters employed in modeling system response: finite impulse response (FIR) filters and infinite impulse response (IIR) filters.

### 4.4.1 FIR Filters

Setting $b_k = 0$ for all $k$ in equation (4.53) results in the MA model and forms what is often designated the all-zero, nonrecursive, or transversal filter. A schematic diagram of a transversal filter is presented in Figure 4.8. The response of the filter can be expressed as follows:

$$y_l(n) = \sum_{j=0}^{J} a_j u_m(n-j) \tag{4.60}$$

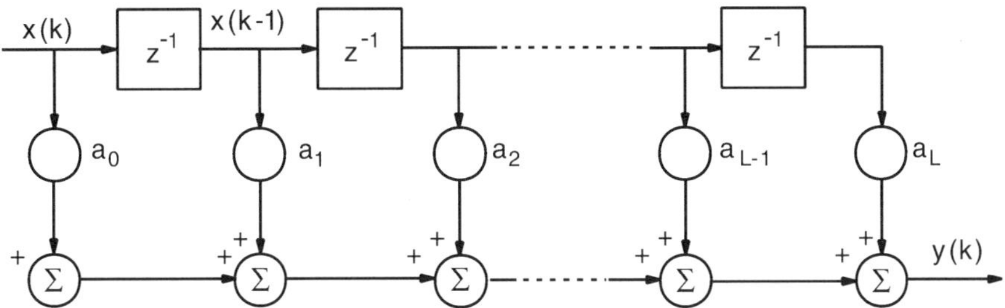


**Figure 4.8:** A transversal filter.

Taking the $z$-transform of the response, one obtains

$$Y_l(z) = U_m(z) \sum_{j=0}^{J} a_j z^{-j} \tag{4.61}$$

which is equivalent to letting $b_k = 0$ in equation (4.55). Thus, the $z$-transfer function of the filter can be expressed as follows, where the designation all-zero filter is apparent:

$$H_{lm}(z) = \sum_{j=0}^{J} a_j z^{-j} \tag{4.62}$$

Recall that the impulse response of the system can be determined from the inverse $z$-transform of the $z$-transfer function. For the transfer function presented in equation (4.62), the inverse $z$-transform is obtained as outlined in Appendix A and is expressed as follows:

$$h_{lm}(n) = \sum_{j=0}^{J} a_j \, \delta(j - n) \tag{4.63}$$

Thus, the impulse response can be expressed in terms of a *finite* series; hence, the designation finite impulse response filter. The coefficients $a_j$ are the values of the impulse response $h_{lm}$.

FIR filters display several characteristics that are worth noting:

1. Always stable for finite valued coefficients
2. Ideal for linear phase applications
3. Capable of approximating arbitrary frequency response characteristics
4. Relatively insensitive to small perturbations in the filter coefficients

In the design and implementation of adaptive structures, FIR filters and adaptive FIR filters are employed in the system identification process as well as in the control process, as discussed in the following sections and in Chapter 8. The FIR filter is attractive for implementation on a digital signal processor due to its simplicity; however, systems with complicated frequency response characteristics might require an excessive number of coefficients to model and/or control the dynamic response of the system. Thus, an alternative to the FIR filter is considered: the IIR filter.

### 4.4.2 IIR Filters

The IIR filter shown in Figure 4.9 is based upon the ARMA model and is often termed a pole-zero or recursive filter. The IIR filter is thus characterized by "feedback," since the response at the current time step is, in general, dependent upon the current and previous inputs as well as previous outputs as indicated below (shown previously in equation (4.53)):

$$y_l(n) = \sum_{j=0}^{J} a_j u_m(n-j) + \sum_{k=1}^{K} b_k y_l(n-k)$$

We have observed earlier that the $z$-transfer function of the IIR filter can be expressed in terms of both zeros and poles upon considering the ARMA model (shown previously in equation (4.56)):

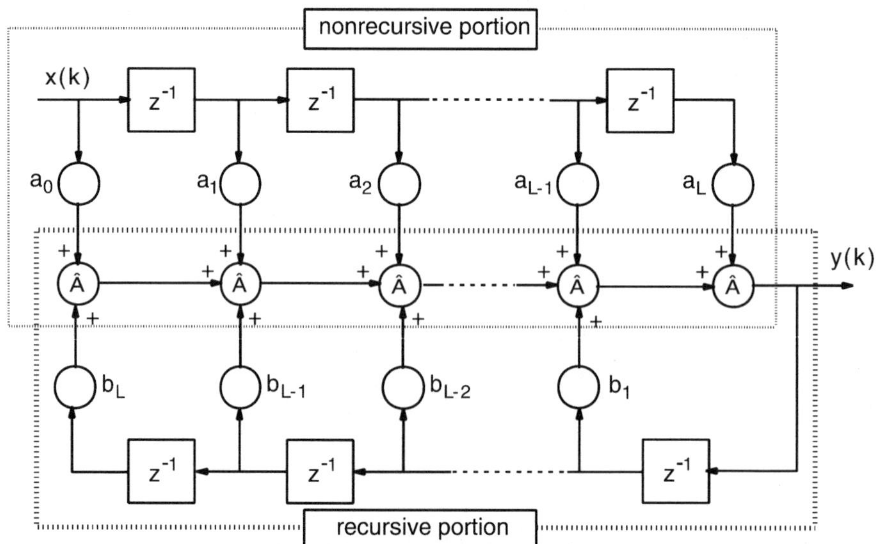


**Figure 4.9:** Schematic of different filter implementations.

$$H_{lm}(z) = \frac{\sum_{j=0}^{J} a_j z^{-j}}{1 - \sum_{k=1}^{K} b_k z^{-k}}$$

Thus, the discrete impulse response of the system can be obtained from the inverse $z$-transform, as outlined previously. The residue theorem is typically employed to compute the inverse $z$-transform. The theorem effectively states that $h_{lm}(n)$ (the impulse response defined in equation (4.63)) is the sum of the residues of the integrand at all of its poles inside the path of integration. To illustrate where the term *infinite* impulse response originated, we present a simple example.

**Example 4.1: Comparison of FIR and IIR Representations of a Simple System** Consider the $z$-transfer function of a simple system:

$$H(z) = \frac{1}{1 - bz^{-1}} \tag{4.64}$$

where $|b| < 1$. Multiply the numerator and denominator by $z$ to obtain

$$H(z) = \frac{z}{z - b} \tag{4.65}$$

Now, perform long division to express the transfer function in terms of a series:

$$H(z) = \sum_{k=0}^{\infty} b_k z^{-k} \tag{4.66}$$

Note the representation of a single coefficient IIR requires an infinite number of coefficients when represented as a series (FIR). The discrete impulse response is obtained from the inverse z-transform and is expressed as follows:

$$h(n) = \sum_{k=0}^{\infty} b_k \delta(k - n),$$

which completes the example.

As indicated in the example problem, the IIR filter results in an impulse response of *infinite* length. Thus one realizes that the impulse response of a reverberant dynamic system (i.e., one that has an exponential decay $\exp(at)$ superimposed on some oscillatory response) can be efficiently modeled with an IIR filter; however, an FIR filter would, in theory at least, require an infinite number of coefficients to represent the response. In practice, a finite number of FIR filter coefficients is used to model a reverberant system, and the truncation of the remaining coefficients constitutes the windowing design procedure. In effect, a rectangular window, or some alternative such as the Hanning window, can be applied to design the appropriate FIR filter.

However, the IIR filter is sometimes more convenient to implement, and it displays several notable characteristics that must be considered:

1. Stable only if the poles all lie within the unit circle in the $z$-plane
2. Relatively sensitive to perturbations in the filter coefficients, especially when some of the poles reside near the unit circle in the $z$-plane
3. Relatively few coefficients required to model the dynamic response of reverberant systems in comparison to the FIR filter
4. Nonlinear phase

Based upon the previously noted characteristics, an IIR filter provides a convenient method of building a discrete-time model of a reverberant dynamic system, since the number of filter coefficients required is greatly reduced in comparison to that of an FIR filter. In addition, the denominator of the filter can be used to estimate the resonant frequencies and damping ratio of the modeled physical system. For systems that are not reverberant (e.g., a single traveling wave), the FIR filter is the representation of choice.

**Example 4.2: FIR/IIR Modeling of a Second-order System** Consider a second-order dynamic system, as detailed in the MATLAB script file **sys2ndid.m**. The system has been designed to have a natural frequency at 50 Hz and a damping ratio of 0.1. A state-variable model is constructed for the purpose of simulating the dynamic response of the system to a prescribed forcing function. A stochastic input force to the system is created and the system response is obtained from simulation. The purpose of generating the discrete-time data set is to illustrate a method of performing system identification. As indicated in the script file, the Fourier transform of the system input and output are computed from the data set, and the autospectrum and cross-spectrum are computed from the Fourier transforms. The data acquisition process and corresponding Fourier analysis are typically performed with a spectrum analyzer in the laboratory. However, the data are simply created from a simple model for the purpose of this example.

Once the cross-spectrum and autospectrum are computed, the FRF can be obtained, as outlined earlier in this chapter. Each step is accurately detailed in the comments of the script file **sys2ndid.m**. Once the frequency response function for the system is obtained, the MATLAB function **invfreqz** can be used to design the FIR or IIR filter. As illustrated in Figure 4.10, the magnitude and phase of the actual FRF are compared to those obtained by designing discrete-time models of the system. Since we know that the system is second-order, a second-order digital IIR filter can be used to accurately model the frequency response, as illustrated in Figure 4.10. If an FIR filter is used, it is obvious from the results presented that

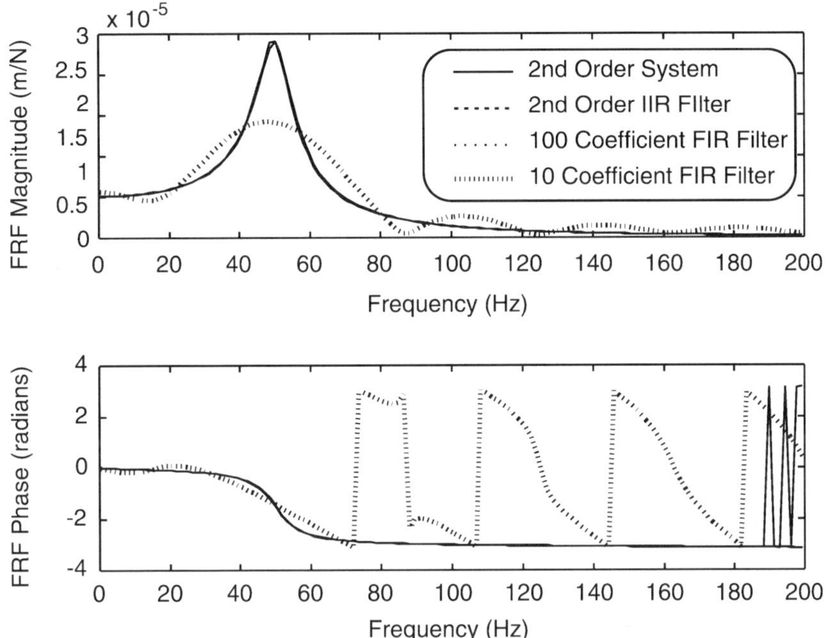


**Figure 4.10:** A comparison between FIR filters and IIR filters for system identification.

increasing the order of the filter results in a better approximation to the actual frequency response of the system. As illustrated, an IIR filter can be implemented with a total of 5 coefficients and yield better performance than a 100-coefficient FIR filter. In fact, the poles of the IIR filter correspond to the poles of the physical plant and can be used to estimate the damping ratio and natural frequency of the system, as indicated in equation (4.59).

For more complex systems, one would proceed by essentially counting the number of complex conjugate pairs of poles and zeros and choosing a reasonable order IIR filter to model the system dynamics. Some iteration may be necessary to improve the quality of the model in terms of matching phase and magnitude of the system frequency response. The resulting discrete-time model can then be used to simulate the system response for control system design and synthesis or to improve the quality of the analytical model used in the design process. More often than not, the designer must rely on models of the system obtained through some form of system identification for successful implementation.

## 4.5 FUNDAMENTALS IN ADAPTIVE FILTER THEORY

An adaptive filter is a self-designing filter that is dependent upon a recursive algorithm and is capable of operating in the presence of slowly (with respect to the adaptation process) time-varying parameters. Adaptive filters are commonly used in prediction, system identification, inverse modeling, and active noise cancellation (ANC). Adaptive filter theory is introduced in this text primarily in terms of its ability to control sound and vibration resulting from some persistent disturbance of a dynamic system. The adaptive filter is implemented to generate appropriate control signals in the presence of time-varying disturbances and operating conditions. Specifics of adaptive filter theory pertaining to control are covered in Chapter 8. The purpose of this section is to introduce the reader to some of the fundamentals in adaptive filter theory.

One should recognize that basically any filter structure can be adapted. For example, one can construct an adaptive IIR filter, an adaptive FIR filter, or an adaptive lattice filter. As one might anticipate, stability of the adaptive IIR filter is of greatest concern, since the poles of the filter must reside within the unit circle during the adaptation process. While the IIR filter is more economical than the FIR filter with respect to the number of filter coefficients required to achieve the same task (as illustrated in Example 4.2), the stability characteristics make it less attractive in the control application. Adaptive lattice filters are also a possible alternative and are attractive since the filter structure orthogonalizes the input signal on a stage-by-stage basis (Cowan and Grant, 1985). Thus, the convergence characteristics can be controlled even when there is a large spread in the eigenvalues of the input correlation matrix. The primary drawback of this approach is the increased complexity in the implementation of the adaptive algorithm.

For the purpose of this text, the focus is on adaptive FIR filters, due to the preferred stability characteristics and simplicity of implementing various adaptive algorithms, such as the least-mean-squares (LMS) algorithm. Adaptive FIR filters are well documented in the literature (Widrow et al., 1975) and have been used extensively in the control of harmonic disturbances since the introduction of the filtered-$x$ LMS algorithm (Widrow and Stearns, 1985) as well as the MIMO filtered-$x$ LMS algorithm (Elliott et al., 1987). The following sections are arranged to provide a basic introduction to the concepts of adaptive filter theory necessary to understand adaptive feedforward control. The reader requiring a more rigorous development is encouraged to study the texts by Cowan and Grant (1985), Widrow and Stearns (1985), and Haykin (1991).

### 4.5.1 The Wiener Filter

Wiener filter theory is typically developed for general complex-valued time series; however, in vibration control applications, one typically deals with inputs and outputs to electromechanical systems for which the sampled-data time series are real-valued. Thus, for the purpose of this text, the development is restricted to real-valued time series. Wiener filters can be described as optimum discrete-time linear filters, and the general filtering process is depicted in Figure 4.11. As illustrated, some input sequence $u(k)$ is passed through the filter and the output $y(k)$ is subtracted from the desired response $d(k)$. The objective

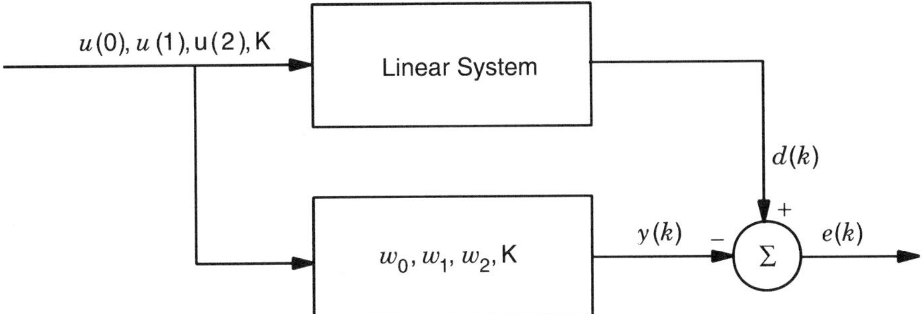


**Figure 4.11:** Schematic diagram of general filtering process.

is to compute the necessary values of the filter to minimize the error $e(k)$. As illustrated in Figure 4.11, the filter is configured for system identification of the linear system. However, the development of the Wiener-Hopf equations are not specific to the system identification process. In general, $d(k)$ is simply some desired response that the filter must be designed to replicate in opposite phase.

***Derivation of the Wiener-Hopf Equations*** At this point, one must choose the type of filter to be employed. The general development of the Wiener-Hopf equations will be performed for an FIR filter such that

$$y(k) = \sum_{n=0}^{\infty} w_n u(k-n), \quad k = 0, 1, 2, \ldots \tag{4.68}$$

The input to the filter and the desired response are assumed to be zero-mean stochastic processes with wide-sense stationarity. To develop an optimal filter design, one must select some *cost function* (i.e., function to be minimized), and the common practice is to choose the mean-square value of the error. This quadratic cost function is most attractive, since a unique minimum is guaranteed. The mean-squared error is defined as follows:

$$\xi = \mathrm{E}[e^2(k)], \tag{4.69}$$

where

$$e(k) = d(k) - y(k) \tag{4.70}$$

To minimize the cost function, one must take the derivative of the function with respect to each filter weight and set the system of equations equal to zero. We thus define the gradient operator such that

$$\nabla_n = \frac{\partial}{\partial w_n}, \qquad n = 0, 1, 2, \ldots \tag{4.71}$$

To minimize the cost function:

$$\nabla_n(\xi) = 0, \qquad n = 0, 1, 2, \ldots \tag{4.72}$$

Substituting equation (4.69) into equation (4.72), one obtains

$$\begin{aligned} \nabla_n(\xi) &= 2\mathrm{E}\left[ e(k) \frac{\partial e(k)}{\partial w_n} \right] \\ &= -2\mathrm{E}[u(k-n)e(k)] \end{aligned} \tag{4.73}$$

As indicated in equation (4.73), the error signal $e(k)$ must be orthogonal to each reference input sample when the optimal filter is employed (Haykin, 1991), since

$$E[u(k-n)e_{\text{opt}}(k)] = 0, \qquad n = 0, 1, 2, \ldots \tag{4.74}$$

where $e_{\text{opt}}(k)$ is the error resulting when the optimal filter is implemented. Equation (4.74) renders the necessary and sufficient conditions for the optimal filter. Since $e_{\text{opt}}(k)$ can be expressed in terms of the desired response and the optimal filter, one obtains

$$e_{\text{opt}}(k) = d(k) - \sum_{i=0}^{\infty} w_{\text{opt}i} u(k-i) \tag{4.75}$$

where $w_{\text{opt}i}$ is the $i$th coefficient of the optimal filter. Substituting equation (4.75) into equation (4.74), one obtains

$$E\left[ u(k-n)\left( d(k) - \sum_{i=0}^{\infty} w_{\text{opt}i} u(k-i) \right) \right] = 0, \qquad n = 0, 1, 2, \ldots \tag{4.76}$$

Thus, upon rearranging the expression,

$$\sum_{i=0}^{\infty} w_{\text{opt}i} E[u(k-n)u(k-i)] = E[u(k-n)d(k)], \qquad n = 0, 1, 2, \ldots \tag{4.77}$$

The expectation on the left-hand side is identified as the autocorrelation with time lag $i - n$ and the expectation on the right-hand side is simply the cross-correlation with time lag $-n$. Hence, one obtains the Wiener-Hopf equations:

$$\sum_{i=0}^{\infty} w_{\text{opt}i} r_{uu}(i-n) = r_{du}(-n), \qquad n = 0, 1, 2, \ldots \tag{4.78}$$

For an FIR filter of size $I$, the Wiener-Hopf equations can be expressed as follows:

$$\sum_{i=0}^{I-1} w_{\text{opt}i} r_{uu}(i-n) = r_{du}(-n), \qquad n = 0, 1, 2, \ldots, I-1 \tag{4.79}$$

The $I$ equations in $I$ unknowns can be written in matrix form:

$$\mathbf{R}\mathbf{w}_{\text{opt}} = \mathbf{P} \tag{4.80}$$

where $\mathbf{R}$ is the input correlation matrix, $\mathbf{w}_{\text{opt}}$ is the optimal filter vector, and $\mathbf{P}$ is the cross-correlation vector. The input correlation matrix can be expressed as follows:

$$\mathbf{R} = E[\mathbf{u}(k)\mathbf{u}^{\mathrm{T}}(k)] \tag{4.81}$$

where

$$\mathbf{u}(k) = [u(k), u(k-1), \ldots, u(n-(I-1))]^{\mathrm{T}} \tag{4.82}$$

The cross-correlation vector is defined in terms of the reference input and the desired response:

$$\mathbf{P} = E[\mathbf{u}(k)d(k)] \tag{4.83}$$

and the optimal filter is

$$\mathbf{w}_{\text{opt}} = [w_{\text{opt}0}, w_{\text{opt}1}, \ldots, w_{\text{opt}(I-1)}]^{\mathrm{T}} \tag{4.84}$$

As long as $\mathbf{R}$ is nonsingular, the optimal Wiener filter can be computed as follows:

$$\mathbf{w}_{\text{opt}} = \mathbf{R}^{-1}\mathbf{P} \tag{4.85}$$

***The Minimum Mean-Squared Error*** To evaluate the performance of the optimal filter design, one is often concerned with the minimum mean-squared error that can be achieved. The minimum mean-squared error can be computed from equation (4.69) by letting $e(k) = e_{\text{opt}}(k)$:

$$\xi_{\min} = E[e_{\text{opt}}^2(k)] \tag{4.86}$$

Utilizing vector notation,

$$e_{\text{opt}}(k) = d(k) - \mathbf{w}_{\text{opt}}^{\mathrm{T}}\mathbf{u}(k) \tag{4.87}$$

Thus,

$$\xi_{\min} = E[d(k)^2 - 2\mathbf{w}_{\text{opt}}^{\mathrm{T}}\mathbf{u}(k) + \mathbf{w}_{\text{opt}}^{\mathrm{T}}\mathbf{u}(k)\mathbf{u}(k)^{\mathrm{T}}\mathbf{w}_{\text{opt}}] \tag{4.88}$$

Substituting equation (4.85) into equation (4.88), applying the appropriate matrix identities, and collecting terms, one obtains

$$\xi_{\min} = E[d^2(k)] - \mathbf{P}^{\mathrm{T}}\mathbf{R}^{-1}\mathbf{P} \tag{4.89}$$

$$= E[d^2(k)] - \mathbf{P}^{\mathrm{T}}\mathbf{w}_{\text{opt}} \tag{4.90}$$

The minimum mean-squared error is thus observed to be a function of the input reference signal and the desired response. Thus, once the filter size is selected, one can compute the minimum mean-squared error that can be achieved before ever computing the optimum Wiener filter through equation (4.89).

***The Performance Surface*** The performance surface can be obtained by varying the coefficients of the filter and plotting the resulting cost function:

$$\xi = E[d(k)^2] - 2\mathbf{w}^{\mathrm{T}}\mathbf{P} + \mathbf{w}^{\mathrm{T}}\mathbf{R}\mathbf{w} \tag{4.91}$$

Since the cost function is quadratic, a single-coefficient filter results in a parabola, a two-coefficient filter yields a paraboloid, and a hyperparaboloid is obtained when more than two coefficients are implemented. The performance surface of an $I$ coefficient filter is thus a hyperparaboloid in $I$ space. Regardless of the filter order, the performance surface is characterized by a unique minimum. The performance surface can be expressed in terms of the minimum mean-squared error by subtracting equation (4.90) from equation (4.91):

$$\xi - \xi_{\min} = \mathbf{P}^{\mathrm{T}}\mathbf{w}_{\text{opt}} - 2\mathbf{w}^{\mathrm{T}}\mathbf{P} + \mathbf{w}^{\mathrm{T}}\mathbf{R}\mathbf{w} \tag{4.92}$$

Recognizing that $\mathbf{P}^{\mathrm{T}} = \mathbf{w}_{\mathrm{opt}}^{\mathrm{T}}\mathbf{R}^{\mathrm{T}}$, one obtains

$$\xi = \xi_{\min} + \mathbf{w}_{\mathrm{opt}}^{\mathrm{T}}\mathbf{R}^{\mathrm{T}}\mathbf{w}_{\mathrm{opt}} - \mathbf{w}^{\mathrm{T}}\mathbf{R}\mathbf{w}_{\mathrm{opt}} - \mathbf{w}_{\mathrm{opt}}^{\mathrm{T}}\mathbf{R}\mathbf{w} + \mathbf{w}^{\mathrm{T}}\mathbf{R}\mathbf{w} \tag{4.93}$$

$$= \xi_{\min} + (\mathbf{w} - \mathbf{w}_{\mathrm{opt}})^{\mathrm{T}}\mathbf{R}(\mathbf{w} - \mathbf{w}_{\mathrm{opt}}) \tag{4.94}$$

The previous simplification is possible since the input correlation matrix $\mathbf{R}$ is real-symmetric. The input correlation matrix can be expressed in modal coordinates such that

$$\mathbf{R} = \mathbf{Q}\mathbf{\Lambda}\mathbf{Q}^{\mathrm{T}} \tag{4.95}$$

where $\mathbf{\Lambda}$ is the diagonal matrix of eigenvalues of $\mathbf{R}$ and $\mathbf{Q}$ is the modal matrix of corresponding eigenvectors. Letting

$$\mathbf{V} = \mathbf{Q}^{\mathrm{T}}(\mathbf{w} - \mathbf{w}_{\mathrm{opt}}) \tag{4.96}$$

the performance surface can be transformed into orthogonal coordinates:

$$\xi = \xi_{\min} + \mathbf{V}^{\mathrm{T}}\mathbf{\Lambda}\mathbf{V} \tag{4.97}$$

This transformation is critical to the study of the stability of various algorithms, as illustrated in the following sections. Since the transformation contains no cross-product terms, the performance surface can be expressed as follows:

$$\xi = \xi_{\min} + \sum_{n=0}^{I-1} \lambda_n v_n^2 \tag{4.98}$$

where $I$ is the order of the FIR filter, $\lambda_n$ is the $n$th eigenvalue of $\mathbf{\Lambda}$, and $v_n$ is the $n$th component of the vector $\mathbf{V}$. One should recognize that $\mathbf{V}$ represents the filter weights in terms of the transformed coordinate system. In addition, as noted by Widrow and Stearns (1985),

$$\frac{\partial^2 \xi}{\partial v_n^2} = 2\lambda_n \tag{4.99}$$

Thus, the eigenvalues of the input correlation matrix yield the second derivatives of the performance surface with respect to the principal axes.

**Example 4.3: System Identification of a Second-order Dynamic System** Consider the second-order dynamic system discussed in Example 4.2. For the purpose of this example, the dynamic system is modeled and a system identifi-

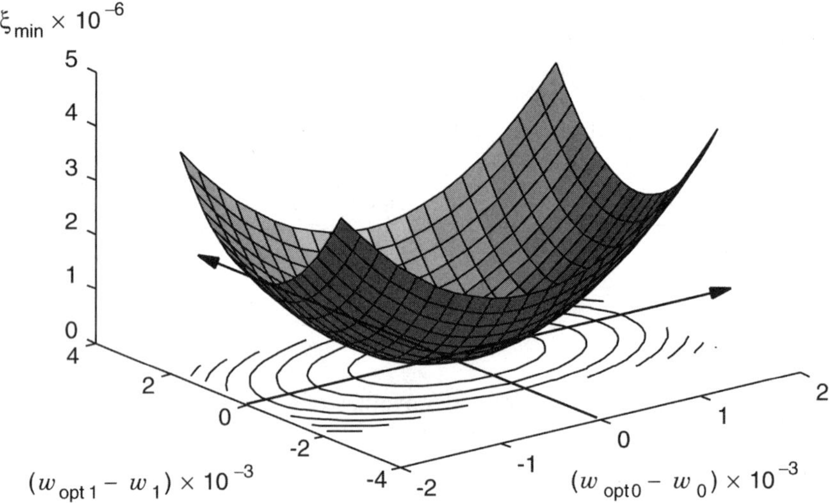


**Figure 4.12:** Performance surface for system identification of a second-order plant with a sample rate of four samples per period.

cation is performed on the plant with a sinusoidal input at a driving frequency of 40 Hz. The optimum Wiener filter is computed as well as the minimum mean-squared error. The script file: **optfilt.m** can be executed to generate the performance surface presented in Figure 4.12. The sample rate was set at 160 Hz, which is four times that of the reference signal, and a two-coefficient FIR filter was chosen for this application. The contour plot facilitates visualizing the orientation of the principle axes. As illustrated, for this example, the principle axes are aligned with the original axes. This result is unique to the chosen sample rate, since *the input correlation matrix is diagonal for a single frequency reference sampled at four samples per period.* Thus, the filter coefficients are orthogonal to each other for this special case.

To illustrate the effect of the sample rate on the shape of the performance surface, execute the script file **optflt5.m**. In this example, the sample rate was changed to five samples per period (i.e., 200 Hz). As illustrated in Figure 4.13, the principal axes are no longer aligned with the axes of the filter coefficients. Interested readers are encouraged to modify the provided script files to study the effect of varying the sample rate on the input correlation matrix and the eigenvalue spread for the input correlation matrix. (To obtain the eigenvalues of **R**, simply execute the MATLAB command **eig**(R) after executing the script files provided.) The disparity in the eigenvalues increases with increasing sample rate.

### 4.5.2 The Method of Steepest Descent

The optimum Wiener filter is typically the objective in adaptive signal processing if the desired response is stationary, but in practice one is confronted with

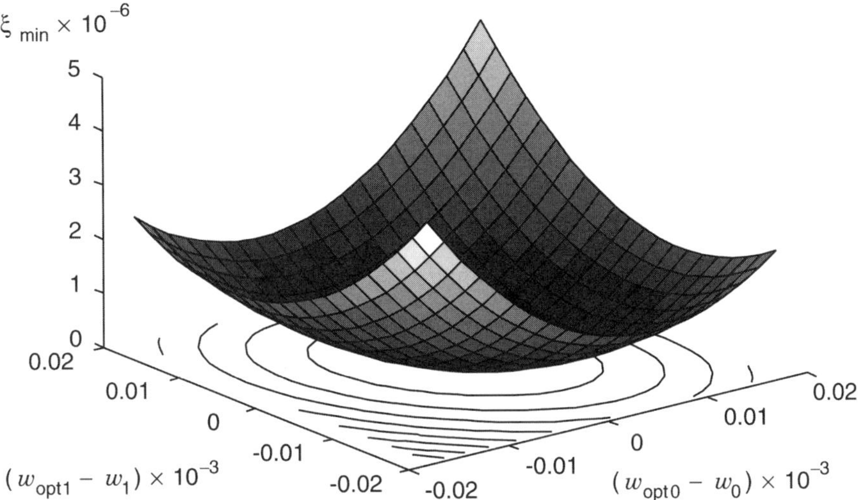


**Figure 4.13:** Performance surface for system identification of a second-order plant with a sample rate of five samples per period.

the challenge of computing the optimum filter in the presence of time-varying parameters and operating conditions. Under these circumstances, the optimum design changes as a function of time, and it is computationally inefficient to compute the solution to the Wiener-Hopf equations by analytical methods. Thus, it is logical to develop an adaptive algorithm to modify the filter as a function of time. One of the oldest and simplest methods of optimization is based upon *the method of steepest descent* (Widrow and Stearns, 1985). The steepest descent algorithm can be expressed simply as follows:

$$\mathbf{w}(k+1) = \mathbf{w}(k) + \tfrac{1}{2}\,\mu(-\boldsymbol{\nabla}(\xi(k))) \tag{4.100}$$

where $\mathbf{w}(k)$ is the value of the filter-weight vector at the $k$th iteration of the algorithm and $\mu$ is a positive-valued real constant. (Some authors multiply $\mu$ by 1 as opposed to $\frac{1}{2}$ in equation (4.100); however, the constant is arbitrary as long as one is consistent.) The gradient vector $\boldsymbol{\nabla}(k)$ at the $k$th iteration is given as follows:

$$\boldsymbol{\nabla}(\xi(k)) = \left[\frac{\partial\xi(k)}{\partial w_0(k)}, \frac{\partial\xi(k)}{\partial w_1(k)}, \ldots, \frac{\partial\xi(k)}{\partial w_{I-1}(k)}\right]^{\mathrm{T}} \tag{4.101}$$

$$= 2\mathbf{R}\mathbf{w}(k) - 2\mathbf{P} \tag{4.102}$$

where the filter-weight vector has $I$ coefficients. To minimize the cost function (i.e., the mean-squared error) with the steepest descent algorithm (Haykin, 1991):

1. Initialize the values of the filter $\mathbf{w}(0)$. The zero vector is an adequate choice.
2. Compute the gradient vector as outlined in equation (4.102).
3. Compute the next estimate of the weight vector as indicated in equation (4.100).
4. Repeat the process starting with step 2.

Step 3 can be simplified by substituting equation (4.102) into equation (4.100) to obtain

$$\mathbf{w}(k+1) = \mathbf{w}(k) + \mu[\mathbf{P} - \mathbf{R}\mathbf{w}(k)], \qquad k = 0, 1, 2, \ldots \tag{4.103}$$

The steepest descent algorithm is obviously recursive, which implies feedback, and thus stability of the algorithm must be considered. A simple schematic diagram of the adaptive algorithm is presented in Figure 4.14. One should recognize that the signal flow diagram is in matrix notation. Since the algorithm is recursive, one might approach the stability analysis as in Appendix A on state-variable models of discrete-time linear systems. For stability, one considers the initial condition response, and thus one lets $\mathbf{P} = \mathbf{0}$. Taking the $z$-transform of equation (4.103) with $\mathbf{P} = \mathbf{0}$, one obtains

$$z\mathbf{w}(z) - \mathbf{w}(0) = \mathbf{w}(z) - \mu\mathbf{R}\mathbf{w}(z) \tag{4.104}$$

Collecting terms,

$$[z\mathbf{I} - (\mathbf{I} - \mu\mathbf{R})]\mathbf{w}(z) = \mathbf{w}(0) \tag{4.105}$$

Expressing in terms of $\mathbf{w}(z)$,

$$\mathbf{w}(z) = [(z-1)\mathbf{I} + \mu\mathbf{R}]^{-1}\mathbf{w}(0) \tag{4.106}$$

The input correlation matrix can be expressed in terms of modal coordinates, as in equation (4.95):

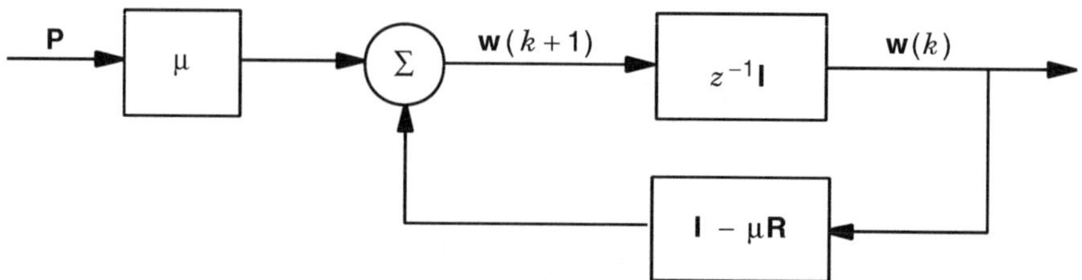


**Figure 4.14:** Schematic diagram of steepest descent algorithm.

$$\mathbf{w}(z) = [(z-1)\mathbf{I} + \mu\mathbf{Q}\mathbf{\Lambda}\mathbf{Q}^{\mathrm{T}}]^{-1}\mathbf{w}(0) \tag{4.107}$$

If the modal matrix is normalized such that $\mathbf{Q}\mathbf{Q}^{\mathrm{T}} = \mathbf{I}$, then

$$\mathbf{w}(z) = [\mathbf{Q}\{(z-1)\mathbf{I} + \mu\mathbf{\Lambda}\}\mathbf{Q}^{\mathrm{T}}]^{-1}\mathbf{w}(0) \tag{4.108}$$

$$= \mathbf{Q}[(z-1)\mathbf{I} + \mu\mathbf{\Lambda}]^{-1}\mathbf{Q}^{\mathrm{T}}\mathbf{w}(0) \tag{4.109}$$

The poles of the discrete-time system are given in equation (4.109), and as discussed in Appendix A, stability is assured if all of the poles reside within the unit circle in the $z$-plane. Thus,

$$-1 < 1 - \mu\lambda_n < 1, \qquad n = 1, 2, \ldots, I \tag{4.110}$$

where $\lambda_n$ is the $n$th diagonal value of the matrix of eigenvalues $\mathbf{\Lambda}$, and $I$ is the order of the filter (i.e., the length of $\mathbf{w}$). Equation (4.110) yields the necessary and sufficient conditions for convergence of the steepest descent algorithm. Solving equation (4.110) in terms of $\mu$, one obtains

$$0 < \mu < \frac{2}{\lambda_n} \tag{4.111}$$

Thus, the maximum value of $\mu$ is obtained when the maximum value of $\lambda_n$ is substituted into equation (4.111):

$$0 < \mu < \frac{2}{\lambda_{\max}} \tag{4.112}$$

Since the poles of the recursive system (i.e., the steepest descent algorithm) are real-valued, the adaptation of each filter weight in transformed coordinates can be described in terms of a first-order, discrete-time system. The homogeneous first-order difference equation can be expressed as follows (Haykin, 1991):

$$v_n(k) = (1 - \mu\lambda_n)v_n(k-1) \tag{4.113}$$

$$= (1 - \mu\lambda_n)^n v_n(0) \tag{4.114}$$

the time constant of which is defined as follows:

$$1 - \mu\lambda_n = \exp\left(-\frac{1}{\tau_n}\right) \tag{4.115}$$

Solving for the time constant in terms of each eigenvalue of the input correlation matrix, one obtains

$$\tau_n = -\frac{1}{\ln(1 - \mu\lambda_n)} \tag{4.116}$$

As discussed by Griffiths (1975), the time constant of the steepest descent algorithm is bounded by the minimum and maximum values of the eigenvalues of the input correlation matrix as follows:

$$-\frac{1}{\ln(1 - \mu\lambda_{\max})} \leq \tau_{SD} \leq -\frac{1}{\ln(1 - \mu\lambda_{\min})} \tag{4.117}$$

where $\tau_{SD}$ is the overall time constant of the steepest descent algorithm. Disparity in the eigenvalues of the input correlation matrix thus yields a wider range for the overall time constant.

The time constant of the mean-squared error is slightly different from that of the adaptive filter. As indicated in equation (4.98), the mean-squared error varies with the square of the filter coefficients in transformed coordinates:

$$\xi = \xi_{\min} + \sum_{n=0}^{I-1} \lambda_n v_n^2$$

Substituting equation (4.114) into the previous expression, one obtains:

$$\xi = \xi_{\min} + \sum_{n=0}^{I-1} \lambda_n (1 - \mu\lambda_n)^{2n} v_n^2(0) \tag{4.118}$$

The time constant of the mean-squared error thus varies as follows:

$$\tau_{n,\xi} = -\frac{1}{2\ln(1 - \mu\lambda_n)} \tag{4.119}$$

The filter weights will thus converge twice as fast as the mean-squared error in the adaptation process. The factor of 2 is directly related to the fact that the mean-squared error varies as a function of the square of the filter coefficients.

### 4.5.3 Newton's Method

Newton's method has long been used in numerical analysis to find the roots of a polynomial. Thus, it is appropriate to investigate its potential as applied to finding the solution of the optimal filter that renders the derivative of the mean-squared error zero. In multidimensional space (hyperspace), Newton's method can be realized as follows (Widrow and Stearns, 1985):

$$\mathbf{w}(k+1) = \mathbf{w}(k) - \tfrac{1}{2}\,\mathbf{R}^{-1}\nabla(\xi(k)) \tag{4.120}$$

For a quadratic error surface such as the mean-squared error, the algorithm yields the optimal filter design in a single iteration; however, $\mathbf{R}^{-1}$ must be computed as well as $\nabla(\xi(k))$, which is not a trivial task.

An alternative to this approach might be considered for the adaptation in multidimensional space; however, let us first consider Newton's method for a single-coefficient filter. The adaptive algorithm can be expressed as follows for the single-coefficient filter:

$$w(k+1) = w(k) - \frac{f(w(k))}{f'(w(k))} \tag{4.121}$$

where $f(w(k)) = \xi'(w(k))$ (Widrow and Stearns, 1985). Replacing $f(w(k))$ with the appropriate derivative of the mean-squared error, one obtains

$$w(k+1) = w(k) - \frac{\xi'(w(k))}{\xi''(w(k))} \tag{4.122}$$

If $\xi'(k)$ and $\xi''(k)$ are not known a priori, the functions must be estimated. A Taylor's series expansion can be performed to compute the estimate of the first and second derivatives using a central difference technique. If a perturbation of the filter coefficient $\Delta w$ is chosen, then the derivatives can be approximated as follows:

$$\xi'(w_k) \approx \frac{\xi(w_k + \Delta w) - \xi(w_k - \Delta w)}{2\Delta w} \tag{4.123}$$

and

$$\xi''(w_k) \approx \frac{\xi(w_k + \Delta w) - 2\xi(w_k) + \xi(w_k - \Delta w)}{(\Delta w)^2} \tag{4.124}$$

where $w_k = w(k)$. *For the quadratic cost function (i.e., the mean-squared error), the derivatives are exact.* This proof is left as an exercise for the reader. However, the proof is relatively simple upon recognizing that the third and higher derivatives of a quadratic function are zero.

Newton's method for the single-coefficient filter can thus be expressed as follows:

$$w(k+1) = w(k) - \Delta w\,\frac{\xi(w(k) + \Delta w) - \xi(w(k) - \Delta w)}{2[\xi(w(k) + \Delta w) - 2\xi(w(k)) + \xi(w(k) - \Delta w)]} \tag{4.125}$$

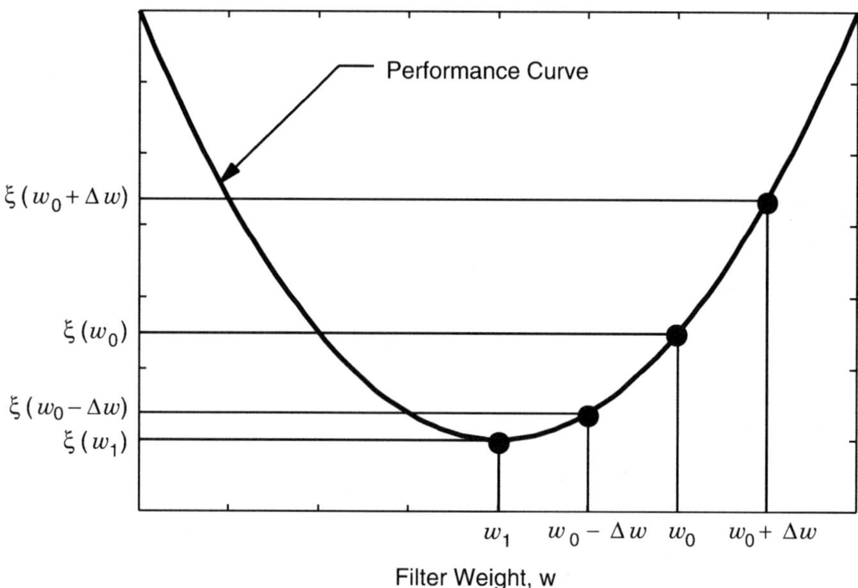


**Figure 4.15:** The single-coefficient implementation of Newton's method and the parabolic performance curve.

**Example 4.4: Newton's Method on a Parabolic Performance Surface** Consider the parabolic performance curve illustrated in Figure 4.15. The minimum of the performance curve and corresponding filter weight illustrated can be found using equation (4.125) as implemented in the script file **newtex.m**. As indicated in the figure, an initial guess at the root of the equation is made (obviously for the purpose of demonstration), and the function is evaluated at this initial guess as well as at some perturbed value of the guess. The optimal solution is then computed in the first iteration based upon equation (4.125). (the figure can be generated with the script file **newtex.m**.)

As indicated in equation (4.125), the filter coefficient is perturbed $\pm\Delta w$ and the mean-squared error is estimated at each perturbation. Thus, some statistics must be performed in taking the expectation of the square of the error to assure a reasonable estimate of the mean-squared error. A single iteration of the algorithm occurs when the next filter coefficient is computed. However, in practice, the actual time required to execute the algorithm is dependent upon the number of samples necessary to estimate the mean-squared error for each perturbation of the filter coefficient. One must also consider the dynamic range of the digital signal processor before selecting the value of the perturbation to assure that a measurable difference in the mean-squared error can be resolved.

The single-coefficient Newton's method can actually be used to adapt the filter coefficients in multidimensional space as follows:

1. Initialize the weight vector and compute the mean-squared error.
2. Perturb a single filter coefficient $\pm\Delta w$ and compute the mean-squared error at each perturbation.
3. Update the filter coefficient using equation (4.125).
4. Proceed to step 2 using the next filter coefficient in the weight vector.

The preceding algorithm cycles through each filter coefficient of the weight vector, one by one, and effectively computes the optimal filter coefficient for the given "slice" of the hyperparabolic surface. By perturbing only a single coefficient, a parabola corresponding to a particular cross section of the hyperparabolic surface is obtained. The algorithm yields the optimal solution for that particular cross section; however, multiple iterations are required to compute the optimal weight vector. The advantage of this approach compared to that of the multidimensional algorithm is the reduction in computational required at each iteration. The disadvantage is the increase in time required to compute the optimal weight vector.

### 4.5.4 The LMS Algorithm

The LMS algorithm (Widrow and Hoff, 1960) is the most commonly used stochastic gradient-based algorithm due to its relative simplicity. The LMS algorithm is a recursive algorithm that can adjust the coefficients of a transversal filter as a function of the measured error $e(k)$, which is the difference between the desired response $d(k)$ and the output of the filter $y(k)$, as illustrated in Figure 4.16. A detailed schematic of a transversal filter has previously been presented

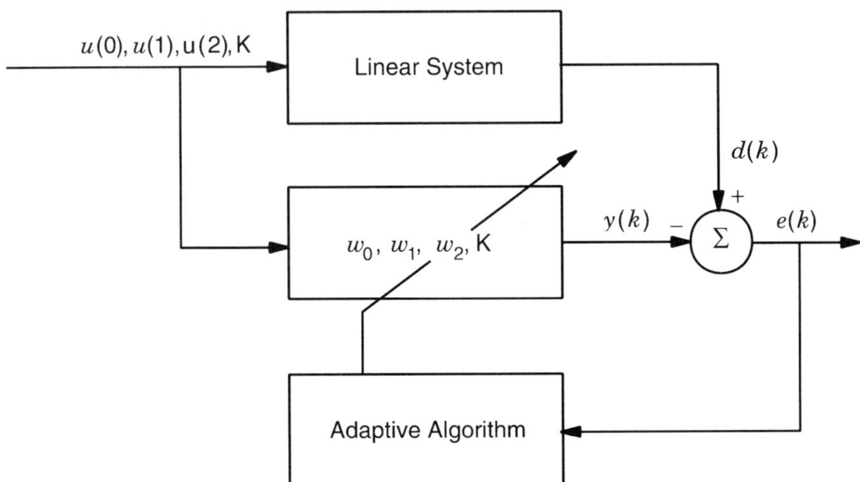


**Figure 4.16:** The adaptive transversal filter.

in Figure 4.8. The basic feature of the LMS algorithm that distinguishes it from other adaptive algorithms is that it does not require any matrix inversions or measurements of the correlation functions in the estimate of the gradient.

To derive the LMS algorithm, we first reconsider the gradient vector that has been presented in equation (4.102). The difficulty rests in estimating the input correlation matrix **R** and the cross-correlation vector **P**. If one selects instantaneous estimates, then

$$\hat{\mathbf{R}}(k) = \mathbf{u}(k)\mathbf{u}^{\mathrm{T}}(k) \tag{4.126}$$

and

$$\hat{\mathbf{P}}(k) = \mathbf{u}(k)d(k) \tag{4.127}$$

Substituting the instantaneous estimates into equation (4.102), one obtains

$$\hat{\nabla}(\xi(k)) = 2\mathbf{u}(k)\mathbf{u}^{\mathrm{T}}(k)\hat{\mathbf{w}}(k) - 2\mathbf{u}(k)d(k) \tag{4.128}$$

However, one should recognize that the filter output is

$$y(k) = \hat{\mathbf{w}}^{\mathrm{T}}(k)\mathbf{u}(k) \tag{4.129}$$

which upon substituting into equation (4.128) and rearranging terms yields

$$\hat{\nabla}(\xi(k)) = -2\mathbf{u}(k)[d(k) - y(k)] \tag{4.130}$$

The instantaneous error is

$$e(k) = d(k) - y(k) \tag{4.131}$$

and thus the estimate of the gradient can be expressed as follows:

$$\hat{\nabla}(\xi(k)) = -2\mathbf{u}(k)e(k) \tag{4.132}$$

Substituting equation (4.132) into equation (4.100), the LMS algorithm is obtained:

$$\hat{\mathbf{w}}(k+1) = \hat{\mathbf{w}}(k) + \mu\mathbf{u}(k)e(k) \tag{4.133}$$

As discussed by Haykin (1991), the LMS algorithm is a multivariable nonlinear stochastic feedback system, which makes the stability analysis somewhat difficult. In general, the LMS algorithm is stable if the convergence parameter $\mu$ is chosen such that the filter weights approach the optimum Wiener filter in the limit and the mean-squared error reaches some steady value in the limit. As detailed by Widrow (1970), the adaptive filter in the LMS algorithm converges to the optimum Wiener filter if

$$0 < \mu < \frac{2}{\lambda_{\max}} \tag{4.134}$$

where $\lambda_{\max}$ is the maximum eigenvalue of the input correlation matrix $\mathbf{R}$. Equation (4.134) provides the necessary and sufficient condition for convergence of the weight vector.

It so happens that the necessary and sufficient condition for convergence of the mean-squared error is more critical. As outlined by Haykin (1991), an additional necessary and sufficient condition is imposed for convergence of the mean-squared error:

$$\sum_{i=1}^{I} \frac{\mu\lambda_i}{2 - \mu\lambda_i} < 1 \tag{4.135}$$

where $I$ is the order of the adaptive filter, and $\lambda_i$ is the $i$th eigenvalue of the input correlation matrix $\mathbf{R}$. Thus, for convergence of the mean-squared error, equations (4.134) and (4.135) provide the necessary and sufficient conditions. Hence, one realizes that convergence of the mean-squared error implies convergence of the weight vector. In equations (4.134) and (4.135), one observes that the *convergence parameter* can be adjusted to control the convergence of the algorithm. However, the convergence parameter also plays an important role as an additional performance measure used to evaluate the LMS algorithm, *the misadjustment.*

The misadjustment $M$ is defined as the ratio of the excess mean-squared error to the minimum mean-squared error that can be achieved (Widrow and Stearns, 1985; Haykin, 1991):

$$M = \frac{\xi(n) - \xi_{\min}}{\xi_{\min}} \tag{4.136}$$

$$= \frac{\sum_i^I \mu\lambda_i/(2 - \mu\lambda_i)}{1 - \sum_{i=1}^I \mu\lambda_i/(2 - \mu\lambda_i)} \tag{4.137}$$

For small $\mu$, the misadjustment can be simplified:

$$M \approx \frac{\mu}{2} \sum_{i=1}^{I} \lambda_i \tag{4.138}$$

$$\approx \frac{\mu I \lambda_{\mathrm{av}}}{2}, \tag{4.139}$$

where $\lambda_{\mathrm{av}}$ is the average of the eigenvalues of the input correlation matrix. To obtain a quick assessment of the level of misadjustment that can be expected, one must simply recall that the trace of the input correlation matrix yields the sum of the eigenvalues:

$$\mathrm{Tr}[\mathbf{R}] = \sum_{i=1}^{I} \lambda_i \tag{4.140}$$

$$= I\lambda_{\mathrm{av}} \tag{4.141}$$

The trace of **R** is simple to compute since **R** is a Toeplitz matrix (Haykin, 1991). Each diagonal entry of **R** is simply the autocorrelation of the reference signal with zero time lag (i.e., the mean-squared value of the input reference). Thus,

$$M \approx \frac{\mu I r_{uu}(0)}{2} \tag{4.142}$$

To minimize the misadjustment, one must decrease the size of the convergence parameter $\mu$. However, decreasing the size of the convergence parameter has the negative effect of increasing the average time constant of the LMS algorithm, which can be approximated as follows:

$$\tau_{\mathrm{av},\xi} \approx \frac{1}{2\mu\lambda_a v} \tag{4.143}$$

$$\approx \frac{1}{2\mu r_{uu}(0)} \tag{4.144}$$

Obviously, one must choose the convergence parameter in accordance with the dynamics of the system to which it is applied. In some cases, rapid convergence might be more important than misadjustment. However, in either case, a quick assessment can be made by simply computing the input power of the reference signal (i.e., $r_{uu}(0)$).

**Example 4.5: System Identification Using the LMS Algorithm** System identification is one of the applications for the LMS algorithm and the process is illustrated in Figure 4.17. As indicated, some known reference input $u(k)$ (typically an applied voltage) is used to drive the dynamic system and the response is measured with some sensor. For the present example, a point force actuator

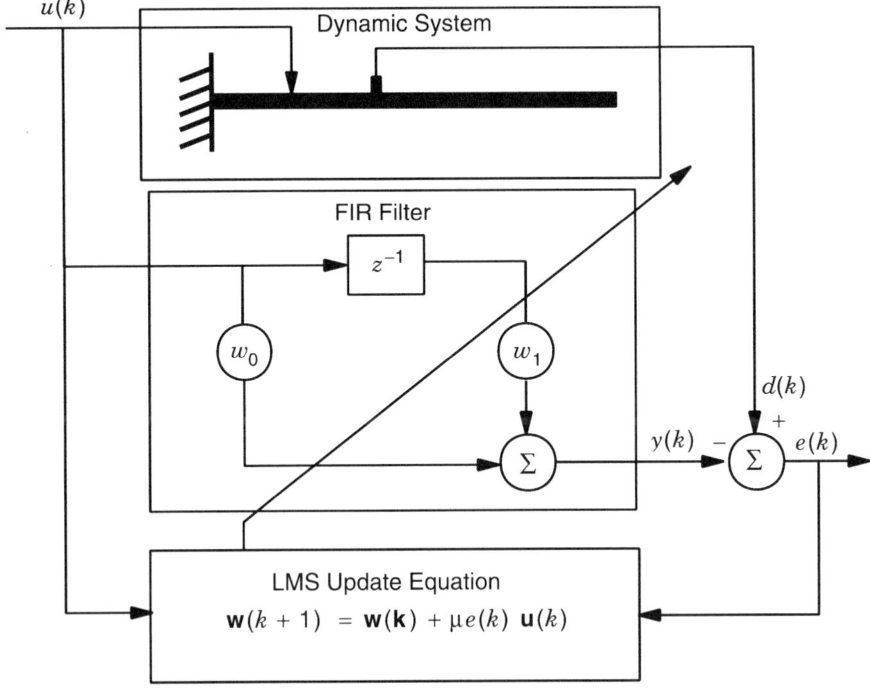


**Figure 4.17:** Schematic diagram of system identification process with the LMS algorithm (two coefficient).

and point displacement sensor are used. The positions of the actuator and sensor, as well as the dimensions of the cantilevered beam used in the simulation, are presented in the script file **bmdriver.m**. The sensor output $d(k)$ becomes the desired response for the LMS algorithm. The same reference signal $u(k)$ is also passed through an adaptive transversal filter. In this case, a single frequency (100 Hz) reference signal is used, and thus a two-coefficient FIR filter suffices. The estimation error $e(k)$ is computed from the output of the transversal filter and the desired response. This estimation error and the reference signal are used to update the filter coefficients at the next time step.

The results from the simulation are illustrated in Figure 4.18. To replicate the results or to experiment with different frequencies, execute or edit the script file **lmsexamp.m**. The desired response is illustrated in the top portion of Figure 4.18 and the estimation error is illustrated in the bottom portion of Figure 4.18. As indicated, the estimation error oscillates for the first second of the adaptation. This oscillation is simply the "ringing" of the reverberant system. One should recognize that this transient response of the system is "uncorrelated" with the reference signal, and thus, the adaptive filter converges in the presence of this uncorrelated "noise." This observation proves important in later chapters devoted to the combination of feedback and adaptive feedforward control.

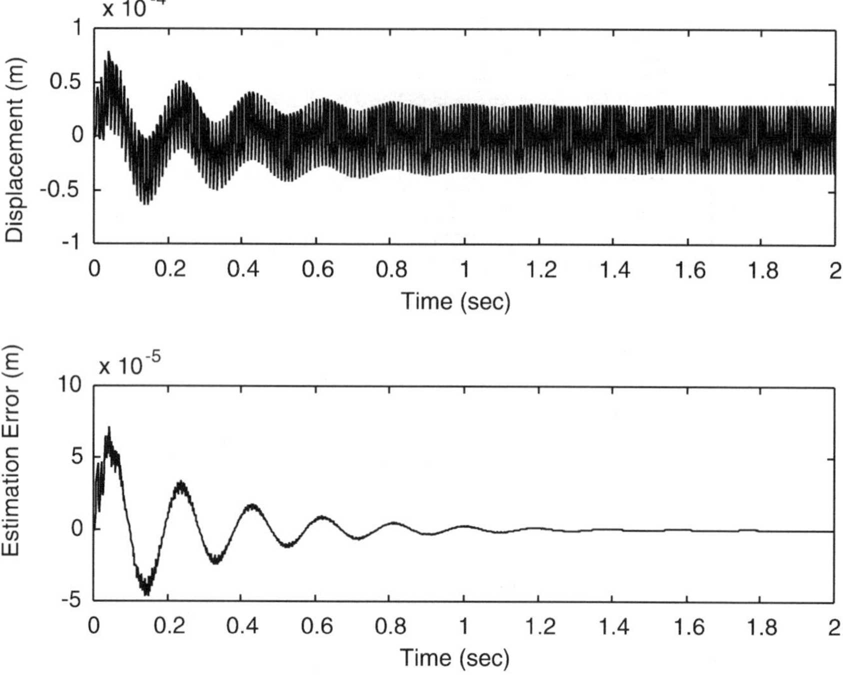


**Figure 4.18:** Simulation of system identification process for single harmonic disturbance on a cantilevered beam.

In Example 4.5, the LMS algorithm has been used for system identification in a single-frequency application. However, more often than not, the reference signal is composed of multiple frequencies, and choosing the appropriate filter size becomes a more complicated issue. In terms of designing the optimum Wiener filter, two filter coefficients are required for each frequency present in the reference (i.e., a reference with two frequencies requires a four coefficient FIR filter). However, as noted by Glover (1977), the order of the filter required when implementing the LMS algorithm is typically much larger, due to the nonlinear feedback associated with the update process. Consider a multifrequency reference input:

$$u(k) = \sum_{m=1}^{M} A_m \cos(\omega_m kT + \theta_m) \tag{4.145}$$

where $A_m$ is the amplitude of the $m$th frequency, $\theta_m$ is the phase associated with the $m$th frequency, and $T$ is the discrete-time sample rate. As outlined by Glover (1977), the $z$-transform of the output of the adaptive transversal filter is a function of a linear time-invariant term and two time-varying terms:

$$Y(z) = \frac{I\mu}{4} E(z) \sum_{m=1}^{M} A_m^2 [U(z \exp(-j\omega_m T))$$
$$+ U(z \exp(j\omega)mT))]$$
$$+ \frac{\mu}{4} \sum_{m=1}^{M} \sum_{n=1}^{M} A_m A_n \beta\left( \frac{\omega_m - \omega_n}{2} T, I \right) [TV], \qquad m \neq n$$
$$+ \frac{\mu}{4} \sum_{m=1}^{M} \sum_{n=1}^{M} A_m A_n \beta\left( \frac{\omega_m + \omega_n}{2} T, I \right) [TV] \tag{4.146}$$

where $[TV]$ is the time-varying term (Glover, 1977), $E(z)$ is the $z$-transform of the error, $U(z)$ is the $z$-transform of the reference, $Y(z)$ is the $z$-transform of the filter output,

$$\beta\left( \frac{\omega_m - \omega_n}{2} T, I \right) = \frac{\sin(I(\omega_m - \omega_n)T/2)}{\sin((\omega_m - \omega_n)T/2)} \tag{4.147}$$

and

$$\beta\left( \frac{\omega_m + \omega_n}{2} T, I \right) = \frac{\sin(I(\omega_m + \omega_n)T/2)}{\sin((\omega_m + \omega_n)T/2)} \tag{4.148}$$

As noted by Glover (1977), when the sinusoids in the reference are closely spaced (in the frequency domain), the only way to eliminate the time-varying terms (i.e., terms multiplied by $[TV]$ in equation (4.146)) is to increase the order of the filter $I$. In some cases, higher harmonic applications in particular, it is possible to find a unique relationship between the frequencies present in the reference, the sample rate, and the filter order to render equations (4.147) and (4.148) zero, as discussed by Clark and Gibbs (1994). For a more extensive discussion, the reader is referred to the references by Glover (1977) and Clark and Gibbs (1994).

### 4.5.5 Choosing the Filter Order and Sampling Rate

In general, the order of the adaptive filter is dependent upon the frequency content of the reference signal as well as the adaptive algorithm used in the filter weight update equation. In theory, with respect to the optimum Wiener filter for multifrequency reference signals, two filter coefficients per frequency are required to minimize the mean-squared error. In terms of the gradient descent algorithm or a time-averaged gradient descent algorithm, two filter coefficients per frequency are sufficient, since the update is based upon statistical estimates of the gradient. In terms of the LMS algorithm, the filter order is dependent upon the sample rate and the relationship between the harmonics (i.e., integer multiples, rational multiples, or irrational multiples). As noted in Section 4.5.4, for the general multifrequency implementation of the LMS algorithm, the order of the filter must be increased to minimize the time-varying terms in the filter output (Glover, 1977). As noted by Clark and Gibbs (1994), for higher harmonic reference signals, if synchronous sampling is employed, certain relationships between the order of the filter and the sample rate will render equations (4.147) and (4.148) zero, eliminating the time-varying terms totally. However, this case is rather unique, and quite often one must resort to a trial and error technique, choosing a filter order and measuring the minimum mean-squared error that can be achieved.

## 4.6 SUMMARY

In Chapter 4, we developed background topics in signal processing required for adaptive signal processing as applied to control of adaptive structures in Chapter 8. Various signal types were reviewed, as well as correlation functions, spectral density functions, windowing, and methods of dealing with finite-length records of signals. Digital filters were introduced, including finite impulse response (FIR) and infinite impulse response (IIR), and, with this background, basic elements of adaptive signal processing were introduced. The Weiner filter, method of steepest descent, Newton's method, and the least-mean-squares (LMS) algorithm were all introduced, including some practical issues with respect to implementation.

## BIBLIOGRAPHY

Bendat, J. S. and A. G. Piersol, 1986. *Random Data*, Wiley, New York.

Clark, R. L. and G. P. Gibbs, 1994. "Implications of Noise on the Reference Signal in Feedforward Control of Harmonic Disturbances," *Journal of the Acoustical Society of America*, **96**(4), 2585–2588.

Cowan, C. F. N. and P. M. Grant, 1985. *Adaptive Filters*, Prentice Hall, Englewood Cliffs, NJ.

Elliott, S. J., I. M. Stothers, and P. A. Nelson, 1987. "A Multiple Error LMS Algorithm and Its Application to the Active Control of Sound and Vibration," *IEEE Transactions on Acoustics, Speech, and Signal Processing*, **ASSP-35**(10), 1423–1434.

Glover, J. R., Jr., 1977. "Adaptive Noise Cancelling Applied to Sinusoidal Interferences," *IEEE Transactions on Acoustics, Speech, and Signal Processing*, **ASSP-25**, 484–491.

Griffiths, L. J., 1975. "Rapid Measurement of Digital Instantaneous Frequency," *IEEE Transactions on Acoustics, Speech, and Signal Processing*, **ASSP-23**, 207–222.

Harris, F. J., 1978. "On the Use of Windows for Harmonic Analysis with the Discrete Fourier Transform," *Proceedings of the IEEE*, **66**(1), 51–83.

Haykin, S., 1991. *Adaptive Filter Theory*, Prentice Hall, Englewood Cliffs, NJ.

Oppenheim, A. V. and R. W. Schafer, 1989. *Discrete-Time Signal Processing*, Prentice Hall, Englewood Cliffs, NJ.

Widrow, B., 1970. "Adaptive Filters," in *Aspects of Network and System Theory*, edited by R. E. Kalman and N. DeClaris, Holt, Rinehart and Winston, New York.

Widrow, B. and M. E. Hoff, Jr., 1960. "Adaptive Switching Circuits," *IRE WESCON Conv. Rec.*, Pt. 4, pp. 96-104.

Widrow, B. and S. D. Stearns, 1985. *Adaptive Signal Processing*, Prentice Hall, Englewood Cliffs, NJ.

Widrow, B., J. R. Glover, J. M. McCool, J. Kaunitz, C. S. Williams, R. H. Hearn, J. R. Zeidler, E. Dong, and R. C. Goodlin, 1975. "Adaptive Noise Canceling: Principles and Applications," *Proceedings of the IEEE*, **63**(12), 1692–1716.

# 5

# TRANSDUCTION DEVICE DYNAMICS AND THE PHYSICAL SYSTEM

## 5.1 INTRODUCTION

One must always consider the effects of transduction device dynamics for successful controller design and implementation in real systems. A schematic diagram of the components required to control a dynamic system is presented in Figure 5.1. The control inputs are typically passed through some form of power amplifier to provide the necessary voltage and current required to drive the actuator transduction devices. The actuators are attached to the surface of the structure or embedded within the structure and used to convey a force, strain, or some other form of dynamic input. The structural response is typically measured with surface-mounted or embedded transduction devices, and the outputs of these devices are likely amplified and filtered by some signal conditioning unit. Thus, control system performance is determined by many different components and their effect on system bandwidth. This composite system, when coupled to a dynamic controller, constitutes the adaptive structure.

The mass, stiffness, and damping of the composite system must be determined. In the case of piezoceramic devices, one must account for the localized modified stiffness of the structural system as well as the localized modified mass. If accelerometers are placed on the surface of the structure, the modified mass of the system must be determined. In some cases, the mass associated with the transduction devices can be neglected, but one must be capable of assessing the validity of the assumption. For moving coil transduction devices such as shakers, the composite system response must be developed in terms of the shaker and structural dynamics. In addition to the dynamics of the transduction devices, filters and amplifiers all have electrical (pole-zero) character-

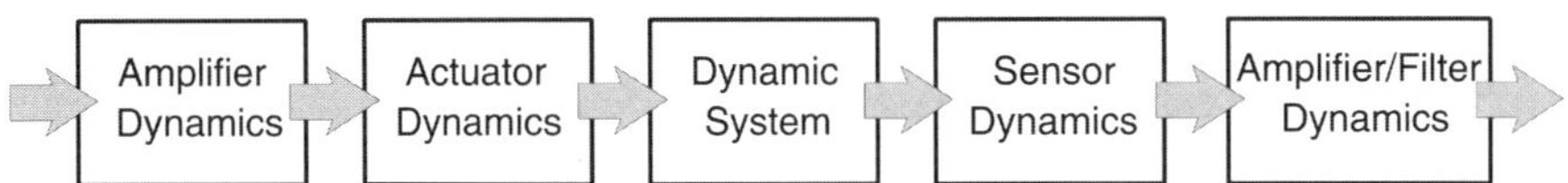


**Figure 5.1:** Schematic diagram of complete system required for control system design and implementation.

istics that affect the temporal response of the system and must be considered to assess stability and performance limitations. Thus, the dynamic response of every component from input to output must be modeled to accurately predict system response.

## 5.2 PRINCIPAL TYPES OF TRANSDUCTION DEVICES

For the purpose of this work, transduction devices have been divided into two categories: active materials and conventional devices. While each category of transduction device can be utilized to achieve an adaptive structure, active materials have been the focus in recent years, since they are readily integrated within the structural design and are analogous to the functional purpose served by muscles, tendons, and the sensory system of biological counterparts in a *very crude sense*. The active materials discussed in this section are piezoelectric devices, magnetostrictive devices, and shape memory alloys. Each of these materials can be loosely described as induced strain devices, capable of electromechanical transduction as an actuator, a sensor, or, in some cases, both. A host of other materials are available for design and implementation, and new materials will continue to evolve. However, the basic techniques of modeling the coupled transduction devices and structural system can be extended from the methods presented in this chapter. Under the category of conventional devices, electrodynamic shakers, strain gages, and accelerometers are discussed. The effect that each device has upon modeling the dynamic response of the structure to which it is coupled depends largely upon the relative mass, stiffness, and dynamics of the device itself. As might be expected, due to their large mass, electrodynamic shakers can have a significant impact on the dynamic response of the coupled structural system. Methods of developing coupled models are presented that demonstrate the importance of including transduction device dynamics before formulating a control law. The basic models presented in this chapter form the foundation for the various examples presented in Chapter 9, which is devoted to control system design and adaptive structure realization.

### 5.2.1 Active Materials

Active materials reviewed in this section include piezoelectric materials, magnetostrictive materials, and shape memory alloys. In general, the bandwidth of the respective materials can be characterized as high for the piezoelectric (well exceeds the bandwidth for structural and acoustic control applications), medium for the magnetostrictive (within the range of structural and accoustic applications), and low for the shape memory alloys (DC to a few hertz due to thermal time constants). Thus, the applications for each material are defined largely by their respective operating bandwidth.

***Piezoelectric Materials*** Materials that become electrically polarized when they are strained are said to display the direct piezoelectric effect, producing an electrical charge at the surface of the material. The converse piezoelectric effect results in a strain in the material when placed within an electric field. The direct and converse piezoelectric effects result in a mechanism for electromechanical transduction, and since the effects are linear over the range of application, a change in the sign associated with the strain or applied field will result in a change in sign of the resulting electrical charge or induced strain, respectively. As the atoms of the material are displaced under deformation, microscopic electric dipoles are formed (see Figure 5.2), and in certain materials, these dipoles yield an average macroscopic behavior: electrical polarization (Auld, 1990). The piezoelectric effect has been observed in a number of materials such as natural quartz crystals, polycrystalline piezoceramic, semicrystalline polyvinylidene polymer, and human bone. For the purpose of this work, the emphasis is placed upon transducers constructed from lead zirconium titanate (PZT) and polyvinylidene fluoride (PVDF).

Before elaborating on the differences between PZT and PVDF, it is appropriate to define the constitutive relations for piezoelectric materials. Stress and strain within material are related as follows:

$$\mathbf{T} = \mathbf{cS} \tag{5.1}$$

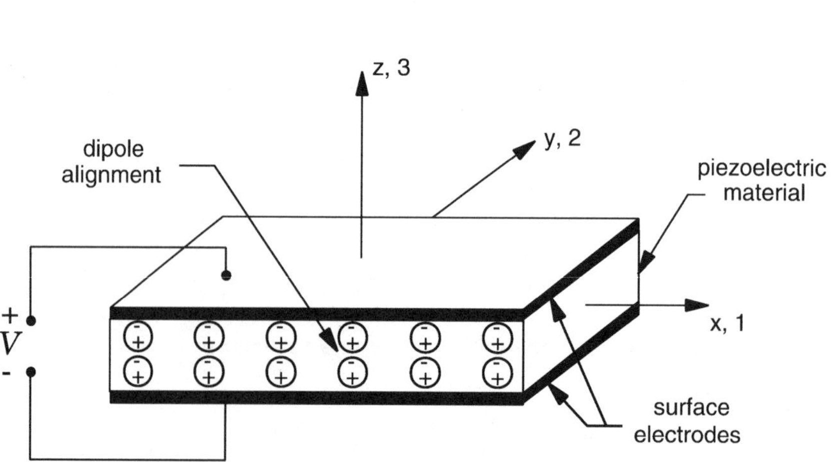


**Figure 5.2:** Schematic diagram of dipole effect induced in piezoelectric material.

where **T** is the vector of material stresses, **c** is the material stiffness matrix, and **S** is the vector of material strains. However, for piezoelectric materials, an additional consideration is required in association with the electrical field:

$$\mathbf{D} = \varepsilon^S \mathbf{E} + \mathbf{eS} \tag{5.2}$$

$$\mathbf{T} = -\mathbf{e}^{\mathrm{T}} \mathbf{E} + \mathbf{c}^{\mathrm{E}} \mathbf{S} \tag{5.3}$$

where **D** is a vector of electrical displacements (charge/area), $\varepsilon^S$ is a matrix of dielectric constants obtained at constant strain (permitivity matrix), **e** is a matrix of piezoelectric material constants relating the voltage to stress by the row and column of the matrix respectively, **E** is a vector of electric fields (volt/meter), and $\mathbf{c}^{\mathrm{E}}$ is the stiffness matrix measured at constant electric field. If each element of the matrix of piezoelectric material constants **e** is designated by $e_{ij}$ where $i$ corresponds to thc row and $j$ corresponds to the column of the matrix, then $e_{ij}$ corresponds to the stress developed in the $j$th direction due to an electric field applied in the $i$th direction. The coordinate system is illustrated in Figure 5.2. More often than not, the piezoelectric strain constants $d_{ij}$, relating the voltage applied in the $i$th direction to a strain developed in the $j$th direction, are provided, as opposed to the stress constants. However, the piezoelectric stress constants can be obtained from the strain constants as follows:

$$\mathbf{e} = \mathbf{dc}^{\mathrm{E}} \tag{5.4}$$

which makes physical sense since stress and strain are typically related through a stiffness matrix. Some material properties for PZT and PVDF are presented in Table 5.1. For greater detail on piezoelectricity or the constitutive relations, the reader is referred to the text by Auld (1990) or the paper by Hagood et al. (1990).

Notable differences exist between PVDF and PZT, guiding the designer in the selection of the one most appropriate for a chosen application. For example,

**TABLE 5.1 Piezoelectric Material Properties**

| | | Values | | |
|---|---|---|---|---|
| Property | Symbols | PVDF | PZT | Units |
| Strain constant | $d_{31}$ | $23 \times 10^{-12}$ | $166 \times 10^{-12}$ | (m/V) |
| | $d_{32}$ | $3 \times 10^{-12}$ | $166 \times 10^{-12}$ | (m/V) |
| | $d_{33}$ | $-30 \times 10^{-12}$ | $360 \times 10^{-12}$ | (m/V) |
| Relative dielectric constant | $K_3$ | 12 | 1700 | |
| Young's modulus | $E_{11}$ | $2 \times 10^9$ | $6.3 \times 10^{10}$ | (N/m$^2$) |
| Density | $\rho$ | 1780 | 7600 | (kg/m$^3$) |

as indicated in Table 5.1, PZT is roughly 4 times as dense, 40 times stiffer, and has a relative permittivity 100 times as great as that of PVDF. Thus, PVDF is much more compliant and lightweight in comparison to PZT, which minimizes the effect of the material on the dynamic properties of a structure constructed from a material such as steel. These properties make this material attractive for sensing applications; however, due to the low relative permittivity, the dielectric constant of PVDF dictates the use of an impedance matching circuit for low-frequency sensing applications (Collins et al., 1990). If one were to decide which material to use as an actuator and which to use as a sensor based upon the piezoelectric strain constants, one would select PZT for actuation and PVDF for sensing applications. As indicated in Table 5.1, the piezoelectric strain constant ($d_{31}$) is roughly five times greater for PZT than PVDF. Thus, a higher value of piezoelectric strain constant implies a greater strain induced for an applied electric field. However, as indicated in the *Kynar Piezo Film Technical Manual* (1987), the dielectric strength of a 28-$\mu m$ thick sheet of PVDF is approximately 70 times greater than that of PZT, which permits the application of much larger electric fields (Collins et al., 1990). If it were not for the low compliance of PVDF in comparison to PZT, the material would make an ideal strain actuator, as the difference in piezoelectric strength provides an increase in strain by an order of magnitude.

A number of practical considerations also factor into the decision of which material to use as an actuator and which to use as a sensor. For example, PVDF can be purchased in large rolls (0.5 meters wide × 6 meters or longer) of material that is readily cut, shaped, and attached to the surface of a structure. However, PZT is manufactured, due to practical constraints, in relatively small elements that are fragile and must be adhered to a flat surface, unless of course the material is manufactured to meet a desired curvature. Secondly, as outlined by Collins et al., (1990), the piezoelectric activity of PVDF is severely degraded in the presence of temperatures exceeding 80°C, which must be considered if the material is to be embedded in composite structures with curing temperatures exceeding this level. In general, the designer must compare the constituent material properties of the structure to that of the active materials and determine the range of actuation desired to decide which material is most appropriate. Limitations on electric fields imposed by the amount of power available to drive the actuator must also be considered in the design process. The choice of material for actuation and sensing is thus dependent upon the application, which reinforces the need for a basic understanding of the physics of the dynamic system before it can be controlled.

***Magnetostrictive Materials*** Magnetostrictive materials, like piezoceramic materials, are induced strain devices; however, transduction occurs through the application of a magnetic field as opposed to an electric field. A schematic diagram of a typical magnetostrictive transducer is depicted in Figure 5.3. As illustrated, the active material, composed of a terbium, dysprosium, iron alloy, is housed in a casing, and a preload is applied to the material. The housing is

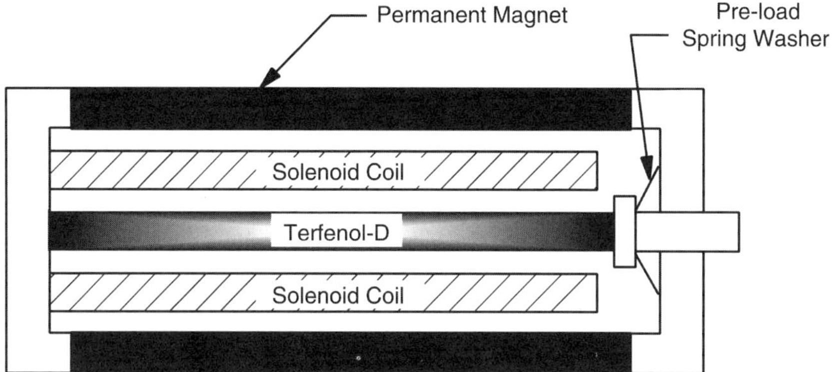


**Figure 5.3:** Schematic diagram of a Terfenol-D actuator.

typically composed of a magnetic casing and contains a wire-wound solenoid, as illustrated in Figure 5.3. Terfenol-D is the trade name commonly used, with "ter" from terfenol, "fe" from iron, "nol" from the Naval Ordinance Laboratory (where it was developed), and "D" from dysprosium. The material is formed such that the magnetic domains are perpendicular to the axis of actuation in the absence of an applied field, and upon applying a prestress, the alignment is refined. As illustrated, the prestress is commonly achieved with a spring washer. Upon inducing a current in the wound solenoid, a strain is induced within the material. In the control of structures, the transducer must "push" and "pull." Thus, to obtain a transducer capable of bidirectional operation, one must apply a biasing field. A DC current applied to the coil in the presence of the permanent magnetic facilitates this objective. As illustrated in Figure 5.4, the strain

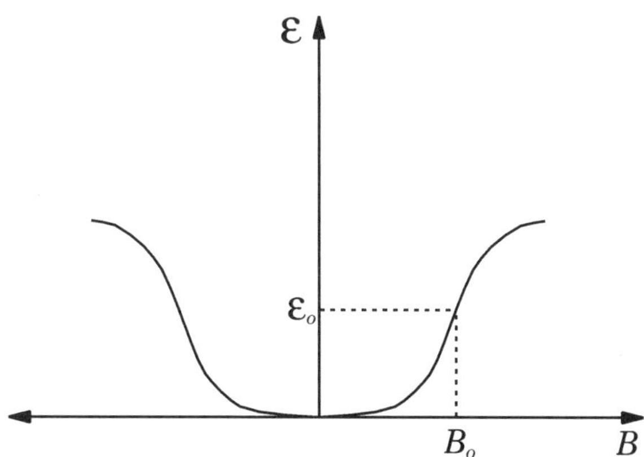


**Figure 5.4:** Schematic diagram of strain versus magnetic field for a magnetostrictive material.

distribution varies as a function of applied magnetic field. The biasing field, with flux density $B_o$, is necessary for bidirectional operation, and the induced strain results from a variation in the flux density about this biased field.

While in the discussion of piezoceramic materials, we noted a significant range of fields over which the transducer response can be assumed linear, such is not the case for magnetostrictive materials, as outlined by Hall and Flatau (1993). In general, the constitutive relations used in modeling such transducers are nonlinear, and the resulting response of a typical transducer to an applied periodic input will result in a dominant oscillatory strain at the period of the input and higher harmonics of that input frequency. Methods of modeling such transducers can be found in the reference by Hall and Flatau (1993) and are not reviewed further herein due to the inherent nonlinearities associated with the transduction device. However, despite the nonlinearities and overhead associated with the size of the transducers, required housing, prestress, and DC current, these transducers find applications in the control of structures and systems requiring large strains or applied forces. However, once housed and prestressed, the transducer is far from fragile, a definite advantage for operation under harsh conditions. For greater details of applications and modeling, the interested reader is directed to the references by Hall and Flatau (1993), Flatau et al. (1993), and Pratt and Flatau (1995).

***Shape Memory Alloys*** Shape memory alloys (SMA) are materials that are characterized by internal crystalline transformations from austenite to martensite, allowing the material to induce large strains (on the order of 10%) upon heating. A contemporary application of the material worth mentioning, due to its novelty, is for the construction of frames used in eyeglasses, the advantage being that the material is capable of recovering from large strains or deformations experienced in everyday use! The most commonly employed material is Nitinol, which is composed of nickel (Ni) and titanium (Ti), and it, like Terfenol-D, was developed at the Naval Ordinance Laboratory (nol) by Buehler and Wiley (1965). It should be noted that, since the internal crystalline transformation occurs as the result of changes in temperature (thermal), the bandwidth of such transducers is rather limited for dynamic control applications. However, with respect to adaptive structures, the transducers have much potential for shape control applications or for globally modifying the effective stiffness of a composite structure. Practically, the material is commonly used in applications for coupling pipes whereby the residual strain in the SMA is used to form a seal between two interconnected pipes (Banks et al., 1996).

The term "memory" is derived from the fact that the alloy is capable of memorizing its original configuration after it has been heated above the characteristic transition temperature; hence the term shape memory effect (SME). A martensitic transformation, named after the German scientist Adolf Martens (Liang and Rogers, 1991), provides the basis for all SMAs. The martensitic transformation is a phase transformation that is accompanied by a shape change. This transformation was first observed in steel after quenching the material while in a high-temperature austenitic phase. The martensitic transformation of SMAs is similar to that of steel, but the transition temperature for the phase transformation is much lower than that of steel. The martensitic fraction is plotted as a function of temperature in Figure 5.5. As illustrated, there are four temperatures to note: martensite finish ($M_f$), martensite start ($M_s$), austenite start ($A_s$), and austenite finish ($A_f$). Upon cooling an SMA in the austenite phase, one reaches the martensite start temperature, $M_s$, at which point the phase transformation begins, and upon complete cooling the martensite finish temperature $M_f$ is reached and the material is in the martensite form. Upon heating the material, the austenite phase transition occurs as illustrated in Figure 5.5. If a stress-induced phase transformation is introduced for which the temperature of the material is less than that of the austenite start $A_s$, the residual strain that remains in the material can be recovered upon heating the material past the austenite finish temperature $A_f$. The recovery of the residual strain is the characteristic SME for which the material was named. Applications for this material continue to evolve and it has played an important role in the field of adaptive structures; however, for the purpose of this text, which is devoted to the dynamic control of structural vibration and acoustic radiation, detailed applications are not explored. For greater details of the SME and SMAs, a number of references are available (Funakubo, 1986; Rogers, 1988, 1990; Liang and Rogers, 1990, 1991).

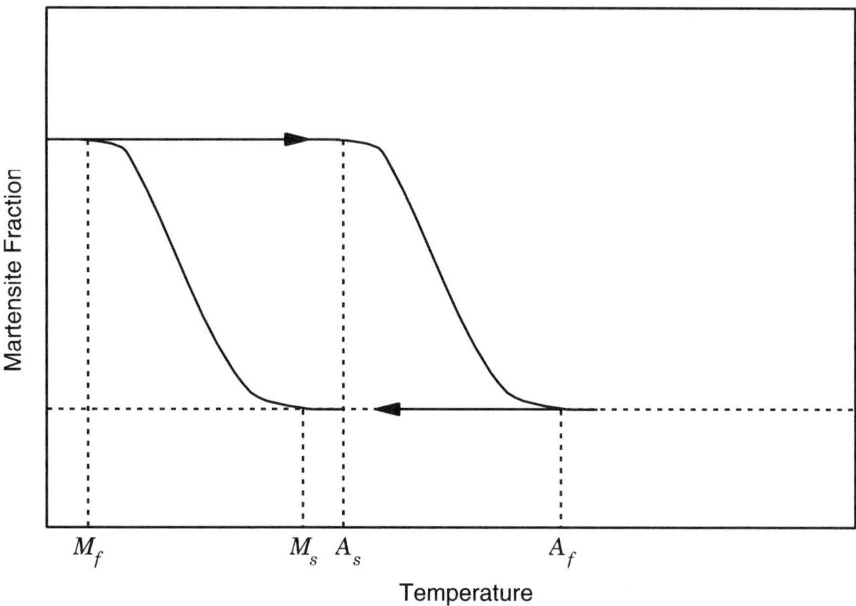


**Figure 5.5:** Schematic diagram of martensite-austenite transformation of a shape memory alloy (after Liang and Rogers, 1991).

***Other Active Materials and Transducers*** There are a number of other materials and transducers that have been used in the field of adaptive structures, including the more general electrostrictive materials (piezoelectric materials are a type of electostrictive material) and optical fiber sensors. Since piezoelectric devices are the more commonly explored of the electrostrictive materials, the discussion herein is limited to piezoceramics. With respect to optical fiber sensors, these transduction devices have generated much interest since they are readily embedded within a composite matrix, do not introduce electromagnetic interference within the structure, and are low-maintenance devices. Preliminary studies have demonstrated potential application as sensors in the control of structural vibration (Murphy et el., 1992). However, since the optical fiber is not used as an actuator in vibration control applications, the authors have chosen not to emphasize the implementation of the transducer in this book, which by no means reflects upon the importance and utility of this device in the emergence of adaptive structures. For an overview of electrostrictive materials and optical fibers, the reader is referred to the manuscript by Banks et al. (1996).

### 5.2.2 Conventional Devices

***Electrodynamic Shakers*** The electrodynamic shaker is one of the most commonly used devices in modal analysis and laboratory experiments. The shaker, like an acoustic loudspeaker, is composed of a permanent magnet, a moving coil through which a current is passed, and a spider (shaker suspension). The method of operation is based upon the principle of induction. Schematic diagrams of a moving coil shaker are depicted in Figure 5.6. Upon applying a voltage across the input terminals, a current is induced in the coil, which forces it to move. The induced force can be computed as follows:

$$F(t) = Bli(t) \tag{5.5}$$

where $B$ is the field strength in tesla, and $l$ is the length of the coil. Once the coil is in motion, a voltage is induced between the ends of the conductor, given by

$$e_{\text{emf}}(t) = Bl\dot{x}(t) \tag{5.6}$$

where $\dot{x}(t)$ is the velocity of the moving coil. The mechanical system can be described as a second-order system when the housing of the shaker is rigidly fixed. A detailed model of the shaker is developed in Section 5.3.

***Strain Gages*** The etched-foil, electrical resistance strain gage is the most commonly employed strain gage in industry, due to its stability, accuracy, small size, bandwidth of operation, relatively low cost, ease of installation, and linear response. A schematic diagram of the device is presented in Figure 5.7. A strain

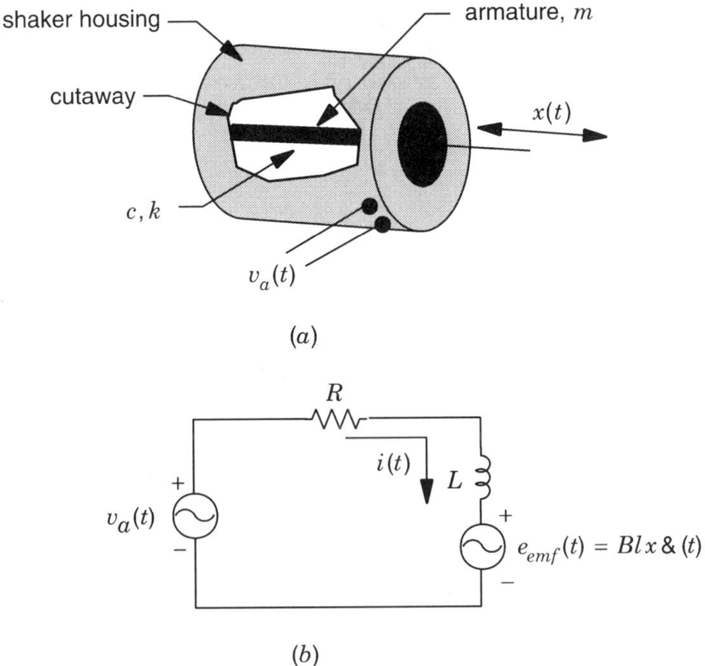


**Figure 5.6:** Schematic diagrams of an electrodynamic shaker. (*a*) Mechanical schematic. (*b*) Electrical schematic.

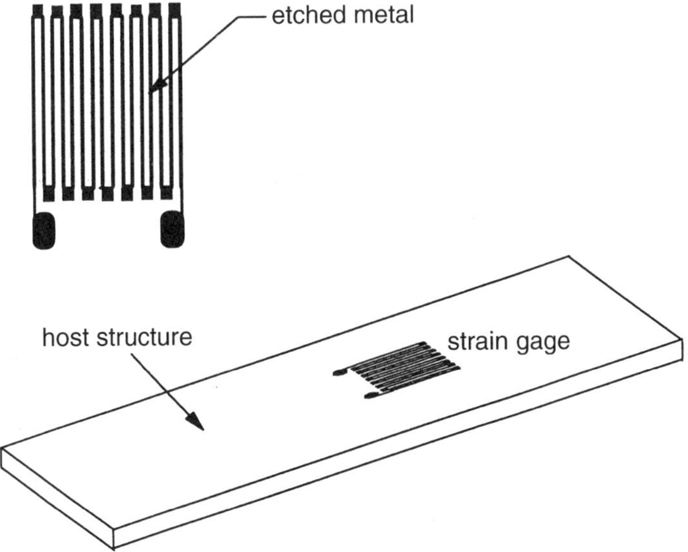


**Figure 5.7:** Schematic diagram of an etched-metal resistive strain gage.

gage is typically bonded to the surface of a structure such that the surface strain can be transmitted to the transducer. As the material is strained, the resistance of the transducer is modified in proportion to the strain. This relationship can be obtained from the expression describing the electrical resistance of a uniform metallic conductor:

$$R = \frac{\rho_R L}{A} \tag{5.7}$$

where $\rho_R$ is the resistivity of the material, $L$ is the length, and $A$ is the cross-sectional area of the conductor. Differentiating the expression for resistance and normalizing with respect to the resistance yields

$$\frac{dR}{R} = \frac{d\rho_R}{\rho_R} + \frac{dL}{L} - \frac{dA}{A} \tag{5.8}$$

Due to the Poisson effect,

$$\frac{dA}{A} \approx -2\nu \frac{dL}{L} = -2\nu\epsilon_a \tag{5.9}$$

where $\nu$ is the Poisson ratio of the material, and $\epsilon_a = dL/L$ (the axial strain in the conductor). Substituting appropriately into equation (5.8) and dividing by the axial strain, one obtains an expression for the sensitivity:

$$\frac{dR/R}{\epsilon_a} = \frac{d\rho_R/\rho_R}{\epsilon_a} + (1 + 2\nu) \tag{5.10}$$

When configured electrically in one arm of a Wheatstone bridge, upon applying a voltage across the diagonal of the bridge as illustrated in Figure 5.8, the output voltage measured across the other diagonal of the bridge is proportional to the change in resistance associated with the strain gage, and thus the voltage is as follows:

$$V_{\text{out}}(t) = \tfrac{1}{4}\, V_{\text{DC}} S_g \epsilon_g(t) \tag{5.11}$$

where $S_g$ is the sensitivity of the strain gage, $V_{\text{DC}}$ is the applied DC voltage across the bridge, and $\epsilon_g(t)$ is the time dependent strain as measured by the gage. More complex bridge configurations and signal conditioning can be employed; however, the basic concepts are evident in the previous development. For a more thorough treatment, the interested reader should consult the reference by Dally et al. (1993).

One important point to be made with respect to this transducer is that the

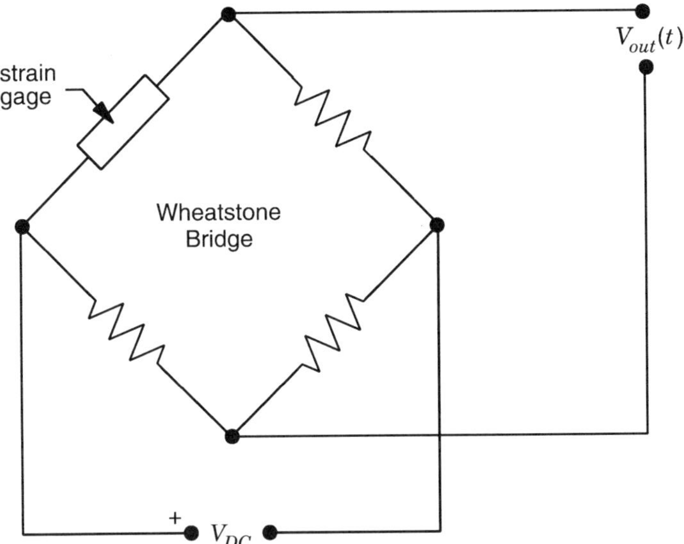


**Figure 5.8:** Schematic diagram of Wheatstone bridge used to obtain the electrical response of the strain gage.

dynamic effect of the device on the structure is negligible for most applications and thus does not need to be modeled in the control system design. Signal conditioning circuitry such as filters and amplifiers required to implement the device, must, however, be included in the final model of the plant. The transducer shape, location, and physical mode of operation all play an important role in the measured output, as will be detailed in Chapter 6.

***Accelerometers*** Accelerometers can be described as seismic acceleration transducers, and are perhaps one of the most commonly employed transducers in the field of vibration analysis and control. As illustrated in Figure 5.9, a single-ended compression accelerometer consists of a seismic mass, attached to the base through a piezoelectric material. (Note that there is a shear force accelerometer design as well, which is less sensitive to bending at the base than the single-ended compression accelerometer.) For all practical purposes, the piezoelectric material serves as a very stiff spring. Thus, when the transducer vibrates, inertia forces result (due to the seismic mass), causing the material to strain, and this strain can be measured through the electromechanical coupling of the material. As long as the attached mass and the body of the transducer move in phase, the output of the transducer is proportional to acceleration. However, if the two bodies move out of phase, indicating that the transducer is being driven above its first mechanical resonance, the output is no longer proportional to acceleration. However, in practice, the fundamental resonance is typically an order of magnitude outside the desired bandwidth of interest.

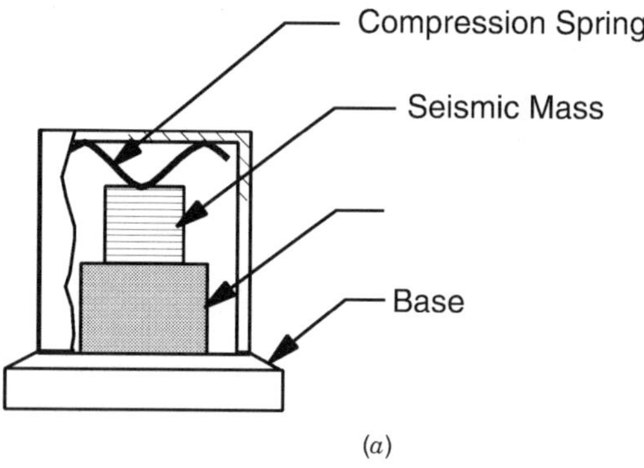


(a)

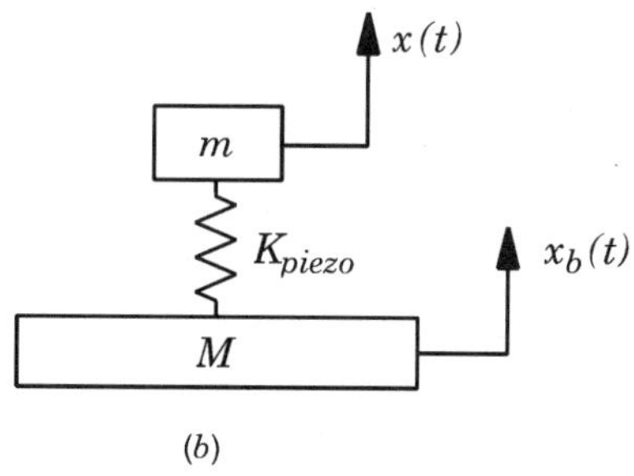


(b)

**Figure 5.9:** (a) Schematic diagram of a single-ended compression accelerometer. (b) Simplified schematic diagram of the same accelerometer.

From the simplified schematic diagram of the single-ended compression accelerometer depicted in Figure 5.9, the equations of motion subject to base motion can be obtained:

$$m\ddot{x}(t) + kx(t) = kx_b(t) \tag{5.12}$$

Redefining the response in relative coordinates such that $y(t) = x(t) - x_b(t)$, one obtains

$$m\ddot{y}(t) + ky(t) = -m\ddot{x}_b(t) \tag{5.13}$$

Thus, the equivalent force presented to the system is the inertia force associated with the acceleration of the base. Applying standard Laplace transform techniques, the system transfer function can be obtained:

$$\frac{y(s)}{x(s)} = -\frac{s^2}{s^2 + \omega_n^2} \tag{5.14}$$

Letting $s = j\omega$, the frequency response function (FRF) is obtained:

$$\frac{y(j\omega)}{x(j\omega)} = \frac{\omega^2}{\omega_n^2 - \omega^2} \tag{5.15}$$

For $\omega << \omega_n$, the system response is proportional to acceleration:

$$y(j\omega) = \frac{\omega^2 x(j\omega)}{\omega_n^2} \tag{5.16}$$

Thus, a very small mass and stiff piezoelectric spring are required to assure that the bandwidth of the transducer far exceeds the bandwidth of the measurement. However, the "spring" cannot be so stiff and the mass so small that the sensitivity of the device is poor, indicating the trade-off associated with the selection of the appropriate transducer. In general, one selects the transducer with the greatest sensitivity and the one that has the least effect on the dynamic response of the system to which it is attached. For a detailed discussion of accelerometers and the effects of attaching such devices to structures, the reference by Ewins (1986) is recommended.

## 5.3 STATE-VARIABLE MODELS OF COUPLED SYSTEM DYNAMICS

Active materials have served as one of the critical components in the recent evolution of adaptive structures. Of all the active materials available, the piezoelectric devices constructed from PZT have been studied and applied to the control of structure-borne sound and vibration more than any other electromechanical material. The majority of the examples presented in this text revolve around the use of this particular material; however, one should recognize that the following development of the model for a piezostructure applies to all induced strain materials in general. Thus, one must simply modify the constituent properties of the materials in the models to vary the transduction devices. To provide alternative examples, the dynamic coupling between moving coil actuators and the structure is presented in addition to the modifications in the structural response due to added masses such as that of accelerometers. We begin the development with the piezostructure.

### 5.3.1 The Piezostructure

A *piezostructure* is defined to be a structure consisting of embedded or surface mounted piezoelectric devices. The dynamic model of the piezostructure includes both electrical inputs and outputs as well as the modified mass and stiffness effects of the structural system due to the additional piezoelectric devices. Hagood et al. (1990) presented the first rigorous development of this model, and the development within this section is consistent with that approach. However, we differ slightly in that only a single component of force and displacement are included in the development to facilitate interpretation, as opposed to the more general vector approach developed by Hagood et al. (1990). Thus, the approach presented in this section can be used to develop models of beams and plates undergoing flexural displacements. Having stated the assumptions associated with the approach presented, the objective is to develop the following electroelastic model (ignoring damping for now) of the piezostructure (Hagood et al., 1990):

$$[\mathbf{M}_s + \mathbf{M}_p]\ddot{\mathbf{r}} + [\mathbf{K}_s + \mathbf{K}_p]\mathbf{r} - \mathbf{\Theta v} = \mathbf{B}_f \mathbf{f} \tag{5.17}$$

$$\mathbf{q} = \mathbf{\Theta}^{\mathrm{T}}\mathbf{r} + \mathbf{C}_p \mathbf{v} \tag{5.18}$$

where $\mathbf{M}_s$ and $\mathbf{M}_p$ are $N \times N$ mass matrices associated with the structure and piezoceramic, respectively; $\mathbf{K}_s$ and $\mathbf{K}_p$ are $N \times N$ stiffness matrices associated with the structure and piezoceramic, respectively; $\mathbf{r}$ is an $N \times 1$ vector of generalized mechanical coordinates; $\mathbf{\Theta}$ is an $N \times M$ electromechanical coupling matrix; $\mathbf{v}$ is an $M \times 1$ vector of generalized electrical coordinates; $\mathbf{B}_f$ is an $N \times L$ matrix of influence functions associated with the $L \times 1$ vector of generalized forces $\mathbf{f}$; $\mathbf{q}$ is an $M \times 1$ vector of applied electrode charges; and $\mathbf{C}_p$ is an $M \times M$ diagonal matrix of piezoelectric capacitances. (Note that the assumed modes approach or a Rayleigh Ritz formulation of the system can be developed to derive the matrices herein, as detailed in Chapter 2.)

The equations of motion describing the electroelastic response of the system can be derived from a modified Rayleigh-Ritz approach (Meirovitch, 1967), which consists of assuming a solution that is a linear combination of admissible functions. (One should also recognize that the assumed-modes method could be used here to obtain the same result.) A generalized form of Hamilton's principle was derived by Crandall et al. (1968) for electromechanical systems:

$$\int_{t_1}^{t_2} \left[\delta(T - V + W_e - W_m) + \delta W\right] dt = 0 \tag{5.19}$$

where $W_m$ is the virtual work due to magnetic terms (negligible for piezoceramics), $T$ is the kinetic energy of the system, $V$ is the potential energy of the system, $W_e$ is the electrical energy of the system, and $W$ represents all other nonconservative work. Each term for the piezostructure can be expressed as follows:

$$T = \frac{1}{2} \int_{V_s} \rho_s \dot{\mathbf{w}}^{\mathrm{T}}(\mathbf{x})\dot{\mathbf{w}}(\mathbf{x}) + \frac{1}{2} \int_{V_{\mathbf{p}}} \rho_{\mathbf{p}} \dot{\mathbf{w}}^{\mathrm{T}}(\mathbf{x})\dot{\mathbf{w}}(\mathbf{x}) \tag{5.20}$$

$$V = \frac{1}{2} \int_{V_s} \mathbf{S}^{\mathrm{T}}(\mathbf{x})\mathbf{T}(\mathbf{x}) + \frac{1}{2} \int_{V_p} \mathbf{S}^{\mathrm{T}}(\mathbf{x})\mathbf{T}(\mathbf{x}) \tag{5.21}$$

$$W_e = \frac{1}{2} \int_{V_p} \mathbf{E}^{\mathrm{T}}(\mathbf{x})\mathbf{D}(\mathbf{x}) \tag{5.22}$$

and

$$\delta W = \sum_{i=1}^{nf} f(\mathbf{x}_i)\delta w(\mathbf{x}_i) + \sum_{j=1}^{nq} q_j \, \delta\varphi(\mathbf{x}_j) \tag{5.23}$$

where $\mathbf{D}(x)$ and $\mathbf{E}(x)$ are vectors of electrical displacements (charge/area) and electrical field (potential/length), respectively, $\mathbf{S}(x)$ and $\mathbf{T}(x)$ are the vectors of material strains (length/length) and stresses (force/area), respectively, $w(\mathbf{x}_i)$ and $f(\mathbf{x}_i)$ are the mechanical displacements and applied forces normal to the surface at field point $\mathbf{x}_i$, respectively; and $q_j$ and $\varphi(\mathbf{x}_j)$ are the applied charge and scalar electrical potential at the $j$th electrode. Based upon this formulation, the equations of motion can be derived for the piezostructure. Rather than repeat the full derivation, the reader is referred to the article by Hagood et al. (1990). However, here we develop each parameter defined in equations (5.17) and (5.18).

First of all, the response of the structure is defined in physical coordinates as a series expansion (vector notation) over the generalized coordinates:

$$w(\mathbf{x}, t) = \mathbf{\Phi}_r(\mathbf{x})\mathbf{r}(t) \tag{5.24}$$

where $\mathbf{\Phi}_r(\mathbf{x})$ is a $1 \times N$ vector of assumed displacement distributions (mode shapes), and $\mathbf{r}(t)$ is an $N \times 1$ vector of generalized displacements. Similarly, the electric potential can be expressed in generalized coordinates as follows:

$$\varphi(\mathbf{x}, t) = \mathbf{\Phi}_v(\mathbf{x})\mathbf{v}(t) \tag{5.25}$$

where $\mathbf{\Phi}_v(\mathbf{x})$ is an $1 \times P$ vector of assumed potential distributions, and $\mathbf{v}(t)$ is a $P \times 1$ vector of generalized electrical coordinates. The displacement and potential distributions must be differentiable to the order of the elastic differential operator used to obtain the material strains and the electrical differential operator used to compute the electric fields, respectively.

Now, we proceed by defining the mass matrices:

$$\mathbf{M}_s = \int_{V_s} \mathbf{\Phi}_r^{\mathrm{T}}(\mathbf{x})\rho_s(\mathbf{x})\mathbf{\Phi}_r(\mathbf{x})d\,V_s(\mathbf{x}) \tag{5.26}$$

and

$$\mathbf{M}_p = \int_{V_p} \mathbf{\Phi}_r^{\mathrm{T}}(\mathbf{x})\rho_p(\mathbf{x})\mathbf{\Phi}_r(\mathbf{x})d\,V_p(\mathbf{x}) \tag{5.27}$$

where $\rho_s(\mathbf{x})$ and $\rho_p(\mathbf{x})$ are the densities of the structure and piezoelectric device, respectively. Note that the mass contribution due to the piezoelectric devices is computed over the volume of all elements deposited on the structure in the global coordinates of the structure ($\mathbf{x}$), determining the spatial aperture of each device. Thus, one computes the mass matrix of the structure and piezoelectric device with respect to the assumed displacement functions in the structural coordinate system.

The stiffness matrix associated with the structure is defined as follows:

$$\mathbf{K}_s = \int_{V_s} [L_w\Phi_r(x)]^{\mathrm{T}} c_s [L_w\Phi_r(x)]d\,V_s(x), \tag{5.28}$$

where $\mathbf{L}_w$ is the elastic differential operator, which for a rectangular plate in flexure can be expressed as follows:

$$\mathbf{L}_w^{\mathrm{T}} = \left[ -z\frac{\partial^2}{\partial x^2} \quad -z\frac{\partial^2}{\partial y^2} \quad -2z\frac{\partial^2}{\partial x \partial y} \right] \tag{5.29}$$

and $\mathbf{c}_s$ is the structural stiffness matrix:

$$\mathbf{c}_s = \begin{bmatrix} \dfrac{E_s}{1-\nu_s^2} & \dfrac{E_s\nu_s}{1-\nu_s^2} & 0 \\ \dfrac{E_s\nu_s}{1-\nu_s^2} & \dfrac{E_s}{1-\nu_s^2} & 0 \\ 0 & 0 & \dfrac{E_s}{2(1-\nu_s^2)} \end{bmatrix} \tag{5.30}$$

$E_s$ and $\nu_s$ are the Young's modulus and Poisson's ratio of the structural material, respectively.

The stiffness matrix for the piezoelectric device can be constructed as follows:

$$\mathbf{K}_p = \int_{V_p} [L_w\Phi_r(x)]^T R_S^T c^E R_S [L_w\Phi_r(x)]d\,V_p(x), \tag{5.31}$$

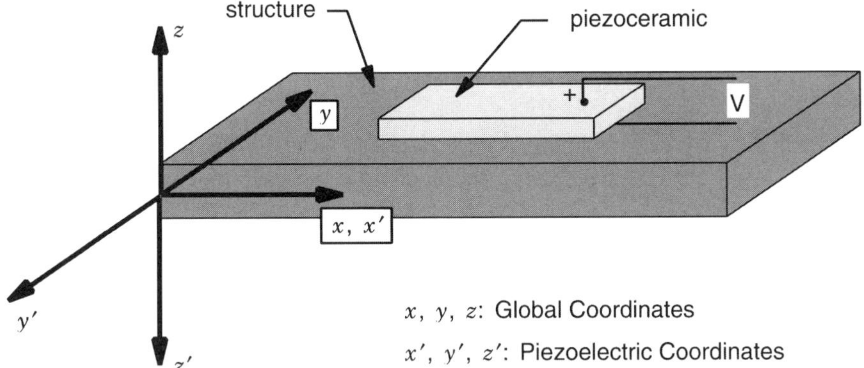


**Figure 5.10:** Schematic diagram of a piezostructure.

where the strain rotation matrix $\mathbf{R}_S$ is defined as follows for the coordinate system depicted in Figure 5.10:

$$\mathbf{R}_S = \begin{bmatrix} 1 & 0 & 0 \\ 0 & -1 & 0 \\ 0 & 0 & -1 \end{bmatrix} \tag{5.32}$$

and the short circuit stiffness matrix of the piezoelectric material is defined as follows:

$$\mathbf{c}^E = \begin{bmatrix} \dfrac{E_p}{1-\nu_p^2} & \dfrac{E_p \nu_p}{1-\nu_p^2} & 0 \\ \dfrac{E_p \nu_p}{1-\nu_p^2} & \dfrac{E_p}{1-\nu_p^2} & 0 \\ 0 & 0 & \dfrac{E_p}{2(1-\nu_p)} \end{bmatrix} \tag{5.33}$$

where $E_p$ and $\nu_p$ are the Young's modulus and Poisson's ratio of the piezoelectric material. Notice that the contribution to the stiffness associated with all piezoelectric devices is computed over the volume of the distributed elements with respect to the coordinate system of the structure (the spatial aperture). In general, each piezoelectric device represents a discontinuity in mass and stiffness over the structural surface in the region of application. Note that the rotation matrix $R_s$ is dependent upon the respective problem.

The electromechanical coupling matrix between the piezoelectric device and the structure can be expressed as follows:

$$\boldsymbol{\Theta} = \int_{V_p} [L_w \Phi_r(x)]^T R_s^T e^T [R_E L_\varphi \Phi_v(x)] d V_p(x) \tag{5.34}$$

Notice that the "mechanical" and "electrical" portions of the integral have been bracketed and are coupled by $\mathbf{e}^{\mathrm{T}}$, the matrix of constants that relate voltage to stress. The electrical field rotation matrix $\mathbf{R}_E$ in equation (5.34) is defined as follows for the coordinate system presented in Figure 5.10:

$$\mathbf{R}_E = \begin{bmatrix} 1 & 0 & 0 \\ 0 & -1 & 0 \\ 0 & 0 & -1 \end{bmatrix} \tag{5.35}$$

and the electrical differential operator can be expressed as follows for polarization in the $z$-direction only:

$$\mathbf{L}_\varphi^{\mathrm{T}} = \begin{bmatrix} 0 & 0 & -\dfrac{\partial}{\partial z} \end{bmatrix} \tag{5.36}$$

The matrix of piezoelectric material constants $\mathbf{e}$ relating voltage applied in the $i$th direction to stress in the $j$th direction (where $i$ corresponds to the row of $\mathbf{e}$ and $j$ corresponds to the column) can be obtained from the matrix of piezoelectric strain constants $\mathbf{d}$ and the short circuit stiffness matrix of the piezoelectric device $\mathbf{c}^E$:

$$\mathbf{e} = \mathbf{d}\mathbf{c}^E \tag{5.37}$$

For the coordinate system defined in Figure 5.10, the matrix of piezoelectric strain constants can be expressed as follows:

$$\mathbf{d} = \begin{bmatrix} 0 & 0 & 0 \\ 0 & 0 & 0 \\ d_{31} & d_{31} & 0 \end{bmatrix} \tag{5.38}$$

Recall that the short circuit stiffness of the piezoceramic device was defined in equation (5.33). The matrix $\mathbf{B}_f$ of influence functions associated with the generalized forces can be computed by evaluating each assumed displacement distribution at the coordinate of the point force due to the spatial aperture associated with a point force: $\delta(\mathbf{x} - \mathbf{x}_i)$. Thus, the $N \times L$ matrix is:

$$\mathbf{B}_f = [\boldsymbol{\Phi}_r^T(\mathbf{x}_{f1}) \quad \ldots \quad \boldsymbol{\Phi}_r^T(\mathbf{x}_{fL})] \tag{5.39}$$

which is simply the displacement distributions (assumed modes) evaluated at the coordinate of each point force. Finally, we must determine the diagonal matrix $\mathbf{C}_p$ of capacitance associated with each piezoelectric device. The capacitance matrix can be computed as follows:

$$\mathbf{C}_p = \int_{V_p} \mathbf{\Phi}_v^{\mathrm{T}}(\mathbf{x})[\mathbf{L}_\varphi^{\mathrm{T}}\mathbf{R}_E^{\mathrm{T}}\varepsilon^S\mathbf{R}_E\mathbf{L}_\varphi]\mathbf{\Phi}_v(\mathbf{x})d\,V_p(\mathbf{x}) \tag{5.40}$$

where the matrix of dielectric constants measured at constant strain can be expressed as follows:

$$\varepsilon^S = \begin{bmatrix} \varepsilon_1^S & 0 & 0 \\ 0 & \varepsilon_1^S & 0 \\ 0 & 0 & \varepsilon_1^S \end{bmatrix} \tag{5.41}$$

In the development presented, we have assumed that a voltage is applied across the piezoelectric device in one dimension only, as illustrated in Figure 5.10 Thus, the integral formulation of equation (5.40) reduces to the following:

$$\mathbf{C}_p = \mathrm{diag}\left[\begin{array}{cccc} \dfrac{\varepsilon_1^S A_{p1}}{h_{p1}} & \dfrac{\varepsilon_1^S A_{p2}}{h_{p2}} & \cdots & \dfrac{\varepsilon_1^S A_{pM}}{h_{pM}} \end{array}\right] \tag{5.42}$$

where $A_{p_m}$ and $h_{p_m}$ are the area and thickness of the $m$th piezoelectric device, respectively. If one chooses to polarize the piezoelectric device in an alternative manner to that presented in Figure 5.10 (i.e., different than the thickness mode) or implement vector forces as opposed to normal forces, see the reference by Hagood et al. (1990).

Now, upon defining all variables, we can cast equations (5.17) and (5.18) in state-variable form:

$$\dot{\mathbf{x}} = \mathbf{Ax} + \mathbf{Bu} \tag{5.43}$$

$$\mathbf{y} = \mathbf{Cx} + \mathbf{Du} \tag{5.44}$$

where

$$\mathbf{A} = \begin{bmatrix} \mathbf{0}_{N\times N} & \mathbf{I}_{N\times N} \\ -[\mathbf{M}_s + \mathbf{M}_p]^{-1}[\mathbf{K}_s + \mathbf{K}_p] & \mathbf{0}_{N\times N} \end{bmatrix} \tag{5.45}$$

$$\mathbf{x}^{\mathrm{T}} = [\mathbf{r}^{\mathrm{T}} \quad \dot{\mathbf{r}}^{\mathrm{T}}] \tag{5.46}$$

$$\mathbf{B} = \begin{bmatrix} \mathbf{0}_{N\times M} & \mathbf{0}_{N\times L} \\ [\mathbf{M}_s + \mathbf{M}_p]^{-1}\boldsymbol{\Theta} & [\mathbf{M}_s + \mathbf{M}_p]^{-1}\mathbf{B}_f \end{bmatrix} \tag{5.47}$$

$$\mathbf{u}^{\mathrm{T}} = [\mathbf{v} \quad \mathbf{f}] \tag{5.48}$$

$$\mathbf{y} = [\mathbf{q}] \tag{5.49}$$

$$\mathbf{C} = [\boldsymbol{\Theta}^{\mathrm{T}} \quad \mathbf{0}_{M\times N}] \tag{5.50}$$

$$\mathbf{D} = [\mathbf{C}_p \quad \mathbf{0}_{M\times L}] \tag{5.51}$$

As outlined above, the resulting charge in the piezostructure is observed; however, one can modify **C** and **D** to observe any set of variables desired.

**Example 5.1: Plate with Pinned Boundaries and Single Distributed Piezoceramic Element** Consider the schematic diagram of a plate presented in Figure 5.11. As illustrated, the plate is configured with two piezoceramic elements mounted symmetrically about the neutral axis. Thus, the spatial aperture of each element in the $x$ and $y$ directions is identical to the other, which effectively leads to a doubling of the mass and stiffness associated with either one of the distributed elements considered alone. The displacement distributions selected for this structure are the eigenfunctions of the homogeneous structure with pinned boundary conditions, as outlined by Meirovitch (1967), since they obviously satisfy the geometric boundary conditions. The vector of displacement distributions can be expressed as follows:

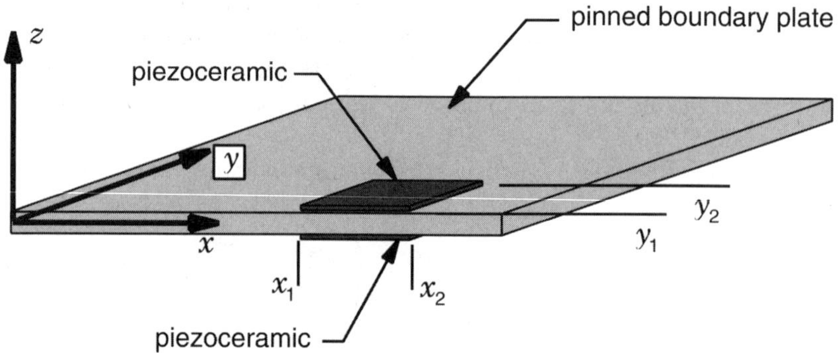


**Figure 5.11:** Schematic diagram of a plate with pinned boundaries configured with a piezoelectric device.

$$\mathbf{\Phi}_r(x, y) = [\phi_1(x, y) \quad \phi_2(x, y) \quad \dots \quad \phi_N(x, y)] \tag{5.52}$$

Substituting into equation (5.26), one obtains an expression for the entry of the $i$th row and $j$th column of the matrix:

$$M_{s_{ij}} = m_s'' \int_0^{L_x} \int_0^{L_y} \phi_i(x, y)\phi_j(x, y) dx\, dy \tag{5.53}$$

where $m_s''$ is the mass per unit area of the structure. Similarly, the mass contribution due to the piezoelectric device for the $i$th row and $j$th column of the mass matrix can be computed as follows:

$$M_{p_{ij}} = m_p'' \int_0^{L_x} \int_0^{L_y} \phi_i(x, y)\phi_j(x, y)[u(x - x_1) - u(x - x_2)] \times [u(y - y_1) - u(y - y_2)] dx\, dy \tag{5.54}$$

where $m_p''$ is the mass per unit area of the piezoceramic element, and $u(x)$ represents a spatial step that can be used to create the spatial aperture associated with the device. This corresponds to limiting the bounds of integration between $x_1$ and $x_2$ in the $x$-dimension and $y_1$ and $y_2$ in the $y$-dimension.

To compute the stiffness matrix, one must compute the stiffness associated with the structure by substituting equation (5.52) into equation (5.28):

$$K_{s_{ij}} = \frac{E_s I_s}{1 - \nu_s^2} \int_0^{L_x} \int_0^{L_y} \left\{ \frac{\partial^2 \phi_i}{\partial x^2} \frac{\partial^2 \phi_j}{\partial x^2} + \frac{\partial^2 \phi_i}{\partial y^2} \frac{\partial^2 \phi_j}{\partial y^2} + \nu_s \left( \frac{\partial^2 \phi_i}{\partial x^2} \frac{\partial^2 \phi_j}{\partial y^2} + \frac{\partial^2 \phi_i}{\partial y^2} \frac{\partial^2 \phi_j}{\partial x^2} \right) + 2(1 - \nu_s) \frac{\partial^2 \phi_i}{\partial x \partial y} \frac{\partial^2 \phi_j}{\partial x \partial y} \right\} dx\, dy \tag{5.55}$$

where $I_s = h_s^3/12$ is the area moment of inertia about the neutral axis, and $h_s$ is the thickness of the structure. Notice that for convenience, the specific dependence of the assumed displacement distributions $\phi_i(x, y)$ on the $x$ and $y$ coordinates was omitted for brevity. A similar expression can be developed for the relative stiffness contribution of the piezoceramic element:

$$K_{p_{ij}} = \frac{E_p I_p}{1-\nu_p^2} \int_0^{L_x} \int_0^{L_y} \left\{ \frac{\partial^2 \phi_i}{\partial x^2} \frac{\partial^2 \phi_j}{\partial x^2} + \frac{\partial^2 \phi_i}{\partial y^2} \frac{\partial^2 \phi_j}{\partial y^2} \right.$$
$$+ \nu_p \left( \frac{\partial^2 \phi_i}{\partial x^2} \frac{\partial^2 \phi_j}{\partial y^2} + \frac{\partial^2 \phi_i}{\partial y^2} \frac{\partial^2 \phi_j}{\partial x^2} \right) + 2(1-\nu_p) \left. \frac{\partial^2 \phi_i}{\partial x \partial y} \frac{\partial^2 \phi_j}{\partial x \partial y} \right\}$$
$$\times [u(x-x_1) - u(x-x_2)][u(y-y_1) - u(y-y_2)] dx\, dy \tag{5.56}$$

where $I_p = h_s^2 h_p/4 + h_s h_p^2/2 + h_p^3/3$ and $u(x)$ is the step function. One should also recognize that the expressions developed in equations (5.54) and (5.56) are for a single piezoceramic device. Thus for symmetrically mounted devices as illustrated in Figure 5.11, each entry for $M_{p_{ij}}$ and $K_{p_{ij}}$ must be doubled.

The capacitance of the device for this example is readily calculated:

$$C_p = \frac{\varepsilon_1^S A_p}{h_p} \tag{5.57}$$

For a parallel wiring configuration typical of symmetrically mounted devices as illustrated in Figure 5.11, the total capacitance is obtained by a parallel equivalent of that expressed in equation (5.57). Thus, the total capacitance used for the device configured in parallel must be doubled.

We proceed by computing the electromechanical coupling matrix for the system presented. Again, for a single piezoceramic element deposited over the surface of the structure as illustrated in Figure 5.11, the electromechanical coupling matrix (a vector for this example) is computed as follows:

$$\mathbf{\Theta} = \int_{x_1}^{x_2} \int_{y_1}^{y_2} \int_{h_s/2}^{(h_s/2)+h_p} \frac{d_{31} E_p}{1-\nu_p} \left\{ \begin{array}{c} \dfrac{\partial^2 \phi_1(x,y)}{\partial x^2} + \dfrac{\partial^2 \phi_1(x,y)}{\partial y^2} \\ \vdots \\ \dfrac{\partial^2 \phi_N(x,y)}{\partial x^2} + \dfrac{\partial^2 \phi_N(x,y)}{\partial y^2} \end{array} \right\}$$
$$\cdot z \frac{\partial \mathbf{\Phi}_v(z)}{\partial z}\, dz\, dy\, dx \tag{5.58}$$

where

$$\mathbf{\Phi}_v(z) = \frac{z - h_s/2}{h_p}, \qquad \text{for} \quad z > 0 \tag{5.59}$$

Notice that one essentially obtains a vector of influence functions in generalized coordinates, functions that couple the electrical response to the mechanical response for each element deposited over the structural surface. As indicated in equation (5.59), a linearly varying voltage profile across the thickness of the piezoceramic is assumed here, and this assumption is consistent with most practical implementations. Again, due to symmetry, the result obtained in equation (5.58) must be doubled. All that remains at this point is to replace the expressions for the eigenfunctions with specific sinusoidal functions and evaluate the integrals presented in this example. For the plate with pinned boundary conditions, the integrals can be evaluated using a set of integral tables. For more complex geometry, numerical integration is, in general, necessary. The evaluation of the integrals is left as an exercise for the reader.

### 5.3.2 Dynamic Actuation Devices and the Structure

A moving coil actuator is one example of a conventional device used to drive a structure. Dynamic shakers are typically constructed from moving coil motors and require external mounting to secure the device. While shakers are often employed in laboratories to perform modal tests, their use for adaptive structures applications is limited by the need for a rigid mount. As an alternative, one can utilize a proof-mass actuator. This device is mounted directly on the structural surface, and the dynamics of this device are typically approximated by a simple second-order spring-mass-damper system. A number of these devices are available commercially, such as the piezoelectric based device manufactured by AVC Incorporated. Whether utilizing the devices to control a structure or for simple tests in laboratory environments, one should recognize how to develop a coupled dynamic model of the structural system, including the dynamics of the actuator. In this section, an electromechanical model of an electrodynamic shaker is developed and coupled to the dynamics of a structure to provide an example.

Consider the schematic diagram of a moving coil shaker presented in Figure 5.6. Analogous to a loudspeaker, a moving coil is placed within the field of a permanent magnet. Upon applying a voltage across the coil, a current is induced. The induced current generates a force at right angles to the plane of the current $i$ and the magnetic field $B$. The process is known as induction, and the force can be expressed as follows:

$$F(t) = Bli(t) \tag{5.60}$$

where $B$ is the field strength in tesla, and $l$ is the length of the coil. However, once the coil begins to move, a voltage $e_{\text{emf}}(t)$ is generated in proportion to the speed of motion $\dot{x}$:

$$e_{\text{emf}}(t) = Bl\dot{x}(t) \tag{5.61}$$

This electromechanical result is commonly known as back emf (electromotive force). Thus, to model the dynamic response of the system, one must couple the electrical dynamics to the structural dynamics. The moving coil can be modeled as a spring-mass-damper system, and we assume that the housing for the permanent magnet is rigidly fixed. A simple first-order differential equation can be developed to describe the electrical circuit, and a second-order differential equation can be constructed for the mechanical system, each depicted in the schematic diagram of Figure 5.6. The equations are expressed as follows:

$$L\frac{di(t)}{dt} + Ri(t) = v_a(t) - Bl\dot{x}(t) \tag{5.62}$$

for the electrical circuit, and

$$m\ddot{x}(t) + c\dot{x}(t) + kx(t) = Bli(t) \tag{5.63}$$

for the mechanical system. Since the differential equations are coupled, the system response can be described as third-order. The dynamic equations describing the response of this transducer can be coupled to that of the structure to describe the fully coupled system.

Consider the beam with pinned boundary conditions illustrated in Figure 5.12. The shaker, which is attached at a point (spatial aperture $\delta(x - x_p)$), is assumed to be rigidly attached to a support structure. The assumed-modes method can be utilized to develop the equations of motion for the coupled dynamic system. The kinetic energy of the mechanical system can be expressed as follows:

$$T(t) = \frac{1}{2}\int_0^{L_x} m'\left[\frac{\partial w(x,t)}{\partial t}\right]^2 dx + \frac{1}{2}\int_0^{L_x} M_{mc}\left[\frac{\partial w(x,t)}{\partial t}\right]^2 \delta(x - x_p)\, dx \tag{5.64}$$

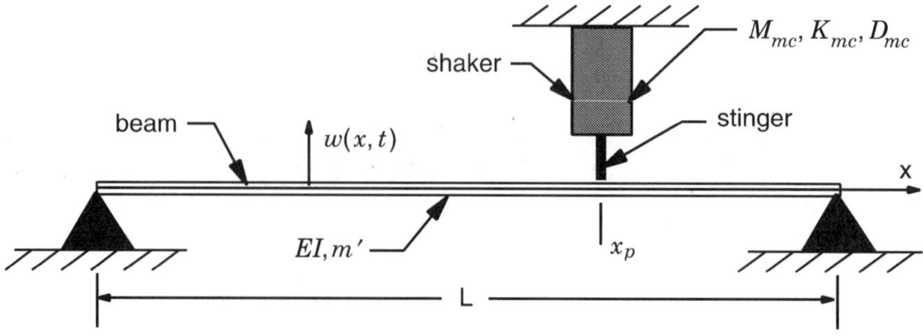


**Figure 5.12:** Schematic diagram of beam being driven by an electrodynamic shaker.

where $m'$ is the mass per unit length of the beam and $M_{mc}$ is the mass of the moving coil armature. Similarly, the potential energy of the mechanical system can be expressed as follows:

$$V(t) = \frac{1}{2}\int_0^{L_x} EI\left[\frac{\partial^2 w(x,t)}{\partial x^2}\right]^2 dx + \frac{1}{2}\int_0^{L_x} K_{mc} w^2(x,t)\delta(x - x_p)dx \tag{5.65}$$

where $E$ is the modulus of elasticity of the structure, $I$ is the area moment of inertia of the structure, and $K_{mc}$ is the stiffness of the spring on the moving coil. The eigenfunctions of the pinned boundary structure are used as the basis for the expansion of the solution:

$$w(x,t) = \sum_{n=1}^{N} \phi_n(x)\eta_n(t) \tag{5.66}$$

Upon substituting equation (5.66) into equations (5.64) and (5.65), the following expressions can be obtained for the mass and stiffness matrices of the mechanical system:

$$m_{ij} = \int_0^{L_x} m'\phi_i(x)\phi_j(x)dx + M_{mc}\phi_i(x_p)\phi_j(x_p) \qquad \text{for } i,j = 1,2,\ldots,N \tag{5.67}$$

and

$$k_{ij} = \int_0^{L_x} EI\,\frac{d^2\phi_i(x)}{dx^2}\,\frac{d^2\phi_j(x)}{dx^2}\,dx + K_{mc}\phi_i(x_p)\phi_j(x_p) \qquad \text{for } i,j = 1,2,\ldots,N \tag{5.68}$$

Notice that the "original" modes of the simply supported beam are now coupled, due to the off-diagonal terms associated with the added mass and stiffness. In general, modifications to the impedance of a reverberant plant lead to modal coupling.

From equations (5.67) and (5.68), the differential equations of the system can be expressed as follows in generalized coordinates:

$$\mathbf{M}\ddot{\boldsymbol{\eta}}(t) + \mathbf{K}\boldsymbol{\eta}(t) = \int_0^{L_x} Bli(t)\delta(x - x_p)\boldsymbol{\Phi}^{\mathrm{T}}(x)dx \tag{5.69}$$

where

$$\mathbf{\Phi}(x) = [\phi_1(x) \quad \phi_2(x) \quad \cdots \quad \phi_N(x)] \tag{5.70}$$

As indicated in equation (5.69), the force resulting from the shaker has been included in the expression for the system response. Of course, the electrical response of the system must be modified as well to include the response of the structure:

$$L\frac{di(t)}{dt} + Ri(t) = v_a(t) - Bl\dot{\boldsymbol{\eta}}^{\mathrm{T}}(t)\int_0^{L_x} \mathbf{\Phi}^{\mathrm{T}}(x)\delta(x - x_p)dx \tag{5.71}$$

The state equations can now be constructed from the preceding development. The state vector is defined as follows:

$$\mathbf{x}^{\mathrm{T}}(t) = [\boldsymbol{\eta}^{\mathrm{T}}(t) \quad \dot{\boldsymbol{\eta}}^{\mathrm{T}}(t) \quad i(t)] \tag{5.72}$$

and the input vector is

$$\mathbf{u}(t) = [v_a(t)] \tag{5.73}$$

We assume that the variable to be observed is the displacement response at the location of the shaker. Thus,

$$\mathbf{A} = \begin{bmatrix} \mathbf{0}_{N\times N} & \mathbf{I}_{N\times N} & \mathbf{0}_{N\times 1} \\ -\mathbf{M}^{-1}\mathbf{K} & \mathbf{0}_{N\times N} & Bl\mathbf{M}^{-1}\mathbf{\Phi}^{\mathrm{T}}(x_p) \\ \mathbf{0}_{1\times N} & -\frac{Bl}{L}\mathbf{\Phi}(x_p) & -\frac{R}{L} \end{bmatrix} \tag{5.74}$$

$$\mathbf{B}^{\mathrm{T}} = \begin{bmatrix} \mathbf{0}_{1\times 2N} & \dfrac{1}{L} \end{bmatrix} \tag{5.75}$$

$$\mathbf{C} = [\mathbf{\Phi}(x_p) \quad \mathbf{0}_{1\times N+1}] \tag{5.76}$$

and

$$\mathbf{D} = [0] \tag{5.77}$$

To demonstrate the impact of coupling the dynamic systems, a MATLAB script file has been created to compare the structural response for a perfect point force input with that of an electrodynamic shaker used to drive the structure. In both cases, the frequency response has been obtained between the displacement response of the structure at the point of application and the applied force. In the case of the shaker, this requires one to construct an observer for the dynamic

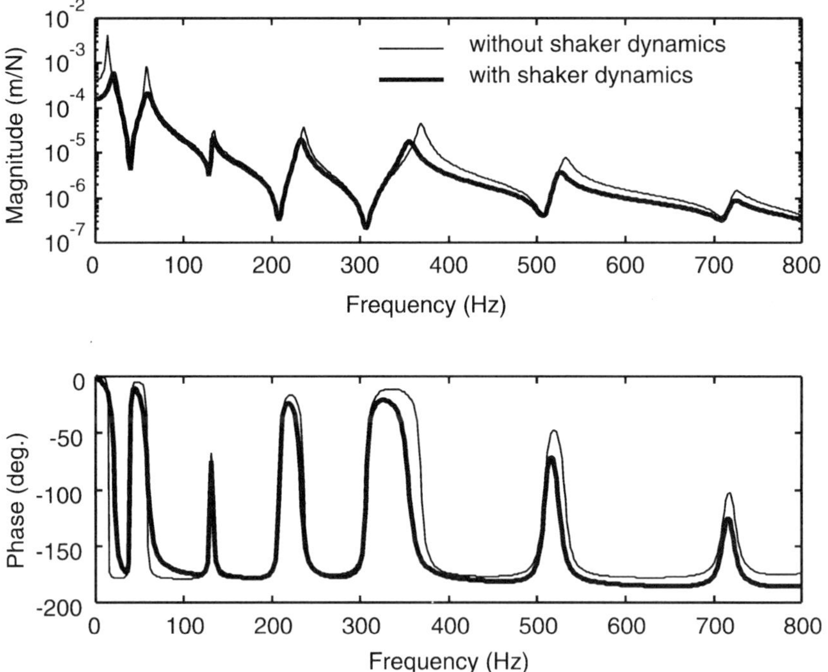


**Figure 5.13:** Frequency response of system with and without added dynamics of the shaker.

force, since the natural input to the system is an applied voltage. Details of the parameters used in the model are defined in the script file **bmshaker.m**. The reader may execute this file to produce the results presented in Figure 5.13. As illustrated, the dynamic coupling associated with the shaker serves to significantly modify the system response at the lower resonance frequencies of the system. The most dramatic effect is the increase in damping observed in the system frequency response. Due to the coupling, some of the structural energy is being dissipated through the damping inherent in the electrodynamic shaker. Whether one is designing a control system or simply comparing experimental results to those obtained from simulations, it is obvious from the example presented here that the dynamic coupling of the transduction device and the physical system is important.

### 5.3.3 Sensor Modeling and the Effect on Dynamic Response

In this section, we consider the effect of attaching transduction devices, such as accelerometers, over the surface of a structure. The use of piezoelectric materials in the design of accelerometers is commonplace to provide the necessary transduction. An introduction to seismic accelerometers has been presented earlier in this chapter and a schematic diagram of the transducer has been presented in Figure 5.9. For all practical purposes, one can replace the accelerometer with an added mass at the point of placement on the structure in developing the system model. The assumption herein is that the first mechanical resonance of the transducer far exceeds the bandwidth of the structural response, and thus the spring-mass dynamics of the transducer are ignored. In feedback control, at modest gains, higher order dynamics such as this can be ignored without having a significant impact on the control system performance. However, if the dynamics of the transducer are within the bandwidth of operation (for example, when a relatively large accelerometer is placed on small test structure), or if large feedback gains are necessary to meet system performance requirements, it may be necessary to develop a coupled dynamic model of the transducer and host structure.

If one simply chooses to model the transducer as an added mass, then the development of the model is consistent with that presented for the piezoelectric device where a set of displacement distributions (modes) are assumed in generalized coordinates, and the added mass serves to modify the expression for the total kinetic energy of the system. Taking an assumed modes approach, one simply modifies the mass matrix of the system as follows:

$$m_{rs} = \int_{V(\mathbf{x})} \rho_s(\mathbf{x})\phi_r(\mathbf{x})\phi_s(\mathbf{x})d\,V(\mathbf{x}) + \int_{V(\mathbf{x})} m_a\phi_r(\mathbf{x})\phi_s(\mathbf{x})\delta(\mathbf{x}-\mathbf{x}_a)dv(\mathbf{x}), \qquad \text{for } r, s = 1, 2, \ldots, N \quad (5.78)$$

where $\delta(\mathbf{x}-\mathbf{x}_a)$ defines the spatial aperture of application. Thus, the mass matrix is populated by off-diagonal terms that serve to couple the response in terms of the assumed modes.

At every level of modeling, one must make assumptions about the dynamic system. For example, the accelerometer *could* be modeled as a beam in longitudinal vibration with an attached mass at the end with a moving boundary. While this would certainly prove more accurate than a simple mass-spring system, it would be overkill. Ultimately, the design engineer must make some assumptions about the necessary level of modeling required, and experience in building models and comparing results between predicted and measured response guides such decisions. If the structure is relatively massive in comparison to the transducers, their effect may in all likelihood be neglected. However, if one is comparing analytical results to experimental results, including the added dynamics of the devices may be important, particularly if the mass of the transducer is significant with respect to that of the host structure. Assumptions are made throughout the engineering process, and the impact of those assumptions on the overall result must be considered.

## 5.4 THE ADAPTIVE PIEZOELECTRIC SENSORIACTUATOR

The *piezoelectric sensoriactuator* is a piezoelectric transducer used simultaneously as a sensor and an actuator, often termed a self-sensing actuator and originally proposed somewhat concurrently by Dosch et al. (1992) and Anderson and Hagood (1992). However, the *adaptive* element was introduced by Cole and Clark (1994) to alleviate problems associated with the drift in capacitance, which rendered compensation difficult. The term *sensoriactuator* was adopted for this work, since the functional application of the transducer is analogous to that of the sensorimotor system in the biological counterpart. In consideration of equation (5.18), Dosch et al. and Anderson and Hagood noted that, if the capacitance of the piezoelectric device is known, one must simply apply the same voltage across an "identical" capacitor and subtract the electrical response from that of the sensoriactuator to resolve the mechanical response of the structure. A schematic diagram of an analog circuit used to resolve the mechanical strain rate from the electrical response is presented in Figure 5.14. As illustrated, two current amplifiers are implemented such that the time derivative of charge (current) can be monitored. There are practical performance limitations of differentiating op-amp circuits, as discussed in more detail below. Charge amplifiers can also be used in place of the current amplifiers to monitor charge, which is proportional to strain. The piezoelectric properties are influenced by variations in environmental conditions and operation, which requires a continual effort to "tune" the circuit of Figure 5.14.

The primary hurdle posed for implementation of the sensoriactuator is accu-

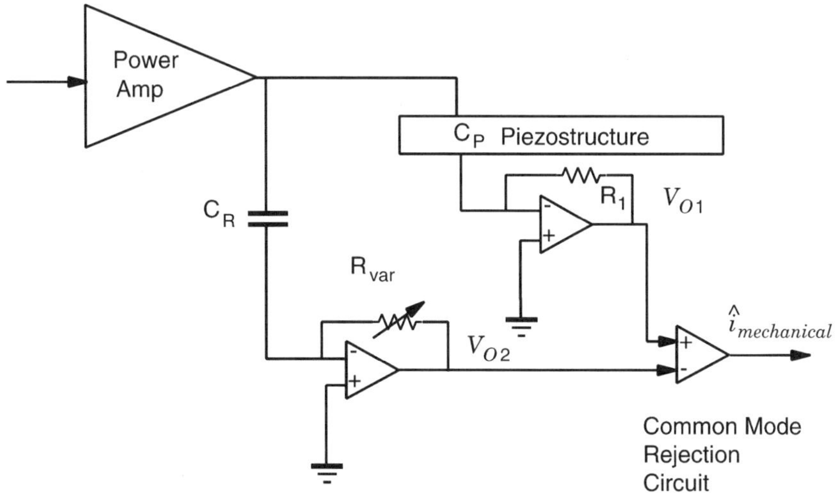


**Figure 5.14:** Schematic of diagram of manually tuned analog compensation circuit (after Anderson and Hagood, 1992).

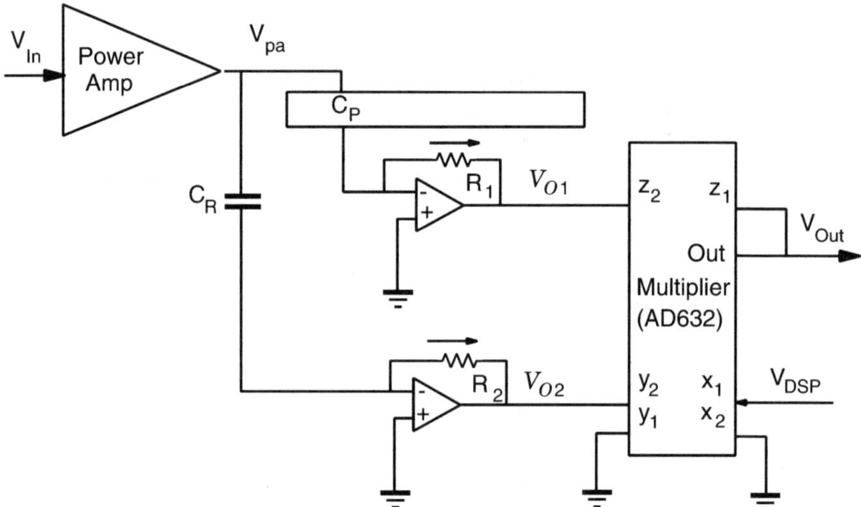


**Figure 5.15:** Schematic of adaptive piezoelectric sensoriactuator circuitry (after Vipperman and Clark, 1995).

rate estimation of the capacitance of the piezoelectric device (Anderson and Hagood, 1992; Dosch et al., 1992; Spangler, 1992). To circumvent this problem, adaptive compensation methods are required. Two approaches have been suggested for the adaptive sensoriactuator implementation. Both are discussed next in detail.

### 5.4.1 Hybrid Analog-Digital Adaptive Sensoriactuator Design

The electrical circuit illustrated in Figure 5.15 is one approach that has been successfully applied to compensate for the electrical response resulting from the applied voltage (Vipperman and Clark, 1995). The blend of analog and digital circuitry is accomplished via the AD632 analog multiplier chip, which has an excellent linear response with a signal to noise ratio (SNR) and common mode rejection ratio (CMRR) of approximately 80 dB. By supplying a DC voltage ($V_{DSP}$) to one of the differential inputs and the signal to be scaled to the other differential input, a voltage-controlled amplifier (VCA) is achieved with the AD632 (see Figure 5.16). Further, the internal output of the VCA can be summed with input $z_2$ of the AD632, removing the need for a separate common mode rejection (CMR) circuit to differentiate the estimated electrical response from the total plant response. The DC voltage is controlled with a digital signal processor and is adapted with the least-mean-squares (LMS) algorithm, as outlined previously by Cole and Clark (1994). As indicated in Figure 5.17, a third leg ($V_{03}$) can be added to differentiate the training signal that is used for adaptation so that the addition of a control signal would not potentially bias the

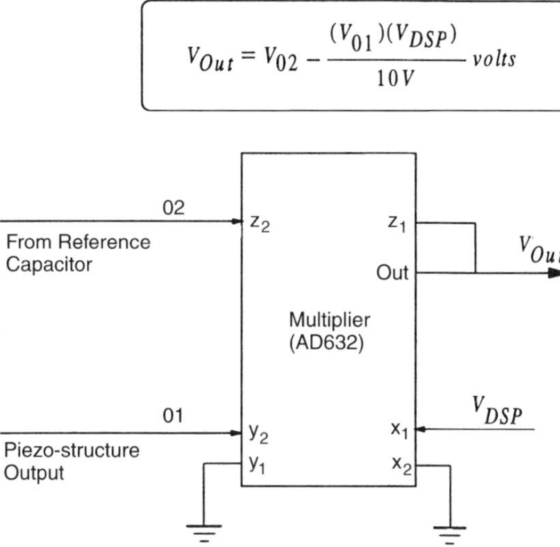


**Figure 5.16:** AD632 Integrated circuit, which acts as a voltage controlled amplifier (VCA) with common mode rejection (CMR) for the adaptive piezoelectric sensoriactuator (APSA) circuit (after Vipperman and Clark, 1995).

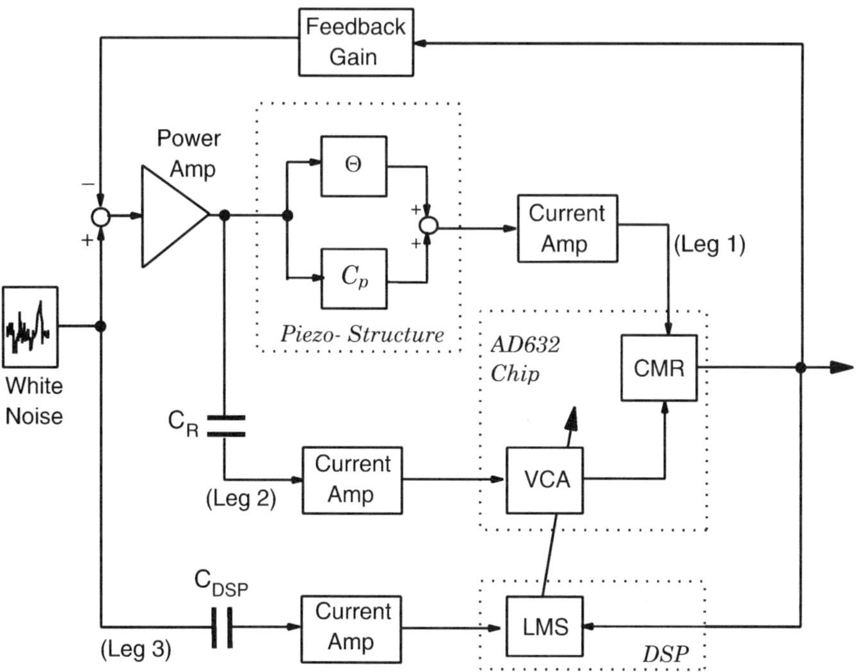


**Figure 5.17:** Schematic of adaptive compensation of a piezoelectric sensoriactuator (after Vipperman and Clark, 1995).

filter solution during closed-loop operation of the adaptive piezoelectric sensoriactuator.

***Analog Circuit Analysis*** After the adaptive filter converges, the dynamic capacitance of the piezostructure can be "identified" (Vipperman, 1997). From Figure 5.15,

$$V_{01}(s) = \mathcal{L}[-R_1 i_{01}(t)] \tag{5.79}$$

$$= \mathcal{L}\left[-R_1\left(C_p \frac{dv_{pa}(t)}{dt} + i_{\text{mech}}(t)\right)\right] \tag{5.80}$$

$$= -sR_1 C_P V_{pa}(s) - R_1 i_{\text{mech}}(s) \tag{5.81}$$

where $\mathcal{L}[\cdot]$ denotes the Laplace transform, $t$ is the time variable, $s$ is the Laplace variable, and all other variables represent electronic components from Figure 5.15. Similarly,

$$V_{02}(s) = -sR_2 C_R V_{pa}(s) \tag{5.82}$$

which leads to

$$V_{\text{Out}}(s) = \frac{V_{DSP}(-sR_2 C_R V_{pa}(s))}{10\text{V}} - sR_1 C_P V_{pa}(s) - R_1 i_{\text{mech}}(s) \tag{5.83}$$

where $10V$ is an internal gain for the AD632 multiplier chip. By inspection, for $V_{\text{Out}}(s)$ to equal only the mechanical response of the piezostructure, $-R_1 i_{\text{mech}}(s)$,

$$V_{DSP} \equiv \frac{10R_1 C_P}{R_2 C_R} \tag{5.84}$$

***Discrete-Time Analysis*** As outlined by Vipperman (1997), the quadratic cost function that is minimized by the LMS algorithm is defined as

$$C(w(k)) = E[V_{\text{Out}}^2] \tag{5.85}$$

where $E[\cdot]$ denotes the expectation operator, $w(k)$ is the digital filter weight, and $k$ is the discrete-time step. The cost is minimized using steepest descent:

$$w(k+1) = w(k) - \mu \nabla C(w(k)) \tag{5.86}$$

$$= w(k) - \mu 2E\left[V_{\text{Out}} \frac{\partial V_{\text{Out}}}{\partial w}\right] \tag{5.87}$$

where $\mu$ is a step size parameter that controls stability and convergence, and

$w(k)$ is the single-coefficient finite impulse response (FIR) filter required to adapt the capacitance, which is equivalent to the output voltage $V_{DSP}$. The LMS algorithm uses a stochastic estimate of the gradient:

$$w(k+1) \approx w(k) + \mu 2 V_{\text{Out}}(k) \frac{V_{02}(k)}{10V} \tag{5.88}$$

The difference here is that the adaptive coefficient is implemented in analog ($V_{DSP}(k+1) \equiv w(k+1)$) such that the controller can be implemented in analog as well to maintain stability guarantees associated with "ideal" colocated rate-feedback control systems (Balas, 1979).

An important result of this work is the ability to dynamically measure the "pseudo-blocked" capacitance in situ. The phrase "pseudo-blocked" was chosen by Vipperman (1997) since the structure to which the transducer is attached has compliance. By knowing the internal gains across the sensoriactuator network and the reference capacitance, the pseudo-blocked capacitance for the sensoriactuator can be determined from equation (5.84) as

$$C_P^s(k) = \frac{R_2 V_{DSP}(k)}{(10\,\text{V}) R_1} C_R \,\mu\text{F} \tag{5.89}$$

where $R_1$, $R_2$ are gain resistors depicted in Figure 5.15, $C_R$ is the reference capacitor, and 10 V is an internal gain in the AD632 multiplier chip.

***Autonomous System Health Monitoring*** Since we are monitoring the capacitance of the transducer in real time via equation (5.89), there is an opportunity to perform basic health monitoring of the piezostructure with respect to the transducer. For example, if a crack develops in the ceramic patch or a lead is disconnected, a change in measured capacitance will result. An edge detector (Oppenheim and Schafer, 1989) or discrete-time differentiator can be used to monitor such a change. The output $y(k)$ of the edge-detector is the difference between the present ($\hat{C}_p(k)$) and past ($\hat{C}_p(k-1)$) estimated discrete dynamic capacitance values, which are proportional to $V_{DSP}(k)$:

$$y(k) = \hat{C}_p(k) - \hat{C}_p(k-1) \tag{5.90}$$

$$y(k) \propto V_{DSP}(k) - V_{DSP}(k-1) \tag{5.91}$$

where $k$ is the sample index. Decimation can be used to monitor the capacitance across many different time scales as desired (Vipperman and Clark, 1995).

### 5.4.2 Nonideal Behavior and Charge Control

Piezoelectric transducers, while typically operated in the "linear range," display nonideal behaviors, such as hysteresis, phase error, and nonlinear magnitude response, if pushed excessively. In addition to the nonideal behavior of the transducer, the reference capacitor $C_R$ used in the adaptive piezoelectric sensoriactuator circuit also displays mild nonideal behavior that can affect the overall performance of the sensoriactuator. To illustrate the significance of a phase mismatch of 0.2 degrees between the reference capacitor and the piezoceramic transducer, Figure 5.18 shows a phasor diagram depicting a complex capacitance vector, the vector due to real compensation with an ideal reference capacitor, the resultant of these two vectors, and a Nyquist plot of the system (Vipperman, 1997). Noting the different scales of the $x$ and $y$ axes, the uncompensated capacitance vector clearly dominates the response. Further, the complex residual from the real compensation is clearly a significant portion of the plant response. The effects of the complex residual and an uncompensated system are compared to a fully compensated system, as indicated in the frequency response plots of Figure 5.19 (Vipperman, 1997). One should note that the phase mismatch between the PZT and reference capacitor varies with different dielectric materials used for the reference capacitor, and a PZT reference capacitor can be used to minimize such phase errors at a slightly increased cost.

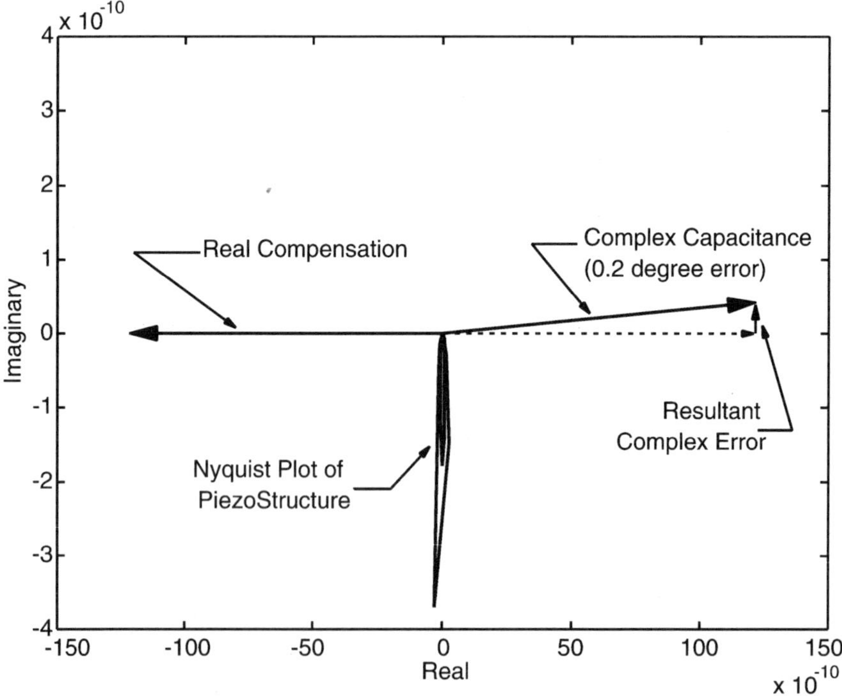


**Figure 5.18:** Phase diagram showing complex capacitance and resultant from real compensation (after Vipperman, 1997).

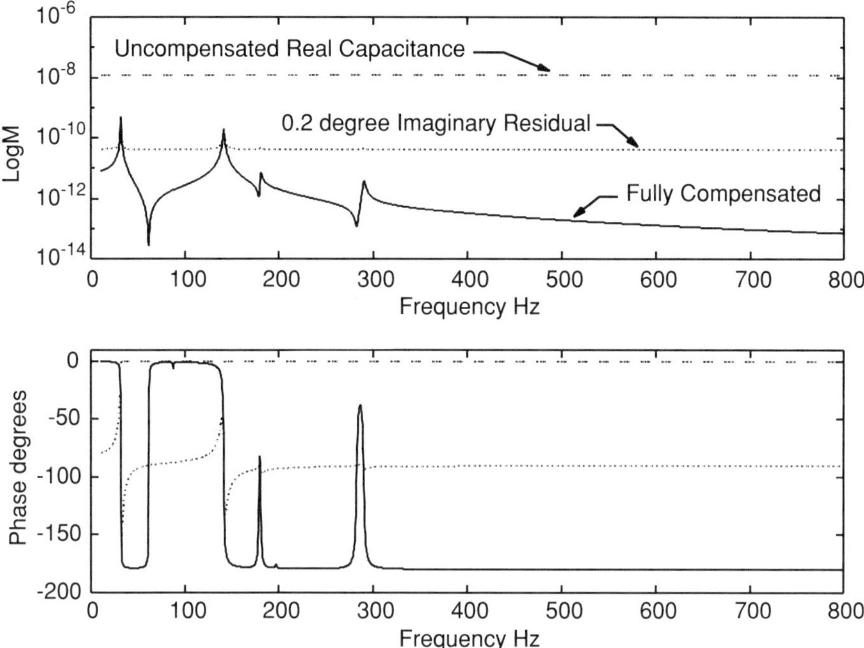


**Figure 5.19:** FRF showing effects of feedthrough and phase error (after Vipperman, 1997).

One method of compensating for nonideal behavior was proposed by Comstock (1976). He discovered that piezoceramics exhibited less hysteresis when they are charge driven as opposed to voltage driven. Piezoelectric charge can be determined by measuring the voltage drop across a large capacitor placed in series with the piezoelectric actuator similar to Figure 5.20. Charge versus displacement curves were shown to have only 4% hysteresis, as compared to 20% hysteresis in a voltage versus displacement curve. Comstock proposed using an active feedback circuit to further reduce hysteresis and partially linearize the charge response (see the circuit in Figure 5.20). Similar charge-control concepts have also been successfully implemented by Newcomb and Flinn (1982), and on an interferometer testbed incorporating sensoriactuator positioner stacks (Spangler, 1992). A simple, passive form of charge control (Kaizuka and Siu, 1988) can be achieved by placing a capacitor $C_{\mathrm{cap}}$ in series with the piezoceramic transducer having a capacitance $C_{\mathrm{piezo}}$ equivalent to that of the transducer, and serving as a natural charge regulator. The disadvantage is that a capacitive voltage divider is created, requiring an increase in the amplifier output by a factor $\gamma$.

$$\gamma = \frac{C_{\mathrm{cap}} + C_{\mathrm{piezo}}}{C_{\mathrm{cap}}} \tag{5.92}$$

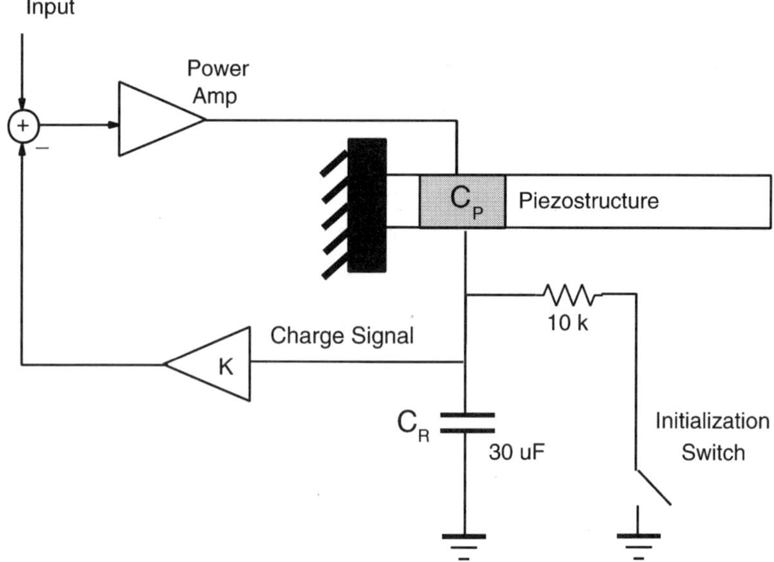


**Figure 5.20:** Active piezoelectric charge control as proposed by Comstock (1976).

to achieve the same voltage across the piezoelectric transducer that is achieved in the absence of the series-capacitor. It is important to note that $(1 - 1/\gamma)$ is approximately the expected reduction of the piezoceramic hysteresis. Therefore, $C_{\text{cap}}$ should be chosen such that $C_{\text{cap}} < C_{\text{piezo}}$, for the most reduction in hysteresis.

### 5.4.3 Purely Analog Adaptive Sensoriactuators

Phasing mismatch and/or the high costs of DSP hardware (although at the writing of this book, DSPs could be purchased in large quantities for under \$3) may be alleviated for purely analog configurations of the sensoriactuator, configurations that rely on the continuous-time LMS algorithm and zero-order or first-order adaptive filter structures (Fannin and Saunders, 1997). The zero-order scheme is identical to the previously discussed hybrid arrangement except that the adaptation mechanism is now analog versus digital. The first-order configuration utilizes the analog adaptive LMS algorithm and a two-weight compensation scheme that leads to adaptation of the capacitance magnitude and the effective corner frequency of the high-pass filter characteristics for the practical sensoriactuator. As shown in Figure 5.21, it is not possible to construct a purely differentiating operational amplifier circuit. Therefore, it is common to intentionally design a high-pass filter that differentiates over the bandwidth of interest (Franco, 1988). In either case, the unknown piezoelectric capacitance leads to uncertainty about the true corner frequency. Two proposed solutions to this problem are discussed next.

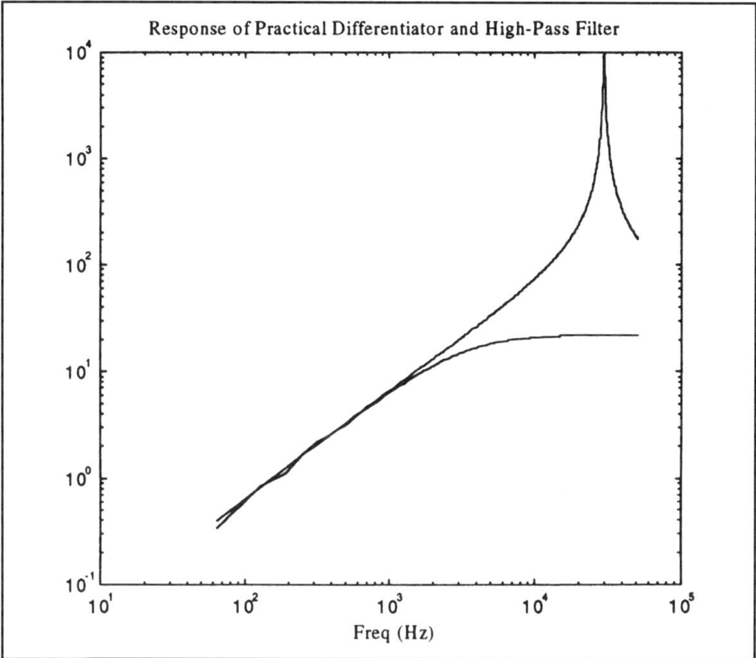


**Figure 5.21:** Higher order dynamics introduced by practical differentiation.

***Analog LMS Algorithm Realization*** A continuous-time development of the LMS algorithm proceeds as discussed by Widrow et al. (1967). It can be shown that the continuous-time LMS algorithm follows directly from the discrete-time case, leading to a weight update equation that involves a convergence parameter $\rho$, the instantaneous values of the error $e(t)$, and a reference signal $X(t)$:

$$\frac{d}{dt} W(t) = \rho e(t) X(t) \tag{5.93}$$

The integral of equation (5.93) yields the analog LMS algorithm:

$$w_i(t) = \rho \int_o^t e(\tau) X_i(\tau) d\tau \tag{5.94}$$

In theory, it can be shown that the system state $W(t)$ is unconditionally stable for arbitrary positive values of the convergence gain (Karni and Zeng, 1989). It is clear from equation (5.94) that the analog LMS block, shown in Figure 5.22 for the signals of interest, includes only a multiplier and an integrator. Therefore, the circuitry requires only one op-amp and one analog multiplier; the convergence gain is determined by the component values of the integrator.

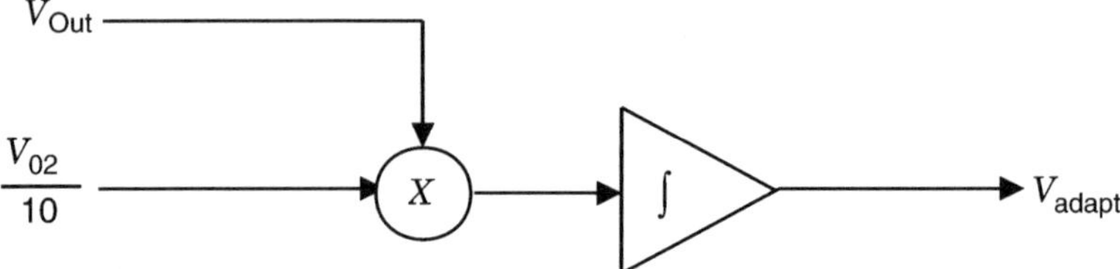


**Figure 5.22:** LMS block used in continuous-time adaptive filter.

Because of the integrator, DC offsets in the circuit will cause the filter coefficient to adapt to the wrong value (Shoval et al., 1995). For this reason, it is very important that the null offset on the integrator op-amp be properly adjusted so as not to add DC offset artificially.

For the sensoriactuator problem, the adaptive signal processing nomenclature confuses matters because we desire the mechanical response as an output. Being careful to avoid confusion, the reference signal $X(t)$ for the adaptive sensoriactuator network is the output of the reference network, defined earlier as $V_{02}$, the error signal $e(t)$ is $V_{\text{Out}}$, and the desired signal is taken to be zero. This was evident in the hybrid adaptive filter described earlier.

***Zero-Order Analog Adaptive Design*** The hybrid adaptive filter presented earlier (Figure 5.15) enabled adaptation of a magnitude condition for the reference leg of the circuit. Because of the expected first-order dynamics associated with the feedthrough signal in the piezoelectric response, it was shown that the capacitor in the reference leg is critical in determining the phase response of the fully adapted system. Stated otherwise, the hybrid configuration provided a single degree of freedom (SDOF) in terms of adaptation. The zero-order analog adaptive filter design reproduces the ealier approach with the addition of the improved differentiator and a tuning potentiometer. The transfer function for the piezoelectric high-pass filter circuit shown in Figure 5.23 is as follows:

$$\frac{V_{01}(s)}{V_{\text{In}}(s)} = -\frac{(R_1/R_{c1})s}{s + (1/R_{c1}C_p)} - R_1 \sum_{r=1}^{\infty} \frac{\Theta^2 s}{s^2 + 2\zeta_r \omega_r s + \omega_r^2} \tag{5.95}$$

The left side of the compensator shown in Figure 5.23 is basically the same as that shown in Figure 5.15, but the adaptive weight is determined by the additional multiplier and integrator, which implement the LMS block shown in Figure 5.22. The adaptive weight provides magnitude adjustment, similar to the hybrid method, but phase matching is provided by tuning the corner frequency of the reference high-pass filter using the potentiometer $RC_2$. This helps to relieve the critical responsibility of perfectly matching the reference capacitor to the piezoelectric capacitance. However, it is difficult to determine when

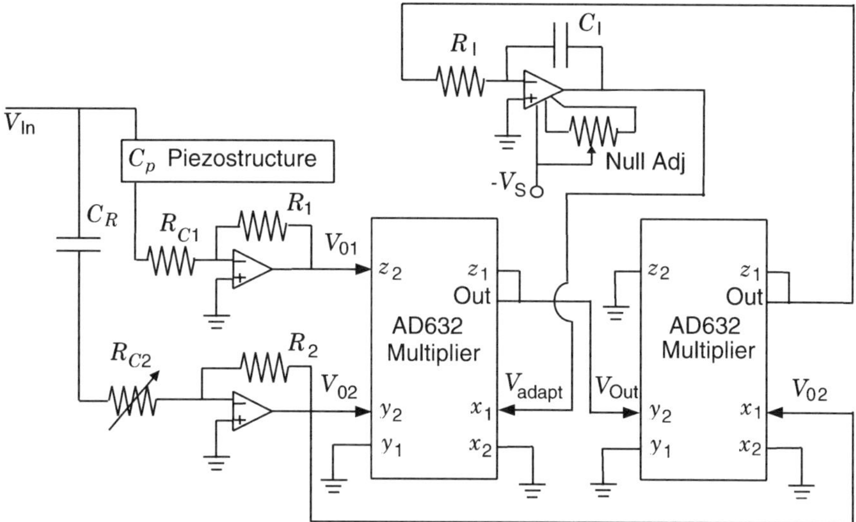


**Figure 5.23:** Zero-order analog adaptive sensoriactuator network.

the phase response of the piezoelectric and reference legs are properly matched, since the mechanical response of the structure is included in $V_{01}$ (see equation (5.95)). Thus, circuit tuning with the potentiometer can prove unreliable.

***First-Order Analog Adaptive Design*** As discussed earlier, the measured feedthrough dynamics (the first term of equation (5.95)) are first-order to some approximation. Given that the coupled piezostructure system retains the first-order dynamics for the feedthrough signal, an adaptable first-order compensation filter can be inverted to achieve perfect compensation of the feedthrough term. This approach is used in the second purely analog adaptive filter design.

The adaptive, high-pass filter requires two integrating op-amps, and two analog multipliers for the LMS algorithm. The summer and integrator in the adaptive filter shown in Figure 5.24 can be constructed using op-amps. Two analog device SSM2164 voltage VCAs operating in a voltage amplification configuration (Figure 5.25) are used to realize the gains blocks $A$ and $B$. Because the VCA gain is proportional to the command voltage, that is, the adaptive filter weights, the LMS algorithm provides a means to make the high-pass filter configuration automatically adaptive. A circuit diagram of the first-order analog adaptive sensoriactuator is shown in Figure 5.26. This configuration is designed for system identification testing.

Tuning of this circuit is very simple. The LMS integrators must be null adjusted to ensure that they are not producing DC offsets that would affect the filter convergence. The null adjust procedure consists of connecting the integrator input to ground, and adjusting the null adjust potentiometer until the integrator output is neither increasing nor decreasing. This procedure should

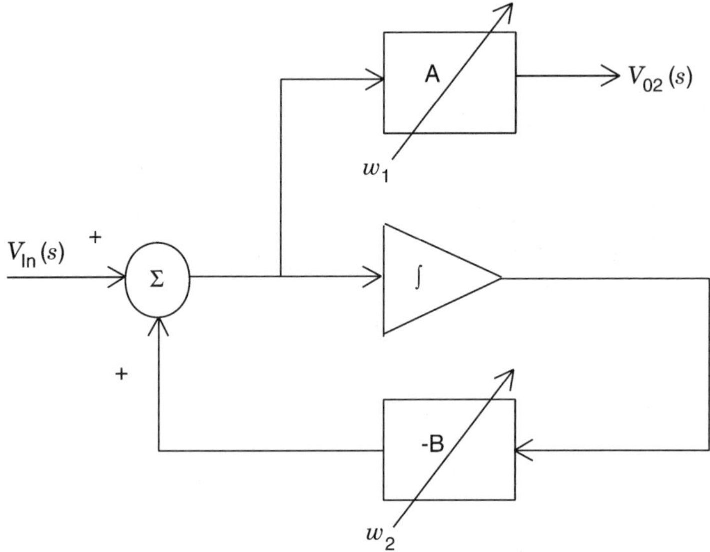


**Figure 5.24:** Adaptive high-pass filter block diagram.

be manually repeated or automated at the beginning of each usage of the circuit. Note that a slightly different configuration of the circuit is required for very high actuation voltages (Fannin and Saunders, 1997).

## 5.5 SIGNAL CONDITIONING

Whether the control system is implemented in analog hardware, digital hardware, or some combination of the two, filters and amplifiers will be necessary to minimize noise, prevent aliasing, increase the gain, and provide the necessary power to implement the control system design. Electrical systems, like

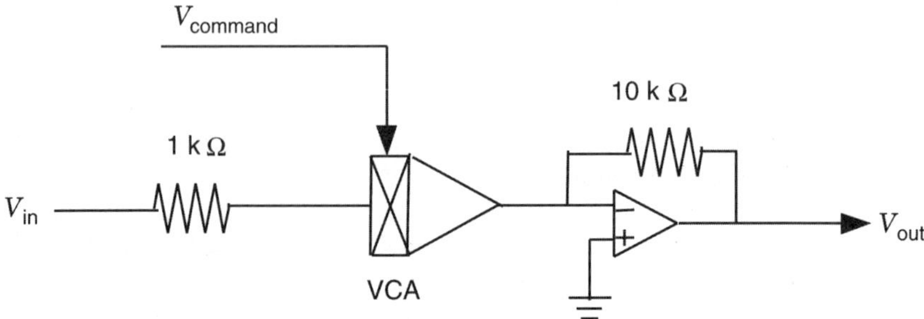


**Figure 5.25:** VCA in voltage amplification configuration.

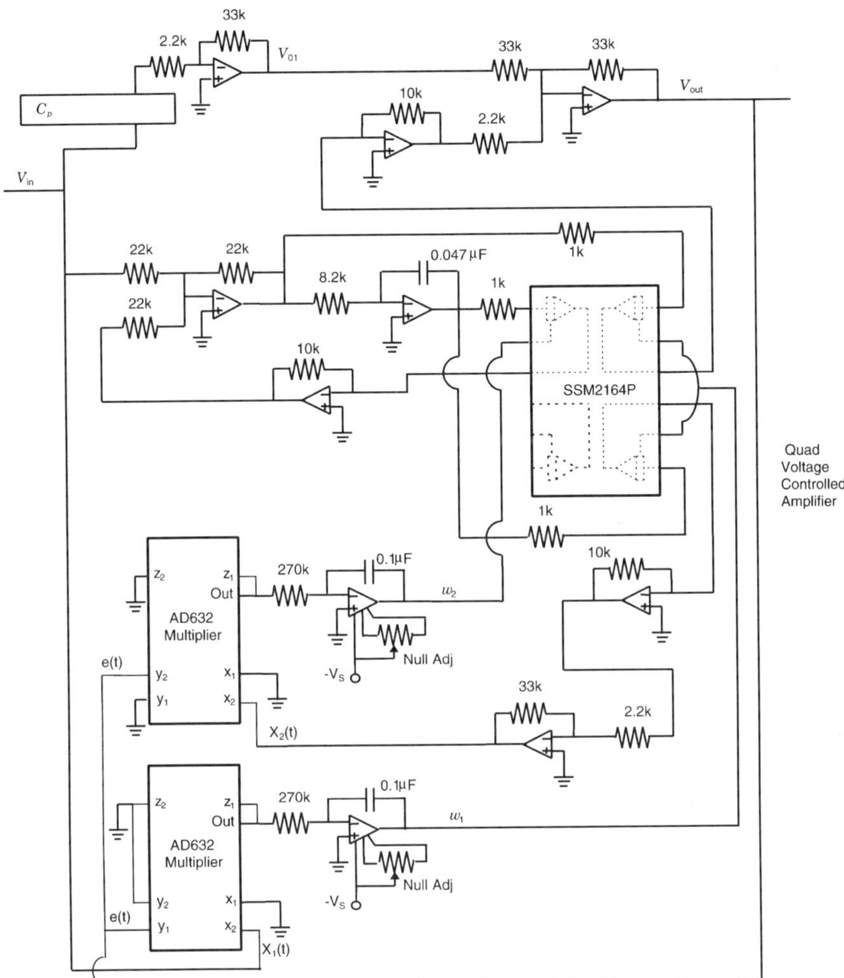


**Figure 5.26:** First-order analog adaptive sensoriactuator network.

mechanical systems, have dynamics that must be included in the development of the system model. The early temptation in the modeling of dynamic systems is to ignore the electrical devices required to implement the control system since the bandwidth of devices such as amplifiers quite often exceeds that of the mechanical system by one or more orders of magnitude. However, as mentioned in the previous section, if high gains are used in feedback, the dynamics of such devices affect stability and will most certainly place finite stability margins on closed-loop systems. Additionally, the current sourcing ability of power amplifiers tend to limit the bandwidth of piezostructures, as does the RC time constant of the system (piezoelectric transducers are characterized by high capacitance).

### 5.5.1 Amplification

There are two basic types of amplification required to implement control system designs: gain and power. With gain amplification, the objective is simply to increase the gain from the transduction devices such as the sensors to compensate for sensitivities of the devices and provide a control signal that is of the appropriate magnitude. To this end, operational amplifier based circuits are typically employed to meet this objective. There are two basic types of gain circuits that can be employed: inverting and noninverting, illustrated in Figure 5.27. One must certainly keep track of the signs imposed by the circuits

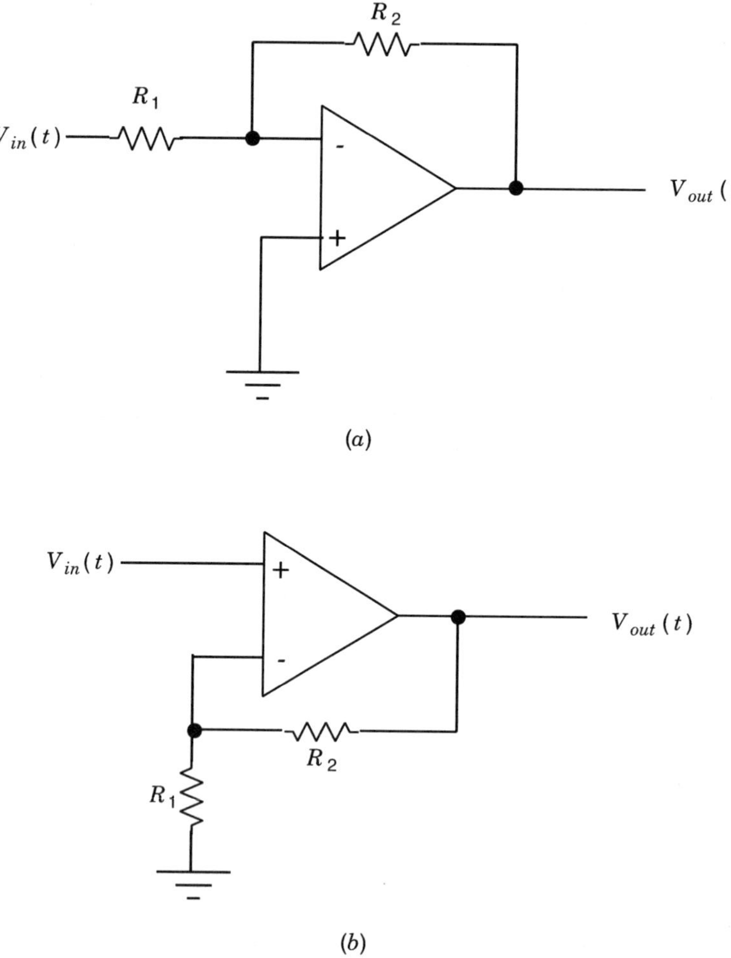


**Figure 5.27:** Schematic diagrams of amplifier circuits. (*a*) Inverting. (*b*) Noninverting.

used to create the gain in feedback control! For the inverting amplifier depicted schematically in Figure 5.27*a*, the gain can be computed as follows:

$$V_{\text{Out}}(t) = -\frac{R_2}{R_1} V_{In}(t) \tag{5.96}$$

and for the noninverting amplifier depicted in Figure 5.27*b*:

$$V_{\text{Out}}(t) = \left(1 + \frac{R_2}{R_1}\right) V_{\text{In}}(t) \tag{5.97}$$

The inverting amplifier can thus be used to attenuate the input as well as increase the input; however, the noninverting amplifier has a lower limit of unity gain, which simply corresponds to a voltage follower. In theory, operational amplifiers are considered to have infinite gain and bandwidth; however, in practice, all operational amplifiers, whether internally compensated (i.e., a 741 op-amp) or uncompensated (i.e., a 748 op-amp), have a finite bandwidth with a 6 dB/octave low-pass rolloff (i.e., a poll at some predetermined frequency). Upon using negative feedback to stabilize the output of the op-amp, the gain is reduced, but the system is stabilized and the bandwidth of operation is extended. For a detailed discussion of practical implementations of operational amplifiers, the interested reader is referred to the reference by Horowitz and Hill (1994). All physical systems will display some rolloff at higher frequencies, and this low-pass filter characteristic must be included in the modeling unless the bandwidth of the amplifiers greatly exceeds that of the dynamic systems and moderate gains are used in feedback. In most cases, it is sufficient to model the dynamics of the electrical system and place it in series with the plant, as illustrated in Figure 5.1.

Power amplifiers are also required in control system realizations to provide the necessary voltage and current required to drive electromechanical devices. Due to inefficiencies in the conversion of electrical energy to mechanical energy, one typically needs more power to drive the control actuators than that which must be supplied to the structure for control. Regardless of the power requirements, like all electrical circuits, amplifiers will have finite bandwidths of operation, and in practice it is best to choose a power amplifier that is flat over a bandwidth extending from DC to an order of magnitude above the bandwidth of the control system. In practice, compensators and filters required in the control system implementation can be implemented into the power and gain amplifier designs. However, for laboratory work, one typically desires a wideband power amplifier to provide the most flexibility for a range of experimental applications.

### 5.5.2 Filtering

Filters are necessary elements in most control system designs, and most compensators in general can simply be classified as filters. If compensators are required in the control system design, then one must obviously model the dynamics of these filters. However, one often places low-pass filters in the loop as well to roll off the response at frequencies well out of the bandwidth to reduce high frequency noise. A schematic diagram of a second-order, Butterworth ($K = 1.586$, Horowitz and Hill, 1994), low-pass filter is illustrated in Figure 5.28. Values of $R$ and $C$ can be chosen to define the corner frequency. A simple analysis of the circuit yields the following transfer function:

$$\frac{V_{\text{Out}}(s)}{V_{\text{In}}(s)} = \frac{K(1/RC)^2}{s^2 + ((3 - K)/RC)s + (1/RC)} \tag{5.98}$$

To convert the transfer function model of the filter into a state-variable model is a simple procedure in MATLAB, requiring the **tf2ss** function. Thus, the filter model can be placed in series with the state-variable model of the piezostructure and sensor if necessary. While a number of manufacturers build programmable filters of various order (i.e., four pole, six pole, and eight pole), one can cascade filters such as that illustrated in Figure 5.28 to create whatever corner frequency and order of filter required. However, the design of low-noise electronic systems is something of an art. The reference by Horowitz and Hill (1994) provides an excellent starting point for designing electronic signal conditioning hardware. If you design your own filters, then modeling the system is simplified since you know how they are built. Most manufacturers will provide you, upon request, with the transfer function of the filters you purchase, and having programmable or selectable corner frequencies certainly adds to efficiency when working on a

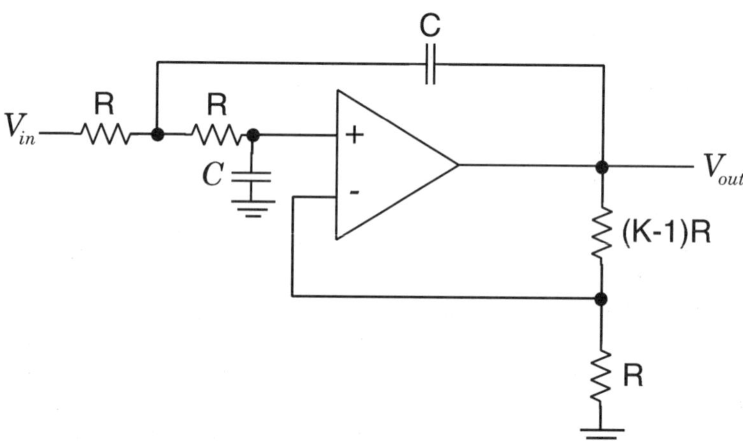


**Figure 5.28:** Schematic diagram of a second-order, low-pass filter.

laboratory test structure. Regardless of whether you design and build your own filters or purchase filters, you must incorporate the dynamics of your signal conditioning electronics into your system model.

## 5.6 SUMMARY

In Chapter 5, we synthesized much of Chapters 2 and 3. Various transduction devices for sensing and actuation were reviewed, including both active materials and conventional devices. State-variable models of dynamic systems, including transduction device dynamics, were developed by the methods outlined in Chapter 2. Various MATLAB models were provided to demonstrate the importance of including the transduction device dynamics. The basic principles required in developing a self-sensing actuator (i.e., the adaptive sensoriactuator) were presented, and both digital and hybrid analog and digital implementations of the transducer in hardware were discussed. Methods of modeling the dynamics of filters, amplifiers, and other signal conditioning hardware components employed in control system realizations were presented.

## BIBLIOGRAPHY


Anderson, E. H. and N. W. Hagood, 1992. "Self-Sensing Piezoelectric Actuation: Analysis and Application to Controlled Structures," *AIAA/ASME/ASCE/AHS/ASC Structures, Structural Dynamics and Materials Conference*, Dallas, TX, pp. 2141–2155.

Auld, B. A., 1990. *Acoustic Fields and Waves in Solids*, Robert E. Krieger Publishing, Malabar, FL.

Balas, M. J., 1979. "Direct Velocity Feedback Control of Large Space Structures," *Journal of Guidance and Control*, **2**(3), 252–253.

Banks, H. T., R. C. Smith, and Y. Wang, 1996. *Smart Material Structures: Modeling, Estimation, and Control*, Masson, Paris, France.

Buehler, W. J. and C. R. Wiley, 1965. Nickel-Based Alloys, US Patent 3,174,851.

Clark, R. L., J. S. Vipperman, and D. G. Cole, 1996. Adaptive Piezoelectric Sensoriactuators, US Patent 5,578,761, Nov.

Cole, D. G. and R. L. Clark, 1994. "Adaptive Compensation of Piezoelectric Sensoriactuators," *Journal of Intelligent Material Systems and Structures*, **5**, 665–672.

Collins, S. A., D. W. Miller, and A. H. von Flotow, 1990. "Sensors for Structural Control Applications Using Piezoelectric Polymer Film," *SERC #12-90*, MIT Space Engineering Research Center, Massachusetts Institute of Technology, Cambridge, MA.

Comstock, R. H., 1976. "Reducing the Strain Uncertainty of Piezoelectric Transducers by the Use of Charge Feedback," *Technical Report C4753*, Charles Stark Draper Laboratory, Cambridge, MA.

Crandall, S. H., D. C. Karnopp, E. F. Kurtz, and D. C. Pridmore-Brown, 1968. *Mechanical and Electromechanical Systems*, Robert E. Krieger Publishing, Malabar, FL.

Dally, J. W., W. F. Riley, and K. G. McConnell, 1993. *Instrumentation for Engineering Measurements*, 2nd ed., Wiley, New York.

Dosch, J. J., D. J. Inman, and E. Garcia, 1992. "A Self-Sensing Piezoelectric Actuator for Collocated Control," *Journal of Intelligent Material Systems and Structures*, **3**, 166–185.

Ewins, D. J., 1986. *Modal Testing: Theory and Practice*, Research Studies Press, Letchworth, England.

Fannin, C. A., and W. R. Saunders, 1997. "Analog Adaptive Piezoelectric Sensoriactuator Design," *Proceedings of the AIAA SDM Conference*, Kissimee, FL.

Flatau, A. B., D. L. Hall, and J. M. Schksselman, "Magnetostrictive Vibration Control Systems" *Journal of Intelligent Material Systems and Structures*, **4**(4), 1993, 560–565.

Franco, S., 1988. *Design with Operational Amplifiers and Analog Integrated Circuits*, McGraw-Hill, New York.

Funakubo, H., 1986. *Shape Memory Alloys*, Gordon and Breach, New York.

Hagood, N. W., W. H. Chung, and A. von Flotow, 1990. "Modelling of Piezoelectric Actuator Dynamics for Active Structural Control," *AIAA-90-1087-CP*.

Hall, D. L. and A. B. Flatau, "Nonlinearities, Harmonics, and Trends in Dynamic Applications of Terfenol-D," Proceedings of SPIE Conference on Smart Structures and Intelligent Materials, vol. 1917, Part 2, 1993, pp. 929–939.

Horowitz, P. and W. Hill, 1994. *The Art of Electronics*, Cambridge University Press, New York.

Kaizuka, H., and B. Siu, 1988. "A Simple Way to Reduce Hysteresis and Creep When Using Piezoelectric Actuators," *Japanese Journal of Applied Physics*, **27**(5), 773–776.

Karni, S. and G. Zeng, 1989. "The Analysis of the Continuous-Time LMS Algorithm," *IEEE Transactions on Acoustics, Speech, and Signal Processing*, **37**(4), 595–597.

*Kynar Piezo Film Technical Manual*, 1987. Pennwalt Corporation. Valley Forge, PA, 65 pp.

Liang, C. and C. A. Rogers, 1990. "A One-Dimensional Thermomechanical Constitutive Relation of Shape Memory Materials," *Journal of Intelligent Material Systems and Structures*, **1**(2), 207–234.

Liang, C. and C. A. Rogers, 1991. "Design of Shape Memory Alloy Coils and Their Applications in Vibration Control," Proceedings of *Recent Advances in Active Control of Sound and Vibration*, edited by C. A. Rogers and C. R. Fuller, Technomic Publishing, Lancaster, PA, pp. 177–198.

Meirovitch, L., 1967. *Analytical Methods in Vibrations*, Macmillan, New York.

Murphy, K. A., A. M. Vengsarkar, B. R. Fogg, M. F. Gunther, and R. O. Claus, 1992. "Spatially-Weighted, Two-Mode Optical Fiber Sensors," Proceedings of *Recent Advances in Adaptive and Sensory Materials and Their Applications*, Blacksburg, VA, edited by C. A. Rogers and R. C. Rogers, Technomic Publishing, Lancaster, PA, 242–253.

Newcomb, C. V. and I. Flinn, 1982. "Improving the Linearity of Piezoelectric Ceramic Actuators," *Electronic Letters*, **18**(11), 442–443.

Oppenheim, A. V. and R. W. Schafer, 1989. *Discrete Time Signal Processing*, Prentice Hall, Englewood Cliffs, NJ.

Pratt, J. and A. B. Flatau, 1995. "Development and Analysis of a Self-Sensing Magnetostrictive Actuator Design," *Journal of Intelligent Material Systems and Structures*, **6**(5), 639–648.

Rogers, C. A., 1988. "Novel Design Concepts Utilizing Shape Memory Alloy Reinforced Composites," *Proceedings of the American Society of Composites 3rd Technical Conference on Composite Materials*, Technomic Publishing, Seattle, WA.

Rogers, C. A., 1990. "Active Vibration and Structural Acoustic Control of Shape Memory Alloy Hybrid Composites: Experimental Results," *Journal of the Acoustical Society of America*, 2803–2811.

Shoval, A., D. A. Johns, and W. M. Snelgrove, 1995. "Comparisons of DC Offset Effects in Four LMS Adaptive Algorithms," *IEEE Transactions on Circuits and Systems*, **42**(3), 176–185.

Spangler, R. L., 1992. *Broadband Control of Structural Vibration Using Simultaneous Sensing and Actuation with Nonlinear Piezoelectric Ceramics*, Ph.D. Thesis, Massachusetts Institute of Technology, Cambridge, MA.

Vipperman, J. S., 1997. *Adaptive Piezoelectric Sensoriactuators for Active Structural Acoustic Control*, Ph.D. Thesis, Duke University, Durham, NC.

Vipperman, J. S. and R. L. Clark, 1995. "Hybrid Analog and Digital Adaptive Compensation of Piezoelectric Sensoriactuators," *AIAA/ASME Adaptive Structures Forum*, New Orleans, LA, pp. 2854–2859.

B. Widrow, P. E. Mantey, L. J. Griffiths, and B. B. Goode, 1967. "Adaptive Antenna Systems," *Proceedings of the IEEE,* **55**(4), 595–597.

# 6

# INTEGRATION OF SPATIAL AND TEMPORAL SIGNAL PROCESSING

## 6.1 INTRODUCTION

In Chapter 5 we have discussed a few of the active materials currently available for research applications in control of vibrating structures. In this chapter, design applications are introduced to demonstrate potential uses and methods of implementation. One should recognize that active materials allow for adaptive structure design; however, a solid understanding of the basic physics of the control problem is required for successful implementation. Plainly stated, the adaptability of the structure is directly related to the ingenuity and creativity of the designer! One must determine if a sensor or actuator is needed for the structure and the material best suited for the application. In many cases, active materials can be utilized as sensors or actuators, and quite often, different active materials can be used for the same application. The designer must choose the material best suited for the application based upon a number of design constraints such as weight of material, complexity in attaching or embedding the material, signal conditioning requirements, and cost of material, to name a few.

In addition to imposed design constraints, one must consider the position, shape, and type of input (e.g., force or moment) and output (e.g. strain or acceleration) of the active materials that create the "nervous system" and "tendons and muscles" of the structure as a function of their specific tasks. The position and shape of active materials can be used to create "spatial windows" with respect to the structure. These "spatial windows" provide a method of performing analog signal processing directly on the structure, often reducing the burden placed upon the digital signal processor in the control application and increasing overall system performance. However, one must remember that events occur in both space and time, and thus temporal signal processing is likely required as well. Some active materials provide a unique method of performing spatial signal processing such as modal filtering; however, time domain signal processing must be performed with digital signal processors. The combination of both *spatial* and *temporal* signal processing results in a tremendous tool for real-time sensing and actuation.

## 6.2 SPATIAL FILTERING FOR SENSING AND ACTUATION

In Chapter 5, we have developed expressions for the electrical response of a piezoelectric material as a function of mechanical strain when bonded to the surface of a structure. The resulting expressions for the sensor output are in units of charge. In addition, the forced response of a structure has been developed as a function of the voltage applied to a piezoelectric actuator bonded to the surface of the structure. In each case, the expressions are dependent upon the piezoelectric properties of the material and the "spatial window" or "aperture" of application. This concept of a spatial window is the topic of discussion in this section.

A one-dimensional test structure, the simply supported beam illustrated in Figure 6.1, is used to convey the concepts by example. Adopting the notation used by Burke and Hubbard (1990), let us assume that the scalar distributed input to a one-dimensional structure at coordinate $\xi$ is described by $u(\xi, t)$ and the scalar output of the system at the spatial coordinate $\mathbf{x}$ is described by $w(\mathbf{x}, t)$. The output can be expressed in terms of the Green's function of the plant as follows:

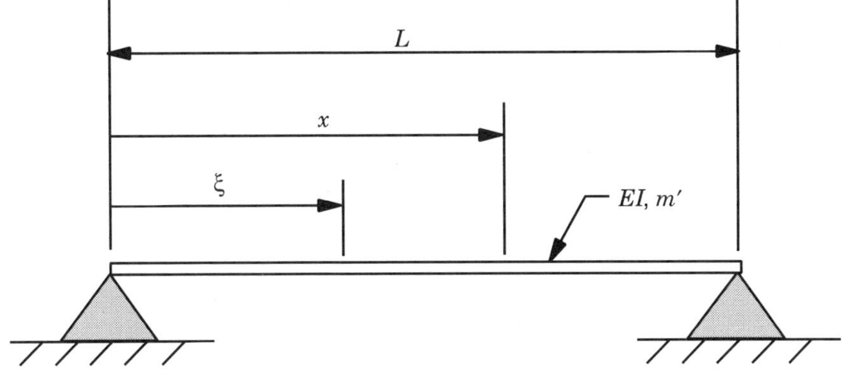


**Figure 6.1:** Schematic diagram of a simply supported beam.

$$w(\mathbf{x}, t) = \int_0^t \int_D h(\mathbf{x}, \xi, t - \tau)u(\xi, \tau)d\xi\, d\tau \tag{6.1}$$

where $h(\mathbf{x}, \xi, t-\tau)$ is the Green's function describing the space and time impulse response of the plant between the specified input and output. As outlined by Burke and Hubbard (1990), the system is considered to be stationary, linear and self-adjoint over the domain $\mathbf{x} \in D$ with boundary $\Gamma$ and homogeneous initial conditions. Taking the Laplace transform of the temporal convolution yields

$$w(\mathbf{x}, s) = \int_D h(\mathbf{x}, \xi, s)u(\xi, s)d\xi \tag{6.2}$$

For self-adjoint systems, $h(x, \xi) = h(\xi, x)$, the impulse response can be expressed as a function of the system eigenfunctions and eigenvalues:

$$h(\mathbf{x}, \xi, s) = \sum_{n=1}^{\infty} \frac{\phi_n(\mathbf{x})\phi_n(\xi)}{\lambda_n(s)} \tag{6.3}$$

where $\lambda_n(s)$ defines the poles of the nth mode $\phi(x)$. A finite number of modes $N$ (in place of $\infty$) can be used to compute the structural response as long as the bandwidth of the input to the structure is well below the resonant frequency of the highest mode included in the expression. In general, one can determine the appropriate number of modes to use in the expansion by monitoring the convergence of the transfer function zeros as they can be used to determine if the compliance at low frequency is accurately captured (Clark, 1997). For convenience, the eigenfunctions are normalized with respect to the mass as outlined by Meirovitch (1967):

$$\int_D m'(\mathbf{x})\phi_m(\mathbf{x})\phi_n(\mathbf{x})d\mathbf{x} = \delta_{mn} \tag{6.4}$$

where $m'(\mathbf{x})$ is the mass normalized with respect to the domain (i.e., per unit length for a beam or per unit area in the case of a plate).

Now, for the simply supported beam illustrated in Figure 6.1, the normalized eigenfunctions presented in equation (6.3) are

$$\phi_n(x) = \sqrt{\frac{2}{m'L}} \sin\left(\frac{n\pi x}{L}\right) \tag{6.5}$$

and the denominator of equation (6.3) can be expressed as follows:

$$\lambda_n(s) = s^2 + 2\zeta_n\omega_n s + \omega_n^2 \tag{6.6}$$

The resonant frequencies of the beam are defined by $\omega_n$ and were presented in Chapter 2. The term $m'$ is used to represent the mass per unit length of the beam, which is defined over the domain $0 < x < L$, and $\zeta_n$ is the proportional damping ratio included to bound the response of the structure.

Returning to our general derivation, one can assume that the input to the structure is separable in space and time:

$$u(x,t) = \sum_{m=1}^{M} L_x^m[\beta_m(x)]u_m(t) \tag{6.7}$$

Taking the Laplace transform,

$$u(x,s) = \sum_{m=1}^{M} L_x^m[\beta_m(x)]u_m(s) \tag{6.8}$$

where the terminology is consistent with that used by Burke and Hubbard (1990). The Laplace transform of the input signal is represented by $u_m(s)$, $\beta_m(x)$ represents the *spatial window* or *aperture* of the $m$th actuator, and $L_x^m$ is the spatial differential operator modeling the $m$th actuator operation (i.e., point force, line moment, etc.). For a point force positioned at $x = x_m$,

$$\beta_m(x) = \delta(x - x_m) \tag{6.9}$$

and

$$L_x^m = \frac{d^0}{dx^0} \tag{6.10}$$

where $\delta(x)$ is the spatial Dirac. If a distributed uniaxial piezoelectric actuator is applied over the structure between $x_1$ and $x_2$, then

$$\beta_m(x) = \mu(x - x_1) - \mu(x - x_2) \tag{6.11}$$

and

$$L_x^m = \frac{d^2}{dx^2} \tag{6.12}$$

where $\mu(x)$ is used to represent the unit step function herein to differentiate it from the input, $u(x,t)$. The above notation has been used in a number of studies to model the spatial window of the chosen actuator (Crawley and de Luis, 1987; Burke and Hubbard, 1990; Clark et al., 1991; Dimitriadis et al., 1991; Gibbs and Fuller, 1992).

A modal expansion can be used to express the system response in terms of the eigenfunctions:

$$w(\mathbf{x}, t) = \sum_{n=1}^{N} \eta_n(t)\phi_n(\mathbf{x}) \tag{6.13}$$

Taking the Laplace transform,

$$w(\mathbf{x}, s) = \sum_{n=1}^{N} \eta_n(s)\phi_n(\mathbf{x}) \tag{6.14}$$

where (from forced equation of motion)

$$\eta_n(s) \equiv \frac{1}{\lambda_n(s)} \sum_{m=1}^{M} u_m(s) \int_D \phi_n(\xi) L_\xi^m[\beta_m(\xi)]d\xi \tag{6.15}$$

Burke and Hubbard (1990) defined the constant $b_{nm}$:

$$b_{nm} \equiv \int_D \phi_n(\xi) L_\xi^m[\beta_m(\xi)]d\xi \tag{6.16}$$

which is the $n$th Fourier coefficient of the $m$th actuator resulting from imposing the spatial window of the actuator with respect to each eigenfunction. These coefficients represent the elements that populate the system input matrix, $B$, for a state space realization. If $b_{mn} \neq 0$, the $n$th mode is controllable by the $m$th actuator. If $b_{mn} \neq 0$ for all $(m, n)$, then the $M$ actuators provide all-mode control (or $N$-mode) control, which is obviously tied directly to controllability.

One can likewise develop an expression for the $p$th error sensor as a function of the spatial window associated with its region of application as well as some temporal filter required to model the dynamics associated with the sensor. (The temporal filter is the topic of discussion in the following sections.) The time-domain response of the $p$th sensor can be expressed in terms of a temporal operator $S_t^p$, a spatial operator $S_{\mathbf{x}}^p$, and a spatial aperture of application $\gamma_p(\mathbf{x})$ as follows:

$$y_p(\mathbf{x}, t) = \int_D S_t^p \left\{ \gamma_p(\mathbf{x}) S_\mathbf{x}^p[w(\mathbf{x}, t)] \right\} d\mathbf{x} \tag{6.17}$$

Taking the Laplace transform,

$$y_p(\mathbf{x}, s) = \int_D k_p(s)\gamma_p(\mathbf{x}) S_\mathbf{x}^p[w(\mathbf{x}, s)] d\mathbf{x} \tag{6.18}$$

where $k_p(s)$ is the Laplace transform of the dynamics imposed by the temporal operator. For a point sensor (e.g., accelerometer) applied at $x = x_p$ on the simply supported beam,

$$\gamma_p(x) = \delta(x - x_p) \tag{6.19}$$

and

$$S_x^p = \frac{d^0}{dx^0} \tag{6.20}$$

One can obtain a measure of displacement, velocity, or acceleration at a point simply by changing the temporal filter $k_p(s)$ (or for that matter, define some other frequency shaped filter). For example, if the sensor output is proportional to velocity, then

$$S_t^p = \frac{d}{dt} \tag{6.21}$$

and thus

$$k_p(s) = s \tag{6.22}$$

Similarly, the model for a uniaxial strain sensor can be developed by letting

$$\gamma_m(x) = \mu(x - x_1) - \mu(x - x_2) \tag{6.23}$$

and

$$S_x^m = \frac{d^2}{dx^2} \tag{6.24}$$

applied between $x_1$ and $x_2$. Substituting the modal expansion of the solution into the expression for the sensor output yields

$$y_p(s) = \sum_{n=1}^{N} \eta_n(s)k_p(s) \int_D \gamma_p(\mathbf{x})S_{\mathbf{x}}^p[\phi_n(\mathbf{x})]d\mathbf{x} \tag{6.25}$$

Another modal influence function is now introduced:

$$c_{pn} \equiv \int_D \gamma_p(\mathbf{x})S_{\mathbf{x}}^p[\phi_n(\mathbf{x})]d\mathbf{x} \tag{6.26}$$

which can also be expressed as

$$c_{pn} \equiv \int_D \phi_n(\mathbf{x})S_{\mathbf{x}}^p[\gamma_p(\mathbf{x})]d\mathbf{x} \tag{6.27}$$

due to the fact that the Green's function, and thus the eigenfunctions, satisfy homogeneous boundary conditions (Burke and Hubbard, 1990). Thus, equation (6.27) serves the same purpose as equation (6.16): the structural modes are spatially weighted according to the window of the sensor or actuator. In fact, these coefficients represent the elements that populate the system output matrix, $C$, for a state space realization. If $c_{pn} \neq 0$, the $n$th mode is observable by the $p$th sensor. If $c_{pn} \neq 0$ for all $(p, n)$, then the $P$ sensors provide all-mode ($N$-mode) observability, which is obviously tied directly to the observability of an LTI system, which was discussed in Chapter 3.

The purpose of presenting this derivation has been to demonstrate that, regardless of the type of actuator or sensor, the spatial window dictates the qualitative operation and the interaction with the structure, including issues such as observability and controllability. Returning to the original example of the simply supported beam, if a point force actuator or sensor is applied at coordinate $x_s$, as illustrated in Figure 6.1, then the coefficients $b_{ns}$ and $c_{sn}$, which contain the effect of the spatial window, are expressed as follows:

$$b_{ns} = c_{sn} = \phi_n(x_s) = \sqrt{\frac{2}{m'L}} \sin\left(\frac{n\pi x_s}{L}\right) \tag{6.28}$$

Similarly, if a rectangular piezoelectric actuator or sensor is positioned between spatial coordinates $x_{1s}$ and $x_{2s}$ as illustrated in Figure 6.1, then the coefficients modeling the effect of the spatial window are

$$b_{ns} = c_{sn} = \frac{d\phi_n(x_{2s})}{dx} - \frac{d\phi_n(x_{1s})}{dx}$$

$$= \sqrt{\frac{2(n\pi)^2}{m'L^3}} \left[\cos\left(\frac{n\pi x_{2s}}{L}\right) - \cos\left(\frac{n\pi x_{1s}}{L}\right)\right] \tag{6.29}$$

Equations (6.28) and (6.29) convey an important concept. For example, if $x_s = L/2$ in equation (6.28), then the structural modes with odd integer values of $n$ are the only modes that can be observed or controlled for the given location. Similar characteristics are observed in equation (6.29); however, there is one distinguishing difference. In the model of the piezoelectric element, the modal influence function is scaled according to the order of the modal index $n$. The scaling term results from the operation of the material since it is a strain sensitive device and must be included for accurate modal filtering. This scaling term tends to place a greater emphasis on higher order modes, which is an important consideration in the design and synthesis of structural controllers. As a general comment, one should recognize that the apertures need not be limited to spatial Diracs or boxcars. The formulation presented is general and applicable to any other shape that can be realized through a general Fourier decomposition of the aperature, influencing both modal controllability and observability.

## 6.3 COLOCATED SENSING AND ACTUATION

As previously outlined in Chapter 5, colocated sensors and actuators are often employed in the design of adaptive structures for active damping, either through direct-velocity-feedback (DVFB) control or positive-position-feedback (PPF) control. In the absence of dynamics associated with the transduction devices and signal conditioning units, enhanced stability characteristics result when actuators and sensors are colocated. In the most general sense, colocation is achieved when the state-space form of (see Chapter 3 for state-space modeling)

$$\mathbf{C}^{\mathrm{T}} = \mathbf{B} \tag{6.30}$$

In other words, the observation matrix is proportional to the transpose of the input matrix in state-variable form when the transduction devices are colocated. Based upon the development in Section 6.2, it is obvious that, if $S^p_{\mathbf{x}} = L^m_\eta$ and $\gamma_p(\mathbf{x}) = \beta_m(\eta)$, then the corresponding actuator and sensor are effectively colocated in their operation, and the devices are of necessity identical. Examples of such colocated sensing and actuation devices has been presented in equations (6.28) and (6.29). However, upon considering equations (6.16) and (6.27), one recognizes that colocation can be achieved if

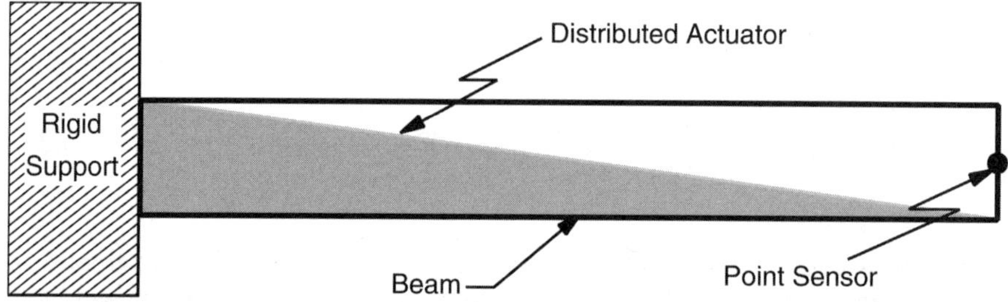


**Figure 6.2:** Schematic diagram of a cantilevered beam with a colocated transducer pair.

$$L_{\xi}^{m}[\beta_m(\xi)] = S_{\mathbf{x}}^{p}[\gamma_p(\mathbf{x})] \tag{6.31}$$

as noted by Burke and Hubbard (1991). In fact, as long as the spatial operation of the sensor and actuator are related by a constant of proportionality, the transduction devices satisfy the necessary condition for colocation.

### 6.3.1 An Example of Different Transducers Implemented as a Colocated Pair

To provide an example of a more general case of colocated transduction devices, consider the cantilevered beam illustrated in Figure 6.2. A distributed piezoelectric actuator is positioned over the length of the beam with a spatial distribution function given as follows:

$$\beta_1(\xi) = u(\xi) - [r(\xi) - r(\xi - L)] \tag{6.32}$$

where $u(\xi)$ is the unit step function and $r(\xi)$ is the ramp function. The spatial aperture of the actuator is illustrated in Figure 6.2. The corresponding spatial operator for the piezoelectric actuator is

$$L_{\eta}^{1} = \frac{d^2}{d\eta^2}$$

Thus,

$$L_{\xi}^{1}[\beta_1(\xi)] = \frac{d^2}{d\xi^2}\,[u(\xi) - \{r(\xi) - r(\xi - L)\}] \tag{6.33}$$

$$= \frac{d\,\delta(\xi)}{d\xi} - \delta(\xi) + \delta(\xi - L) \tag{6.34}$$

The distributed piezoelectric device results in a point slope input at the root as well as a point displacement input at the root and tip (Burke and Hubbard, 1991). Due to physical constraints, the point displacement and slope inputs at the root can be ignored, yielding:

$$L_{\xi}^{1}[\beta_1(\xi)] = \delta(\xi - L) \tag{6.35}$$

If a point sensor is placed at the tip of the beam such that

$$\gamma_1(x) = \delta(x - L) \tag{6.36}$$

and

$$S_x^1 = \frac{d^0}{dx^0} \tag{6.37}$$

then

$$S_x^1[\gamma_1(x)] = \delta(x - L) \tag{6.38}$$

and the colocation condition is satisfied even though the sensor and actuator have totally different spatial operators and apertures! Although this is a special case, it serves to demonstrate an important concept: *one does not have to place actuators and sensors at the same spatial coordinates to achieve colocation.* As illustrated in Chapter 9, colocated transduction devices simplify the control system design and can result in significant performance advantages for adaptive structures while implementing the design in physical coordinates as opposed to modal coordinates.

### 6.3.2 Effects of Spatial Aperture and Spatial Differential Operator

In the design of adaptive structures, the spatial aperture and spatial differential operator are the two elements most critical to the integration of the transducer. The spatial aperture and differential operator control the low-frequency and high-frequency asymptotes of the frequency response for colocated transducer pairs. They also serve to control the corner frequency of the dereverberated structure (i.e., the structural response in the absence of reflections associated with boundary conditions). Burke and Hubbard (1991) detailed the effects of the spatial aperture and differential operator on the frequency response of colocated transducers, and in a more recent study, McCain and Crawley (1995) investigated the asymptotic limits of the structural frequency response of beams with various boundary conditions, spatial apertures, and differential operators. The results of these studies are reviewed here, since understanding the interaction of the transducer design and structural dynamics is paramount to the realization of adaptive structures.

Consider a beam with pinned boundary conditions whose response can be described from Bernoulli-Euler beam theory:

$$EI\frac{\partial^4 w(x,t)}{\partial x^4} + m'\frac{\partial^2 w(x,t)}{\partial t^2} = f(x,t) \tag{6.39}$$

where $EI$ is the flexural rigidity, $m'$ is the mass per unit length, $w(x,t)$ is the displacement response, and $f(x,t)$ is the applied force. One can solve this differential equation through a modal or wave approach as outlined in Chapter 2. Since asymptotic limits are of interest in this example, the wave approach is used to eliminate the need for retaining an excessive number of terms in the series expansion. For harmonic motion, the solution can be expressed as follows:

$$w(x,t) = (A_1\exp(jkx) + A_2\exp(-jkx) + A_3\exp(kx) + A_4\exp(-kx))\exp(j\omega t) \tag{6.40}$$

where the structural wavenumber is defined as

$$k^4 = \frac{m'\omega^2}{EI} \tag{6.41}$$

The solution is thus composed of left and right traveling waves, as well as left and right evanescent (decaying near-field) waves. The spatial apertures under consideration in this example are a Dirac, a rectangle, and a triangle, as illustrated in Figure 6.3.

If a colocated point actuator-sensor pair (i.e., Dirac aperture) is incorporated at the center of the beam, one can express the frequency response as follows:

$$H_D(\omega) = -\frac{\exp(kl/2)(\sin(kl/2) - \cos(kl/2)) + \exp(-kl/2)(\sin(kl/2) + \cos(kl/2))}{4EIk^3\cos(kl/2)(\exp(kl/2) + \exp(-kl/2))} \tag{6.42}$$

where it is understood that $k$ is a function of $\omega$, and $l$ is the length of the beam.

A script file **colocate.m** has been generated to plot the frequency response of various actuator-sensor pairs centered on the beam with pinned boundaries. Dimensions of the structure and material properties are included in the script file. The magnitude of the frequency response of the beam for the actuator-sensor pair described by a spatial Dirac is presented in Figure 6.4 and can be generated with the script file. Both the low-frequency and high-frequency asymptotes of the structural frequency response are plotted with the reverberant frequency response.

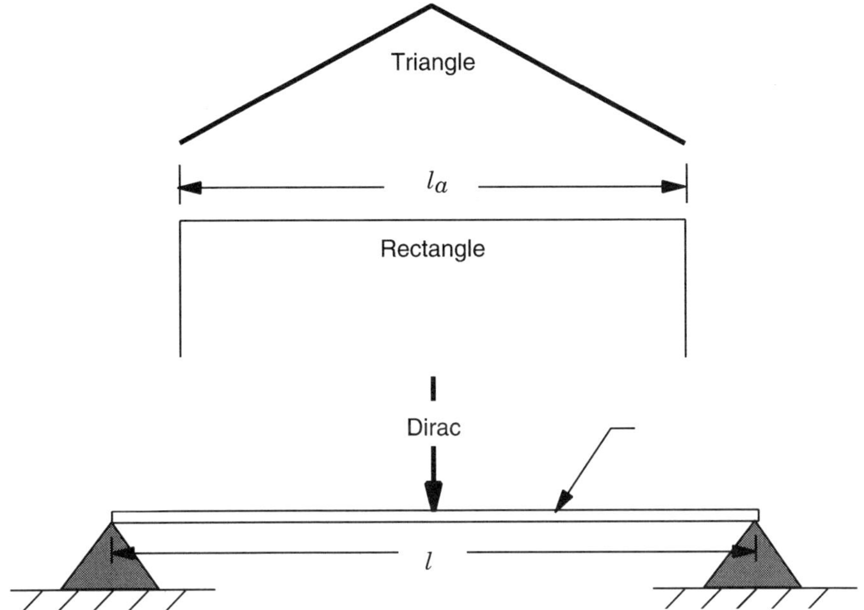


**Figure 6.3:** Schematic diagram of various transducer pair spatial apertures.

The low-frequency asymptote for the actuator-sensor pair centered on the beam can be obtained by taking the limit of equation (6.42) as $k \to 0$. The result is $-(l^3/48EI)$. Similarly, the high-frequency asymptote for the actuator-sensor pair centered on the beam can be obtained by taking the limit as $k \to \infty$. The result is $-(1/2EIk^3)$ (McCain and Crawley, 1995). Although not shown in equation (6.42), the low-frequency asymptote varies with position. As the

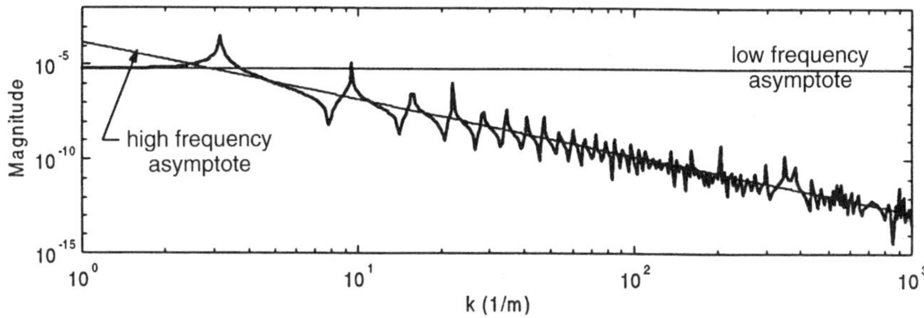


**Figure 6.4:** Magnitude of frequency response for a beam with pinned boundary conditions and a colocated point-force/point-displacement transducer pair centered at beam midpoint.

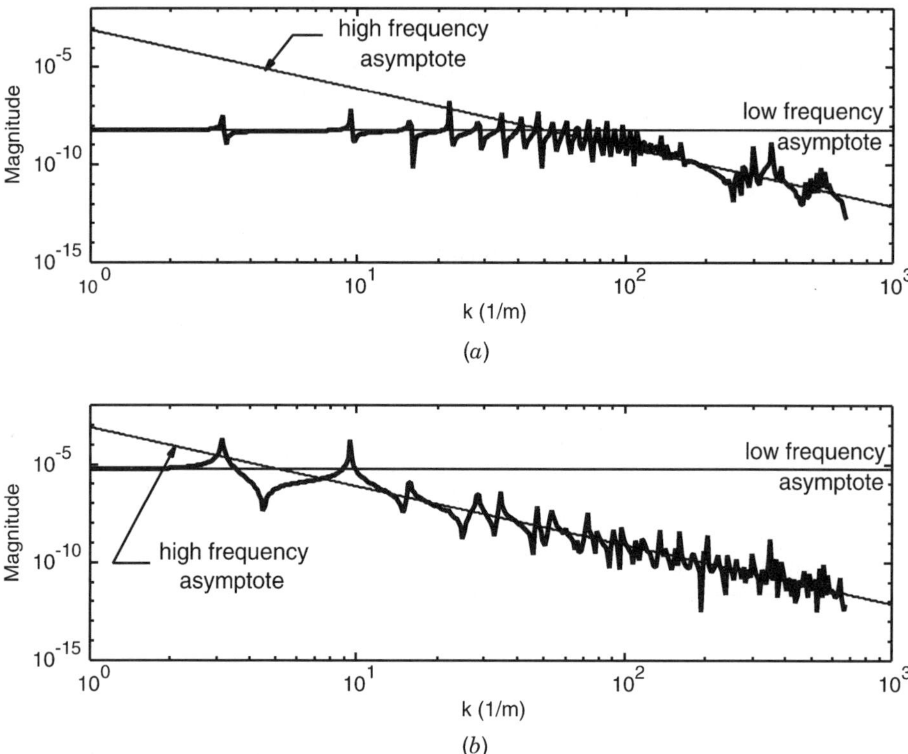


**Figure 6.5:** Magnitude of frequency response for a beam with pinned boundary conditions and a colocated rectangular transducer pair centered at beam midpoint. (*a*) Transducer covering 6% of the beam. (*b*) Transducer covering 60% of the beam.

actuator-sensor pair is moved closer to the boundary, the low-frequency asymptote decreases in magnitude, due to an increase in stiffness closer to the structural boundary. The intersection of the low-frequency and high-frequency asymptotes dictate the "corner frequency" of the structural frequency response, and thus as one moves the actuator-sensor pair closer to the boundary, the corner frequency will increase. The high-frequency asymptote does not change with position and has a rolloff proportional to $k^{-3}$.

A similar expression (to that of equation (6.42)) for the frequency response associated with a colocated rectangular transducer pair centered on the beam can be developed (McCain and Crawley, 1995), and the appropriate equation is incorporated in the script file **colocate.m**. The script file can be used to generate the frequency response functions illustrated in Figure 6.5. As in the previous example, the low-frequency and high-frequency asymptotes are plotted with the reverberant frequency response for the actuator-sensor pair. For this aperture, the low-frequency asymptote is $(l_a^3/EI$ and the high frequency asymptote is $(1/2EIk)$ where $l_a$ is the length of the rectangular aperture. Notice that the high-frequency asymptote has a rolloff proportional to $k^{-1}$. The high-frequency asymptote remains unaffected by the aperture; however, the low-frequency asymptote is observed to increase with increasing aperture length. This increase can be viewed as a decrease in the stiffness as seen through the actuator-sensor pair.

The magnitude of the frequency response of a rectangular transducer pair covering 6% of the beam is presented in Figure 6.5*a* and the magnitude of the frequency response of a transducer pair covering 60% of the beam is illustrated in Figure 6.5*b*. Notice the increase in pole-zero separation at low frequency with increasing aperture size. Physically, the aperture-dependent characteristic is a function of the ratio of the actuator length to that of the standing waves on the structure. As the actuator spans a length equal to or greater than a spatial wavelength of a standing wave, the transducer pair is no longer capable of efficiently coupling to that standing wave. The corner frequency associated with the intersection of the asymptotes thus decreases as the length of the rectangular aperture is increased. The implication is that larger rectangular apertures are required to control long-wavelength, low-frequency, structural modes.

The final aperture under consideration is the triangular aperture, which is a spatial Bartlet window. The equation required to compute the frequency response of the structure with a colocated, triangular, transducer pair centered on the structure is included in the script file **colocate.m**. Executing this script file produces plots of the magnitude of the frequency response for apertures extending over 6% and 60% of the beam, as illustrated in Figure 6.6*a* and Figure 6.6*b*, respectively. For the triangular aperture, the low-frequency asymptote is $(l_a^3/12EI)$, and the high-frequency asymptote is $(5/2EIk^3)$ (McCain and Crawley, 1995). As was the case for the colocated point force/sensor pair, the high-frequency rolloff associated with the colocated triangular transducer pair is proportional to $k^{-3}$. However, like the colocated rectangular transducer, the low-frequency asymptote is controlled by the length of the aperture and the ratio of this aperture to standing waves on the structure, as can be seen by example upon comparing Figure 6.6*a* to Figure 6.6*b*. Like the rectangular, colocated, transducer pair, the triangular actuator-sensor pair can be moved to various locations on the structure without affecting the analytical value of the asymptotes (McCain and Crawley, 1995).

From the examples provided, it is obvious that the low-frequency asymptote, high-frequency asymptote, and the corner frequency associated with the intersection of the asymptotes can be modified by choice of aperture, position, and differential operator associated with the physics of the transducer pair. As indicated by the frequency response functions (FRFs) presented, the spacing between the poles and zeros of the colocated induced strain actuator pairs (rectangular and triangular apertures) is relatively small if the ratio of the aperture to that of the structure is small. As a result, very little performance can be achieved with direct output feedback control (static gain) when the ratio of the transducer aperture to that of the standing waves on the structure is small. However, when the ratio is large (i.e., of order of the structural dimension), the low-frequency

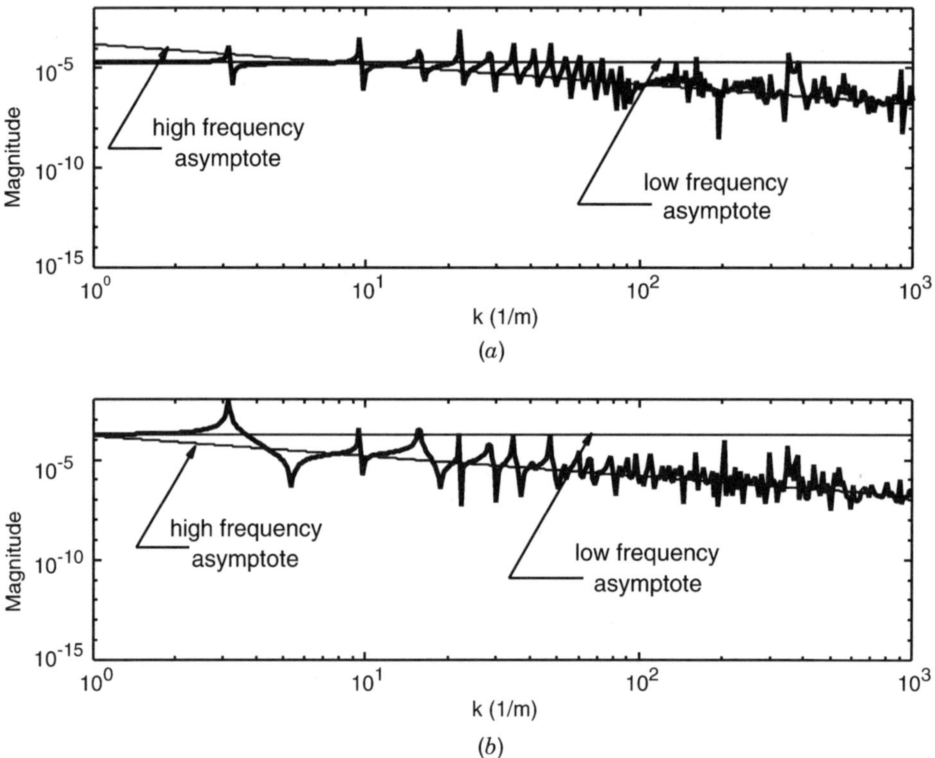


**Figure 6.6:** Magnitude of frequency response for a beam with pinned boundary conditions and a colocated triangular transducer pair centered at beam midpoint. (*a*) Transducer covering 6% of the beam. (*b*) Transducer covering 60% of the beam.

response of the structure is emphasized, and the pole-zero separation increases, as was illustrated in Figures 6.5*b* and 6.6*b*. For such distributed transducer pairs, colocated output-feedback control can be used to control the low-frequency response with reasonable expectations for performance. However, it may not be practical to extend an actuator-sensor pair over the entire surface of a structure, in which case dynamic compensation will be a necessity to enhance the low-frequency performance of the system. These issues are addressed in Chapter 9; however, they are mentioned here to motivate the reader to consider all aspects of adaptive structure design, as they will ultimately impact the complexity of the control system design and realization.

## 6.4 MODAL SENSING AND ACTUATION

Modal sensors and actuators are a special class of spatial sensors and actuators. The uniqueness of such devices rests in the fact that the structural response is measured in the modal coordinates, as opposed to the natural coordinates. Meirovitch and Baruh (1983) first developed the concept of independent modal space control (IMSC); however practical realization often proves difficult due to the problems associated with accurately estimating the modal response of structures with a discrete array of transducers. Until recently (Freudinger, 1990; Shelly, 1991), methods of calculating modal filter vectors for arbitrary structures were not available. The objective in modal filtering is to design a device that operates on single or multiple modes of the structure. Two different approaches to this problem are outlined in this section: modal filtering with a single distributed element and modal filtering with an array of discrete elements.

### 6.4.1 Modal Filtering with a Distributed Element

Upon considering equation (6.16) of the previous section, one observes that in order to selectively actuate structural modes, the integral equation must equal zero for all values of $m \neq n$. The task is thus to find a spatial window that is orthogonal to all other structural modes. Alternately, to selectively observe structural modes, the integral of equation (6.27) must equal zero for all values of $p \neq n$. For now, we limit our discussion to the propagation of flexural waves on a two-dimensional structure. As shown by Lee and Moon (1990), the response of a piezoelectric sensor deposited over the surface of a thin rectangular plate can be expressed as follows:

$$q(t) = \frac{h_p + h_s}{2} \int_0^{L_x} \int_0^{L_y} \Gamma(x, y) \left( e_{31}^0 \frac{\partial^2 w}{\partial x^2} + e_{32}^0 \frac{\partial^2 w}{\partial y^2} + e_{36}^0 \frac{\partial^2 w}{\partial x \partial y} \right) dy\, dx \tag{6.43}$$

where $q(t)$ is the sensor output charge; $w(x, y, t)$ is the displacement field of the plate; $e_{31}^0$ and $e_{32}^0$ denote the piezoelectric's charge per unit area in the $x$ and $y$ directions, respectively; $e_{36}^0$ is the charge per unit area in the cross-axis (skew) direction; $\Gamma(x, y)$ is the piezo's shape and polarization profile as a function of area; $h_p$ is the thickness of the plate; $h_s$ is the thickness of the piezoelectric transducer; and $L_x$ and $L_y$ are the $x$ and $y$ dimensions of the plate. We assume $e_{36}^0 = 0$, for example, the sensor's piezoelectric axes are coincident with the geometric axes of the plate (no skew). With respect to the terminology introduced in the previous section, $S_t^p = 1$ and

$$\gamma(\mathbf{x}) S_{\mathbf{x}}^p = \frac{h_p + h_s}{2} \Gamma(x, y) \left( e_{31}^0 \frac{\partial^2}{\partial x^2} + e_{32}^0 \frac{\partial^2}{\partial y^2} + e_{36}^0 \frac{\partial^2}{\partial x \partial y} \right) \tag{6.44}$$

If we assume the solution of the plate equations is separable, consistent with the presentation in Chapter 2 (which is only rigorously true if at least two opposing sides of the plate are simply supported; otherwise the expansion holds as a limit in the mean (Love, 1944)), then the structural response can be expressed as follows:

$$w(x, y, t) = \sum_{m=1}^{M} \sum_{n=1}^{N} W_{mn}\varphi_m(k_x^{(m)}x)\psi_n(k_y^{(n)}y)\exp(j\omega t) \tag{6.45}$$

where $k_y^{(n)}$ and $\psi_n$ are eigenvalues and eigenfunctions of the separable solution in the $y$ direction and $k_x^{(m)}$ and $\varphi_n$ are eigenvalues and eigenfunctions in the $x$ direction.

Consistent with the presentation of Clark and Burke (1996), one can develop an expression for a sensor varying in the $y$ dimension as a function of $x$ and extending over the $x$ dimension of a panel with pinned boundary conditions at $x = 0$ and $x = L_y$, as follows:

$$q(t) = \frac{h_p + h_s}{2} \int_0^{L_x} \int_{a-bS(x)}^{a+bS(x)} \left\{ \sum_{m=1}^{M} \sum_{n=1}^{N} W_{mn} \sin\left(\frac{m\pi y}{L_y}\right) \right. \\ \times \left[ e_{31}^0 \frac{\partial^2 \varphi_m(k_x^{(m)}x)}{\partial x^2} - e_{32}^0 \left(\frac{n\pi}{L_y}\right)^2 \varphi_m(k_x^{(m)}x) + \right] \left. \exp(j\omega t) \right\} dy\, dx \tag{6.46}$$

where the sensor aperture is shaped along the $x$ direction, as illustrated in Figure 6.7, such that

$$S(x) = \sum_{k=1}^{K} S_k \gamma_k(x) \tag{6.47}$$

Substituting equation (6.47) into equation (6.46), and assuming that the $y$ direction mode shapes are sinusoids, one obtains:

$$q(t) = -2\exp(j\omega t)b(h_p + h_s) \sum_{k=1}^{K} \sum_{m=1}^{M} \sum_{n=1}^{N} \left\{ W_{mn} S_k \sin\left(\frac{n\pi a}{L_y}\right) \right. \\ \cdot \int_0^{L_x} \left[ e_{31}^o \varphi_m''(k_x^{(m)}x) - e_{32}^0 \left(\frac{n\pi}{L_y}\right)^2 \varphi_m(k_x^{(m)}x) \right] \gamma_k(x)dx \left. \vphantom{\int} \right\} \tag{6.48}$$

Hence, if the objective is to observe a single mode, then the following integrals must be zero for $m \neq k$:

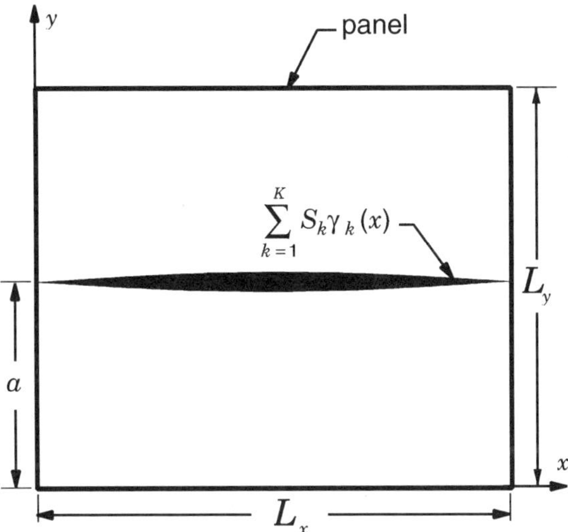


**Figure 6.7:** Schematic diagram of a distributed transducer with spatial aperture described by $S(x)$.

$$\int_0^{L_x} \varphi''_m(k_x^{(m)}x)\gamma_k(x)dx \tag{6.49}$$

and

$$\int_0^{L_x} \varphi_m(k_x^{(m)}x)\gamma_k(x)dx \tag{6.50}$$

Unless $\varphi_m(k_x^{(m)}x)$ is a sinusoidal function, it will not be orthogonal to its Laplacian. As a result, the integral of equation (6.48) will include *cross terms* for $m \neq k$. This result indicates that for nonsinusoidal mode shapes in $x$, the modal aperture shaping for a plate cannot be rigorously achieved. This does not mean that an aperture yielding a good approximation to the desired modal filter cannot be designed, but it does serve to place a design constraint on the realization of modal sensors for plate structures.

However, for structure whose response can be described as a function of a single dimension (i.e., beams), the design criterion can be concisely expressed as follows:

$$\int_0^{L_x} \varphi''_m(k_x^{(m)}x)\gamma_k(x)dx = \delta_{mk} \tag{6.51}$$

For beams with pinned boundary conditions, the Laplacian of the eigenfunction is proportional to the eigenfunction and hence the sensor aperture can be designed accordingly. However, for other beams, the aperture should be designed according to the curvature of the desired eigenfunction.

With respect to the terminology introduced in the previous section, the sensor aperture is defined by the following relationship, where the function is made proportional to the mass ($m'(x)$) to satsify equation (6.4):

$$S_{\mathbf{x}}^{m}[\gamma_m(x)] = m'(x)\varphi_m(k_x^{(m)}x) \tag{6.52}$$

Hence, in general, the operation of the material used to create the sensor (i.e., $S_{\mathbf{x}}^{m}$) must be incorporated into the design. For a piezoelectric material, the linear operator can be represented by a second-order differential, as indicated in equation (6.12), since the material is sensitive to strain. Thus, the spatial window for this material is obtained by integrating twice:

$$\gamma_m(x) = \int\int m'(x)\varphi_m(k_x^{(m)}x)dx\, dx \tag{6.53}$$

While the example problems in this text are purposely restricted to systems with analytical solutions to convey the basic concepts, the formulation for designing modal sensors and actuators is applicable for structures that must be characterized experimentally or numerically. As previously discussed, distributed modal sensors cannot be rigorously generalized to two-dimensional structures, but approximations to desired modal filters can be achieved, as has been demonstrated by Sullivan et al. (1996). Even if it were possible to design the exact aperture for two-dimensional structures, realization of such apertures through variations in polarization profile presents a different hurdle. As illustrated in Figure 6.8, a modal sensor for a one-dimensional structure, such as the simply supported beam, can be realized by shaping the sensor to replicate the desired spatial window, in this case, the first structural mode. However, for a two-

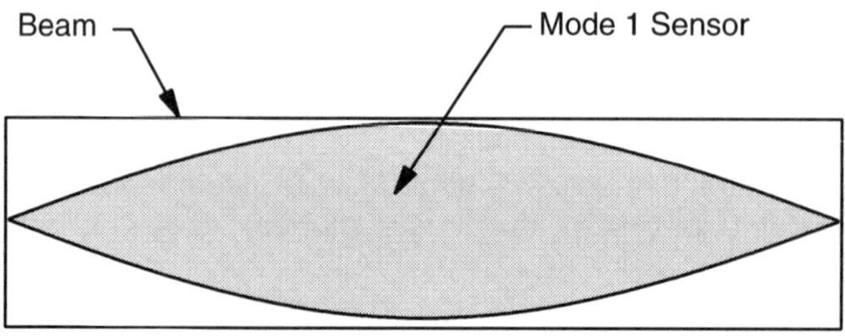


**Figure 6.8:** Schematic diagram of a first structural mode sensor for a simply supported beam.

**Figure 6.9:** Schematic diagram of a first structural mode sensor for a simply supported plate.

dimensional structure such as the simply supported plate (a structure for which the modal sensor can be designed), the spatial window varies as a function of both dimensions and must be realized by effectively varying the polarization profile or thickness of the material as a function of the out-of-plane displacement. A typical modal sensor can be designed to observe the first structural mode of the simply supported plate, with the thickness (or other properties) varying across the plate with the shape shown in Figure 6.9. A number of possible methods for achieving such shapes have been suggested (Burke and Hubbard, 1991); however, practical methods of constructing such spatial windows are currently under development (Burke and Sullivan, 1995).

### 6.4.2 Modal Filtering with Discrete Elements

Due to the limitations outlined with respect to distributed modal filtering, alternative methods must be considered. Modal filtering with an array of discrete elements was first introduced by Meirovitch and Baruh in 1983, along with the concept of IMSC. The objective in modal filtering is to identify a vector such that the inner product of that vector with a vector containing the response of an array of sensors applied in physical coordinates accurately represents the response of the structure in modal coordinates. A procedure for determining the modal filter vector is presented here for sensors with arbitrary spatial windows and operation. Recalling equation (6.18),

$$y_p(\mathbf{x}, s) = \sum_{n=1}^{N} \eta_n(s) k_p(s) \int_D m'(\mathbf{x}) \gamma_p(\mathbf{x}) \phi_n(\mathbf{x}) d\mathbf{x}$$

and substituting the result of equation (6.27), one obtains

$$y_p(\mathbf{x}, s) = \sum_{n=1}^{N} \eta_n(s) k_p(s) c_{pn} \tag{6.54}$$

Thus the response at the $p$th error sensor can be expressed in terms of the $N$ modal coordinates and the selected temporal filtering $k_p(s)$. The operation of the selected sensor (i.e., strain, displacement, etc.) is contained in the modal influence function $c_{pn}$, as indicated in equation (6.27). The spatial window created by the sensor is contained in this equation as well. In general, one needs at least as many discrete error sensors as there are modes present in the structural response to accurately estimate the modal response. Thus if the response of $N$ modes is desired, $M$ sensors are required where $M \geq N$ (one must obey the sampling theorem). If the temporal filter associated with equation (6.54) is the same for all sensors, then one can express the response of $M$ sensors in vector notation:

$$\mathbf{y}(s) = k(s) \mathbf{C}' \eta(s) \tag{6.55}$$

where $\mathbf{y}(s)$ is an $M \times 1$ vector of sensor outputs, $k_p(s) = k(s)$ for all $M$ sensors, $\mathbf{C}'$ is an $M \times N$ matrix of modal influence coefficients with $c_{pn}$ as the element of the $p$th row and $n$th column, and $\eta(s)$ is an $N \times 1$ vector of the modal response. The solution of these equations can be obtained by taking the pseudo-inverse:

$$\eta(s) = [\mathbf{C}'^{\mathrm{T}} \mathbf{C}']^{-1} \mathbf{C}'^{\mathrm{T}} \frac{\mathbf{y}(s)}{k(s)} \tag{6.56}$$

where the $n$th row of $[\mathbf{C}'^{\mathrm{T}} \mathbf{C}']^{-1} \mathbf{C}'^{\mathrm{T}}$ constitutes a least-mean-squares (LMS) solution for the $n$th modal filter vector. (Methods of designing discrete modal filters and the performance degradation resulting from the use of a finite array of sensors was detailed by Morgan (1991)).

An alternative approach to the design of the appropriate modal filter vectors was provided in a recent study by Shelly et al. (1993). In that study, the adaptive LMS algorithm (see Chapter 4) was used to determine the modal filter vector by constructing the error such that

$$\mathbf{e}_n(k) = \phi_n(k) - \varphi_n^{\mathrm{T}}(k) \mathbf{y}(k) \tag{6.57}$$

where $\phi_n(k)$ is the true discrete time response of the $n$th modal coordinate at the $k$th time step, $\varphi_n(k)$ is the $k$th iterate of the $n$th modal filter vector, and $\mathbf{y}(k)$ is a vector containing the output of an array of sensors at the $k$th iteration. Taking the expected value of the error squared yields a quadratic cost function with a unique minimum, as discussed in Chapter 4. Implementing the LMS algorithm, one obtains

$$\varphi_n(k+1) = \varphi_n(k) + \mu \mathbf{e}_n(k)\mathbf{y}(k) \tag{6.58}$$

Upon converging, the coefficients $\varphi_n(k)$ of the modal filter vector will be identical to the coefficients of the $n$th row of $[\mathbf{C}'^{\mathrm{T}}\mathbf{C}']^{-1}\mathbf{C}'^{\mathrm{T}}$.

In both of the techniques described above, one must have a-priori knowledge of the structural mode shapes obtained from some form of system identification to design the modal filter vectors. Shelly et al. (1993) attempted to reduce the level of identification required by driving a simple second-order system with the same force used to control the structure; however, the modal participation vector was needed to effectively tranform the force to modal coordinates. Since the modal participation vector is essentially what you are trying to determine, a true adaptive modal filter has not yet been demonstrated.

As a final note, the discussion in this section has focused primarily on sensing; however, if an array of discrete actuators identical in shape and form is used to control the structure, one can use the same approach to design the modal filter vector for actuation. In this case, the modal filter vector is used to selectively excite structural modes with a discrete array of actuators. In fact, if the operation of the actuator is identical to that of the sensor (for example, a rectangular piezoelectric element), then the same modal filter vector used to monitor the response of independent modes can be used to control the response of independent modes.

## 6.5 SOME EXAMPLES OF SPATIAL AND TEMPORAL SIGNAL PROCESSING

If vibration control of resonant structures is the primary objective, then spatial filtering for actuation and sensing may suffice. However, a number of control problems require *both spatial and temporal* signal processing. For example, if the objective is to control the structural acoustic response of the system, one must account for the interaction between structural modes as a function of frequency. Thus, both spatial and temporal signal processing are required to predict and control the far-field sound radiation. As another example, consider the infinite beam in flexure, as discussed in Chapter 2. To control the vibration in this type of structure, one can minimize the response of the traveling waves due to some disturbance input. The approach is to assume a solution of the waveform that can be expressed in terms of real and complex exponentials or an infinite Fourier series. For infinite beams, or systems displaying such characteristics (i.e., where the reflected wave is very small), the response is best modeled with exponentials since the expression can be reduced to four terms: the left and right propagating waves and the left and right evanescent or decaying (near-field) waves. To control the traveling waves in the structure, one must design a sensor that can monitor the amplitude of the waves. Such a sensor requires the combination of spatial and temporal signal processing. This section is devoted to presenting practical examples of such systems.

### 6.5.1 Spatial and Temporal Signal Processing to Estimate Traveling Waves in a Nondispersive Medium

Let us consider waves traveling in a nondispersive medium, such as extensional waves in beams (or one-dimensional acoustic plane waves). Extensional motion in beams consists of only traveling waves (decaying waves are not solutions to the differential equation of motion). In this case, the response of the beam can be expressed as follows:

$$U(x,t) = [E\exp(jk_e x) + F\exp(-jk_e x)]\exp(-j\omega t) \tag{6.59}$$

where $U(x,t)$ is the total in-plane displacement, $k_e$ is the extensional wavenumber defined in Chapter 2, $E$ is the positive traveling extensional wave, $F$ is the negative traveling extensional wave, and $\omega$ is the circular frequency of excitation. Let us position four rectangular piezoelectric elements on the surface of the beam in symmetric pairs, as illustrated in Figure 6.10. The sensors are separated by a distance $L_{es}$ and are of length $L_s$. Recalling the expression for the response of a piezoelectric element to uniform strain in one dimension, one can express the output of each of the four sensors as follows:

$$q_1(t) = -2je_{31}^0 L_y \sin\left(\frac{k_e L_s}{2}\right)[E+F]\exp(-j\omega t) \tag{6.60}$$

$$q_2(t) = -2je_{31}^0 L_y \sin\left(\frac{k_e L_s}{2}\right)[E+F]\exp(-j\omega t) \tag{6.61}$$

$$q_5(t) = -2je_{31}^0 L_y \sin\left(\frac{k_e L_s}{2}\right)[E\exp(jk_e L_{es}) + F\exp(-jk_e L_{es})]\exp(-j\omega t) \tag{6.62}$$

$$q_6(t) = -2je_{31}^0 L_y \sin\left(\frac{k_e L_s}{2}\right)[E\exp(jk_e L_{es}) + F\exp(-jk_e L_{es})]\exp(-j\omega t) \tag{6.63}$$

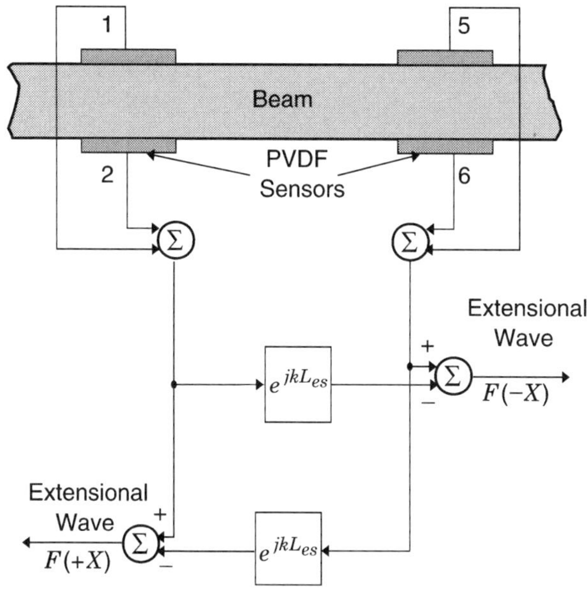


**Figure 6.10:** Schematic diagram of extensional wave sensor.

where the subscript indicates the position of the sensor as illustrated in Figure 6.10. To resolve the extensional motion from the output of the sensors, one must sum the outputs from the symmetrically positioned sensors:

$$q_{e12} = q_1(t) + q_2(t) \tag{6.64}$$

$$q_{e56} = q_5(t) + q_6(t) \tag{6.65}$$

This assures that the response of the flexural waves is eliminated as long as the sensors have the same sensitivity and are colocated at symmetric positions on the structural surface. (If the sensitivities of the sensors are not the same, they can be matched by adjusting hardware signal conditioner gains.) Substituting equations (6.60)–(6.63) into equations (6.64)–(6.65), one obtains

$$\begin{bmatrix} q_{e12} \\ q_{e56} \end{bmatrix} = -4e_{31}^0 L_y \sin\left(\frac{k_e L_s}{2}\right) \begin{bmatrix} 1 & 1 \\ \exp(jk_e L_{es}) & \exp(-jk_e L_{es}) \end{bmatrix} \begin{bmatrix} E \\ F \end{bmatrix} \exp(-j\omega t) \tag{6.66}$$

The objective is to solve the system of equations such that the response of the right and left traveling waves are resolved in real time. This must be accomplished with temporal signal processing. As demonstrated by Gibbs (1993), if

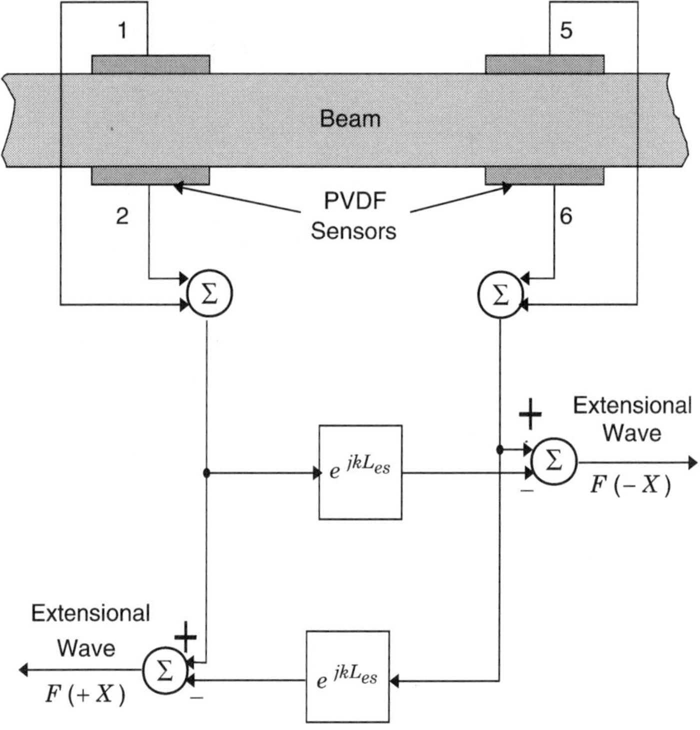


**Figure 6.11:** Flow chart of extensional wave sensor implementation.

the sensors are positioned (i.e., the spatial filtering) a full wavelength apart at the sample rate frequency, then a pure time delay (i.e., the temporal filtering, represented by a [0 1] finite impulse response (FIR) filter) yields the desired solution to equation (6.66). The solution to the system of equations is presented in the schematic of Figure 6.11. Mathematically, one can solve the system of equations by subtracting a "spatially delayed" version of each respective sensor from the other:

$$\begin{bmatrix} q_{e12} - q_{e56}\exp(jk_eL_{es}) \\ q_{e56} - q_{e12}\exp(jk_eL_{es}) \end{bmatrix} = -4e_{31}^0 L_y \sin\left(\frac{k_eL_s}{2}\right)\begin{bmatrix} 1-\exp(j2k_eL_{es}) & 0 \\ 0 & j\sin(k_eL_{es}) \end{bmatrix} \times \begin{bmatrix} E \\ F \end{bmatrix}\exp(-j\omega t) \tag{6.67}$$

The system of equations is now in diagonal form, but maintains frequency dependence through the extensional wavenumber term $k_e$. However, if one lets

$$\exp(jk_eL_{es}) = \exp(j\omega T) \tag{6.68}$$

then a temporal filter (i.e., the [0 1] FIR filter) can be incorporated to effectively model the physical time delay inherent in the propagation of the wave in the structure. This is readily observed upon substituting equation (6.68) into equation (6.67):

$$\begin{bmatrix} q_{e12} - q_{e56}\exp(j\omega T) \\ q_{e56} - q_{e12}\exp(j\omega T) \end{bmatrix} = -4e_{31}^0 L_y \sin\left(\frac{k_eL_s}{2}\right)\begin{bmatrix} 1-\exp(j2k_eL_{es}) & 0 \\ 0 & j\sin(k_eL_{es}) \end{bmatrix} \times \begin{bmatrix} E \\ F \end{bmatrix} \exp(-j\omega t) \tag{6.69}$$

Thus, one can resolve a signal proportional to the right and left traveling extensional waves of the beam in real time since

$$\begin{aligned} E &\propto q_{e12}(k) - q_{e56}(k-1) \\ F &\propto q_{e56}(k) - q_{e12}(k-1) \end{aligned} \tag{6.70}$$

If the exact amplitude of the traveling waves is desired, one can incorporate the physical dimensions of the structure, sensors and the spatial separation between the sensors along with the associated piezoelectric stress constant to calibrate the output.

It is convenient at this point to reflect upon equation (6.69) and discuss the physics associated with the spatial filter. It is apparent that the matrix on the right-hand side of the equation must be nonsingular. This requirement dictates that

$$0 < \frac{k_eL_s}{2} < \pi \tag{6.71}$$

and

$$0 < k_eL_{es} < \pi \tag{6.72}$$

Physically, this means that the separation distance between the sensors ($L_{es}$) and the length of the sensors ($L_s$) is determined by the shortest structural wavelength over the frequency range of application. For example, if the maximum frequency of interest is $f_{\max}$, then the sample rate must be

$$f_s = 2f_{\max} \tag{6.73}$$

to satisfy the Nyquist criterion. Substituting this expression into equation (6.71) and equation (6.72) yields:

$$0 < L_s < \frac{c}{f_{\max}} \tag{6.74}$$

and

$$0 < L_{es} < \frac{c}{2f_{\max}} \tag{6.75}$$

Thus, the sensors must be separated by less than half a wavelength of the maximum frequency of interest to resolve the traveling waves, and each sensor must measure less than a full wavelength at the maximum frequency. Obviously, if the sensors measured a full wavelength, then the integral of the strain over the length of the sensor would be zero, and thus the wave could not be resolved. The purpose of this discussion was to demonstrate that the spatial filtering provided by the sensors is tied to the temporal filtering and is of equal importance in resolving the traveling waves.

The novelty in the previous formulation rests in the observation that a simple time delay FIR filter in combination with spatial measurements of strain provides a direct transformation to the coordinates of the traveling waves. The outlined approach is applicable for broadband measurements up to the Nyquist frequency of the discrete-time sample rate. While the example presented is specific to the propagation of extensional waves in a beam, it is applicable in general to any nondispersive media.

### 6.5.2 SPATIAL AND TEMPORAL SIGNAL PROCESSING TO ESTIMATE TRAVELING WAVES IN A DISPERSIVE MEDIUM

In the example of this section, we consider the propagation of a flexural wave in a one-dimensional beam. For this example, we choose to position the sensors on the structure a distance far enough from discontinuities to assure that the evanescent waves are no longer present to any measurable degree. Thus, the response of the beam can be expressed in terms of the left and right traveling flexural waves, as outlined in Chapter 2.

$$W(x,t) = [A\exp(jk_f x) + B\exp(-jk_f x)]\exp(-j\omega t) \tag{6.76}$$

where $W(x,t)$ is the total out-of-plane displacement, $k_f$ is the flexural wavenumber, and $\omega$ is the circular frequency of excitation. Again, four piezoelectric sen-

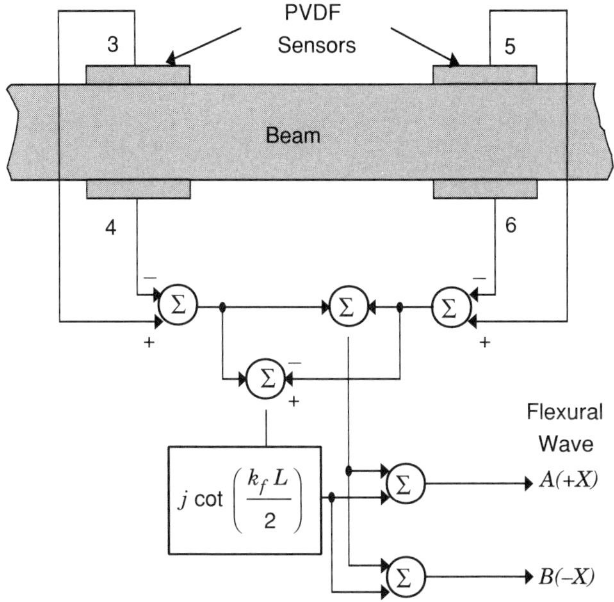


**Figure 6.12:** Schematic diagram of flexural wave sensor.

sors are mounted to the beam in symmetric pairs and spaced a distance $L_{fs}$ apart, as illustrated in Figure 6.12. The output of each piezoelectric sensor can be expressed as a function of the right and left traveling waves upon substituting the above expression into the one-dimensional sensor equation presented in Chapter 2. The response at each sensor illustrated in Figure 6.12 can be expressed as follows:

$$q_3(t) = -2\left(h + \frac{T_s}{2}\right) e_{31}^0 L_y k_f \sin\left(\frac{k_f L_s}{2}\right) [A + B] \exp(-j\omega t) \tag{6.77}$$

$$q_4(t) = 2\left(h + \frac{T_s}{2}\right) e_{31}^0 L_y k_f \sin\left(\frac{k_f L_s}{2}\right) [A + B] \exp(-j\omega t) \tag{6.78}$$

$$q_5(t) = -2\left(h + \frac{T_s}{2}\right) e_{31}^0 L_y k_f \sin\left(\frac{k_f L_s}{2}\right) [A \exp(jk_f L_{fs}) + B \exp(-jk_f L_{fs})] \exp(-j\omega t) \tag{6.79}$$

$$q_6(t) = 2\left(h + \frac{T_s}{2}\right) e_{31}^0 L_y k_f \sin\left(\frac{k_f L_s}{2}\right) [A \exp(jk_f L_{fs}) + B \exp(-jk_f L_{fs})] \exp(-j\omega t) \tag{6.80}$$

where $q_i(t)$ is the output of the $i$th sensor and $L_s$ is the length of the sensor. In the previous example, the colocated sensors were summed to eliminate the contribution of the flexural motion and develop a signal proportional to the extensional motion only. However, if the objective is to develop a sensor whose signal is proportional to the flexural wave, one must take the difference between the outputs of the colocated sensors. The extensional wave is thus eliminated from the derived error. The "flexural" sensor signals are thus created as follows:

$$q_{f34}(t) = q_3(t) - q_4(t) \tag{6.81}$$

$$q_{f56}(t) = q_5(t) - q_6(t) \tag{6.82}$$

The system of equations can be expressed in matrix notation:

$$\begin{bmatrix} q_{f34} \\ q_{f56} \end{bmatrix} = -4\left(h + \frac{T_s}{2}\right) e_{31}^{0} L_y k_f \sin\left(\frac{k_f L_s}{2}\right) \begin{bmatrix} 1 & 1 \\ \exp(jk_f L_{fs}) & \exp(-jk_f L_{fs}) \end{bmatrix} \times \begin{bmatrix} A \\ B \end{bmatrix} \exp(-j\omega t) \tag{6.83}$$

The solution to this system of equations is obtained by Gaussian elimination, as outlined by Gibbs (1993), and is designated the cotangent method. A schematic diagram of the flexural wave sensing algorithm is presented in Figure 6.12, and the system of equations corresponding to the solution is

$$\begin{bmatrix} (q_{f34} + q_{f56}) + j\cot\left(\frac{k_f L_{fs}}{2}\right)(q_{f34} - q_{f56} \\ (q_{f34} + q_{f56}) - j\cot\left(\frac{k_f L_{fs}}{2}\right)(q_{f34} - q_{f56}) \end{bmatrix} = -16\left(h + \frac{T_s}{2}\right) e_{31}^{0} L_y k_f \sin\left(\frac{k_f L_s}{2}\right) \times \cos\left(\frac{k_f L_{fs}}{2}\right) \begin{bmatrix} \exp\left(\frac{jk_f L_{fs}}{2}\right) & 0 \\ 0 & \exp\left(-\frac{jk_f L_{fs}}{2}\right) \end{bmatrix} \begin{bmatrix} A \\ B \end{bmatrix} \exp(-j\omega t) \tag{6.84}$$

The solution to this system of equations is dependent upon the relationship of the frequency of the wave to the dimension of the sensor and the separation distance between the sensors. The critical requirements are

$$0 < \frac{k_f L_{fs}}{2} < \frac{\pi}{2} \tag{6.85}$$

and

$$0 < \frac{k_f L_s}{2} < \pi \tag{6.86}$$

Thus for the flexural wave sensors, the dimension of the sensor $L_s$ must be less than a full wavelength of the smallest structural wave to resolve, as was the case for the extensional wave sensors for the same reason: the integral of the strain over a full wave is zero. On the other hand, the distance between the sensors must be less than a quarter wavelength of the smallest structural wave to assure that the cosine term in equation (6.84) is greater than zero. Again, the relationship between the temporal and spatial filtering are bound together by the wave speed in the medium.

Since flexural waves are dispersive (i.e., the speed of sound in the medium $c$ is frequency dependent), the time delay method used to resolve the extensional wave presents significant difficulties in the design of a temporal filter to achieve the same task for the flexural wave. Formulating the solution as in equation (6.84) simplifies the design of the temporal filter. As indicated in Figure 6.12, the flexural wave amplitudes are obtained from simple addition and subtraction as well as by filtering the difference between the sensors with the cotangent function in the rectangular box. This cotangent function is frequency dependent and can be approximated with a three-coefficient infinite impulse response (IIR) filter.

As an example, consider an aluminium beam configured with polyvinylidene fluoride (PVDF) material for sensing. The dimensions and material properties utilized in the example are presented in Table 6.1. Based upon the tabulated material properties and dimensions, the cotangent filter illustrated in Figure 6.12 can be expressed as a function of frequency over the desired range of excitation. For this example, the desired response is identified between 500 and 1500 Hz, and the phase and magnitude of the function are computed in 10-Hz increments.

**TABLE 6.1 Material Properties of Piezoelectric Sensor and Beam**

| Name | Symbol | Value | Units |
|---|---|---|---|
| Stress coefficient | $e_{31}$ | 65.3E − 3 | m/V |
| Density | $\rho_s$ | 1780 | kg/m$^3$ |
| Modulus | $E_{11}$ | 2.0E + 9 | N/m$^2$ |
| Thickness | $T_s$ | 28E − 6 | m |
| Length | $L_s$ | 1.0 | cm |
| Density | $\rho_b$ | 2700 | kg/m$^3$ |
| Modulus | $E_b$ | 7.1E + 10 | N/m$^2$ |
| Thickness | $2h$ | 3.175 | mm |
| Width | $L_y$ | 7.62 | cm |

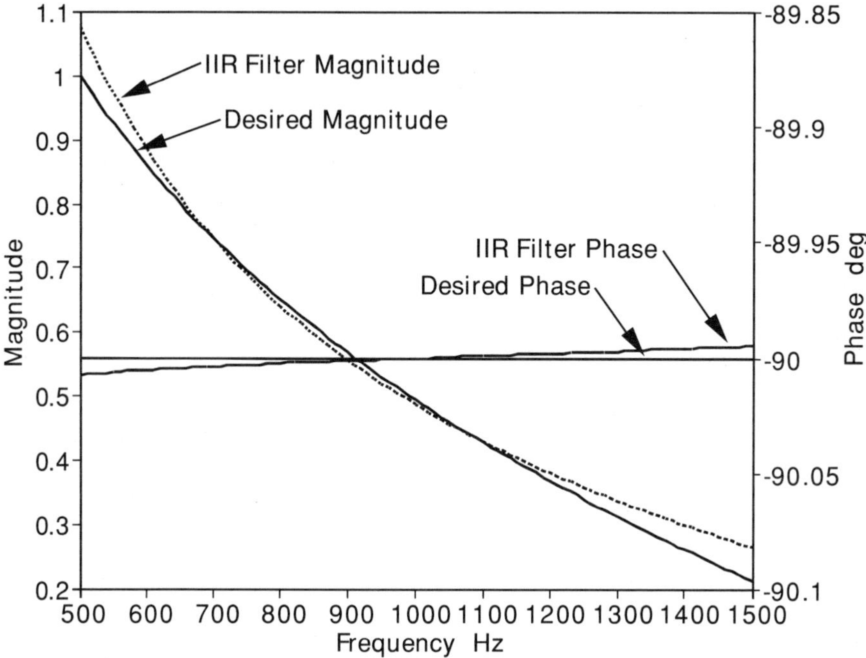


**Figure 6.13:** Comparison of desired FRF and designed FRF.

The **invfreqz** function provided with MATLAB has been used to design the three-coefficient IIR filter to represent the cotangent function over the desired frequency range, and the sample rate has been set at 5300 Hz. The design results in the following IIR filter represented by the $z$-transform:

$$G(z) = \frac{0.32835 + 0.32842z^{-2}}{1 - 0.9999z^{-2}} \tag{6.87}$$

As illustrated in Figure 6.13, the IIR filter response agrees with that of the cotangent function to within 0.2 dB. The program used to design the IIR filter for this example problem is called **flexiir.m**, and it can be executed to replicate the results presented in Figure 6.13.

### 6.5.3 Spatial and Temporal Signal Processing in Structural Acoustics

Research in active structural acoustic control (ASAC) has prompted a number of studies in the design of structural sensors that yield a signal proportional to the far-field sound radiation. The design of structural based sensing techniques is important for two reasons: (1) often it is impractical to place microphones in the acoustic far-field to provide estimates of sound radiation and (2) causality prevents the active control of sound radiation resulting from random excitation due to the time delay associated with the propagation of the sound wave in the acoustic medium (i.e., by the time the sound reaches the error sensor, the structural response cannot be adjusted to compensate). The difficulty in designing structural based sensors is illuminated upon considering the Rayleigh integral, presented in Appendix C:

$$p(\mathbf{r}, t) = \frac{j\omega\rho_o}{2\pi} \int_S \frac{\dot{w}_n(\mathbf{r}_s)\exp(-jkR)}{R} \, dS \exp(j\omega t)$$

where $R = |\mathbf{r} - \mathbf{r}_s|$ and the remainder of the variables are defined in Appendix C. The Rayleigh integral essentially breaks the vibrating structure into an infinite number of volumetric sources, and the acoustic pressure resulting from each infinitesimal source is summed (i.e., the integral) to compute the pressure at any field point. For now, we express the acoustic pressure as a function of the structural modes of a two-dimensional, rectangular structure:

$$p(\mathbf{r}, t) = \frac{j\omega\rho_o}{2\pi} \sum_{n=1}^{N} \dot{\eta}_n(t) \int_0^{L_x} \int_0^{L_y} \Phi_n(x, y) \, \frac{\exp(-jkR(x, y))}{R(x, y)} \, dydx \tag{6.88}$$

where $\dot{\eta}_n(t)$ is the velocity response of the $n$th generalized coordinate, and $\Phi_n(x, y)$ is the $n$th eigenfunction of the system. A modal acoustic influence function can be defined from the previous expression, as outlined previously by Fuller and Burdisso (1991) and Clark et al. (1993), such that

$$\Gamma_n(k, x, y) = \int_0^{L_x} \int_0^{L_y} \Phi_n(x, y) \, \frac{\exp(-jkR(x, y))}{R(x, y)} \, dydx \tag{6.89}$$

This modal acoustic influence function is frequency dependent, as indicated by the presence of the acoustic wavenumber. Thus, to compute the pressure in the acoustic field, the structural response must be filtered in both space and time. If the sound pressure is estimated from the above equation, one must perform modal vector filtering or distributed modal sensing and then design an IIR filter for each respective mode to incorporate the temporal filtering. An alternate form of the Rayleigh integral offers a more elemental approach to the problem of predicting the sound radiation.

Consider the Rayleigh integral expressed as a function of acceleration over the planar surface:

$$p(\mathbf{r}, t) = \frac{\rho_o}{2\pi} \int_S \frac{\ddot{w}_n(\mathbf{r}_s)\exp(-jkR)}{R} \, dS \exp(j\omega t) \tag{6.90}$$

For $\mathbf{r} >> \mathbf{r}_s$, the integral can be formulated as follows (Fahy, 1985):

$$p(\mathbf{r}, t) = \frac{\rho_o}{2\pi r} \int_S \ddot{w}_n(\mathbf{r}_s) \exp(-jkR) dS \exp(j\omega t) \tag{6.91}$$

Applying the piston approximation as outlined by Maillard and Fuller (1994), and discretizing the surface of the structure into rectangular elements, the acoustic response can be expressed as the superposition of a sum of radiating elements:

$$p(\mathbf{r}, t) = \frac{\rho_o}{2\pi r} \sum_{p=1}^{P} \ddot{w}(\mathbf{r}_p) \int_{S_p} \exp(-jkR) dS \exp(j\omega t) \tag{6.92}$$

where $P$ is the number of rectangular elements used in the discretization, $\mathbf{r}_p$ is the vector to the center of the $p$th rectangular element, and the acceleration of the surface of the $p$th element is approximated by the acceleration measured at the $p$th spatial coordinate:

$$\ddot{w}(\mathbf{r}) \approx \ddot{w}(\mathbf{r}_p) \quad \text{on } S_p \tag{6.93}$$

The far-field pressure can be constructed by summing the radiated pressure from each of the $P$ pistons based upon the assumed acceleration over the elemental surface. For low-order structural modes, the expression can be further simplified:

$$p(\mathbf{r}, t) = \frac{\rho_o}{2\pi r} \sum_{p=1}^{P} S_p \ddot{w}(\mathbf{r}_p) \exp(-jk|\mathbf{r} - \mathbf{r}_p|) \exp(j\omega t) \tag{6.94}$$

where $S_p$ is the elemental area of the $p$th *monopole* radiator discussed by Maillard and Fuller (1994). The far-field pressure can thus be approximated based upon a filtered summation of the acceleration measured at discrete points on the structural surface. The filter takes the following form:

$$\Lambda_p(\omega) = \frac{\rho_o}{2\pi r} S_p \exp\left(-j\omega \frac{|\mathbf{r} - \mathbf{r}_p|}{c}\right) \tag{6.95}$$

where the scalar multiple is the same for each element and the exponential effectively models the half-space Green's function between each elemental radiator and the far-field.

With the assumption of far-field sound radiation, the following approximation is valid, as outlined by Wallace (1972) and discussed in Appendix C:

$$|\bar{r} - \bar{r}_p| \approx r - x_p \sin(\theta)\cos(\phi) - y_p \sin(\theta)\sin(\phi) \tag{6.96}$$

Substituting equation (6.96) into equation (6.95), one obtains

$$\Lambda_p(\omega) = \frac{\rho_o \exp(-jkr)}{2\pi r} S_p \exp\left( j\omega \frac{x_p \sin(\theta)\cos(\phi) + y_p \sin(\theta)\sin(\phi)}{c} \right) \quad (6.97)$$

The exponential terms in the above equation are constant magnitude (i.e., unity) and linear phase; thus a simple FIR filter can be used to model the dynamics. The term $|\mathbf{r} - \mathbf{r}_p|/c$ serves as a time delay in the filter, relating the dynamics of the sound propagation of the $p$th structural element to the far-field coordinate defined by $\mathbf{r}$. If we pause to consider the implications, the equations are physically plausible, since the time required to propagate sound from a structural element to a coordinate in the far-field is defined solely by the distance from that coordinate and the wave speed in the acoustic medium, $c$, which is a constant.

To provide a specific example of the previous formulation, we consider a simply supported beam, surrounded by rigid baffle extending over an infinite half-space, with a point force located at the spatial coordinate $x_f$ as indicated in Figure C.1. The sound radiation from such a structure is derived in Appendix C. For the beam, equation (6.97) can be expressed as follows:

$$\Lambda_p(\omega) = \frac{\rho_o \exp(-jkr)}{2\pi r} S_p \exp\left( j\omega \frac{x_p \sin(\theta)\cos(\phi)}{c} \right) \quad (6.98)$$

where the $y$ direction dependence is eliminated from the filter since the acceleration is assumed uniform over the $p$th element of the beam. Ignoring the scaling terms in equation (6.98), the filter design is reduced to the following expression:

$$\Lambda_p(\omega) = \exp\left( -j\omega \left( \Delta\tau - \frac{x_p \sin(\theta)\cos(\phi)}{c} \right) \right) \quad (6.99)$$

where

$$\Delta\tau = \frac{r}{c} - \kappa T \quad (6.100)$$

$\kappa$ is a *positive integer* (due to the discrete-time implementation), and $T$ is the discrete-time sample period. The objective is to select $\kappa$ such that the physical delay imposed by the distance to the far-field coordinate from the center of the beam is removed from the expression. This serves to reduce the order of the FIR filter required to approximate the temporal characteristics as outlined by Maillard and Fuller (1994). In addition, the causality issue is addressed in this formulation since one can predict what the acoustic response of the structure will be a priori. For a causal filter,

$$\Delta\tau \geq \frac{x_p \sin(\theta)\cos(\phi)}{c}, \qquad p = 1, 2, \ldots, P \tag{6.101}$$

As indicated in equation (6.101), one must select the time delay $\Delta\tau$ based upon the maximum value of $x_p$ to guarantee a causal filter design.

For the purpose of this example, we “instrument” the analytical structure with four equally spaced point acceleration transducers. Hence, the beam is segmented into four discrete sections of length $L_x/4$. The output of each transducer is passed through a separate FIR filter designed as outlined above, and the output of each filter is summed to yield an estimate of the farfield sound pressure at an acoustic field point with polar coordinates $(r, \theta = 45°, \phi = 0°)$ as illustrated in Figure C.1 of Appendix C. The choice of $r$ is arbitrary, since the distance from the structure affects the magnitude of the response only. The filters were designed using the **invfreqz** function from MATLAB, and the program **discpres.m** can be used to design the filters.

The FRFs between predicted pressure based upon the FIR filter design and the input force, as well as those between the theoretical acoustic pressure and the input force, are presented in Figure 6.14. As illustrated, the approximation is valid over a frequency range corresponding to the first three structural resonances and tends to deviate for higher order structural modes. This observation

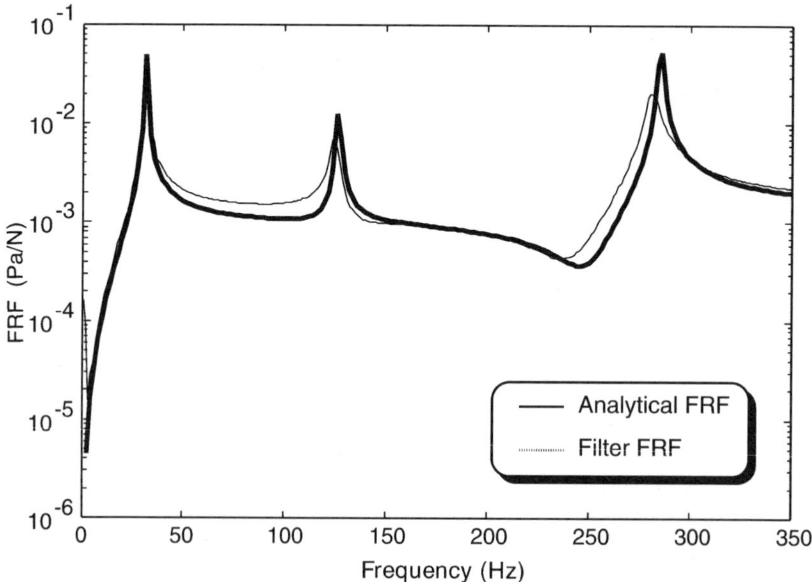


**Figure 6.14:** Comparison of structural acoustic FRF and designed FRF from the discretized array of sensors.

is a result of the spatial filtering that takes place due to the number of sensors used in the design process. One must have at least one discrete sensor per anti-node of the highest structural mode of interest in formulating the acoustic response.

## 6.6 SUMMARY

In Chapter 6, we have presented the starting point for adaptive structure design: the combination of spatial and temporal signal processing. The importance of transducer apertures in the design of adaptive structures was demonstrated, and concepts associated with colocated transducers were reviewed. Modal sensing and actuation were also reviewed in terms of developing an "ideal" form of spatial signal processing. Various examples combining elements of spatial and temporal signal processing to facilitate the design and implementation of adaptive structures were presented.

## BIBLIOGRAPHY

Burdisso, R. A. and C. R. Fuller, 1992. "Theory of Feedforward Control System Eigenproperties," *Journal of Sound and Vibration* **153**(3), 437–451.

Burke, S. E. and J. E. Hubbard, Jr., 1990. "Distributed Transducers, Collocation, and Smart Structural Control," *SPIE's 1990 Technical Symposium on Optical Engineering and Photonics in Aerospace Sensing*, Orlando, FL.

Burke, S. E. and J. E. Hubbard, Jr., 1991. "Distributed Transducer Vibration Control of Thin Plates," *Journal of the Acoustical Society of America*, **90**(2), 937–944.

Burke, S. S. and J. M. Sullivan, 1995. "Distributed Transducer Shading Via Spatial Gradient Electrodes," in *Smart Structures and Materials 1995: Smart Structures and Integrated Systems*, Inderjit Chopra, Ed., Proceedings SPIE 2443, pp. 716–726.

Clark, R. L., 1997. "Accounting for Out-of-Bandwidth Modes in the Assumed Modes Approach: Implications on Colocated Output Feedback Control," accepted to *Journal of Dynamic Systems, Measurement, and Control.*

Clark, R. L. and S. E. Burke, 1996. "Practical Limitations in Achieving Shaped Modal Sensors with Induced Strain Materials," *ASME Journal of Vibration and Acoustics*, **118**, 668–675.

Clark, R. L., C. R. Fuller, and A. L. Wicks, 1991. "Characterization of Multiple Piezoelectric Actuators for Structural Excitation," *Journal of the Acoustical Society of America*, **90**(1), 346–357.

Clark, R. L., R. A. Burdisso, and C. R. Fuller, 1993. "Design Approaches for Shaping Polyvinylidene Fluoride Sensors in Active Structural Acoustic Control (ASAC)," *Journal of Intelligent Material Systems and Structures*, **4**, 354–365.

Collins, S. A., D. W. Miller, and A. H. von Flotow, 1990. "Sensors for Structural Control Applications Using Piezoelectric Polymer Film," *SERC #12-90, MIT Report*, Massachusetts Institute of Technology, Cambridge, MA.

Crawley, E. F. and J. de Luis, 1987. "Use of Piezoelectric Actuators as Elements of Intelligent Structures," *AIAA Journal*, **25**(10), 1373–1385.

Dimitriadis, E. K., C. R. Fuller and C. A. Rogers, 1991. "Piezoelectric Actuators for Distributed Vibration Excitation of Thin Plates," *Journal of Vibration and Acoustics*, 100–107.

F. Fahy, 1985. *Sound and Structural Vibration*, Academic Press, New York.

Freudinger, L., 1990. *Analysis of Structural Response Data Using Discrete Modal Filters*, M.S. Thesis, University of Cincinnati, Department of Mechanical Industrial, and Nuclear Engineering, Cincinnati, OH.

Fuller, C. R. and R. A. Burdisso, 1991. "A Wavenumber Domain Approach to the Active Control of Structure-Borne Sound," *Journal of Sound and Vibration*, **148**(2), 355–360.

Gibbs, G. P. 1993, *Simultaneous Active Control of Flexural and Extensional Power Flow in Thin Beams*, Ph.D. Dissertation, Virginia Polytechnic Institute and State University, Department of Mechanical Engineering, Blacksburg, VA.

Gibbs, G. P. and C. R. Fuller, 1992. "Excitation of Thin Beams Using Asymmetric Piezoelectric Actuators," *Journal of the Acoustical Society of America*, **92**(6), 3221–3227.

*Kynar Piezo Film Technical Manual*, 1987. Pennwalt Corporation, Valley Forge, PA, 65 pp.

Lee, C. K. and F. C. Moon, 1990. "Modal Sensors/Actuators," *ASME Journal of Applied Mechanics*, **57**, 434–441.

Love, A., 1944. *A Treatise on the Mathematical Theory of Elasticity*, Dover, New York.

Maillard, J. and C. R. Fuller, 1994. "Advanced Time Domain Wave-Number Sensing for Structural Acoustic Systems. I. Theory and Design," *Journal of the Acoustical Society of America*, **95**(6), 3252–3261.

McCain, A. J. and E. F. Crawley, 1995. "Shaped Actuators and Sensors for Local Control of Intelligent Structures," *SERC #3-95, MIT Report*, Massachusetts Institute of Technology, Cambridge, MA.

Meirovitch, L., 1967. *Analytical Methods in Vibrations*, Macmillan, New York.

Meirovitch, L. and H. Baruh, 1983. "Control of Self-Adjoint Distributed-Parameter Systems," *Journal of Guidance and Control*, **5**(1), 60–66.

Morgan, D. R. 1991. "An Adaptive Modal-Based Active Control System," *Journal of the Acoustical Society of America*, **89**(1), 248–256.

Shelly, S. J., 1991. *Investigation of Discrete Modal Filters for Structural Dynamic Applications*, Ph.D. Dissertation, University of Cincinnati, Department of Mechanical, Industrial, and Nuclear Engineering, Cincinnati, OH.

Shelly, S. J., R. J. Allemang, G. L. Slater, and J. F. Schultze, 1993. "Active Vibration Control Utilizing an Adaptive Modal Filter Based Modal Control Method," *Proceedings of the 11th International Modal Analysis Conference*, pp. 751–758.

Sullivan, J. M., J. E. Hubbard, and S. E. Burke, 1996. "Modeling Approach for Two-Dimensional Distributed Transducers of Arbitrary Spatial Distribution," *Journal of the Acoustical Society of America*, **99**(5), 2965–2974.

Wallace, C. E., 1972. "Radiation Resistance of a Baffled Beam," *Journal of the Acoustical Society of America*, **51**(3), 936–945.

# 7

# CLASSICAL CONTROL FOR ADAPTIVE STRUCTURES

## 7.1 INTRODUCTION

The application of classical control design methods to *single-input, single-output* (SISO) structural plants provides many important insights into the more complex behavior of multivariable adaptive structures subjected to numerous types of disturbances or command inputs. Controllers designed using classical control methods result in fixed-gain, feedback controllers that are not adaptive in the truest sense; their effects on structures result in *constrained adaptive structures*, which means that the structures can adapt their open-loop behavior within the constraints of the control system dynamics. In this chapter, we review SISO analytical models for a structure, system performance objectives, compensator design, and closed-loop analysis techniques from classical control theory in the context of constrained adaptive structures.

Structures are categorized as a class of *reverberant plants* because there are always boundary reflections of structural waves originating from any forces or moments (i.e., disturbances or control inputs) applied to the structure. Analytically, the modeling of reverberant plants can be approached using wave formulations (Cremer, 1988) or posed as a Sturm-Liouville problem (Churchill and Ward, 1978). Wave solutions are treated in Chapters 2 and 9, where it is shown that wavenumber models can be obtained from discrete models and lead to unique controller designs for finite-dimensional structures. The Sturm-Liouville boundary value problem refers to a broad class of two-point boundary value problems that guarantee that the governing partial differential equation results in a countable infinity of eigenvalues (structural resonances) and eigenfunctions (modeshapes). Together, they are termed the *eigenstructure* of the system. In turn, as has been demonstrated in Chapter 2, the solution to the partial differential equation can be expressed in terms of a linear expansion of assumed modes and generalized coordinates:

$$w(\mathbf{x}, t) = \sum_{m=1}^{\infty} \phi_m(\mathbf{x}) q_m(t) \tag{7.1}$$

where $\phi_m(\mathbf{x})$ is the $m$th assumed mode, and $q_m(t)$ is the $m$th corresponding generalized coordinate. In general, we limit the expansion to a finite number of assumed modes, and the importance of this result to design and analysis of SISO controlled adaptive structures is that we can develop finite models to describe the structural response. From an engineering perspective, this is very helpful. Analytical, numerical, and experimental models can all provide very accurate estimates of the structural dynamic system as outlined in the following section. The accuracy of those estimates begins to degrade with increasing frequency. However, the application of active control is typically required at low frequencies, and thus the decreasing accuracy of the model at high frequencies can be treated with uncertainty models (Skogestad and Postlethwaite, 1996). When using finite models, the designer must always be aware of truncation effects that occur as we limit the number of assumed modes included in the expansion of equation (7.1) or alternatively the number of discrete elements included in a finite element model. These effects can be nearly eliminated in the control bandwidth by including a large number of out-of-bandwidth modes in the system models or by a simple static correction as discussed by Clark (1997). For more details, the equivalence of discrete and distributed parameter structural models is treated thoroughly in numerous structural dynamics or vibrations textbooks, for example, Inman (1994).

### 7.1.1 SISO Modeling

As outlined in Chapter 2, given equation (7.1), we can develop a system of equations that describe the dynamic response for $n$ generalized coordinates such that

$$\mathbf{M}\ddot{\mathbf{q}}(t) + \mathbf{C}\dot{\mathbf{q}}(t) + \mathbf{K}\mathbf{q}(t) = \mathbf{Q}(t) \tag{7.2}$$

where $\mathbf{q}(t) \in R^{n\times 1}$; $\mathbf{Q}(t) \in R^{n\times 1}$; and the mass, damping, and stiffness matrices can be developed from analytical, numerical, or experimental methods. In Chapter 2, we have focused on the development of such models based upon analytical methods. For now, we assume that the damping matrix $\mathbf{C}$ is proportional to the mass and stiffness matrices through some ratio $\mathbf{C} = \alpha\mathbf{M} + \beta\mathbf{K}$, which means that the eigenvectors of the discrete system are real valued. Note that $w(\mathbf{x}, t)$ and $f(\mathbf{x}, t)$ represent the response and applied force in physical coordinates, respectively. The generalized forces are obtained from the applied force as follows:

$$\mathbf{Q}(t) = \int_D f(\mathbf{x}, t)\mathbf{\Phi}(\mathbf{x})dD \tag{7.3}$$

where $\mathbf{\Phi}(\mathbf{x})$ is a $n \times 1$ vector of the assumed modes $\phi_n(\mathbf{x})$. Following the format described in Chapter 6, if we let $f(\mathbf{x}, t) = F(t)\delta(\mathbf{x} - \mathbf{x}_j)$, then

$$\mathbf{Q}(t) = \mathbf{\Phi}(\mathbf{x}_j)F(t) \tag{7.4}$$

For the SISO system, the response vector is composed of a single element:

$$y(t) = w(\mathbf{x}_i, t) \tag{7.5}$$

$$= \mathbf{\Phi}^{\mathrm{T}}(\mathbf{x}_i)\mathbf{q}(t) \tag{7.6}$$

Taking the Laplace transform of equation (7.2), one obtains

$$[\mathbf{M}s^2 + \mathbf{C}s + \mathbf{K}]\, \mathbf{q}(s) = \mathbf{Q}(s) \tag{7.7}$$

where we have assumed zero initial conditions. Equations (7.2) and (7.7) form the basis for the description of a SISO dynamic relationship between the applied force (*input*) and the displacement response (*output*).

If we define the system matrix

$$\mathbf{B}(s) = [\mathbf{M}s^2 + \mathbf{C}s + \mathbf{K}] \tag{7.8}$$

then we can compactly write

$$\mathbf{q}(s) = \mathbf{B}^{-1}(s)\mathbf{Q}(s) \tag{7.9}$$

Recognizing that $w(s) = \mathbf{\Phi}^{\mathrm{T}}(\mathbf{x}_i)\mathbf{q}(s)$ and $\mathbf{Q}(s) = \mathbf{\Phi}(\mathbf{x}_j)F(s)$, one can write

$$w(s) = \mathbf{\Phi}^{\mathrm{T}}(\mathbf{x}_i)\mathbf{B}^{-1}(s)\mathbf{\Phi}(\mathbf{x}_j)F(s) \tag{7.10}$$

The *transfer function*, $P(s) = \mathbf{\Phi}^{\mathrm{T}}(\mathbf{x}_i)\mathbf{B}^{-1}(s)\mathbf{\Phi}(\mathbf{x}_j)$, is obtained from the inverse of the system matrix, that is, the elements of the assumed modes at the excitation and response locations, and is guaranteed to exist for self-adjoint systems. The matrix is a ratio of polynomials of order $2n$, defined by

$$P(s) = \frac{\mathbf{\Phi}^{\mathrm{T}}(\mathbf{x}_i)\mathrm{adj}[\mathbf{B}(s)]\mathbf{\Phi}(\mathbf{x}_j)}{\det[\mathbf{B}(s)]} \tag{7.11}$$

$$= \frac{N_P(s)}{D_P(s)} \tag{7.12}$$

The roots of the denominator polynomial $D_P(s) = \det[\mathbf{B}(s)]$ are referred to as the *poles* of the transfer function and the roots of the numerator polynomial $N_P(s)$ are called the *zeros* of the transfer function. The physical interpretation of both quantities is quite important and follows from equation (7.10). At frequencies that correspond to zeros of the transfer function, the structural response is "blocked." No vibration is possible, and subsequently cannot be measured at those frequencies. The poles are those frequencies where the response is very large, bounded only by the damping of the structure. Expansion of the numerator using the adjoint matrix description results in explicit descriptions of the structural transfer function zeros. However, it is more direct to use a change of coordinates, as shown next.

Since we are now working with a discrete model of the dynamic system, we can perform a similarity transform on the system matrix. The columns of the matrix $\mathbf{\Psi} = \mathbf{\Psi}(\Delta\mathbf{x})$, used in this similarity transform, are composed of the eigenvectors of the discrete eigenvalue problem resulting from the finite-dimensional mass and stiffness matrices of equation (7.2). Thus, if we let

$$\mathbf{q}(t) = \mathbf{\Psi}\boldsymbol{\eta}(t) \tag{7.13}$$

where $\boldsymbol{\eta}(t)$ is the response in orthogonal coordinates, and $\mathbf{\Psi}$ is a matrix of the eigenvectors of the discrete model. Taking the Laplace transform and substituting into equation (7.7), one obtains

$$[\mathbf{M}s^2 + \mathbf{C}s + \mathbf{K}]\mathbf{\Psi}\boldsymbol{\eta}(s) = \mathbf{Q}(s) \tag{7.14}$$

Premultiplying by $\mathbf{\Psi}^{\mathrm{T}}$,

$$\mathbf{\Psi}^{\mathrm{T}}[\mathbf{M}s^2 + \mathbf{C}s + \mathbf{K}]\mathbf{\Psi}\boldsymbol{\eta}(s) = \mathbf{\Psi}^{\mathrm{T}}\mathbf{Q}(s) \tag{7.15}$$

However, since the eigenvectors are usually normalized with respect to the mass matrix for the discrete problem, equation (7.15) can be reduced to the following:

$$\mathrm{diag}[s^2 + 2\zeta_n\omega_n s + \omega_n^2]\boldsymbol{\eta}(s) = \mathbf{\Psi}^{\mathrm{T}}\mathbf{Q}(s) \tag{7.16}$$

Thus, we now have an expression in orthogonal coordinates, which can be used to express the response in physical coordinates as follows:

$$w(s) = \mathbf{\Phi}^{\mathrm{T}}(\mathbf{x}_i)\mathbf{\Psi}\ \mathrm{diag}[s^2 + 2\zeta_n\omega_n s + \omega_n^2]^{-1}\mathbf{\Psi}^{\mathrm{T}}\mathbf{\Phi}(\mathbf{x}_j)F(s) \tag{7.17}$$

If we define $\mathbf{\Gamma}$ such that

$$\mathbf{\Gamma}(\mathbf{x}_j) = \mathbf{\Psi}^{\mathrm{T}}\mathbf{\Phi}(\mathbf{x}_j) \tag{7.18}$$

and

$$\mathbf{\Gamma}^{\mathrm{T}}(\mathbf{x}_i) = \mathbf{\Phi}^{\mathrm{T}}(\mathbf{x}_i)\mathbf{\Psi} \tag{7.19}$$

then we can express equation (7.17) as follows:

$$w(s) = \mathbf{\Gamma}^{\mathrm{T}}(\mathbf{x}_i)\mathrm{diag}[s^2 + 2\zeta_n\omega_n s + \omega_n^2]^{-1}\mathbf{\Gamma}(\mathbf{x}_j)F(s) \tag{7.20}$$

Thus, the system transfer function can be expressed compactly as follows:

$$P(s) = \frac{w(s)}{F(s)} = \mathbf{\Gamma}^{\mathrm{T}}(\mathbf{x}_i)\mathrm{diag}[s^2 + 2\zeta_n\omega_n s + \omega_n^2]^{-1}\mathbf{\Gamma}(\mathbf{x}_j) \tag{7.21}$$

Assuming harmonic response, each entry of the transfer function presented in equation (7.23) can be represented in terms of the following expression:

$$P(j\omega) = \frac{w(j\omega)}{F(j\omega)} = \sum_{n=1}^{N} \frac{\Gamma_n(\mathbf{x}_i)\Gamma_n(\mathbf{x}_j)}{(\omega_n^2 - \omega^2) + j2\omega\zeta_n\omega_n} \tag{7.22}$$

$$= \sum_{n=1}^{N} \frac{\Gamma_n(\mathbf{x}_i)\Gamma_n(\mathbf{x}_j)}{(j\omega - p_n)(j\omega - p_n^*)} \tag{7.23}$$

where, for the underdamped system, $\omega_n$ is the natural frequency of the $n$th orthogonal coordinate, $\zeta_n$ is the damping ratio of the $n$th orthogonal coordinate, and $\Gamma_n(\mathbf{x}_j)$ is the $n$th entry of the vector $\mathbf{\Gamma}(\mathbf{x}_j)$. The poles of the system are expressed in complex conjugate pairs such that

$$p_{n(1,2)} = -\zeta_n\omega_n \pm j\omega_n\sqrt{1 - \zeta_n^2} \tag{7.24}$$

This expression can be written in several different ways and has been the subject of intense discussion by researchers in the field of modal testing and analysis. Our purpose is to describe the influence of the poles and basis functions on the zeros for the structure so we choose a partial fraction expansion of equation (7.23) to produce a pole-residue equation of the form

$$P(j\omega) = \sum_{n=1}^{N} \left[ \frac{R_n}{j\omega - p_n} + \frac{R_n^*}{j\omega - p_n^*} \right] \tag{7.25}$$

where

$$R_n = \frac{\Gamma_n(\mathbf{x}_i)\Gamma_n(\mathbf{x}_j)}{j2\omega_n\sqrt{1-\zeta_n^2}} \tag{7.26}$$

and $R_n^* = R_n$ is valid for proportional damping only ($^*$ is the complex conjugate operator). For the case of nonproportional damping, complex modes must be used in the modal expansion using slightly different analysis techniques (Klosterman, 1984).

If we examine the expression in equation (7.25) for the case of $N = 3$ and a single position measurement $w_k(j\omega)$, it can be shown that

$$\begin{aligned}\frac{w_k(j\omega)}{f_l(j\omega)} = {} & \frac{(j\omega - p_2)(j\omega - p_2^*)(j\omega - p_3)(j\omega - p_3^*)\Gamma_1(\mathbf{x}_i)\Gamma_1(\mathbf{x}_j)}{\prod_{n=1}^{3}(j\omega - p_n)(j\omega - p_n^*)} \\ & + \frac{(j\omega - p_1)(j\omega - p_1^*)(j\omega - p_3)(j\omega - p_3^*)\Gamma_2(\mathbf{x}_i)\Gamma_2(\mathbf{x}_j)}{\prod_{n=1}^{3}(j\omega - p_n)(j\omega - p_n^*)} \\ & + \frac{(j\omega - p_1)(j\omega - p_1^*)(j\omega - p_2)(j\omega - p_2^*)\Gamma_3(\mathbf{x}_i)\Gamma_3(\mathbf{x}_j)}{\prod_{n=1}^{3}(j\omega - p_n)(j\omega - p_n^*)}\end{aligned} \tag{7.27}$$

This expression explicitly indicates the nature of the zeros, that is, the roots of the numerator polynomial, for a receptance measurement. The zero frequencies are determined by the elements of the basis functions at the input and output locations described by $\boldsymbol{\Gamma}(\mathbf{x}_i)$ and $\boldsymbol{\Gamma}(\mathbf{x}_j)$, respectively, as well as the poles of the system. The $N - 2$ order of the numerator polynomial is for position measurements only; velocity measurements lead to order $N - 1$ and acceleration measurements to order $N$. Equation (7.29) also indicates the potential problem for exact zero estimation for analytical models that are truncated to low order ($N < \infty$). The zero frequencies depend on the entirety of poles—a low-order truncation cannot adequately describe the interactive effects. This is addressed in more detail in Chapter 9.

State-variable methods can also be used to illuminate similar relationships between transfer function poles, zeros, structural resonances, damping values, and modeshapes. As indicated above, homogeneous solutions for the $n$-dimensional problem of equation (7.2) yield $n$ discrete *eigenvalues* $\lambda_n$, with corresponding *eigenvectors* $\boldsymbol{\Psi}_n$, *natural frequencies* $\omega_n$, modal *damping ratios* $\zeta_n$, and $n$ discrete assumed modes $\phi_n(\mathbf{x})$ of the structure.

Consistent with the presentation in Chapter 3, we develop the state-variable representation of the system for comparison with the transfer function representation. From equation (7.15), we utilized a similarity transform to decouple the equations of motion, which in the time domain results in the following expression:

$$\begin{bmatrix} 1 & 0 & \cdots & 0 \\ 0 & 1 & \cdots & 0 \\ & \ddots & & \\ 0 & 0 & 0 & 1 \end{bmatrix} \ddot{\boldsymbol{\eta}}(t) + \begin{bmatrix} 2\zeta_1\omega_1 & 0 & \cdots & 0 \\ 0 & 2\zeta_2\omega_2 & \cdots & 0 \\ & \ddots & & \\ 0 & 0 & 0 & 2\zeta_n\omega_n \end{bmatrix} \dot{\boldsymbol{\eta}}(t)$$

$$+ \begin{bmatrix} \omega_1^2 & 0 & \cdots & 0 \\ 0 & \omega_2^2 & \cdots & 0 \\ & \ddots & & \\ 0 & 0 & 0 & \omega_n^2 \end{bmatrix} \boldsymbol{\eta}(t) = \boldsymbol{\Psi}^{\mathrm{T}}\mathbf{Q}(t) \tag{7.28}$$

Now, as discussed in Chapter 3, the state variable is defined as

$$\mathbf{X}(t) = [\eta_1(t) \ \cdots \ \eta_n(t) \ \vdots \ \dot{\eta}_1(t) \ \cdots \ \dot{\eta}_n(t)]^{\mathrm{T}} \tag{7.29}$$

The state-space dynamic equation for the structural response can be written as

$$\dot{\mathbf{X}}(t) = \mathbf{A}\mathbf{X}(t) + \mathbf{B}F(t) \tag{7.30}$$

and the single output $y(t)$, produced by the single input $F(t)$, can be expressed in terms of the system states and the applied input:

$$y(t) = \mathbf{C}\mathbf{X}(t) + DF(t) \tag{7.31}$$

Recall that the *state* matrix $\mathbf{A} \in \mathcal{R}^{2n\times 2n}$ is of the form:

$$\mathbf{A} = \begin{bmatrix} \mathbf{0} & \mathbf{I} \\ \mathrm{diag}[-\omega_n^2] & \mathrm{diag}[-2\zeta_n\omega_n] \end{bmatrix}$$

The *input* vector $\mathbf{B} \in \mathcal{R}^{n\times 1}$ weights the force at the position of the applied input according to the orthogonal coordinate transformation:

$$\mathbf{B} = [\mathbf{0} \mid \mathbf{\Gamma}(\mathbf{x}_j)]^{\mathrm{T}}$$

The product of this vector and the forcing magnitude produces the force in the orthogonal coordinates. Equation (7.33) is called the output equation. The contents of the *output* vector **C** and *feedthrough* coefficient $D$ depend on user-selected units of the output vector $y(t)$. For the example used in this discussion, the displacement response $w(\mathbf{x}_j, t)$ is selected, and **C** can be expressed as:

$$\mathbf{C} = [\mathbf{\Gamma}(\mathbf{x}_i) \mid \mathbf{0}]$$

The feedthrough matrix $D$ will be zero. (What are the forms of **C** and $D$ for velocity outputs? Acceleration outputs?) For position outputs, notice that the output equation is equivalent to the discrete modal expansion theorem:

$$w(\mathbf{x}, t) = \mathbf{\Gamma}^{\mathrm{T}}(\mathbf{x})\boldsymbol{\eta}(t) \tag{7.32}$$

and therefore produces the displacement in physical coordinates that is formed by the choice of **C** and $D$ in the output equation. Manipulation of the different forms for the state equations is required as we encounter different design objectives for adaptive structures.

Recalling equation (3.62), the system transfer function is obtained as follows:

$$\frac{w(s)}{F(s)} = [\mathbf{C}(s\mathbf{I} - \mathbf{A})^{-1}\mathbf{B} + D] \tag{7.33}$$

From the definitions of the state-space realization (**A**, **B**, **C**, **D**), and the inverse of a matrix, we see that

$$P(s) = \frac{\mathbf{C}\ \mathrm{adj}[s\mathbf{I} - \mathbf{A}]\mathbf{B} + D}{\det[s\mathbf{I} - \mathbf{A}]} \tag{7.34}$$

Upon comparison of equation (7.11) and equation (7.34), it again becomes clear, based upon the derivation of equation (7.34), that the structural transfer function zeros depend on the modeshapes and the transfer function poles. It can be shown that the magnitudes of the complex poles, for underdamped cases, equal the natural frequencies of the structure. Because of our assumptions for equations (7.2) and (7.7), the system poles occur in complex conjugate pairs. As detailed in equation (7.24), the complex conjugate pair of poles for each mode can be represented as follows:

$$s_{n1,2} = -\zeta_n\omega_n \pm i\omega_n\sqrt{1-\zeta^2} \tag{7.35}$$

$$= \sigma_n \pm i\omega_d \tag{7.36}$$

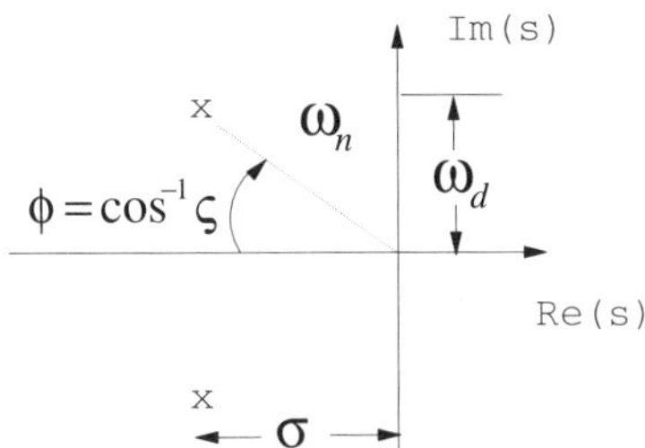


**Figure 7.1:** Reverberant system poles in the $s$-plane.

where $\sigma_n$ is the projection of the pole on the real axis and is an approximate measure of the structural mode's speed of response to force inputs. The imaginary term $\omega_d$ is called the *damped natural frequency*. For a mobility transfer function ($\dot{w}/F$), the structural response magnitude displays a peak at the damped natural frequency. Figure 7.1 shows a small region of the $s$-plane that encloses a single pair of complex poles. Both poles are located on a circle about the origin of radius $\omega_n$. The perpendicular distance from the pole to the real axis corresponds to the damped natural frequency. Finally, the angle between the pole's position vector (from the origin) and the real axis defines the damping ratio $\zeta_n$. Notice that poles are located on the imaginary axis for structural models that neglect damping.

The zeros of a transfer function are equally as important as the poles and are extremely important to any detailed analysis of the phase behavior. Zeros are classified into two primary categories, minimum phase zeros and nonminimum phase zeros. Minimum phase zeros are located in the left-half Laplace plane, where the structural and transducer dynamic poles are located. They are given this name because it can be shown that colocated transfer functions of reverberant systems exhibit alternating poles and zeros that remain in the left-half Laplace plane. The phase of colocated transfer functions is alternately controlled by the influence of the poles (decreasing phase) in the same half plane, and by the increasing phase of its zeros so that the phase remains within a 0 to $\pi$ boundary for all frequencies. This is the minimum phase boundary that can be realized for any structural transfer function, or for any reverberant system transfer function.

Nonminimum phase zeros are located in the right-half plane so that they exhibit decreasing phase with increasing frequency, identical to the phase changes observed for left-half plane poles. The right-half plane zeros present two fundamental problems to adaptive structure design. First, if an inverse filter of a particular path through an adaptive structure is required for compensation, it will be noncausal if there are right-half plane zeros in the respective transfer function. Second, those zeros are generally destabilizing to a closed-loop, feedback system, as they attract structural poles in the presence of feedback. This is discussed in more detail in Section 7.3.

It is helpful to understand the physical significance of left-half and right-half plane zeros from the perspective of the structure's dynamic response. Minimum phase response occurs for drive point (colocated) transfer functions $w_j(s)/F_j(s)$, because each mode is observed to be in phase with the input force at that location. In fact, this must be the case if the transducer is truly colocated with the input. As the excitation frequency moves past each resonant frequency, the phase between the structural displacement and the force changes abruptly (depending on the damping) by 180 degrees. However, the structural response due to the next higher structural resonance must necessarily be in phase with the force, as the resonant frequency has not been reached. This positive adjustment in phase is the same phase contributed by the zero from a control system perspective. Of course, the phase decreases again as the next resonant frequency is surpassed, and so on. In reality, the transducers are no longer ideally colocated when the frequency gets high enough such that the ratio between the structural wavenumber and the transducer aperture reach the order of $\pi$. Therefore, if the measured response is not a true drive point response, the resulting transfer function from applied force to measured output exhibits nonminimum phase behavior. This subject is treated in detail by Tohyama and Lyon (1989).

The simple models described in this section need to be generated to examine some of the fundamental characteristics of adaptive structures. To facilitate this, the toolbox titled **BLevTb.m** (developed by Russell Groves, Duke University) can be used to develop FRF plots for different beam boundary conditions and selectable input/output locations. Students are encouraged to become comfortable with magnitude and phase characteristics for varying beam structures. For the remainder of this chapter, the closed-loop response characteristics are sought using the tools of classical control system theory. The SISO structural plant displays characteristics similar to many other second-order plants, for which analysis tools like Bode's and Nyquist's frequency response, and Evans' root-locus diagrams, have proven quite helpful in describing closed-loop response. Next, we introduce some nomenclature—followed by an overview of root-locus diagrams and frequency response analyses—that help us to analyze the performance of SISO adaptive structures.

## 7.2 SYSTEM NOMENCLATURE

Classical control system design is traditionally concerned with the effects of *feedback* loops between exogenous input signals and selected system output signals. Control systems engineers analyze dynamic relationships between input and output signals; structural dynamicists discuss input forces and torques, and structural velocities, strains, or accelerations as outputs. In practice, inputs may be from exogenous disturbances or control actions intentionally introduced to the system using actuators. The output signals are generated by appropriate sensors that measure the physical quantities important to the problem. In Chapter 5, the wide variation in transducer dynamics has been discussed and the

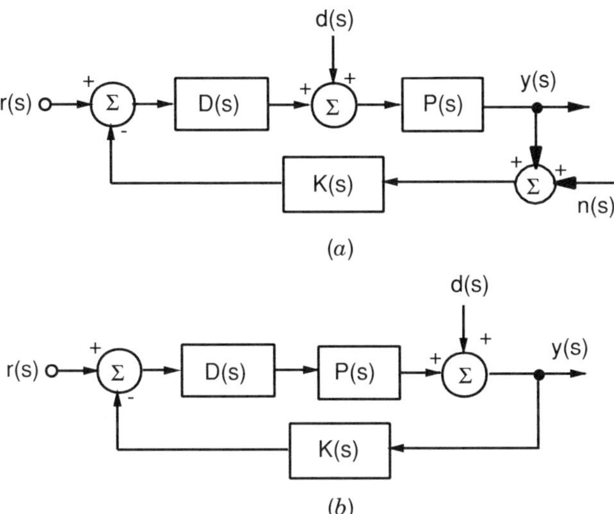


**Figure 7.2:** Output feedback with reference input and (*a*) disturbance input to the plant, (*b*) disturbance input to measured output.

impact of those dynamics on the output signal cannot be overemphasized. The control system modeler must include those dynamics in either the compensator or the plant descriptions, because they will surely affect the closed-loop performance and stability. In this section, we introduce some general notation that will be used to describe the input and output signals, as well as the transfer functions needed to describe an adaptive structure's closed-loop response.

Two general topologies for a single-loop feedback system with negative feedback are shown in Figure 7.2 where inputs and outputs are defined as

| | |
|---|---|
| $d(s)$ | Disturbance input |
| $n(s)$ | Measurement noise |
| $r(s)$ | Reference or command input |
| $y(s)$ | Measured output |
| $e(s) = r(s) - y(s)$ | Error |

and the dynamic processes are

| | |
|---|---|
| $D(s)$ | Series compensator transfer function |
| $P(s)$ | Plant transfer function |
| $K(s)$ | Feedback compensator transfer function |

The inputs and outputs can be single variables or vectors of variables; the dynamics would be represented using transfer function matrices in the latter case (see Chapter 3).

SISO control loops exhibit some different properties, and are easier to understand, than the multiple-input, multiple-output (MIMO) control descriptions necessary for vector inputs and outputs. We reserve detailed analyses of multivariable feedback systems until Chapter 8. Multivariable control systems are also discussed in more detail by Maciejowski (1989), Skogestad and Postlethwaite (1996) and Zhou et al. (1996), for general reverberant systems.

For adaptive structures, the transfer functions of the closed-loop system represented by Figure 7.2 are dependent on the structure's dynamics $P(s)$ and the compensator dynamics in the forward path, $D(s)$, as well as the feedback path $K(s)$. Our emphasis is to interpret well-known relationships for feedback control systems in regards to adaptive structure (closed-loop) systems. Beginning at the left upper corner of Figure 7.2*a*, there is a "reference" or "command" signal, $r(s)$, which the output attempts to follow. Thus, units of the input must be consistent with the feedback signal $K(s)y(s)$. For the unity feedback case, the reference input must also have consistent units with the output $y(s)$. Then, $r(s)$ is selected to be the desired value of $y(s)$ for the closed-loop system. For nonunity feedback, the units of the compensator $K(s)$ or the prefilter (if necessary) translate the units for equivalency at the summation between the reference and feedback signals. The *forward path* contains a *series compensator* $D(s)$ and the structural plant dynamics $P(s)$. Often, the model $P(s)$ includes transducer dynamics, as indicated by the series connections introduced in Figure 5.1. Disturbances on the system must also be induced strains, forces, and moments. In fact, all exogenous inputs, with the exception of the measurement noise signal $n(s)$, must be of these types. The noise signal is most helpful for *robustness* and *stability* analyses of the closed-loop system. It is often acceptable to ignore the noise input if the structure's measurement transducers are located in high response regions of the structure. Continuing, the feedback compensator $K(s)$ is used to close the loop around the structure. As mentioned above, this can also be achieved with only a wire between the output and the summer. In that case, we refer to the system as a *unity feedback system*.

The response $y(s)$ for the case of negative feedback shown in Figure 7.2a can be written as

$$y(s) = P(s)d(s) + D(s)P(s)(r(s) - K(s)y(s) - K(s)n(s)) \tag{7.37}$$

leading to three closed-loop transfer functions between the inputs and the output $y(s)$:

$$\frac{y(s)}{r(s)} = \frac{P(s)D(s)}{1 + P(s)D(s)K(s)} \tag{7.38}$$

$$\frac{y(s)}{d(s)} = \frac{P(s)}{1 + P(s)D(s)K(s)} \tag{7.39}$$

and

$$\frac{y(s)}{n(s)} = -\frac{P(s)D(s)K(s)}{1 + P(s)D(s)K(s)} \tag{7.40}$$

It should be noted that a positive summation of the feedback signal changes the sign of the denominators in these equations and changes the closed-loop properties significantly. We concentrate on negative feedback only. The transfer function $y(s)/e(s)$, called the error transfer function, is used to analyze the command tracking performance for shape control problems. It is covered in Section 7.6. The reciprocal of the denominator of the three closed-loop transfer functions is commonly called the *sensitivity function*: $S(s) = (1 + P(s)D(s)K(s))^{-1}$. The rational function $P(s)D(s)K(s)$ is called the *loop gain*, and it is clear that the sensitivity is small for high loop gain. $P(s)D(s)K(s)$ is also referred to as the loop transfer function. The closed-loop transfer function $y(s)/r(s)$ can then be written as $P_c(s) = S(s)P(s)D(s)$, and the closed-loop transfer function $y(s)/n(s)$ can be expressed as $T(s) = S(s)P(s)D(s)K(s)$. The remaining transfer functions can also be written more compactly in this notation, so that the closed-loop output becomes

$$y(s) = S(s)P(s)d(s) + P_c(s)r(s) - T(s)n(s) \tag{7.41}$$

For adaptive structures, we are often interested in the effects of measurement errors and disturbance signals on the output $y(s)$. From equation (7.40), we can see that there is a tradeoff that must be made if the closed-loop system is intended to minimize the effects of noise and disturbances. This is most easily appreciated for the case where the disturbance enters at the output (Figure 7.2*b*) and the feedback transfer function $K(s)$ is unity, leading to the equality

$$S(s) + T(s) = 1 \tag{7.42}$$

This expression is the basis for naming $T(s)$ the *complementary sensitivity*. If we make the sensitivity low to reduce the effects of disturbances, the complementary sensitivity approaches unity, which is equivalent to very little filtering of measurement noise. The balance of equation (7.42) is the basis for frequency-domain loop-shaping. At low frequencies, disturbance signal powers are typically much larger than noise signal powers. Therefore, we make $S(s)$ small (i.e., large loop gain), and thus $T(s)$ is large. At high frequencies, where noise signal powers dominate, the gains are reversed. The transition point depends on the specific design problem. This design trade-off is a classical result of control theory, but the adaptive structure application leads to slight exceptions to this result.

The traditional loop-shaping trade-off only exists for certain classes of adaptive structure problems. Adaptive structures that point and track are two such classes. For example, the adaptive structure might be a robot moving or holding a tool, or repositioning its grip on a tool. These are tracking and regulator problems that must be robust in the face of high-frequency disturbances and measurement noise. Combined slewing and vibration control of a robotic link is another example where the designer must consider vibration suppression and regulation. Two common exceptions to the desired loop shapes described above result from active vibration suppression and active structural acoustic control of a structure. Both of these problems are disturbance rejection problems that do not require simultaneous command tracking.

In general, the interpretation of the sensitivity and complementary sensitivity are quite different for SISO vibration or structural acoustic control of reverberant structures, specifically when the disturbance enters through an alternative path of the structure (i.e., it is physically located at a different point on the structure from the control input). To provide a motivating example, consider a rectangular plate configured with a disturbance force entering at one point on the structure and a control force entering at another point on the structure. The dimensions of the plate and locations of the disturbance and control forces are documented in the script file **plt_lead.m**. The performance measure is chosen as the displacement response at the location of the disturbance, and the sensor for feedback is chosen to be displacement colocated with the control force.

In the traditional servocontrol problem, the disturbance typically enters in the same loop as the control input $u(s)$, and the sensed variable $y(s)$ is typically the same as the error or performance variable $z(s)$. However, in the design of adaptive structures for rejection of vibration or structural acoustic radiation, the performance variables are frequently not accessible with transducers. Thus, the *performance path* is defined between the disturbance $d$ and some physically motivated measure or estimate of performance $z(s)$. However, the *control path* is defined between the control input $c$ and the sensed variable $y(s)$. Hence, the sensitivity and complementary sensitivity are very effective in evaluating performance for the SISO servocontrol system (i.e., the tracking or reference command problem); however, they are less informative in the disturbance rejection problem for adaptive structures, particularly when the disturbance and error are not colocated with the control and sensor.

A block diagram of the example discussed is depicted in Figure 7.3. As illustrated, the system transfer matrix between the applied inputs and outputs can be represented as follows:

$$\mathbf{P}(s) = \begin{bmatrix} P_{zw}(s) & P_{zu}(s) \\ P_{yw}(s) & P_{yu}(s) \end{bmatrix} \tag{7.43}$$

where $P_{zw}(s)$ is the transfer function between the disturbance $w(s)$ and the performance output $z(s)$. The remaining transfer functions are defined similarly. If the control input is defined such that

$$u(s) = K(s)y(s) \tag{7.44}$$

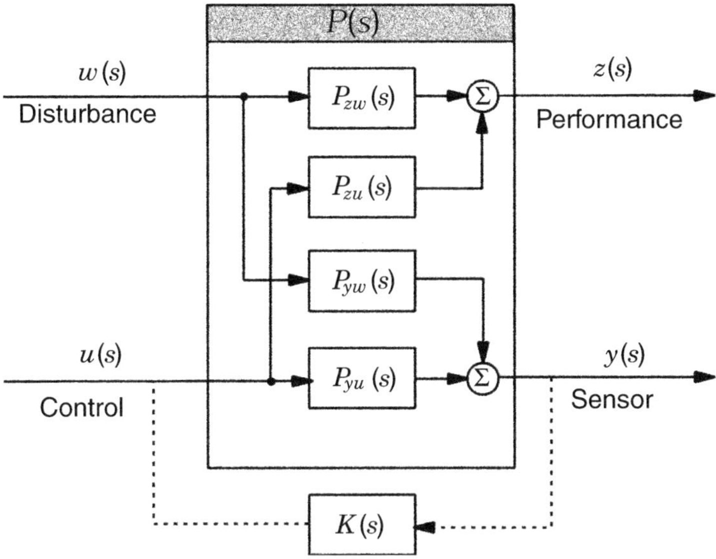


**Figure 7.3:** Block diagram of system with an independent disturbance-to-performance path and control-to-sensor path.

for scalar inputs and outputs, then one can derive an expression for the closed-loop transfer function from $w(s)$ to $z(s)$, $T_{zw}(s)$, as outlined by Hong and Bernstein (1997):

$$T_{zw}(s) = F(K(s))S(s) \tag{7.45}$$

where

$$F(K(s)) \equiv P_{zw}(s) - [P_{zw}(s)P_{yu}(s) - P_{zu}(s)P_{yw}(s)]K(s) \tag{7.46}$$

The sensitivity function is defined as follows:

$$S(s) = (1 - P_{yu}(s)K(s))^{-1} \tag{7.47}$$

Thus, as noted by Hong and Bernstein, the closed-loop response in the performance path is dependent upon the dynamics of the control path, the performance path, the associated cross transfer functions between the two paths, and the dynamic compensator. A reduction in $F(K(s))$ or the sensitivity $S(s)$ results in a reduction of the corresponding closed-loop response $T_{zw}(s)$. However, one must be selective in choosing where to reduce the sensitivity, due to the Bode sensitivity integral constraint (Bode, 1945):

$$\int_0^\infty \log |S(j\omega)| d\omega = 0 \tag{7.48}$$

The obvious result of this constraint is that, if the sensitivity is reduced over some frequency bandwidth, it will be increased over some other bandwidth. Thus, while one would like the sensitivity to be less than unity over the entire bandwidth to provide disturbance rejection, this is not physically possible.

Given this constraint, how should we design the loop shape? Obviously, we wish to reduce the sensitivity over the bandwidth where $F(K(s))$ is a maximum and allow the sensitivity to increase where $F(K(s))$ is a minimum. Thus, the objective is not merely to reduce the sensitivity, but rather to reduce the sensitivity in selective bandwidths, likely in the proximity of resonances for lightly damped reverberant plants. However, as the modal density increases for reverberant structures, open-loop shaping becomes quite involved because the dynamics of the compensator must typically increase. In fact, the difficulty presented by this problem provides motivation for the application of closed-loop design procedures using $\mathcal{H}_2$ and $\mathcal{H}_\infty$ control synthesis techniques (a topic discussed in Chapter 8).

However, before proceeding with the example in this section, it is worthwhile to demonstrate two special cases, which were discussed by Hong and Bernstein (1997). Consider the case when the control actuator is colocated with the disturbance *or* the case when the sensor is colocated with the performance measure. For both of these cases,

$$F(K(s)) = P_{zw}(s) \tag{7.49}$$

Thus, the closed-loop response for the performance path can be evaluated as follows:

$$T_{zw}(s) = P_{zw}(s)S(s) \tag{7.50}$$

Now, the magnitude of the frequency response can be expressed as follows:

$$|T_{zw}(j\omega)| = |P_{zw}(j\omega)||S(j\omega)| \tag{7.51}$$

From Equation (7.51), the effects of spillover are obvious. As the sensitivity is reduced, so is the closed-loop response of the performance path, in direct proportion. However, in frequency bandwidths where the sensitivity is greater than unity, and it will be due to equation (7.48), the closed-loop response of the system exceeds that of the open-loop response. Thus, it is obvious that if spillover is to occur, as a designer, you want to force the spillover to occur in bandwidths where the open-loop response is at a minimum relative to the remainder of the bandwidth. Additionally, it is better to smear the spillover over the bandwidth than to cause a sharp transition at any particular frequency. While this design procedure is quite manageable for low-order dynamic systems, it is quite challenging for high modal density systems. The difference in complexity reflected in equations (7.45) and (7.50) serves to demonstrate why loop shaping becomes more difficult when the disturbance and the performance measure are not colocated with the control input and sensor.

To continue with the example, consider the frequency response functions (FRFs) presented in Figure 7.4. These FRFs can be generated with the script file **plt_lead.m**. In Figure 7.4*a*, the open- and closed-loop responses between the control path ($u$ to $y$) and the performance path ($d$ to $z$) are presented. Proportional control is used, and as illustrated, the closed-loop frequency response of the control path ($y/u$) decreases significantly, although the closed-loop response corresponding to the performance path ($z/d$) results in very little attenuation. However, the resonance frequency of the fundamental mode increased, due to the added stiffness resulting from the SISO feedback loop Upon considering the magnitudes of the sensitivity ($S$ in the legend), complementary sensitivity ($T$ in the legend), and loop gain illustrated in Figure 7.4*b*, one observes that

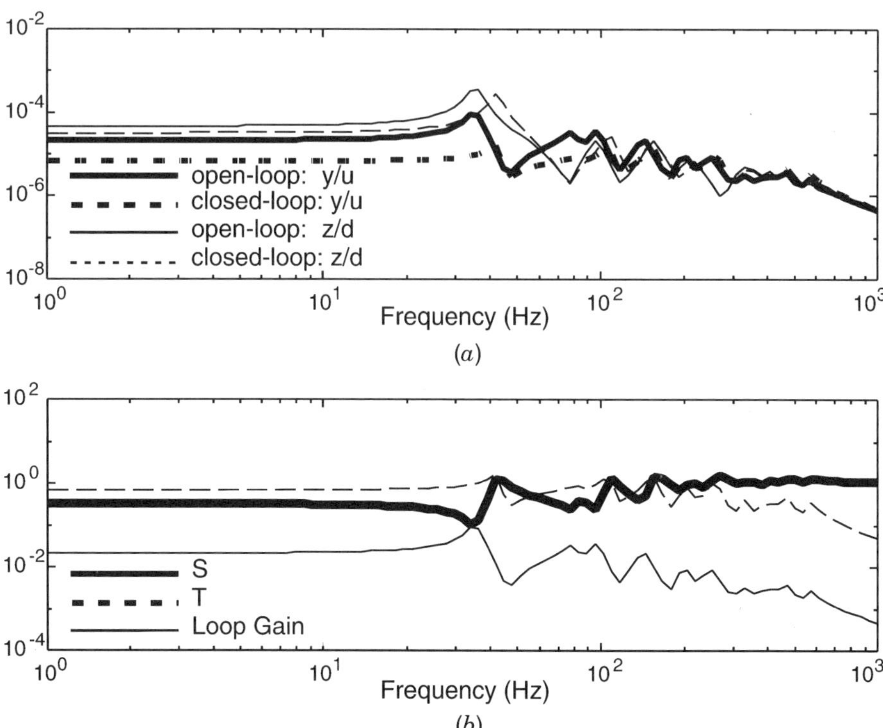


**Figure 7.4:** (*a*) Open- and closed-loop FRFs for proportional feedback on a reverberant plate model. (*b*) Sensitivity, complementary sensitivity, and loop gain for the same plant.

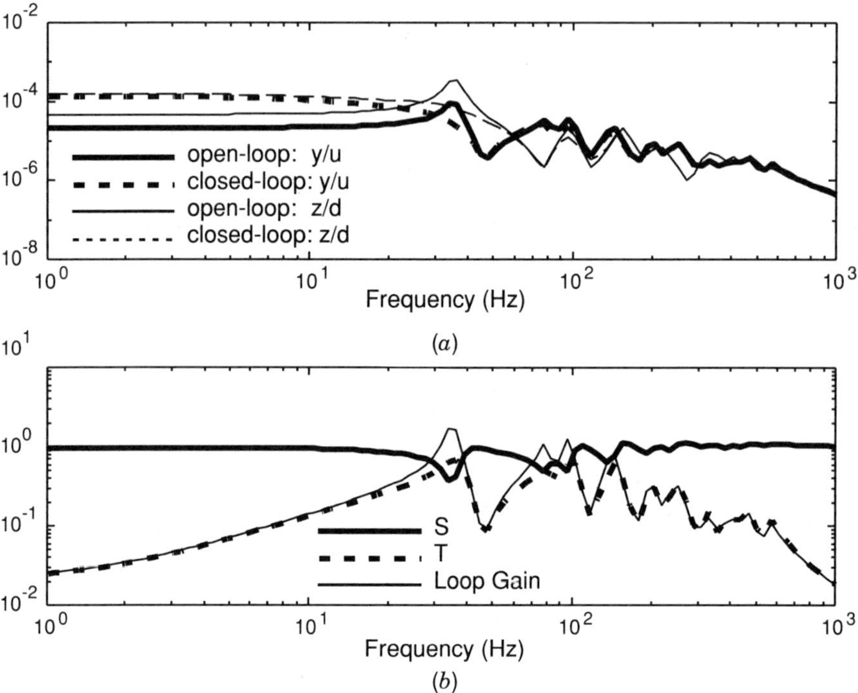


**Figure 7.5:** (*a*) Open- and closed-loop FRFs for lead compensation feedback on a reverberant plate model. (*b*) Sensitivity, complementary sensitivity, and loop gain for the same plant.

the sensitivity is less than unity up to approximately 50 Hz, at which point it approaches unity, and the complementary sensitivity begins to roll off. However, as illustrated in Figure 7.4*a*, the benefit of this increased sensitivity is observed primarily in the control path from $u$ to $y$.

Now, consider the same structural plant, except in this case, we implement a lead compensator, which is discussed in greater detail later in this chapter. The location of the pole and zero for the lead compensator are provided in the script file **plt_lead.m**. The resulting open- and closed-loop frequency responses for the control path and the performance path are illustrated in Figure 7.5*a*. In this case, the closed-loop response in both paths of the plant have similar trends. The compliance of the structure is increased at low frequency; however, significant levels of attenuation result near the resonance corresponding to the fundamental mode. The lead compensator is selected to effectively accomplish dissipative control in this bandwidth. Upon considering the plots of the sensitivity function, complementary sensitivity function, and loop gain in Figure 7.5*b*, one recognizes that at the frequency corresponding to the resonance of the fundamental mode, the sensitivity is decreased; however, the sensitivity is very near unity over the remainder of the bandwidth. Thus, the compensator is used to target a particular bandwidth and particular type of control (rate-feedback).

As can be seen from the two examples provided, the sensitivity and complementary sensitivity do not necessarily provide a good indication of the closed-loop performance when the disturbance path is different from the control path. This is also the case for multivariable control system applications. However, if the sensed variable is the same as the desired performance variable, then they are quite useful in the analysis of the closed-loop performance. For such problems, it is only important that the loop transfer function $P(s)D(s)K(s)$ be large in the bandwidth of the disturbance so that the sensitivity is less than some design value $S^*$:

$$\left| \frac{1}{1 + P(s)D(s)K(s)} \right| \leq S^* \tag{7.52}$$

The bandwidth of disturbances may or may not be at very low frequencies, making this problem somewhat different than those typically encountered by designers of SISO servocontrol systems. Because of the importance of disturbance rejection (associated with disturbances presented at the plant input) for adaptive structure design, this topic is treated with significant detail in Section 7.6.

Finally, looking back at equations (7.37) through (7.40), we highlight two important features of the different closed-loop transfer functions. First, the denominators are the same for each path through the structure. The form of the denominator $1 + P(s)D(s)K(s)$ indicates that the roots of the open-loop system, described by the factors of $P(s)D(s)K(s)$, are changed in the presence of feedback. The roots of the denominator polynomial are called the *poles* of the system. Through our selection of $D(s)$ and $K(s)$, we can affect exactly how the closed-loop poles are modified. Second, the numerators are different for each case. The roots of each numerator are called the *zeros* of the system. We see that the open-loop zeros still remain in the closed-loop system, although there are additional zeros contributed by the compensators $D(s)$ and $K(s)$. These are two classical, well-known features of closed-loop systems, and form the cornerstone for design of SISO adaptive structures.

## 7.3 ROOT LOCUS AS A DESIGN TOOL

The design of adaptive structures often requires iterative simulation and experimentation of the open-loop system during the design phase of the closed-loop control system. In this design phase, it is usually necessary to work with high-order dynamic models to achieve acceptable correlation between test data and model data from the structural prototype. Once the design model has achieved reasonable fidelity, the first adaptive structure design consideration will be related to closed-loop *stability*. One way to test for stability is to factor the closed-loop characteristic equation $1+P(s)D(s)K(s)$ and determine the location of the closed-loop poles. It is difficult, and unnecessary, to factor the high-order, rational function approximations used to represent the structure's controlled dynamic response. The *root-locus method* is a graphical means of identifying the closed-loop poles and zeros of a SISO adaptive structure configuration for all values of compensator gains while requiring only one factorization of the open-loop characteristic equation. Beginning with various forms of the characteristic equation, dependencies of the closed-loop system poles on arbitrary system parameters (usually the feedback gain) can be plotted in the complex plane. The power of the method is that all of the closed-loop information can be obtained by dealing with the open-loop transfer function. This allows the designer to quickly assess the system's stability over the operational range of the variable parameter.

Root-locus techniques prove to be particularly helpful as we demonstrate the closed-loop performance of SISO adaptive structures with and without modeling effects of transduction device dynamics. Today, many computational programs can sketch root-locus diagrams much faster than one can complete such diagrams using the rules originally established by Evans. We assume that the reader has already had an introduction to the sketching rules, and we use the MATLAB root-locus commands for the majority of the examples presented.

Root-locus diagrams are discussed in virtually all feedback control theory textbooks, including Melsa and Schultz (1969), Hale (1977), and Franklin et al. (1994), to name a few. These references present discussions explaining *why* the root-locus technique works; our purpose is to present a review of *how* to sketch root-locus diagrams and use the diagrams to understand essential differences between different adaptive structure designs. Section 7.3.1 is devoted entirely to a basic review of sketching root-locus diagrams. In Section 7.3.2 we analyze the movement of an adaptive structure's closed-loop poles as proportional feedback is applied to position, velocity, and acceleration sensors. Section 7.3.3 deals with changes in the closed-loop pole locations as transducer dynamics increase the numbers of poles and zeros in the Laplace left-half plane.

### 7.3.1 Basic Review

The root-locus method begins with the following representation of a linear system's characteristic equation:

$$1+kP(s)D(s)K(s)=0 \tag{7.5.3}$$

where $k$ is a variable gain that has been factored out of the open-loop transfer function. The closed-loop poles occur at those values of $s$ for which the characteristic equation is equal to zero. We can then write

$$kP(s)D(s)K(s) = -1 = |1|\angle 180 \tag{7.54}$$

Equating magnitude and phase terms in the previous equation, we can write

$$|kP(s)D(s)K(s)| = 1 \tag{7.55}$$

$$\angle kP(s)D(s)K(s) = 180 \text{ degrees} \tag{7.56}$$

The root-locus plot for the system is the locus of all values of $s$ that satisfy the phase-angle condition, independent of the magnitude condition. Construction of the locus proceeds by substituting a specific value of $s$ into the phase-angle criterion and determining if the equation is satisfied. Algorithms are easily written to accomplish this checking in a numerically efficient manner (see **rlocus.m** in the Control Systems Toolbox).

Consider the characteristic equation in polynomial form

$$\nabla(s) = a_n s^n + a_{n-1} s^{n-1} + \cdots + a_0 = 0 \tag{7.57}$$

If we wish to develop a root locus based on changes in the coefficient $a_{n-1}$, we divide equation (7.57) by all terms except the term with that coefficient. Consider as a simple example the unforced, SDOF spring-mass-damper equation of motion in the Laplace domain with zero-initial conditions:

$$ms^2X(s) + csX(s) + kX(s) = 0 \tag{7.58}$$

Dividing equation (7.58) appropriately, the root-locus form is written as

$$1 + \frac{cs}{ms^2 + k} = 0 \tag{7.59}$$

Thus, the roots of the equation change as $c$ varies from zero to infinity

For most cases, rearranging the characteristic equation is not necessary. Referring back to equation (7.38), the denominator expression is already in the correct form, that is:

$$\nabla(s) = 1 + kP(s)D(s)K(s) \tag{7.60}$$

Finally, we note that the loop transfer function $kP(s)D(s)K(s)$ is a ratio of polynomials, so that the characteristic equation is:

$$1 + kP(s)D(s)K(s) = 1 + k\,\frac{N_P(s)N_D(s)N_K(s)}{D_P(s)D_D(s)D_K(s)} = 0 \tag{7.61}$$

where it can be seen that for $k = 0$, the roots of the expression correspond to the open-loop poles and as $k$ approaches infinity, the roots approach the zeros of the open-loop system.

Referring back to equation (7.59), we can now examine how the poles change as the viscous damping constant varies. The zero location is $s = 0$ and is marked with an O. The pole locations are identical to the undamped system poles, given by

$$p_{1,2} = \pm j\omega_n, \tag{7.62}$$

and marked with X's in Figure 7.6. The real axis root locus exists for all negative real values because of the zero at the origin. As indicated in Figure 7.6, both poles have departure angles of 180 degrees. The loci eventually re-enter the real axis, as one pole moves toward negative infinity and the other pole moves toward the zero at the origin. Of course, we recognize that the point of re-entry corresponds to critical damping so that the value of the damping constant should be equal to unity at this location. As $c$ continues to increase, the poles become overdamped, as is evident by their location on the real axis. The plot of the root locus in Figure 7.6 confirms all of these features.

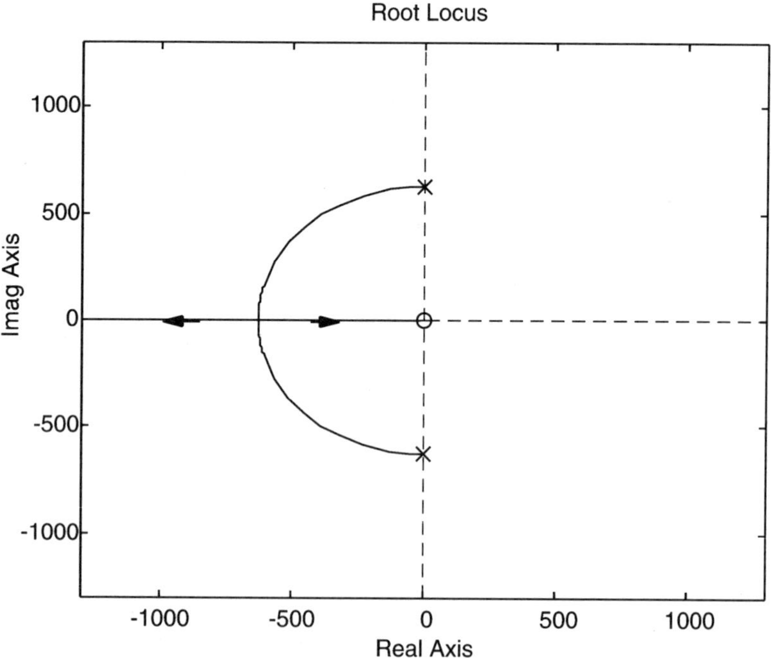


**Figure 7.6:** Root-locus diagram for increasing damping coefficient; SDOF mass-spring-damper system.

### 7.3.2 Root-Locus Diagrams for Structures with Proportional Feedback

Constrained adaptive structure root-locus diagrams change markedly for different selections of feedback variables. We begin with a SISO receptance model, as introduced earlier for the homogeneous problem:

$$ms^2y(s) + csy(s) + ky(s) = F(s) \tag{7.63}$$

For proportional feedback, $F(s) = -K_f y(s)$, the closed-loop transfer function is

$$\frac{y(s)}{F(s)} = \frac{1}{[s^2 + (c/m)s + ((k/m) + (K_f/m))]} \tag{7.64}$$

Defining $\omega^2_{n,cl} = (K_f/m)$ and recalling definitions of the damping ratio and natural frequency from vibration theory, the poles of the closed-loop system can be written as

$$s_{1,2} = -\zeta\omega_n \pm \sqrt{(\zeta\omega_n)^2 - (\omega^2_n + \omega^2_{n,cl})} \tag{7.65}$$

Almost always, we are interested in the underdamped case ($\zeta < 1$), so that

$$s_{1,2} = -\zeta\omega_n \pm j\omega_n\sqrt{1 + \frac{\omega^2_{n,cl}}{\omega^2_n} - \zeta^2} \tag{7.66}$$

Recall that the damped natural frequency is the imaginary part of each complex conjugate pole. Equation (7.64) shows that the closed-loop damped natural frequency increases as $K_f$ increases, but the real part of the pole is identical to the open-loop value.

Using root-locus sketching rules, it is straightforward to determine that there are two branches, no real-axis root locus, and two asymptotes, for the system of equation (7.64). The departure angles are ±90 degrees. The root locus is shown in Figure 7.7*a*. The behavior of the loci agrees with the conclusions made in the previous paragraph. Both poles move toward zeros at infinity, which causes the damping to decrease and the closed-loop natural frequency to increase as $K_f$ becomes higher. Using the trigonometric relations defined in Section 7.1, it can be shown that the closed-loop damping ratio decreases as the proportional gain increases according to the relationship

$$\zeta_{cl} = \zeta_{ol}\,\frac{k^{1/2}}{K_f^{1/2}} \tag{7.67}$$

This adaptive structure result is in line with proportional controllers for other second-order dynamic plants; that is, use of proportional (position) feedback is

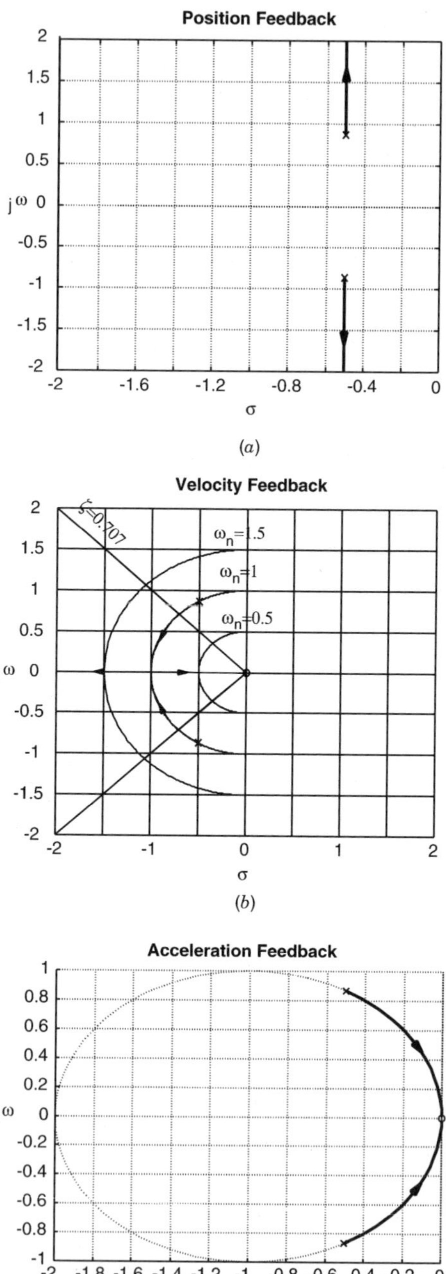


**Figure 7.7:** Root-locus diagrams for proportional output feedback, SDOF mass-spring-damper system. (*a*) Position. (*b*) Velocity. (*c*) Acceleration.

not a favorable approach if the design objective is active damping. One should also note that the decrease in damping with increasing $K_f$ will be constrained by physical limits of the force actuation process being applied to the structure.

Similar analyses can be completed for proportional feedback of velocity (sometimes referred to as rate feedback) or acceleration feedback. One significant change that occurs with different selection of output variables is that the transfer function acquires one or two zeros at the origin, for velocity and acceleration feedback, respectively. In addition, we see a change in the dependence of the real and imaginary parts of the system poles on the feedback gain. Velocity feedback implies that $F(s) = -C_f s y(s)$ and the acceleration feedback force becomes $F(s) = -M_f s^2 y(s)$. Beginning with equation (7.63), the closed-loop poles can be obtained as before. This is left for the reader to complete. As an alternative, the root loci can be sketched for both closed-loop systems as shown in Figure 7.7(*b*) and (*c*). Negative rate feedback moves the poles along a constant radius circle in the $s$-plane until they re-enter the real axis, implying that the natural frequency is unaffected as damping is introduced by the feedback force proportional to velocity. For negative acceleration feedback, the poles move along inverse circular paths and converge on the two zeros at the origin. Therefore, the closed-loop natural frequency and the damping ratio decrease as the feedback gain $M_f$ is raised. The designer's satisfaction with these results depends, of course, on the design objectives. If the purpose is to add damping, the SDOF analysis indicates that velocity outputs must be selected for feedback.

Theoretically, infinite dimensional order for real structures adds tremendous complexity to the simple analyses introduced above. However, the SDOF results do provide important generalities about output feedback for constrained adaptive structures. Perhaps the most important is that negative velocity feedback is the most "natural" loop-closure approach if one wishes to increase the damping in a structure. An interesting root-locus analysis of a higher order model, characterized by six modes that are very lightly damped, is shown in Figure 7.8. The output variable is once again velocity and is for a colocated input-output transfer function (i.e., minimum phase). Notice that the path of the pole moving to the real axis root locus is a circle, as before. However, it is no longer the fundamental mode that follows that particular locus. The student can access the file **highpz.m** to continue the investigation of the root-locus behavior for higher order systems. Using the file, it is a simple matter to examine the root loci as a function of modal density, pole-zero spacing, and output variable units (e.g., velocity vs. acceleration outputs). In fact, the results of this script file provide extremely insightful information about the closed-loop behavior for any reverberant systems characterized by high modal densities.

### 7.3.3 Effects of Transducer Dynamics

The root locus is also an excellent analysis tool for adaptive structures with higher dimensional dynamics caused by dynamics of the actuators and sensors

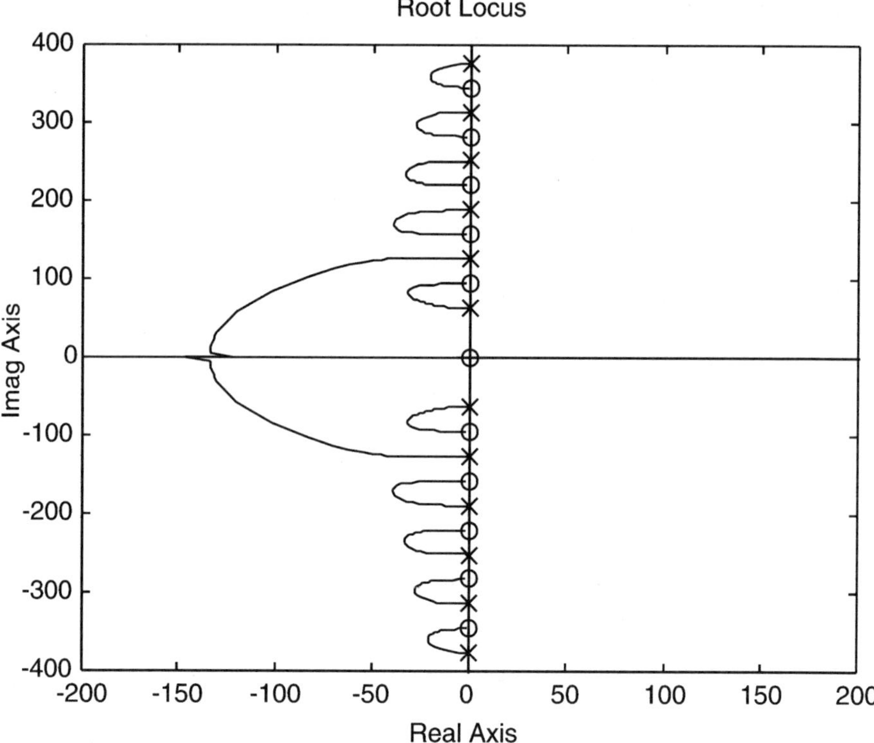


**Figure 7.8:** Root locus for a six-mode beam model, velocity feedback.

that are required for practical control problems. Let us assume that we wish to generate a closed-loop controller that can regulate the shape of a structure. This is classified as a command tracking problem. For simplicity, we consider using a linear actuator to affect the structure and a displacement probe to measure the position at some reference point on the structure. These transducers are not of the embedded, integrated types typically discussed for adaptive structures but emphasize the effects of transducer dynamics for a structural control problem. Consider that the structural plant is a simple SDOF model.

If we assume that the linear actuator can be modeled approximately as a second-order system, the transfer function from the applied voltage to the lead screw position is

$$\frac{y(s)}{v_a(s)} = \frac{A}{(\tau_1 s + 1)(\tau_2 s + 1)} \tag{7.68}$$

The displacement probe has a time lag represented by a first-order pole, leading to a total of three additional poles in the closed-loop system. We assume that $\tau_2 \gg \tau_1$ and that the simple pole from the probe dynamics is approximately

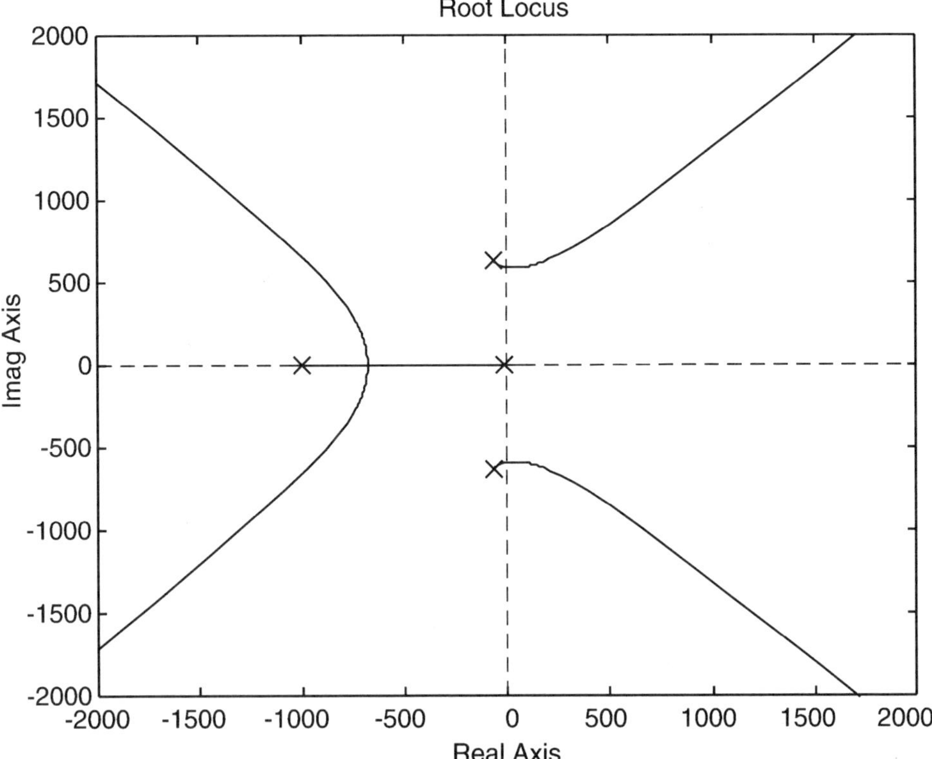


**Figure 7.9:** Effects of additional transducer dynamics on a SDOF system with position feedback.

1000 rad/sec. Thus, only two poles are in the vicinity of the two complex structural poles. (The actuator pole associated with $\tau_1$ is far to the left.) If we examine the root locus for the open-loop transfer function, shown in Figure 7.9, we see that now the system quickly goes unstable for proportional gain feedback. This was not the case for the hypothetical proportional control discussed above. The conclusion is clear: for real systems, the adaptive structure plant dynamics must include the transduction device dynamics. Root-locus analyses, performed using the correct plant dynamics (plant + transducers + electronics), provide accurate predictions of the adaptive structure's stability bounds and approximations of the closed-loop transient performance.

## 7.4 FREQUENCY RESPONSE ANALYSIS AND DESIGN

Frequency response methods are widely used in the design of control laws for SISO systems, with adaptive structures as no exception. Frequency domain design methods are very popular for several very practical reasons. First, feedback compensators can be generated directly from knowledge of certain experimental FRFs, which are simply related to the corresponding transfer functions by

$$\frac{y(j\omega)}{u(j\omega)} = \left.\frac{y(s)}{u(s)}\right|_{s=j\omega} \tag{7.69}$$

FRFs are relatively easy to acquire and often serve as adequate substitutes for pole-zero mathematical models in the design process. Second, the designer can assess the eventual closed-loop stability by examining the Bode form and/or the Nyquist form of the open-loop frequency response function. Finally, we have seen in Section 7.2 that there are frequency-dependent specifications for the gain of a closed-loop system depending on the performance objectives. For example, some designs require high loop gain $|P(j\omega)K(j\omega)|$ for reduced sensitivity to disturbances, but lower gain where noise signal energy controls the frequency response. Other closed-loop system characteristics, such as command tracking and transient response, are also easily assimilated into the FRF format. Detailed frequency response design methods are the topic of "loop-shaping," a design approach that relies on FRF measurements of the plant and knowledge of the FRF characteristics for any compensators that will be added to the plant. Loop shaping methods are discussed in Chapter 8, where it is shown that frequency domain design can also be used for multivariable control applications.

In this section, frequency response analysis and design methods for SISO paths through the adaptive structure are emphasized. The link between FRFs and transfer functions is discussed, followed by a description of the data acquisition and signal processing procedures that must be used to generate experimental FRFs. The Bode plot, a display technique that presents logarithmic frequency versus logarithmic magnitude and linear phase of the FRF data, is used to present important information—minimum and nonminimum phase behavior, stabilizability, structural damping estimates, and so on—that assists in the controller design process for the structure. Similar information can be obtained from the Nyquist displays, and examples are provided. Finally, the important issue of colocated versus noncolocated actuator-sensor pairs is illustrated using the frequency response information for a multimodal model of a SISO structure.

### 7.4.1 The Frequency Response Function

Frequency response concepts are well understood by most structural dynamicists and a large number of engineers in varied specialty areas including vibrations and acoustics, as well as civil, and structural engineering. However, the use of frequency response data for control system design relies on certain viewpoints that are not typically encountered in these other fields. We wish to integrate the frequency domain concepts of control theory with the signal processing/structural dynamics perspective that adaptive structure designers need to master before designing the closed-loop system for performance and stability.

The FRF is directly related to the transfer functions described earlier, both theoretically and experimentally. Knowledge of the FRF follows directly if the transfer function of the structure is known. The general notation for a transfer function is

$$H(s) = \frac{X(s)}{F(s)} \tag{7.70}$$

and for an FRF, it is written

$$H(j\omega) = \frac{X(j\omega)}{F(j\omega)} \tag{7.71}$$

It can be shown that the frequency response may be obtained from the transfer function simply by letting $s = j\omega$. Visually, this is illustrated in Figure 7.10, which is derived from a Matlab mesh plot of a second-order transfer function:

$$G(s) = \frac{\omega_n^2}{s^2 + 2\zeta_n\omega_n s + \omega_n^2}$$

where $\omega_n \simeq 4$ rad/sec. The FRF is valid along the $j\omega$-axis only. The numerical value of $\zeta_n$ is small for the example, as indicated by the large amplitude of the transfer function and the FRF in the vicinity of the poles. It is easy to think of similar plots for higher dimensional transfer functions (i.e., more poles and zeros) if one imagines sketching a pole-zero map of the $s$-plane on a flat membrane. Then, the membrane is pulled up at the pole locations and pushed down at the zero locations. To visualize the corresponding FRF, the transfer function is "sliced" along the imaginary axis of the Laplace plane or the $j\omega$ axis. The FRF is then the elevation of the transfer function plot along the positive portion of the $j\omega$ plane, as shown in the figure.

We recall that the time responses of singularities (i.e., poles) on the $j\omega$ axis are undamped sinusoids, $y(t) = \exp(j\omega t)$. This is indicative of the true meaning for the FRF, namely that it provides the system response to discrete sinusoidal inputs of frequency $j\omega$. Conversely, the transfer function can be obtained from the FRF by replacing $\omega$ by $s/j$. As discussed in Chapter 3, the transfer function provides the response to arbitrary inputs. Therefore, we conclude that the FRF of a linear system uniquely defines the impulse response of the system, and thus the time response, to arbitrary inputs.

The frequency response of a system can be obtained experimentally or developed numerically from several forms of linear models. There are many factors that affect whether one should use experimental FRF models or numerical FRF models. Experimental Bode plots are created by acquiring specific input-output

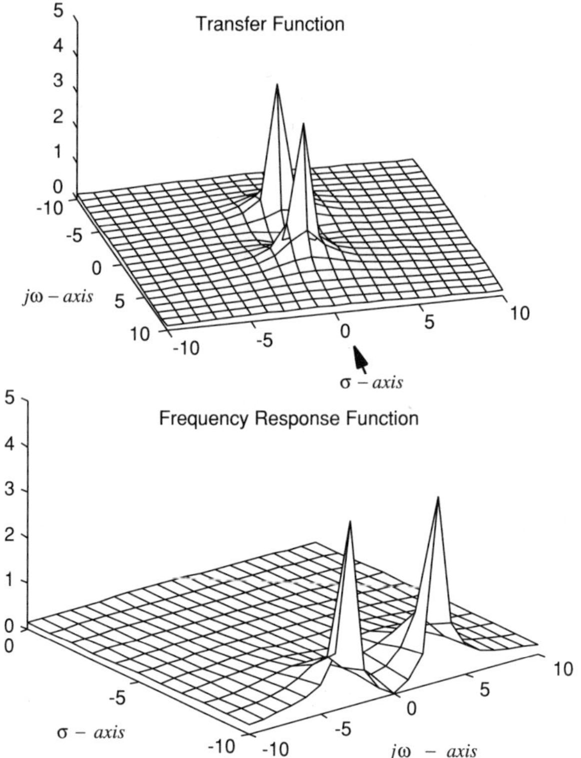


**Figure 7.10:** Relationship of the FRF to the transfer function.

measurements and then postprocessing the data using signal processing methods as discussed by Bendat and Piersol (1984) or Newland (1975). Even for stationary linear systems, there are always accuracy issues that arise because of electrical noise on the measurement transducers, inaccurate calibrations of transducers, poor dynamic range settings of the analog to digital converters, dynamic range of the processor used to create the frequency domain power spectra, and linear dependencies of the selected plant input (i.e. unmeasured inputs) that may not be obvious to the designer of the experiments. The accuracy of Bode plots generated from numerical models is determined by the uncertainty of the modeling process; electrical noise, dynamic range, and so on, are not factors. Instead, the designer must consider truncation effects, unmodeled dynamics, uncertainty of the modeled dynamics, and inaccuracy of input/output disturbance models, all of which can degrade the Bode plot fidelity. Because of the popularity of Bode plots in controller design practice, we briefly review the essential steps that lead to Bode plots using experimental and analytical-numerical methods. Some of the information presented below was introduced earlier in Chapter 4; however, this topic is so important to SISO control system design for real structures that it is reviewed again below.

### 7.4.2 Experimental Development of Bode Plots

Bode plots are equivalent to FRF measurements made between particular input-output locations on a structure. Even though FRFs are strictly associated with sinusoidal inputs, it can be shown that the linear interdependence of time-frequency relationships allows the determination of an FRF to *any* known time input. This feature represents a tremendous opportunity to reduce the complexities that arise if a system FRF is acquired by stepping through sinusoidal inputs of varying frequencies. The sinusoidal excitation methods require swept-sine analyzers or burdensome acquisition and processing software that are not always readily available. Instead, the FRFs can usually be measured with good accuracy (given careful planning), using random or chirp inputs whose bandwidth matches the bandwidth of interest. There are certainly a wide number of factors that can affect the ultimate accuracy of the FRF estimates (Ewins, 1986). The reader is encouraged to learn or review proper structural testing procedures before embarking on the experimental path. For our purposes, we wish only to review the synthesis of FRF estimates from time ensembles of input and output measurements. We assume that the data has been acquired with proper attention paid to the structural test support system, transducer mounting, power amplification, signal conditioning, windowing, and other potential sources for experimental error. Once test conditions are adequate, accurate adaptive structure FRFs, and hence Bode plots, can be acquired by using professional signal analyzers or by developing appropriate software that implements the operations summarized below.

Typical estimation of a system's FRF begins with fast Fourier transform (FFT) calculations to generate frequency domain representations, specifically the discrete Fourier transforms (DFT), of input $f_r$ and output $x_r$ time sequences. The transforms can be written

$$\mathcal{F} = \frac{1}{N} \sum_{r=0}^{N-1} f_r e^{-i(2\pi k r/N)} \qquad K = 0, 1, 2, \ldots, (N-1) \tag{7.72}$$

$$\mathcal{X} = \frac{1}{N} \sum_{r=0}^{N-1} x_r e^{-i(2\pi k r/N)} \qquad K = 0, 1, 2, \ldots, (N-1) \tag{7.73}$$

where the DFTs are complex-valued vectors whose resolutions depend on the length $N$ of the discrete sequences $f_r$ and $x_r$. Estimates of the spectral densities lead to expressions for the one-sided autospectra:

$$G_{xx}(\omega) = \mathcal{X}\mathcal{X}^* \tag{7.74}$$

$$G_{ff}(\omega) = \mathcal{F}\mathcal{F}^* \tag{7.75}$$

of the response and excitation, respectively. These are real-valued at all frequencies. The cross-spectra $G_{xf}(\omega) = G_{fx}(\omega)$ are obtained from the DFTs according to

$$G_{xf}(\omega) = \mathcal{X}\mathcal{F}^* \tag{7.76}$$

$$G_{fx}(\omega) = \mathcal{F}\mathcal{X}^* \tag{7.77}$$

For linear systems, the FRFs are obtained from direct relationships between the autospectra and the cross-spectra. In the following descriptions, we ignore the treatment of noise signals on the input and output measurements, but the experimenter should be very careful to understand the noise sources in the test setup. The noise can be electrical in nature, or caused by unrecognized couplings between inputs and additional processes. Not accounting for noise inputs, we can write three important relationships:

$$G_{xx}(\omega) = |H(\omega)|^2 G_{ff}(\omega) \tag{7.78}$$

$$G_{fx}(\omega) = H(\omega)G_{ff}(\omega) \tag{7.79}$$

$$G_{xx}(\omega) = H(\omega)G_{xf}(\omega) \tag{7.80}$$

Equations (7.78) through (7.80) can then be manipulated to give measured estimates of the FRF $H(\omega)$. It can be seen that $H(\omega)$ may be obtained by either of two independent calculations. These two estimates are typically referred to as the "$H_1$" and "$H_2$" estimators, given by

$$H_1(\omega) = \frac{G_{fx}(\omega)}{G_{ff}(\omega)} \tag{7.81}$$

and

$$H_2(\omega) = \frac{G_{xx}(\omega)}{G_{xf}(\omega)} \tag{7.82}$$

The $H_1$ estimate is based on the assumption that the output measurements are noisy. The $H_2$ estimate assumes noise on the input. For the remainder of this book, we assume that any FRF measurements correspond to the $H_1$ FRF estimator. The reader should consult the references to appreciate the differences between the $H_1$ and $H_2$ estimates when noise exists on both the input and output signals (which, in practice, is *always* the case).

Once the FRF estimates have been measured, using equations (7.78) through (7.80), the experimental Bode plots are created using logarithmic displays of the frequency axes and magnitude ordinate axis. The phase plots are usually presented using linear degrees or radians on the ordinate. The capability to generate these Bode plots in the laboratory or field test environments begins with the data acquisition process. One can purchase any number of data acquistion boards today. However, be aware that the acquisition system should use a simultaneous sample/hold amplifier and multiplex through the analog-to-digital conversion. A sequential sample/hold configuration will result in erroneous phase information. After sampling, spectrum estimates and plotting operations are easily implemented using commands from the MATLAB Signal Processing and Control Systems Toolboxes. So, the FRF estimates for the structure can be acquired quite economically. Additionally, commercial signal analyzers are typically configured to provide Bode plots of the output-to-input FRFs (based on the $H_1$ estimator).

For example, Figure 7.11 shows an experimental Bode plot for a free-free aluminum beam excited by a piezoceramic element located at $0.2L$. The output measurement is acquired using the same piezoceramic element, wired in a sensoriactuator configuration as discussed in Chapter 5. The Bode plot shows the characteristic alternating poles and zeros associated with colocated inputs and outputs. This is evident in the alternating peaks and valleys of the magnitude plot and the restriction of the phase boundaries between $\frac{\pi}{2}$ and $-\frac{\pi}{2}$ radians. The units of the experimental FRF are shown in volt/volt. This is often satisfactory

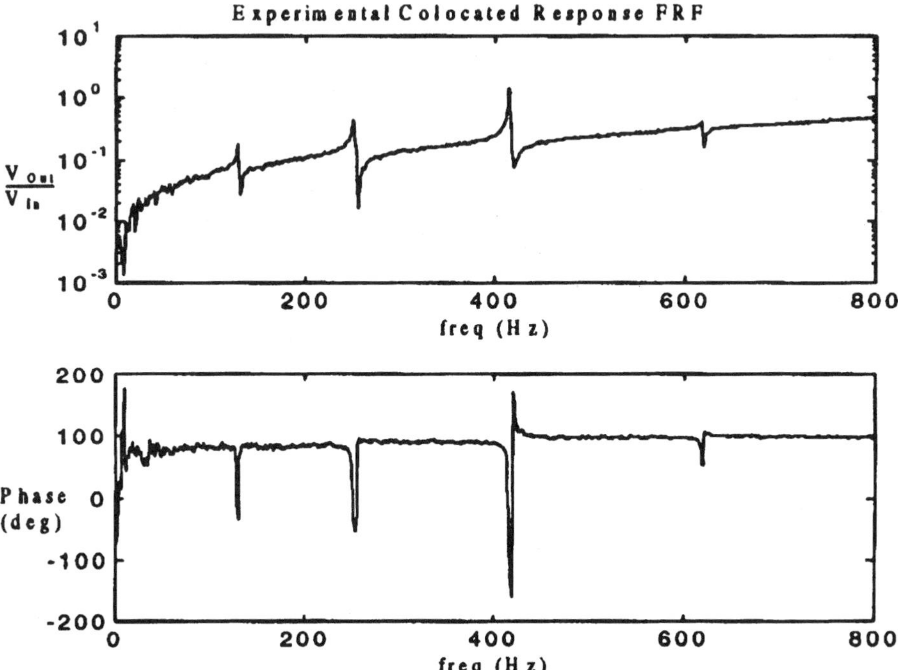


**Figure 7.11:** Bode diagram for an experimental FRF; free-free aluminum beam.

for the controller design, but the dynamic sensitivity can be used to convert to physical units if a comparison or validation of an analytical-numerical FRF is desired.

Experimental measurement of Bode plots for the open-loop transfer function of a structure is the first step in the design process if no analytical or numerical model is available. It is shown in Section 7.5 that the compensator can be designed directly from this information. However, it is always preferable to conduct a parallel modeling effort during the design of an adaptive structure since the physics of the dynamic system are typically better understood by the designer during this process.

### 7.4.3 Analytical-Numerical Development of Bode Plots

Bode plots of the adaptive structure can also be created from analytical or numerical models created using techniques presented elsewhere in the text. To summarize, it is only necessary to have some dynamic description of the structure in order to generate the Bode plot. If the dynamics are available as an analytical transfer function or FRF, the Bode diagram is straightforward. If the dynamic model is a differential equation, one can take the Laplace transform and proceed to the Bode plot. If the dynamic model is based on wave theory, the designer may transform into pole-zero descriptions and then generate Bode plots. If the dynamic model is a finite element model or is based on statistical energy analysis (SEA), it may be possible to construct a state-space model or a modal model for the former, or simply evaluate the output to varying frequency sinusoidal inputs that span the desired control bandwidth.

The MATLAB software generates Bode plots using the **bode.m** command. The model may be in state-variable, transfer function, or pole-zero-gain format. For example, Figure 7.12 is a Bode plot created for a numerical model of the free-free beam whose experimental frequency response was shown in the previous section (Figure 7.11). As before, this represents the response of a self-sensing (sensoriactuator) piezoceramic located at $0.2L$. The file **ffbeam.m** may be used by the student to create additional Bode diagrams that correspond to different actuator and sensor locations, different frequency bandwidths, and so on, for the free-free beam.

The benefits of introducing Bode plots in our discussion of adaptive structures are not limited strictly to controller design. There are great advantages to the designer when there is clear understanding of the physical parameters that control different frequency regimes of the FRF representations. This is best illustrated by considering the *Bode form* of an FRF, given for a general second-order system by

$$\frac{1/\omega_n^2}{[(j\omega/\omega_n)^2 + 2\zeta(j\omega/\omega_n) + 1]} \tag{7.83}$$

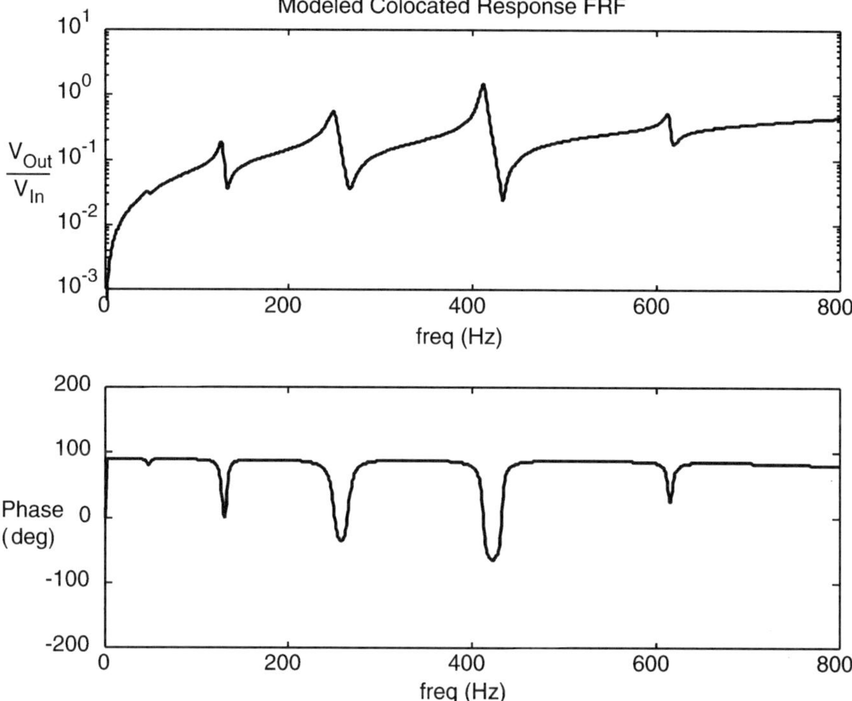


**Figure 7.12:** Bode diagram for a numerical FRF; free-free aluminum beam.

This Bode form is valid for arbitrary second-order transfer functions. When we write the viscously damped SISO receptance $(x/F)$ in the Bode form, we find

$$\frac{x(j\omega)}{F(j\omega)} = \frac{1/k}{[(j\omega/\omega_n)^2 + 2\zeta(j\omega/\omega_n) + 1]} \tag{7.84}$$

which indicates that the DC gain (defined as the system gain when $\omega = 0$) is inversely proportional to the structural stiffness. Stated otherwise, equation (7.84) demonstrates the well-known result that the static structural displacement (rotation) per unit force (torque) decreases as the stiffness increases. A similar expression can be written for the accelerance $(\ddot{x}/F)$ as

$$\frac{-\omega^2 x(j\omega)}{F(j\omega)} = \frac{-\omega^2/k}{[(1 + (j\omega_n/\omega)^2) + 2\zeta(j\omega_n/\omega)]} \tag{7.85}$$

which shows that, at high frequencies, the accelerance is inversely proportional to the structural mass. These dependencies are illustrated in Figure 7.13. When

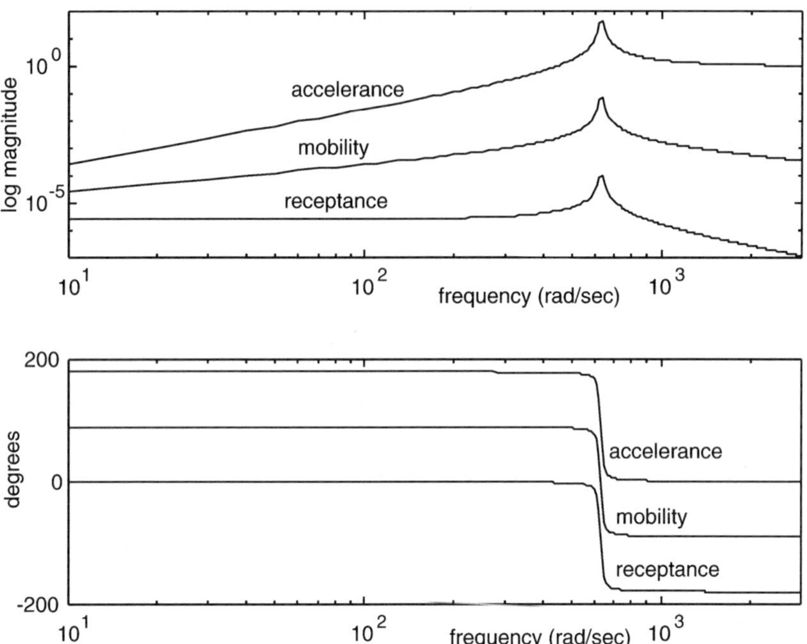


**Figure 7.13:** Bode diagram of receptance, mobility, and accelerance FRFs of a SDOF system.

feedback controllers are designed for the adaptive structure, it is usually informative to examine the closed-loop FRF in the Bode form for a clear description of the controller's effect on the structural stiffness, damping, and mass. These concepts are demonstrated later in Section 7.6.

### 7.4.4 Alternative Representations

The complex-valued FRF is often displayed on a *polar plot* that shows the magnitude versus the phase angle for each frequency of the FRF using polar coordinates. It is also common to display only the real part or imaginary part versus frequency, although this format is used more for modal analysis than for control system analysis. We wish to examine polar plots, which have a strong history in both structural modal analysis and SISO closed-loop control analysis where they form the basis for the *Nyquist stability criterion.* For the following discussion, we defer to the label of polar plots instead of Nyquist plots unless the plots are used for stability analysis of the closed-loop adaptive structure. Details of the procedures used to generate the polar representations are not included here but can be found in classical control theory textbooks (e.g., Ogata, 1970).

Dynamics of typical SISO transfer functions for adaptive structures can be factored into integral, derivative, first-order, and quadratic factors. The Bode form for the general case can be written

$$P(j\omega) = \frac{K(j\omega)^{\alpha}(1 + j\omega T_1)(1 + j\omega T_2)\ldots}{(j\omega)^{\beta}(1 + j\omega\tau_1)(1 + j\omega\tau_2)\ldots} \tag{7.86}$$

Corresponding polar plots for type 0 ($\beta = 0$) and type 1 ($\beta = 1$) second-order transfer functions, with no numerator dynamics, are shown in Figure 7.14. The type 0 polar plot is theoretically interesting because it is identical to the receptance of a single, lightly damped mode for a structure with no transducer dynamics. Viscous damping is assumed. Kennedy and Pancu (1947) were the first to identify the relationship between the polar plot of a point receptance and the structural parameters $\omega_n$, $\zeta_n$, *and* $R_n$ [see equations (7.25) and (7.26)]. The modal analysis community adopted these relationships for frequency domain system identification of structures with light damping and sparse modal density. Figure 7.15 displays polar plots for the receptance and the mobility of a structural mode. Both appear circular over the frequencies used in the analysis, but it can be shown that only the mobility plot yields an exact circle. If we denote the mobility as

$$Y(\omega) = \frac{j\omega}{(-\omega^2 m + k) + j\omega c} \tag{7.87}$$

then the equation for mobility circle can be written as

$$\left(\mathrm{Re}(Y(\omega)) - \frac{1}{2c}\right)^2 + (\mathrm{Im}(Y(\omega)))^2 = \left(\frac{1}{2c}\right)^2 \tag{7.88}$$

Therefore, the polar plot of $Y(\omega)$ traces out a circle of radius $1/2c$ and is centered at $(1/2c, 0)$, where $c$ is the viscous damping coefficient. A similar result can be shown for the receptance transfer function using an assumption of hysteretic damping. The resonant frequency is located at the bottom of the mobility circle, and the bandwidth is also available from these types of plots. As the frequency bandwidth is extended to include the entire frequency axis, the polar plots become significantly more complicated at first appearance. With practice, the adaptive structure designer can use such plots to determine whether a measured FRF is minimum phase, a drive-point measurement, or a transfer-point measurement, and if the damping is proportional or nonproportional. Additionally, expected margins of stability can be assessed when a compensator is integrated with the dynamics of the structure.

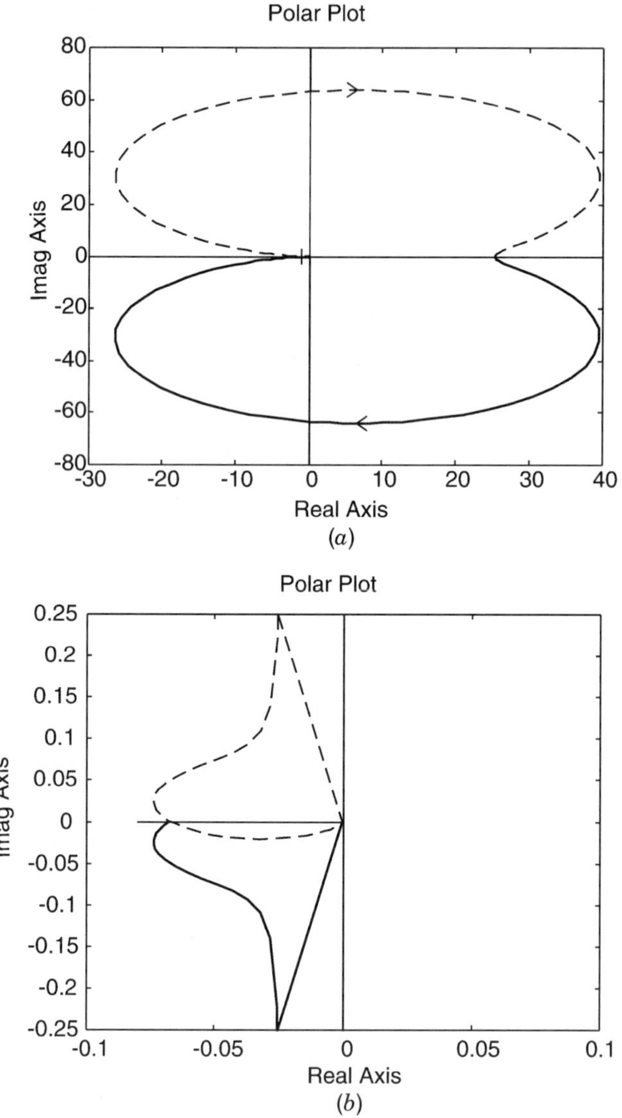


**Figure 7.14:** Polar diagrams for second-order FRFs. (*a*) Type 0. (*b*) Type 1.

### 7.4.5 Stability Margins

It has been mentioned earlier that colocated configurations for adaptive structures are stabilizing to closed-loop adaptive structure designs because these structures tend to be minimum phase systems and display robust stability over practical ranges of compensator gain. (The construction of minimum phase adaptive structures is discussed in Chapter 9.) For minimum phase

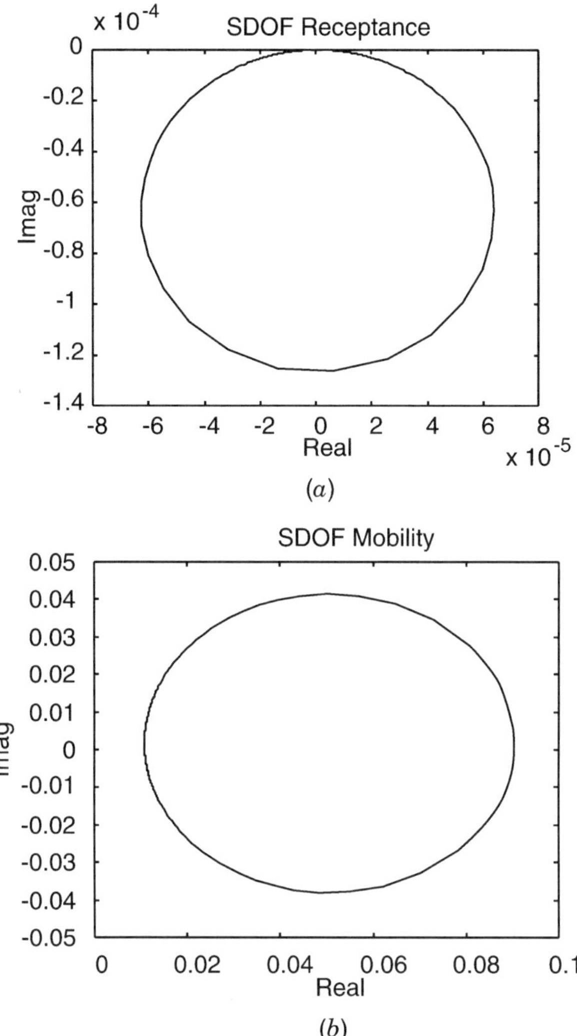


**Figure 7.15:** Polar diagrams of FRFs. (*a*) SDOF receptance. (*b*) SDOF mobility.

systems, the designer can use well-known expressions that predict the *stability margins* from the Bode magnitude and phase plots. *Gain margin* (GM) and *phase margin* (PM) are two metrics of closed-loop stability that can be used during design of SISO adaptive structural plants. Unless the designer intentionally builds colocation into the adaptive structure, it is very likely that any arbitrary SISO transfer function for MIMO structure design will be nonminimum phase. In that case, Nyquist stability analysis must be used to determine the gain and phase margins. In this section, we wish to review the interpretation of stability margins

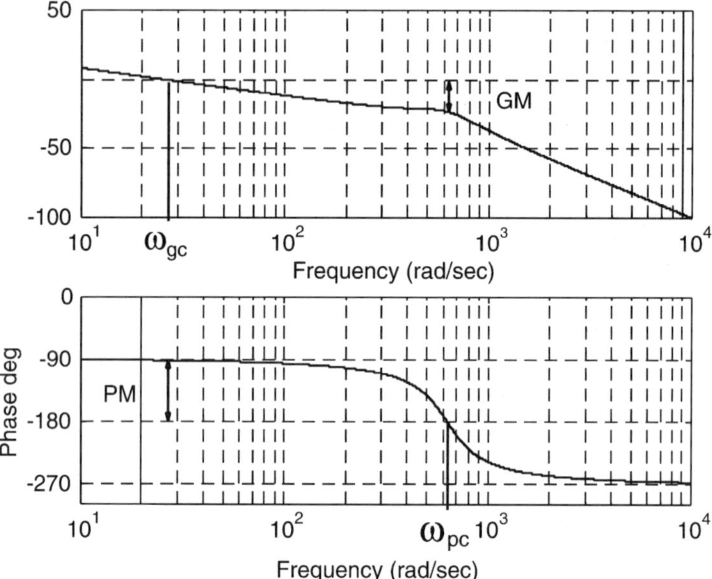


**Figure 7.16:** Definition of gain margin (GM) and phase margin (PM) using Bode diagrams.

and discuss the treatment of multiple crossover frequencies that occur for real structural plants.

***Gain and Phase Margins*** Figures 7.16 and 7.17 show a Bode plot and the corresponding polar plot for a type 1 system with second-order dynamics (see equation (7.86)). The stability of the resulting closed-loop systems can be analyzed directly from either of these FRF displays. Referring to the Bode plot of Figure 7.16, we observe that there are two ways to predict the degree of instability:

*Gain Margin:* The gain margin is defined as the reciprocal of the linear magnitude, or the decibel magnitude difference from 0 dB, at the frequency where the Bode phase is $-180$ degrees. This provides expressions for the linear and logarithmic gain margin (GM), respectively:

$$\text{GM} = \frac{1}{|G(j\omega)|} \tag{7.89}$$

$$= -20 \log |G(j\omega)| \text{ dB} \tag{7.90}$$

The logarithmic gain margin is positive valued when the system is stable; its decibel value represents how much the gain can be increased before a minimum phase system will become unstable.

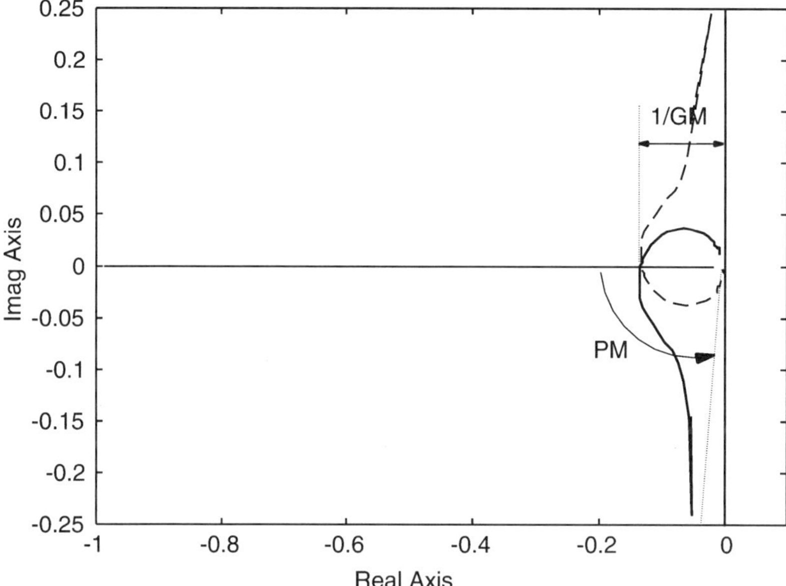


**Figure 7.17:** Definition of gain margin (GM) and phase margin (PM) using Nyquist diagrams.

*Phase Margin:* The phase margin is equal to the phase at the gain crossover frequency ($|G(j\omega_{gc})| = 1$) added to 180 degrees, that is,

$$\mathrm{PM} = 180 \text{ deg} + \arg[G(j\omega_{gc})] \tag{7.91}$$

The phase margin represents the amount of phase lag that can be accomodated at the gain crossover frequency $\omega_{gc}$ before a minimum phase system becomes unstable.

To demonstrate the concepts graphically for a plant with multiple resonances, consider the plot of the magnitude and phase of the FRF presented in Figure 7.18. The measured FRF corresponds to that of a speaker and microphone located in a rectangular enclosure, as detailed by Clark et al. (1996). The multiple resonances are characteristic of any reverberant plant, and adaptive structures, in general. To determine the stability margins, we must consider the frequencies at which the gain crosses unity and the frequencies at which the phase passes through multiples of $-180$ degrees to obtain the phase and gain margins, respectively. As illustrated, for a plant with multiple modes, there are many crossover frequencies due to the multiple resonances. The borders of the shaded bars in Figure 7.18 have been generated to show the frequencies at which the phase drops below $-180$ degrees, and the gain margin must be evaluated at each of these frequencies. The *worst case* (i.e., smallest gain) corresponding to

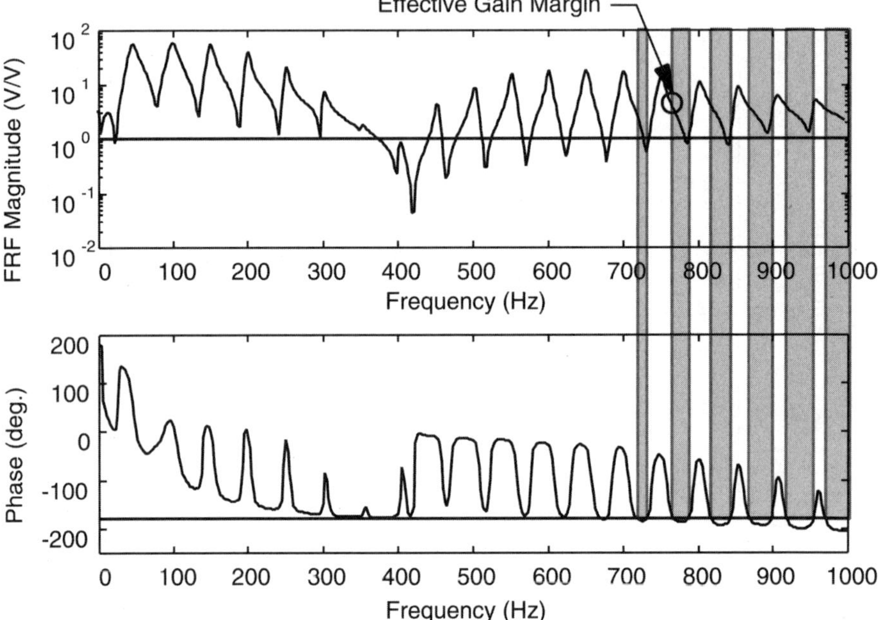


**Figure 7.18:** Gain margin analysis for a reverberant plant with multiple resonances in the bandwidth of interest.

this family of gain margins is considered to be the overall gain margin of the system. This example serves to emphasize the general difference in the level of complexity associated with the application of classical control system design approaches to high-order dynamic plants. In general, such complexity motivates us to consider closed-loop shaping, which can be cast from the general $\mathcal{H}_2$ or $\mathcal{H}_\infty$ control sytem design approach. This is discussed in greater detail in Chapter 8.

We emphasize that stability margins can be determined from Bode plots *only* if the open-loop system is minimum phase. Correct determination of the margins for nonminimum phase systems are generally not possible from Bode frequency response plots. For example, a nonminimum phase system can have negative phase and gain margins. The exact behavior of gain and phase margins for nonminimum phase systems is highly dependent on the specific transfer function of interest. Analysis procedures are available and are discussed by Shahian and Hassul (1993) for specific examples. Adaptive structure designs that rely on noncolocated transducer placements have a high probability of occurence of right-half plane zeros. Time delays in the open-loop system are equivalent to right-half plane zeros also. Although these systems remain closed-loop stable for very small gains, the closed-loop poles must move to instability (recall that the closed-loop pole locations eventually equal the open-loop zero locations as the closed-loop gain increases). The reader is encouraged to determine through analysis or experiments whether a particular adaptive structure design is minimum phase before using the guidelines proposed in this section. Unfortunately, identification of nonminimum phase zeros is very challenging. Interested readers should review the treatment of this subject by Tohyama and Lyon (1989).

For most cases, the designer should rely on Nyquist stability criteria if the goal is prediction of closed-loop stability. Referring back to Figure 7.17, we comment that the polar plots shown are equivalent to Nyquist diagrams when they are used for stability analysis. They are created by mapping a closed contour that encompasses the entire right-half plane to a conformal plane that is the evaluation of the open-loop transfer function for all values of $s$ along the contour. Then, the closed-loop stability can be predicted using the Nyquist stability criterion:

$$Z = N + P \tag{7.92}$$

where $P$ equals the number of right-half plane poles in the open-loop transfer function, $N$ is the number of clockwise encirclements of the point $s = -1 + j0$ in the Laplace frequency domain, and $Z$ represents the number of unstable closed-loop roots of the closed-loop characteristic equation. This stability test is valid for any transfer function and is the preferred analysis for closed-loop stability of systems whose models are not well known, or perhaps unknown entirely. This happens when the system has only been measured experimentally. Many modern frequency analyzers have the capability of producing a Nyquist diagram for the measurements made during system identification tests. Then, the designer need only determine the control system design that keeps $Z = 0$ and stay below that gain for the controller being investigated.

### 7.4.6 Bode Diagrams for Colocated and Noncolocated Designs

To conclude this section on frequency response analysis, the issue of colocated and non-colocated FRFs is revisited using the Bode plots shown in Figures 7.19 and 7.20. The first Bode plot is for the colocated FRF, and displays the minimum phase characteristics discussed earlier in Section 7.1.1. This type of Bode plot is always observed for so-called "drive-point" responses, that is, for input-output measurements where the input force and sensed output are at the same spatial location on the structure. If minimum phase characteristics are not observed for a drive point measurement, then the experimenter must determine the reasons why. Possible reasons include the presence of transducer dynamics in the bandwidth of interest, time delay in the measurement path, or inadequate spatial matching of the actuator and sensor footprint. It has been shown in Chapter 6 that the requirements for colocation can be achieved for a state-space realization when $\mathbf{B}^{\mathrm{T}} = \mathbf{C}$. This effectively conveys that the modal participation at the inputs and outputs must be the same in a relative sense (obviously scaling can occur depending upon transducer sensitivities). Figure 7.20 illustrates an example of a noncolocated

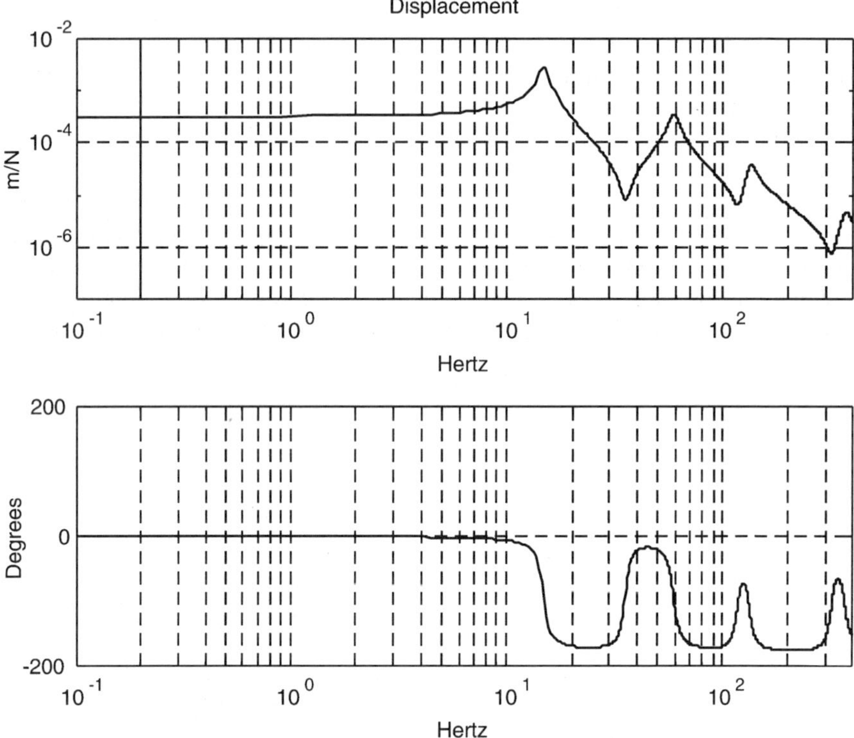


**Figure 7.19:** Bode diagram for colocated FRF; six-mode beam model.

FRF. The phase could fall off even more rapidly than shown for heavily damped structures with significant transducer dynamics in the bandwidth. Based on the previous discussions of stability margins, it should be clear that the control of structures with non-colocated actuator/sensor pairs is more challenging than the colocated case with respect to ensuring stability.

## 7.5 SERIES COMPENSATOR DESIGN

In classical control theory, compensator designs that are located in the forward path of the closed-loop system are called *series compensators*. Referring back to Figure 7.2, $D(s)$ represents the series compensators that introduce additional dynamics in order to ensure certain properties of the closed-loop transfer function. In this section, we discuss the impact of commonly used series compensators on adaptive structure dynamics.

There are some important distinctions that affect the design of SISO compensators for adaptive structures versus SISO compensators that are found in process control, navigation, and other types of controlled plants. First, the use

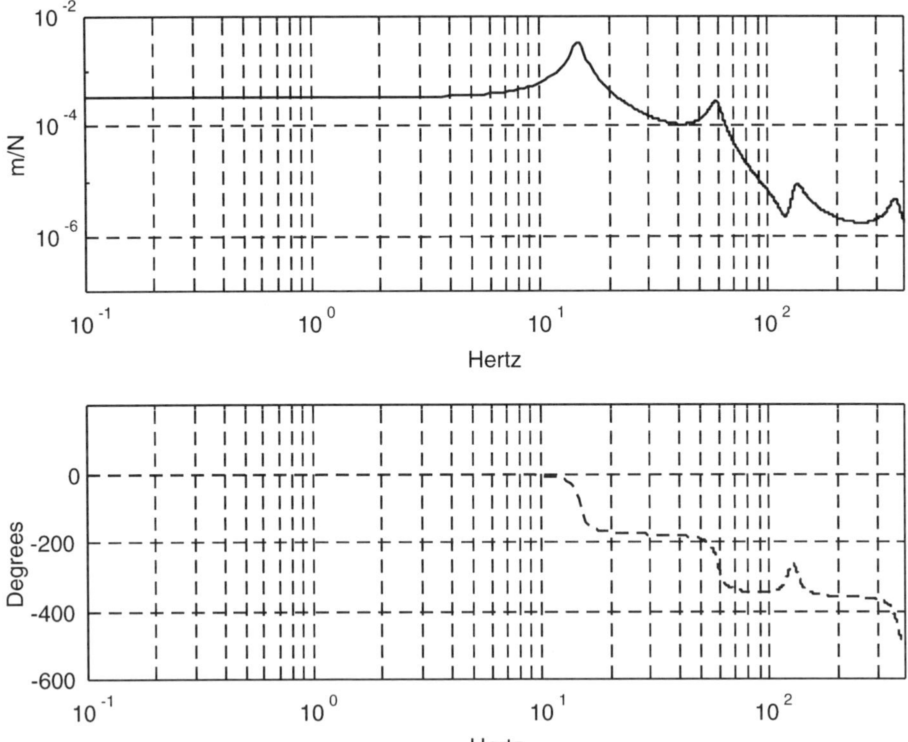


**Figure 7.20:** Bode diagram for noncolocated FRF; six-mode beam model.

of forward path compensation for most SISO closed-loop systems is usually related to a need for excellent command tracking and transient performance. The compensator design attempts to provide fast speed of response to command signals (step, ramp, programmed trajectories), while minimizing the transient overshoot and the steady-state errors of the plant output. For the adaptive structure plant, the need for tracking and fast transient response is encountered far less often than the need for enhanced structural damping or suppression of disturbance forces on the structure. Second, the order of the plant dyamics for adaptive structures (or any reverberant plant) is extremely large compared to many other plants that rely on series compensation for good performance. This results in relatively low control bandwidths for series compensated adaptive structures, as seen in the following sections.

There are advantages in designing traditional series compensators for the structural plant, especially for certain applications. For example, robot structures are often required to move an arm to a precise location with maximum speed. The flexibility of the arm is often the speed-limiting parameter. In that situation, the oscillatory response is usually dominated by several low-frequency modes that can be effectively compensated using familiar architectures such as lead, lead-lag, and proportional-integral-derivative (PID) designs. These compensators are easy to realize in analog or digital form and may be the most economical, cost-effective solutions for certain applications, particularly when there is not a sufficient number of sensors to provide critical multivariable response measurements required by more sophisticated compensators.

The following sections investigate the effects of series compensator designs on the nominal adaptive structure. Design equations are presented using frequency domain representations, and root-locus diagrams are used extensively to illustrate the structure's closed-loop performance. We do not attempt to synthesize compensators based on desired performance objectives because it is well known that this class of SISO compensation techniques generally cannot meet all of the performance objectives we demand of adaptive structures (see Section 7.6). Multivariable controllers are required for optimal performance. The multivariable compensator designs are discussed in Chapter 8.

### 7.5.1 General Considerations for Adaptive Structure Series Compensation

The inclusion of $D(s)$ in the system of Figure 7.2 provides $m$ additional open-loop zeros and $n$ additional open-loop poles ($m \leq n$) that act to modify the roots of the closed-loop structure. For series compensators, the open-loop zeros are also closed-loop zeros for command-output transfer functions but not for disturbance-output transfer functions. (This suggests that compensator zeros might be used to cancel structural poles in some command-tracking designs, but it is not recommended.) There is no theoretical limitation on the order of the compensator's numerator and denominator dynamics, but implementation issues typically limit the number of added zeros and poles to less than a dozen for SISO designs. This is because the structure's high modal density causes the closed-loop response to be highly sensitive to exact placement of the compensator dynamics. This is particularly true for very lightly damped structures. Therefore, the designer must synthesize the compensator in a top-down design approach or somehow be able to assess the best impact of many compensator zeros and poles all at once (the precise objective of optimal control!). For those reasons, we postpone the discussion of high-order series compensation and concentrate on the effects of well-known, simpler forms of series compensators.

The relative order of the compensator dynamics and the structural dynamics is extremely influential on the structure's response to series compensation. The number of compensator states is selectable, of course, but is always orders of magnitude less than the structure's order. As a result, the compensator dynamics cannot optimally modify the structure's dynamics over wide frequency ranges. This single fact must be appreciated in order to determine when it is possible to successfully use simple series compensation on a structure. To illustrate, we consider the transfer function of a general series compensator:

$$D(s) = \frac{Z(s)}{P(s)} = k_D \frac{(s+z_1)(s+z_2)\ \ldots\ (s+z_m)}{(s+p_1)(s+p_2)\ \ldots\ (s+p_n)} \tag{7.93}$$

with numerator order $m$ and denominator order $n$. Proceeding with the knowledge that our series compensator designs will be limited to low dimensions, we wish to reach some conclusions about their actions on the plant. Figures 7.21–7.23 show the root-locus diagrams for proportional feedback of structural displacement, velocity, and acceleration, respectively. This type of compensation is referred to as gain-compensation. The plant model corresponds to a steel, simply supported beam with colocated piezoceramic elements used for the sensing and actuation. The figures are similar to the SDOF and higher order root-locus diagrams introduced in Figures 7.7 and 7.8. There are some important conclusions to draw from these diagrams. Recall that the denominators are identically of order $2N$ (where $N$ is the number of modeled modes) for all three structural frequency response functions, but that the order of the numerators are different according to

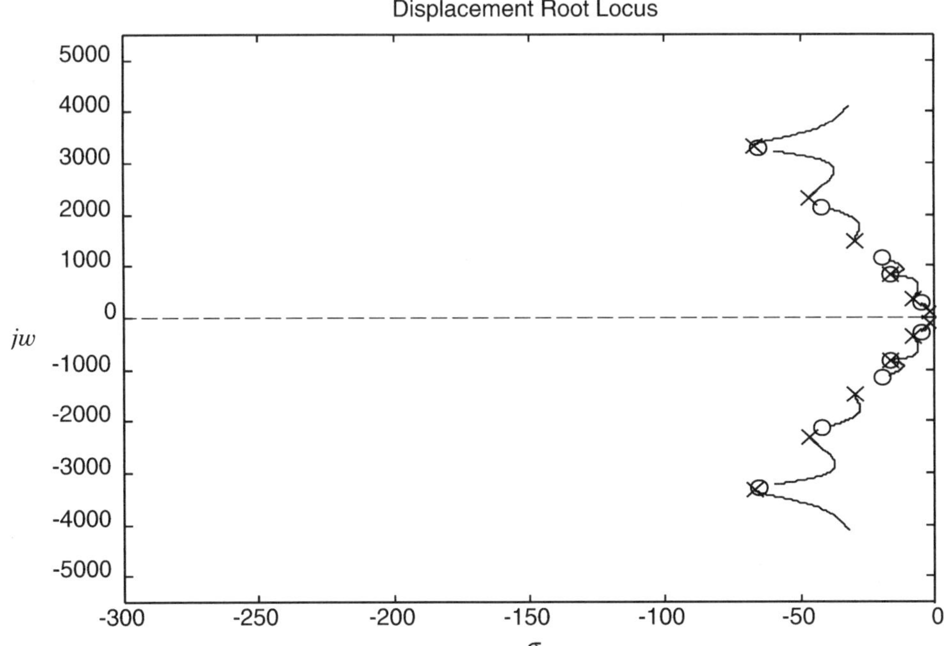


**Figure 7.21:** Root-locus diagram for proportional compensation, displacement feedback.

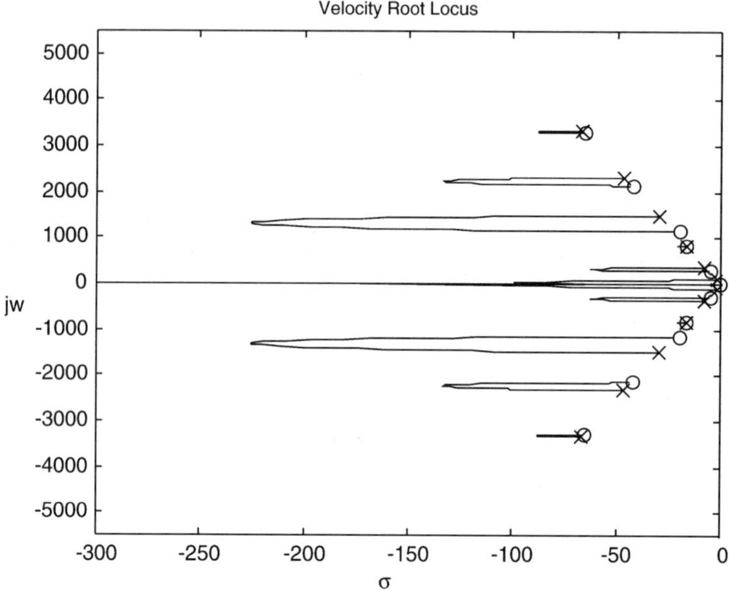


**Figure 7.22:** Root-locus diagram for proportional compensation, velocity feedback.

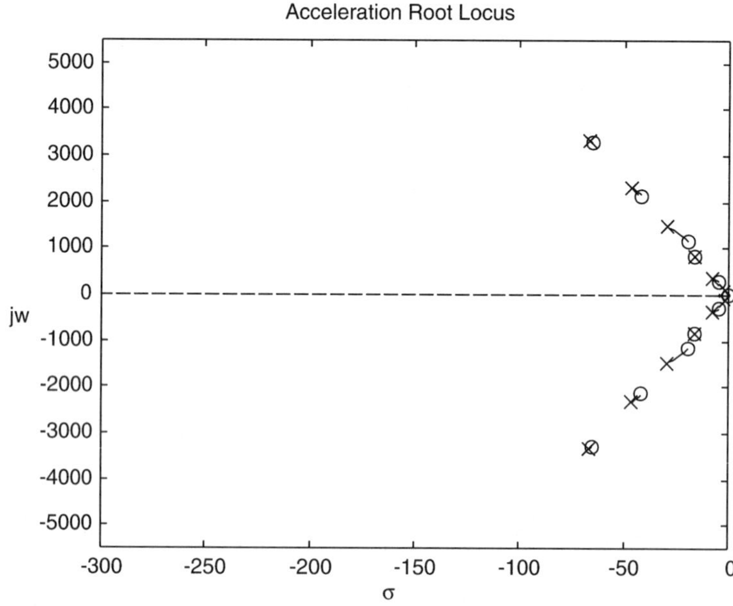


**Figure 7.23:** Root-locus diagram for proportional compensation, acceleration feedback.

$$\frac{x_k(j\omega)}{f_l(j\omega)} = \frac{\prod_{m=1}^{N-2} (j\omega - z_m)(j\omega - z_m^*)}{\prod_{n=1}^{N} (j\omega - p_n)(j\omega - p_n^*)} \tag{7.94}$$

$$\frac{\dot{x}_k(j\omega)}{f_l(j\omega)} = (j\omega) \frac{\prod_{m=1}^{N-2} (j\omega - z_m)(j\omega - z_m^*)}{\prod_{n=1}^{N} (j\omega - p_n)(j\omega - p_n^*)} \tag{7.95}$$

$$\frac{\ddot{x}_k(j\omega)}{f_l(j\omega)} = (j\omega)^2 \frac{\prod_{m=1}^{N-2} (j\omega - z_m)(j\omega - z_m^*)}{\prod_{n=1}^{N} (j\omega - p_n)(j\omega - p_n^*)} \tag{7.96}$$

Since the numerators differ by either one or two zeros at the origin, the corresponding root-locus plots differ also. The most significant differences between the three cases can be described using the real-axis root-locus rule and the calculation of departure angles of the poles. The velocity transfer function, with a single zero at the origin, is the only structural transfer function to exhibit a real-axis root locus. This provides a mechanism for multi-degree-of-freedom FRFs to display similarities to the SDOF case shown in Section 7.3. The fundamental mode of the system moves along a circular arc to the real axis, where it becomes overdamped for increasing values of gain, as seen in Figure 7.22. Velocity feedback, as seen earlier, leads to the most damping in the closed-loop system. Position and acceleration feedback, similar to the SDOF solution, cannot move poles far to the left under the action of proportional gain only, as can be seen in Figures 7.21 and 7.23, respectively. Thus, one important conclusion is that we simply need to add derivative action, that is, a zero at the origin, if we are to successfully feed back a structural displacement signal. It would be equally valid to add integral action for acceleration signals. A second observation that is evident in the figures is that the effect of a zero at the origin can be stabilizing for the higher order modes but has little effect on the system damping at high frequencies. This is seen in greater detail in Section 7.6. For now, the conclusion is that performance effects from the low-frequency zero are limited to low frequency dynamics, but the effects are generally dependent upon the modal density since the relative pole-zero spacing determines the poles affected most by the presence of feedback.

Widely studied series compensator designs typically have fewer than three zeros ($m < 3$) and/or poles ($n < 3$) in their FRF representations. These include the lead, lag, and lead-lag compensators, as well as proportional-integral (PI), proportional-derivative (PD), and proportional-integral-derivative (PID) compensators. The following sections examine the impact each of these compensation schemes have when used to control an adaptive structure.

***Lead Compensation for the Structural Plant*** The FRF of a lead compensator can be written in several ways. To highlight the locations of poles and zeros, we use the description

$$D(j\omega) = \alpha \frac{j\omega + \omega_c}{j\omega + \alpha\omega_c}, \qquad \alpha > 1 \tag{7.97}$$

It should be pointed out that this realization cannot be achieved using passive analog circuitry because of the gain $\alpha$ at higher frequencies. This does not present a problem for digital implementations (or active filter implementations), and there is a simple rearrangement of the compensator equation that conforms to the passive lead analog circuit shown in Figure 7.24*c*. The corresponding frequency response function is

$$D(j\omega) = \frac{1}{\alpha} \frac{j\omega\alpha\tau_c + 1}{j\omega\tau_c + 1}, \qquad \alpha > 1 \tag{7.98}$$

where the gain is now attenuated below the corner frequency rather than ampli-

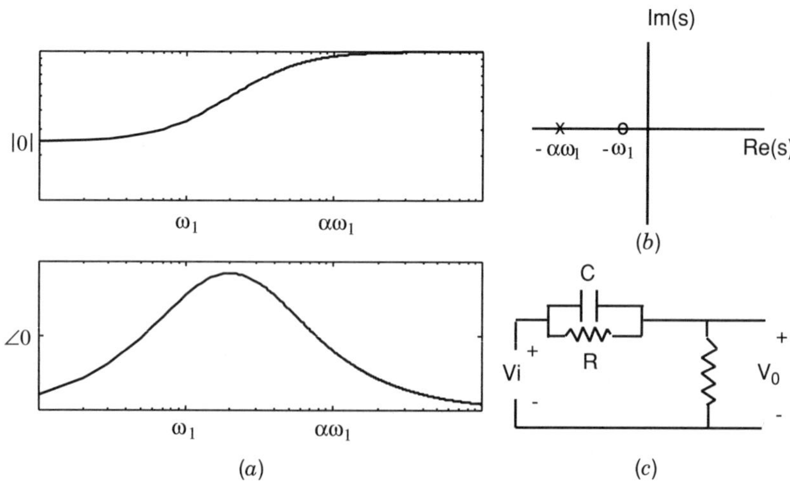


**Figure 7.24:** Lead compensation. (*a*) Bode diagram. (*b*) Pole-zero plot. (*c*) Passive circuit realization.

fied above the corner frequency. Of course, digital implementations may use either equation (7.97) or equation (7.98) to generate the difference equations or discrete-time state-space model for the compensator (Kuo, 1963).

Equation (7.97) and Figure 7.24 demonstrate that the frequency of the zero is always lower than the pole frequency for $\alpha > 1$. The frequency response plot illustrated shows the expected features of the zero and pole as frequency increases. At DC, the gain is unity but increases to $\alpha$ at high frequencies. Usually, the lead compensator is included when it is desirable to add phase lead at or near the phase crossover frequency for the purpose of improving the gain margin in the closed-loop system. This will also hold true for the structural design problem, but the modally dense response of the structure makes the conventional design approach quite challenging. Instead, we prefer to focus on design using root-locus diagrams for the lead-compensated structure.

Figure 7.25 shows the effects of lead compensation on the same structure that was gain-compensated as in Figures 7.21–7.23. The compensator's zero location was designed at $2\pi$ radians/second and the pole location was placed at $100\pi$ radians/second, almost two frequency decades higher. Large separation of the zero and pole frequencies was selected to illustrate the effect of lead compensation on displacement feedback. It is clear that the lead compensator is able to provide substantial damping to the closed-loop response at the lower frequency modes, in stark contrast to the proportional gain controller. This can be explained by examining the departure angles for the poles of the open-loop

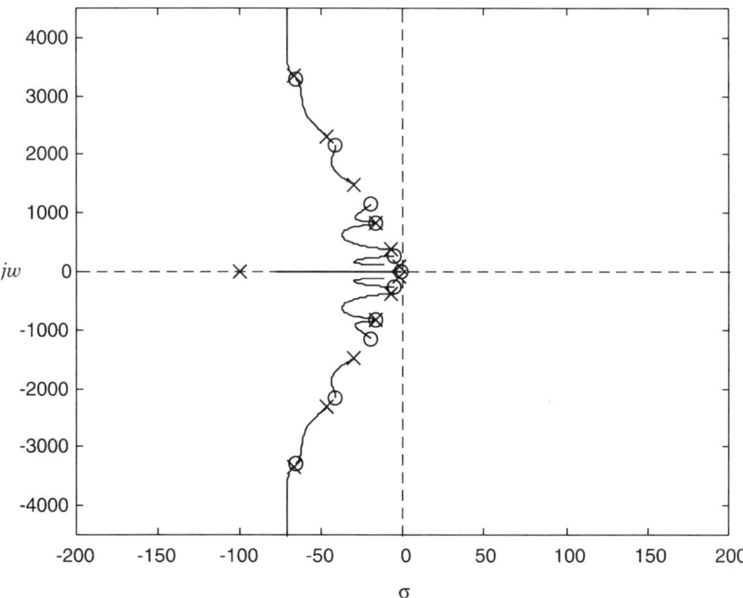


**Figure 7.25:** Root-locus diagram for lead compensation, displacement feedback.

transfer function. From the angle criteria, a general expression for the departure angle of the largest magnitude pole in a MDOF colocated adaptive structure can be found for the case of light damping and proportional compensation:

$$\theta_d^\circ(x) = ((n-2) * 90^\circ - (n-1) * 90^\circ - 180^\circ) \tag{7.99}$$

where $n$ is the number of poles in the transfer function. For this six-mode model of the structure, the calculated departure angle is 90 degrees. The remaining poles in the system display approximately the same departure angles. Increases in damping reduce the departure angles slightly. Similar expressions for velocity and acceleration feedback are shown below:

$$\theta_v^\circ(\dot{x}) = ((n-1) * 90^\circ - (n-1) * 90^\circ - 180^\circ) \tag{7.100}$$

$$\theta_a^\circ(\ddot{x}) = ((2n) * 90^\circ - (n-1) * 90^\circ - 180^\circ) \tag{7.101}$$

These expressions show that the departure angles are 180 degrees and −90 degrees, respectively. Clearly, it is advantageous to generate closed-loop systems that exhibit 180 degree departure angles if the goal is higher damping in the system. It is now obvious that the action of the lead compensator is to modify the departure angles of the displacement feedback adaptive structure so that it appears to behave as a rate feedback system, that is, $\theta_d(x) = 180$ degrees. This will be true as long as the additional pole from the compensator is fast enough that its angle to the root locus is close to zero for all frequencies of interest.

The effect of the lead compensator on the velocity feedback plant is not advantageous. The additional low-frequency zero at $\omega = 2\pi$ radians/second acts to transform the velocity signal to an acceleration signal, thereby destroying the favorable departure angle seen for the case of displacement feedback. In other words, the system effectively becomes acceleration feedback. The compensator pole restores the velocity feedback behavior at higher frequencies, but there are few positive effects from the simple compensator at these higher frequencies. The reader is encouraged to support these observations by using the files discussed below.

Similar analyses can be made for the case of lead-compensated acceleration feedback. (*Does lead compensation provide beneficial closed-loop behavior in this case?*) However, the goal of this discussion is to introduce the distinct differences from compensation of different feedback signals. The reader should remember that the analysis above has been specified for colocated, that is, minimum phase, SISO configurations of the structure. You may use the MATLAB script file **leadrl.m** to examine different compensator corner frequencies for displacement, velocity, or acceleration feedback. It is also possible to change the number of modes in the structural transfer function and to examine the corresponding root-locus diagrams for noncolocated sensor/actuator pairs. Note that transducer dynamics are not included in that specific script file.

***Lag Compensation of the Structural Plant*** The transfer function for the lag compensator is

$$D(s) = \frac{1}{\alpha} \frac{s + \alpha\omega_c}{s + \omega_c}, \qquad \alpha > 1 \tag{7.102}$$

which shows that the lag compensator simply reverses the locations of the lead compensator zero and pole on the negative real axis. This causes attenuation of the open-loop FRF amplitude above the first corner frequency, as shown in Figure 7.26*a*. This property of lag compensators is used primarily to retain steady-state accuracy provided by integral action of the plant while attenuating the higher frequency magnitude in an attempt to improve the stability margins. The usual strategy is to use the attenuation to lower the gain crossover frequency in order to recover a more favorable phase margin. A passive analog circuit that implements the lag compensator effects for a real system is indicated in Figure 7.26*c*. For digital implementations, the designer would simply construct a digital filter with the frequency domain properties corresponding to equation (7.102).

For structural control, it is informative to examine the root-locus diagrams

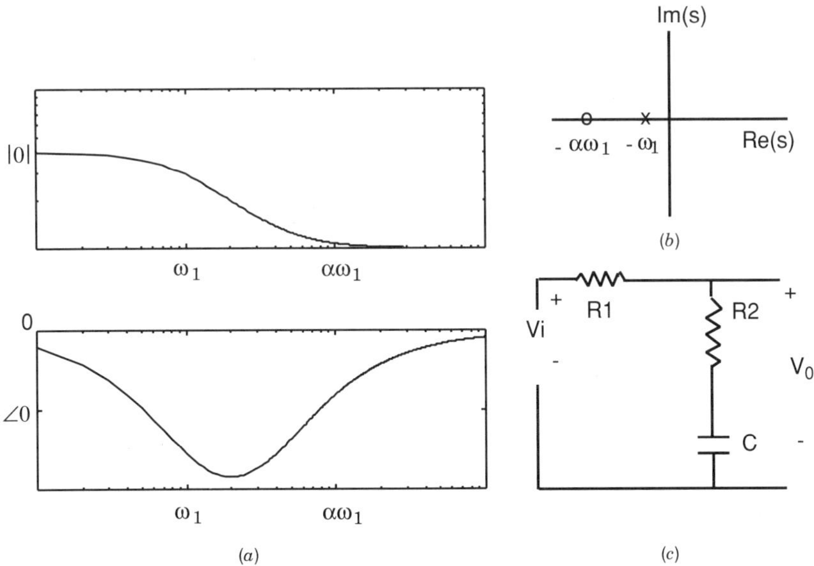


**Figure 7.26:** Lag compensation. (*a*) Bode diagram. (*b*) Pole-zero plot. (*c*) Passive circuit realization.

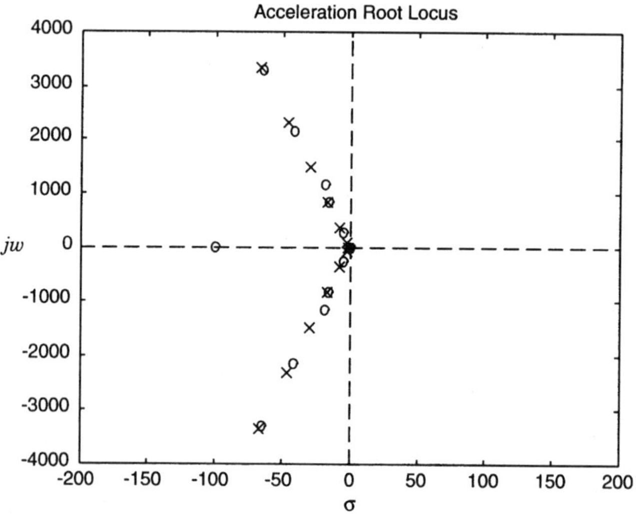


**Figure 7.27:** Root-locus diagram for lag compensation, acceleration feedback.

again. The lag compensator's low-frequency pole acts nearly like an integrator, just as the lead compensator's low-frequency zero acts to differentiate the plant. Thus, the lag compensator would be expected to be beneficial for acceleration feedback configurations. This is the case, depending on the actual resonant frequencies of the structure. Figure 7.27 shows the closed-loop poles where the same frequencies have been used in the compensator design, simply reversed for the pole and zero. Of course, there is a problem for the lag compensated displacement feedback design. The additional low-frequency pole causes the departure angle to approach $\theta_d(x) = 0$ degrees, a destabilizing result for even moderately damped structures. The file **lagrl.m** may be used to further examine the effects of the lag compensator on the root locus for the closed-loop plant.

The three compensators discussed in the remainder of this section are the PD, PI, and PID designs. The PD and PI forms have only slight differences in their effects, when compared to the lead and lag designs above. The PID compensator is higher order than the others, providing slightly more flexibility in the adaptive structure design.

***PD Compensation of the Structural Plant*** Proportional-plus-derivative (PD) compensators have a transfer function defined by

$$D(s) = K_p T_d \left( s + \frac{1}{T_d} \right) \tag{7.103}$$

where $K_p$ is the proportional gain and $T_d$ is the additional gain of the derivative control action through the plant. The compensator equation is often cast in the form $D(s) = K_p(T_d s + 1)$ also. This compensator implies pure derivative action above the corner frequency $\omega_c = 1/T_d$, a requirement that cannot be entirely realized by any analog circuit because all operational amplifiers exhibit non-ideal behaviors. Attempts to build pure differentiators are not recommended as they will generate unstable circuits; instead, the derivative action should be rolled off at some acceptable frequency (Franco, 1992). In addition, derivative control is "noisy," which means that it amplifies high-frequency noise because of montonically increasing gain with increasing frequency.

The PD compensatation effects on the structural plant are very similar to the lead-compensation effects discussed earlier if the compensator zero $s = -1/T_d$ is low frequency. Because any practical implementation also adds a high frequency pole, this compensator can be viewed as identical to the lead compensator for our interests. The reader may use the file **leadrl.m** to examine closed-loop pole locations and show that cascading a lead network effectively provides PD compensation for small $\omega$.

***PI Compensation of the Structural Plant*** Proportional-plus-integral (PI) compensators have a transfer function defined by

$$D(s) = K_p \left( \frac{s + (1/T_I)}{s} \right) \tag{7.104}$$

where $K_p$ is the proportional gain and $T_I$ is the gain of the integral control action through the plant. The PI compensator is quite similar to the lag compensator, the only difference being that the low-frequency pole is always found at the origin.

As with the PD compensator, specific effects of the PI compensator on the structural plant are not examined here, although the reader is encouraged to use the script file **lagrl.m** to show that cascading a lag network effectively provides PI compensation for large $\omega$.

***PID Compensation of the Structural Plant*** The final series form we consider is called a proportional-integral-derivative (PID) compensator. The transfer function for PID control is

$$D(s) = K_p \left( 1 + \frac{1}{T_I s} + T_d s \right) \tag{7.105}$$

where all variables have been previously defined. Equation (7.105) can be rearranged to show the zero and pole locations more clearly:

$$D(s) = K_p \left( \frac{T_d s^2 + s + (1/T_I)}{s} \right) \tag{7.106}$$

(Note that the compensator as presented is acausal. However, a high-frequency pole can be used in a causal realization.) This compensator simply blends the three fundamental control actions—gain control, derivative control, and integral control—into a single compensator. One advantage of the PID compensator for adaptive structures is the addition of two zeros, instead of only one zero as supplied by the lead and PD compensators. The analog implementation of the PID compensator is best accomplished using active filter methods. A simplified circuit schematic for a PID compensator that can be used to drive relatively low power actuators (up to 25 W) is shown in Figure 7.28. The corresponding PID gains are given by

$$K_p = \frac{R_2}{R_1} \tag{7.107}$$

$$T_I = R_{i_1} C_i \tag{7.108}$$

$$T_d = R_{d_2} C_{d_2} \tag{7.109}$$

The digital implementation follows by taking the $Z$-transform of equation (7.106) and creating the difference equation. Discrete-time methods are not included in this treatment of adaptive structures. The interested reader can refer to Kuo (1963), Shahian and Hassul (1993), Franklin et al. (1994), and Santina et al. (1994).

Referring to equation (7.106), the quadratic roots in the numerator lead to substantially more variations on possible closed-loop pole locations when compared to the other series compensators. The compensator zeros are determined by

$$s_{1,2} = -\frac{0.5}{T_d} \left( 1 \pm \sqrt{1 - 4\frac{T_d}{T_I}} \right) \tag{7.110}$$

so that the zeros lie on the negative real axis for $4(T_d/T_I) < 1$ and are complex conjugates in the left-half plane for $4(T_d/T_I) > 1$. The wide range of design choices for placing these zeros makes it more difficult to make concise statements about the effects of PID compensation on the structural plant. However, it is informative to examine several results. Figure 7.29 shows the root-locus diagram for the same structural plant used earlier with displacement feedback. The structure is controlled by a PID compensator with $T_d = 0.25$ and $T_I = 1.5$. This ratio of derivative and integral gain leads to compensator zeros given by $s_1 = -3.0$ radians/second and $s_2 = -0.1$ radians/second. In this case, the

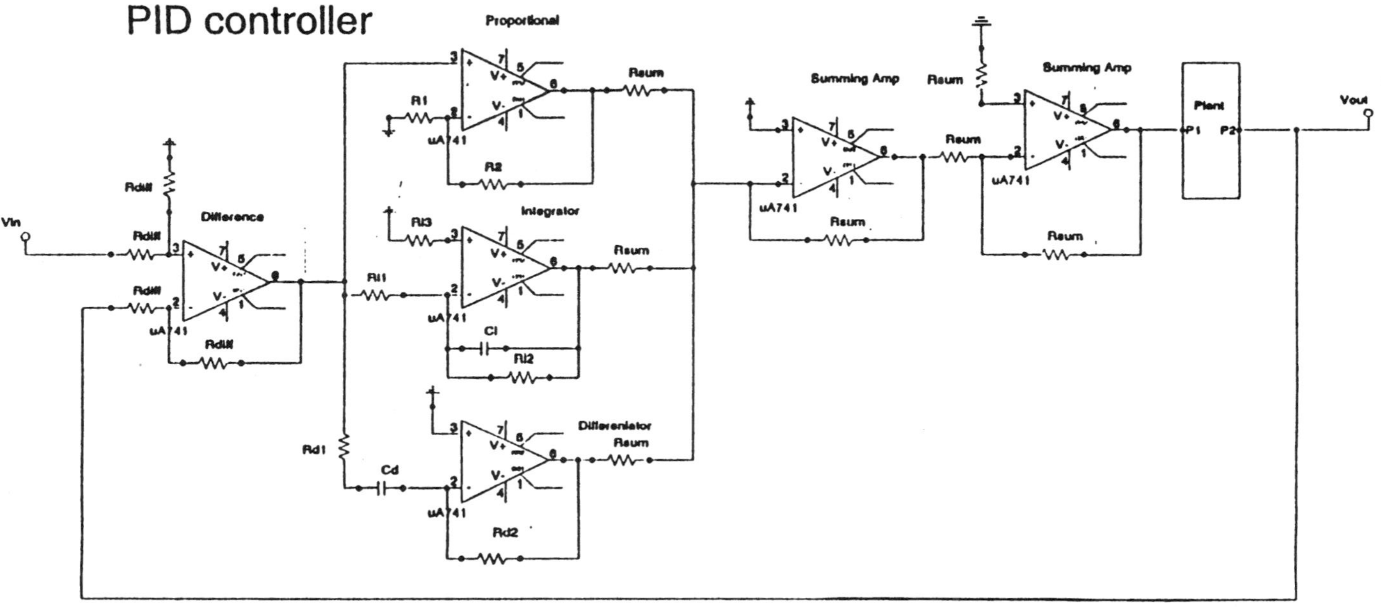


**Figure 7.28:** Analog circuit realization for a PID compensator.

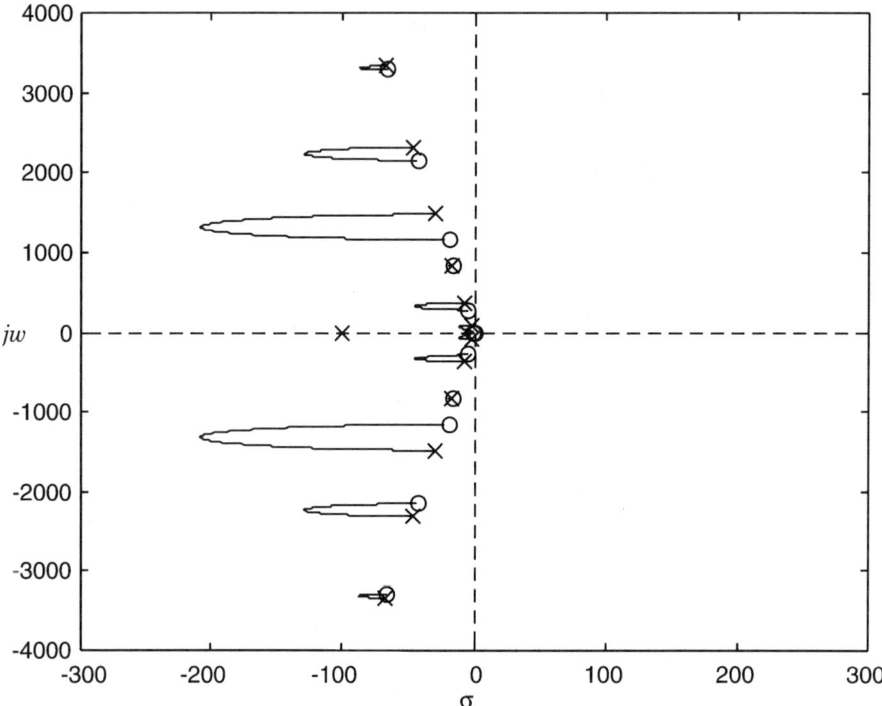


**Figure 7.29:** Root-locus diagram for PID compensation, displacement feedback.

compensator pole at the origin is effectively canceled by the low-frequency zero, leaving only the zero at $s = -3$ to affect the closed-loop system poles. As before, the placement of this zero close to the origin causes the displacement feedback configuration to behave like velocity feedback, with excellent options for closed-loop pole locations. It appears that this solution is comparable to that of the lead compensator but requires extra hardware to construct. However, if transient considerations are important, the PID-compensated structure is now a type 1 system, which leads to better steady-state error properties than the lead-compensated structure that is type 0.

Selection of PID gains that produce complex conjugate compensator zeros in the open-loop transfer function presents several challenges in adaptive structure design. These types of zeros have the ability to attract structural poles to good damping regions, but since there are only two zeros, there will be limited effects. This results in low-frequency control system bandwidths, as has been the case for the other series compensators. The designer may wish to cascade any number of PID compensators, with varying $T_D$ and $T_I$ gains, so that all dominant structural poles may be affected. This is possible and is reminiscent of the positive position feedback (PPF) design approach offered by Goh and Caughey (1985) and investigated by numerous other researchers. Compen-

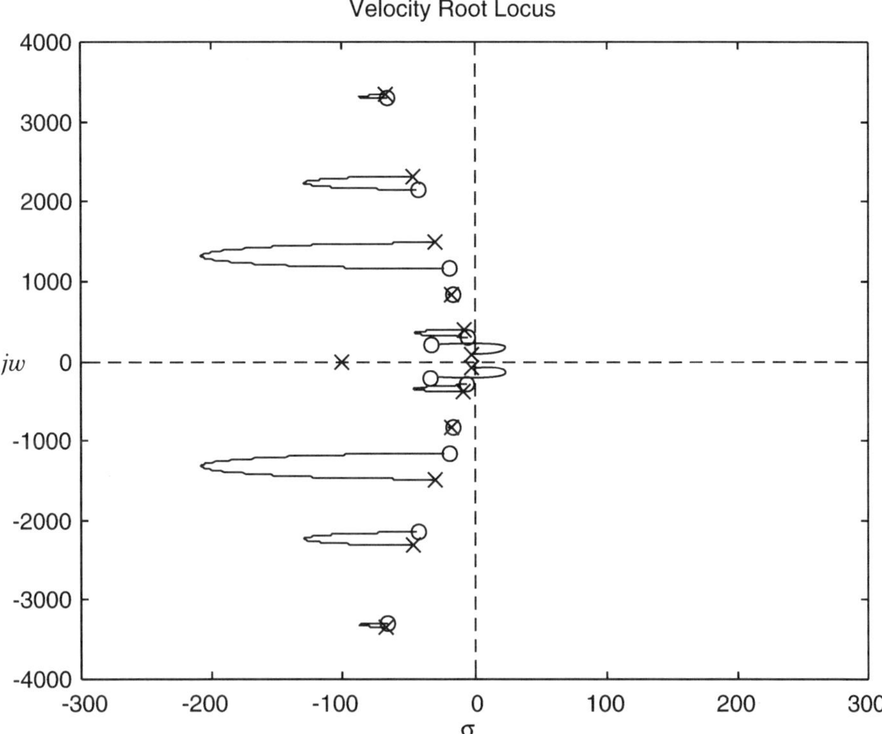


**Figure 7.30:** Root-locus diagram for PID compensation ($s_{1,2} = -50 \pm j218$), displacement feedback.

sators such as the PPF and series PID styles are generally referred to as tuning filters, as they are carefully placed to tune the closed-loop response according to some desired performance metrics. To date, there are no fixed design rules for the series PID tuning filter approach with adaptive structures. Figures 7.30 and 7.31 show the root-locus diagrams for displacement and velocity feedback using $T_d = 0.01$ and $T_I = 0.002$, which yields compensator zeros as $s_{1,2} = -50 \pm j218$. The figures emphasize that precise selection of the zero locations is required for good adaptive structure closed-loop properties and that the compensator design must be selected based on the available feedback signals.

***Summary Comments*** To conclude this section, it is important to emphasize some issues that accompany series compensation design for adaptive structures. All of the examples presented above have been based on colocated transducers. The closed-loop performance changes significantly with noncolocation, and each case should be analyzed separately. The discussions above have illustrated how different compensators can effectively modify a structure's response to displacement feedback so that it behaves as if it were velocity feedback. It is

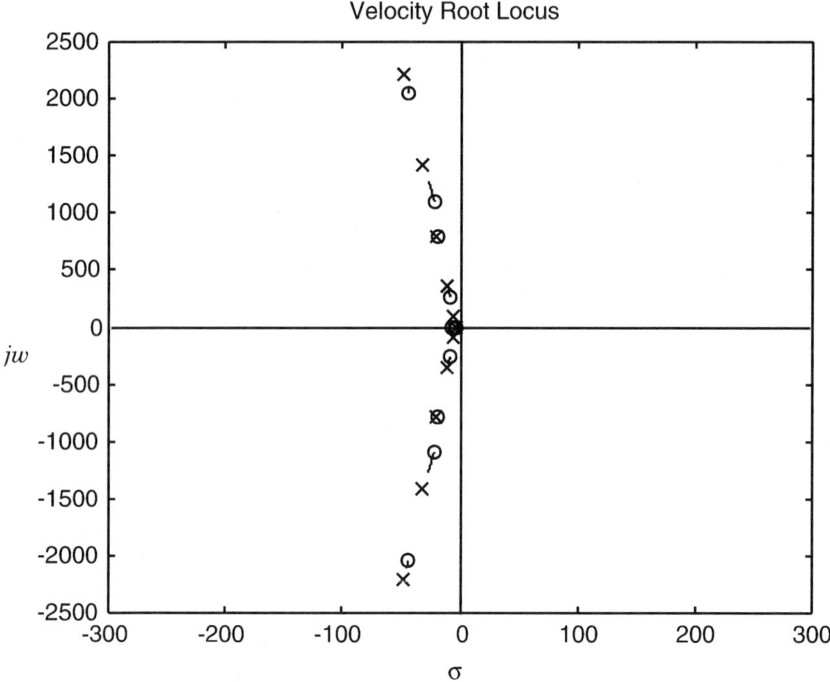


**Figure 7.31:** Root-locus diagram for PID compensation ($s_{1,2} = -50 \pm j218$), velocity feedback.

equally possible to degrade the positive effects of true velocity feedback when the wrong compensator design is selected. The designer must be aware of the types of signals that will be available for feedback and choose the appropriate compensators. Notice that the presence of transducer dynamics sometimes act to modify the closed-loop response favorably or unfavorably, also (Cannon and Rosenthal, 1984; Inman et al., 1990). In that regard, it is not unreasonable to treat the transducer dynamics as compensator dynamics and use the types of analyses presented in this chapter to analyze the effects of the transducers on the adaptive structure's performance. Alternatively, one may choose to "design" the dynamics of the transducer in an integrated approach to achieve the required compensation without adding additional compensator dynamics or electronics.

## 7.6 CONTROLLER DESIGN OBJECTIVES APPLIED TO THE STRUCTURAL PLANT

Feedback control design for dynamic systems usually *begins* with a mathematical model, which drives the early stages of the design process. We have been considering linear models that can be developed analytically or numerically. Using these models, it is possible to exploit the methods of classical control system design for the adaptive structure. Therefore, one of the first steps in adaptive structure design is to design (or test) for linearity (see Chapter 3). It is possible that nonlinearities may be detectable, most often related to the active materials and transducers that are integral to the adaptive structure. Linearization techniques, such as Taylor series expansions, small perturbation methods, describing functions, and so on, can be used in these situations for a first-order remedy. More complex nonlinear analyses may be justified for more complex dynamics, such as aeroelastic flutter or high amplitude acoustic excitation of structures. Fortunately, structural response linearity is observed for many situations, especially if the application limits the frequency bandwidth and ranges of input amplitudes. In this section, we examine adaptive structure designs, using classical control theory and linear SISO models with and without consideration of augmented transduction device dynamics. Historical uses for classical control design can be easily associated with adaptive structure performance objectives, as shown in the list below.

FEEDBACK CONTROL PROPERTIES FOR ADAPTIVE STRUCTURES

| Control System Objectives | Adaptive Structure Application |
|---|---|
| 1. Accurate tracking of a command input | Structural shape control<br>Rigid body control |
| 2. Reduced sensitivity to disturbance inputs | Active vibration control<br>Active structural damping<br>Active constrained layer damping |
| 3. Modification of system eigenproperties | Active structural damping<br>Active structural acoustic control<br>Structural shape control |
| 4. Reduced sensitivity to uncertainties | Robust adaptive structures |

The left column lists the primary effects of feedback control, which are also design objectives for either SISO or MIMO adaptive structure configurations. It could be argued that command or reference input tracking is the single most important use for simple feedback controllers. This is not actually the case when designing adaptive structures, although there are two important applications of command regulation. Structural *shape control* requires position sensing and position actuation, which can be implemented with strain gages, displacement probes, or similar sensors. Actuation can be successfully applied using a variety of schemes, some of which were discussed in Chapter 5. Examples of shape control might include pointing an antenna or prescribing a specific deflection of a structural member, for example a span of a bridge. The second type of regulation problem includes rigid body control applications, which are most common in spacecraft attitude control, and have been covered in detail by Junkins and Kim (1993).

The remaining feedback control objectives are more applicable to the adaptive structure designs being considered here. *Active vibration control* (AVC) or *active structural acoustic control* (ASAC) are two important adaptive structure functions that rely on reducing sensitivity to force disturbances. Typically, the goal is to reduce the effects of forces entering the structure in locations that cannot be modified through fundamental design. For example, a marine ship requires that the propulsion shaft be terminated by a thrust block, a structure that transfers the propulsion forces to the hull. By design, this structure must have high stiffness and mass; therefore, it also transfers the superimposed oscillating forces to the hull, where they then create undesirable vibrations and noise. AVC is a broad label but is used here to refer to minimization of structural vibrations through an increase in the closed-loop system sensitivity (defined in the next section) or placement of zeros where the disturbance signal power is high. Mechanically, this may be interpreted as changing the impedance of the structure at the disturbance location. Similarly, ASAC is a *selective minimization* of structural vibrations that efficiently radiate acoustic energy away from (or through) the structure. Mitigation of disturbances can be accomplished using feedback, feedforward, or combined feedback-feedforward architectures. SISO feedback applications as discussed in this chapter; MIMO control methods are discussed in the remaining chapters.

The third design objective, modification of system eigenproperties, or eigenstructure assignment, is mostly associated with *active damping* of a structure. Output feedback, or state feedback, can often be designed so that structural poles move to regions of the $s$-plane associated with higher damping as compared to the open-loop pole locations. This was illustrated in the previous section for series compensated structures. From a system perspective, this can be accomplished in several ways: pole placement, ad hoc gain stabilization, or optimal methods. The end result is the same for any of these active damping methods—closed-loop structural poles are more heavily damped than the open-loop structural poles. Active damping relies exclusively on feedback control to increase the dissipation of a structure's vibrational energy through the use of installed actuator-sensor-controller components. The active structural damping is associated with the transformation of vibrational energy to heat and mechanical losses in the attached control hardware. As might be expected, there are limits as to how much additional damping can be achieved with feedback control. A second use of eigenstructure assignment is reshaping of the system modeshapes. This technique can be considered to be zero placement and requires large numbers of sensors and actuators, making it a MIMO design method; it is not discussed in this section.

Finally, reduced sensitivity to uncertainties is a positive benefit of any type of feedback control system. This is related to the topic of *robust control*, which is treated in detail by Skogestad and Postlethwaite (1996). For an in-depth treatment of robust stability and robust performance for various types of uncertainties encountered for SISO control system design, the interested reader is encouraged to review the suggested reference. Chapter 9 introduces demonstration problems that illustrate the design of robust adaptive structures.

### 7.6.1 Command Tracking

One of the most discussed topics in classical feedback control is command tracking or regulation. The control system designer must be able to predict how well the output will follow a reference input of arbitrary statistics. Command tracking is assessed using steady-state error analyses that monitor the error signal

$$e(s) = r(s) - y(s) \tag{7.111}$$

For the adaptive structure, $r(s)$ is the reference input, $y(s)$ is the measured structural displacement, and $e(s)$ is always derived from the difference between the input and the output. Of course, $e(s)$ should have a steady-state value equal to zero for ideal command tracking. Figure 7.32 shows the closed-loop block diagram for the two different cases of unity feedback and a compensator $K(s)$ in the feedback path. For unity feedback configurations, the error signal is the same as the output of the comparator that sums the reference and feedback signals. Feedback compensator dynamics and any transducer dynamics can always be moved into the forward path of the closed-loop systems, so we discuss adaptive structure command tracking using only the unity feedback system. Taking the Laplace transform of $e(s)$ and dividing by $r(s)$, the error-to-reference transfer function becomes

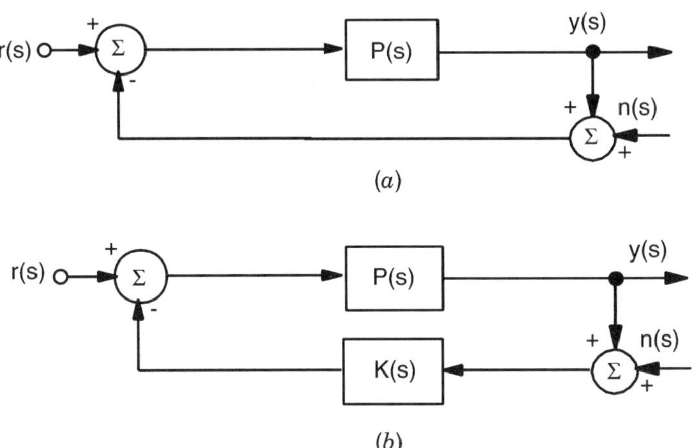


**Figure 7.32:** The system block diagram for command tracking. (*a*) Unity feedback. (*b*) Feedback compensation.

$$\frac{e(s)}{r(s)} = 1 - \frac{y(s)}{r(s)} \tag{7.112}$$

$$= 1 - \frac{K(s)P(s)}{1 + K(s)P(s)} \tag{7.113}$$

$$= \frac{1}{1 + K(s)P(s)} \tag{7.114}$$

The steady-state error can then be found from the final value theorem (with the condition that all poles of $se(s)$ are in the left-half plane):

$$e(t)|_{t \to \infty} = \lim_{s=0} \frac{sr(s)}{1 + K(s)P(s)} \tag{7.115}$$

This equation allows us to examine steady-state error behavior using well-known definitions of signal input type and system type (Franklin et al., 1994). Readers who are not familiar with this terminology can browse the referenced material, but the system type is numerically equal to the number of free integrators in the system and the input type refers to the exponent $t$ in $r(s) = 1/s^t$. It is clear from equation (7.115) that system poles at the origin *may* cause the steady-state error to remain finite or become unbounded, depending on the input type. Here, we are specifically interested in the accuracy of command tracking for certain inputs when the plant transfer function $P(s)$ represents a dynamic structure.

We begin with a SDOF, mass-spring-damper model, as shown in Figure 7.33. The Laplace transform of the force balance equation leads to a transfer function between position and force input described by

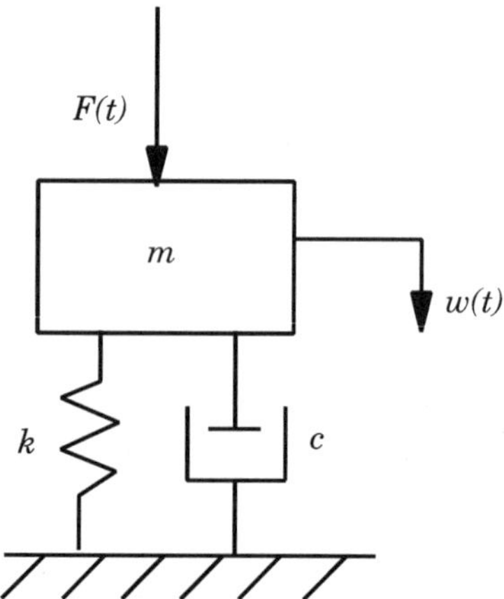


**Figure 7.33:** Spring-mass-damper with force for position control.

$$\frac{Y(s)}{F(s)} = \frac{1/k}{(s^2/\omega_n^2) + (2\zeta/\omega_n)s + 1} \tag{7.116}$$

Thus, the plant transfer function $P(s)$ for the SDOF vibration model is type 0; that is, it has no poles at the origin. For a step input $r(s) = 1/s$ and a series compensator $K(s) = N_K(s)/D_K(s)$, the steady-state error

$$e(t)|_{t\to\infty} = \lim_{s=0} \frac{s(1/s)}{1 + [N_K(s)/D_K(s)][(1/k)/((s^2/\omega_n^2) + (2\zeta/\omega_n)s + 1)]} \tag{7.117}$$

must have at least one integration in $D_K(s)$ if it is to be zero. If the compensator is a pure gain $K$, *the steady-state error for a structural step response will always be a finite value* that depends on the magnitude of the gain as described by

$$e(t)|_{t\to\infty} = \lim_{s=0} \frac{1}{1 + K[(1/k)/((s^2/\omega_n^2) + (2\zeta/\omega_n)s + 1)]} \tag{7.118}$$

Note that the previous discussion holds true for velocity or acceleration output measurements also. It should also be mentioned that transducer dynamics will not provide free integrators to the structural plant model. Therefore, this result holds true for structures with and without transducer dynamics.

Real structures have infinitely many degrees of freedom but the system type remains the same as the simple SDOF example. Position inputs are the most likely to be encountered for shape control of adaptive structures; therefore, one will always have to accept a finite error in the accuracy of these command tracking applications if the system is limited to proportional compensation. (The same problems arise for control of rigid body dynamics under the action of ramp, or velocity, inputs.) For shape control, this is an important issue. If the control system is asked to deform a point on the reverberant structure to a reference position (essentially a step input), the response will display typical rise time, overshoot, and steady-state error behavior associated with many other types of second-order systems. This is because the control actuator(s) is attempting to act upon the local dynamic impedance, which imposes second-order dynamics on the response. For perfect shape control, series compensators must be used to introduce the integral action that transforms the structure to a type 1 system, thereby providing zero steady-state error. The lag compensator and the PID compensator, discussed in the previous section, can both satisfy this requirement. Selection of the specific compensator also needs to be based on the damping requirements for the problem.

### 7.6.2 Reduced Sensitivity to Disturbances

The topic of disturbance sensitivity is extremely important in adaptive structure design. That is because many adaptive structures are designed to minimize vibration or acoustic response to numerous forms of external force disturbances on the structure. There are several uniquely different methods that can be used to suppress the effects of disturbances—disturbance rejection, active damping, and loop shaping. All of these methods lower the structural response magnitude caused by the disturbances with varying degrees of success. This section focuses primarily on disturbance rejection, which is specifically intended for reduction of disturbance effects. We assume that the disturbances are *dynamic*; constant input disturbances can be treated simply with integral feedback (Kailath, 1980) and are encountered far less often in adaptive structure applications. Active damping reduces the effects of steady disturbances but has wider ranging effectiveness and is discussed in Section 7.6.3 on system eigenproperties. Loop shaping techniques for disturbance rejection are accomplished using optimal control methods, such as $\mathcal{H}_2$ or $\mathcal{H}_\infty$ control, and are discussed in Chapter 8 and demonstrated in Chapter 9.

Disturbance rejection, sometimes called disturbance compensation, can be distinguished from active damping through the following example. Suppose that the effects of a *dynamic* disturbance force $d(s)$ (i.e., an undesired input) on a structure are to be controlled by a single control force $u(s)$. In state-variable form, we can express this situation (omitting $s$ for now) as

$$\dot{\mathbf{X}} = \mathbf{AX} + \mathbf{B}u + \mathbf{L}d \tag{7.119}$$

$$y = \mathbf{CX} + D_u u + D_d d \tag{7.120}$$

where dimensions of the states $\mathbf{X} \in R^{2N \times 1}$ and the corresponding state matrices do not change the following analysis. The open-loop poles and zeros for the two transfer functions $y_u/u$ and $y_d/d$ can be determined from the numerator and denominator terms of the expressions

$$y_u = [\mathbf{C}(s\mathbf{I} - \mathbf{A})^{-1}\mathbf{B} + D_u]\, u \tag{7.121}$$

$$y_d = [\mathbf{C}(s\mathbf{I} - \mathbf{A})^{-1}\mathbf{L} + D_y]\, d \tag{7.122}$$

indicating that the poles are identical for both transfer functions (as to be expected) but the zeros are different (also to be expected). Now, there are two possible mechanisms that can be used to reduce the amplitude of $y_d$ at frequencies of the disturbance signal $d$. First, the pole residues can be reduced. Recalling equation (7.26), the residues depend on the eigenvectors and the pole locations. The residues are decreased by decreasing the magnitude of the eigenvectors and by increasing the damping of the poles. These changes must be such that they can affect $y_d$ in the bandwidth of the disturbance. The second approach is to cancel the disturbance. If equation (7.120) were for a scalar case (i.e., one state), exact cancellation of the disturbance would require that

$$Bu = -Ld \tag{7.123}$$

Then the control signal $u = -(L/B)d$ would prevent any response of $y_d$ to $d$ at all frequencies. Adaptive structures are generally not scalar in dimension, and the order of **X** leads to matrices $B \in R^{2N \times 1}$ and $L \in R^{2N \times 1}$ for SISO situations. Therefore, the solution of the disturbance cancellation signal is overdetermined and can only be solved for in a least-squares sense given by

$$u = -(\mathbf{B}^{\mathrm{T}}\mathbf{B})^{-1}\mathbf{B}^{\mathrm{T}}\mathbf{L}d \tag{7.124}$$

The expression $(\mathbf{B}^{\mathrm{T}}\mathbf{B})^{-1}\mathbf{B}^{\mathrm{T}}$ is called the pseudoinverse of $B$. Substitution into equation (7.120), assuming that $D_u$ and $D_d$ equal zero, yields

$$\dot{\mathbf{X}} = \mathbf{AX} - \mathbf{B}(\mathbf{B}^{\mathrm{T}}\mathbf{B})^{-1}\mathbf{B}^{\mathrm{T}}\mathbf{L}d + \mathbf{L}d \tag{7.125}$$

$$y = \mathbf{CX} \tag{7.126}$$

which shows that the effects of $d$ on the structure may not be completely annihilated but can be reduced. This is referred to as disturbance rejection. As shown by the previous discussion, the structural poles do not play a role in this process. Instead, the closed-loop system is arranged in such a way that the effect of the disturbance input(s) is blocked. This blocking action can be achieved at all frequencies if there is a robust solution to the pseudoinverse. Unfortunately, this does not always occur. Instead, it is possible to place blocking zeros in the transfer function $y_d/d$ at selective frequencies where the signal energy is high. This is a topic of much importance to adaptive structure design, as discussed next.

Disturbance rejection receives little attention in many textbooks devoted to classical control. When it does, the concept of disturbance rejection is usually related to the *internal model principle*. The internal model principle states that there must be a model of the disturbance dynamics in the closed-loop system in order to achieve disturbance rejection (and robust tracking) with no steady-state error. This is usually achieved by designing a feedback compensator whose poles include the appropriate disturbance dynamics. For example, if a structure were subject to a sinusoidal disturbance $d(j\omega)$ at frequency $\omega_o$, a feedback compensator $K(s) = k/(s^2 + \omega_o^2)$ in the system of Figure 7.32b would produce the following transfer function between the output and the disturbance input:

$$\frac{y}{d} = \frac{s^2 + \omega_o^2}{s^2 + \omega_o^2 + P(s)k} \tag{7.127}$$

which shows that the disturbance input will be blocked at frequency $\omega_o$ because of the complex conjugate zeros that originate in the feedback compensator. (Recall that the closed-loop zeros of the disturbance-to-output transfer function are the sum of the open-loop zeros and the poles of the feedback compensator.)

To generalize this result, the following discussion relies heavily on the presentation of Sievers and von Flotow (1990) for narrowband disturbance rejection.

We have already determined that the output of a linear structure subject to disturbance inputs, as shown in Figure 7.2(a), depends on the sensitivity function $S(s)$ as

$$y(s) = [I + K(s)P(s)]^{-1}P(s)d(s) \tag{7.128}$$

$$= S(s)P(G)d(s) \tag{7.129}$$

As mentioned earlier, our design goal is to decouple the disturbance from the output in the frequency band of the disturbance. This must be performed so that the closed-loop response is also stable. Therefore, $K(s)$ is selected for both stability and disturbance suppression characteristics. It is clear from equation (7.129) that the minimum sensitivity $S(s)$ corresponds to the minimum impact of the disturbance on $y(s)$. For SISO control system designs, the maximum achievable attenuation corresponds to the Bode magnitude of the sensitivity $S(s)$ in the bandwidth of the disturbance. MIMO control systems must use singular value analysis to determine the maximum disturbance rejection in any particular direction. This is covered in Chapter 8.

According to equation (7.127), disturbance rejection can be achieved by judicious zero placement, which has a direct appeal to our intuition that single frequency disturbances can be "canceled." A second viewpoint is that the open-loop transfer function $K(s)P(s)$ must be large wherever $S(s)$ is small. Assuming a narrowband disturbance of frequency $\omega_o$, a quality factor of

$$Q = \frac{\omega_2 - \omega_1}{\omega_o} > 10 \tag{7.130}$$

can be achieved by designing the compensator as shown in Figure 7.34. The compensator is a bandpass filter with center frequency $\omega_o$ and a passband (magnitude greater than 0 dB) defined by the lower cutoff $\omega_1$ and upper cutoff $\omega_2$ frequencies. In other words, the compensator looks identical to the narrowband disturbance compensator, as called for by the internal model principle. At the frequency $\omega_o$, the open-loop transfer function gain approaches infinity, so the sensitivity approaches zero, implying nearly exact cancellation of the disturbance. The attenuation above and below the passband of the compensator ensures that the closed-loop system has a favorable gain margin at all frequencies outside the compensator bandwidth. Therefore, the designer must only ensure that the compensator is phase-stabilized in the bandwidth of the compensator.

Phase stabilization of the disturbance rejection system can be understood by examining the form of the compensator transfer function, given by

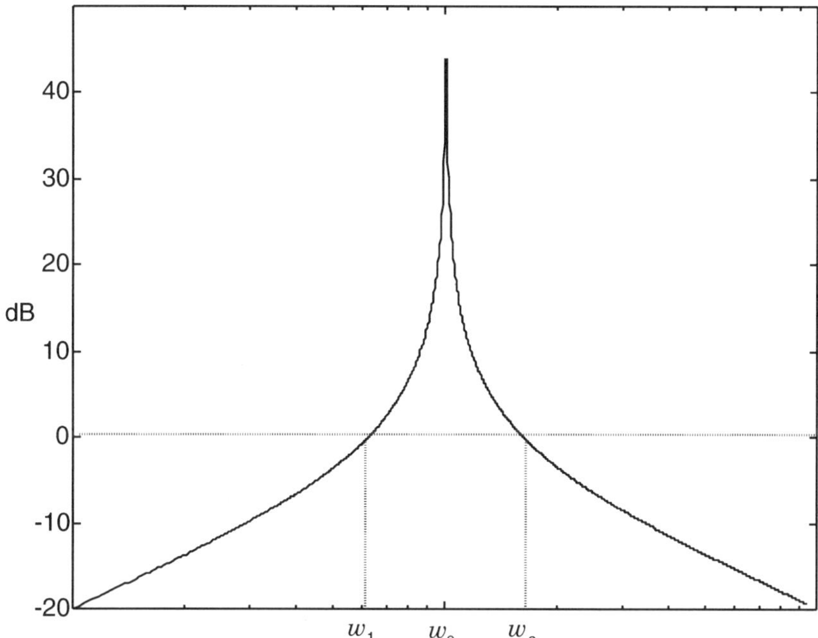


**Figure 7.34:** Typical compensator magnitude for narrowband disturbance rejection.

$$K(s) = \frac{k_c(\tau_c s + 1)}{s^2 + \omega_o^2} \tag{7.131}$$

The complex conjugate poles at the disturbance frequency $\omega_o$ provide the high gain required for complete attenuation of the disturbance. The Bode gain $k_c$ can adjust the compensator bandwidth, depending on the desired out-of-band characteristics. For example, low DC gains result in poor command tracking while higher DC gains lead to less stable systems. The zero location $\omega_c = 1/\tau_c$ is responsible for the phase stabilization in the compensator bandwidth. As discussed earlier, the root-locus diagram for a dissipative system shows that lightly damped poles of the open-loop transfer function, in this case from the compensator or the structure, should have departure angles of 180 degrees. This implies maximum damping of the closed-loop poles and a "velocity-like" closed-loop transfer function. It has been shown earlier that velocity feedback leads to the best phase characteristics from the viewpoint of stability. When the plant dynamics are well known, it is possible to design the compensator zero so that the departure angles are 180 degrees. When this is not possible, the compensator poles should be damped by some amount in order to avoid high gain instabilities that might occur for uncertain phase in the compensator bandwidth.

Rejection of narrowband exogenous disturbances using feedback control is always accomplished by placing closed-loop zeros at the frequencies of the disturbances. There are many variations on the actual implementation of this zero placement, as seen in the remaining chapters. Disturbance estimation can be built into an optimal estimator (i.e., Kalman filter) so that the disturbance rejection design is integrated into a linear quadratic Gaussian (LQG) optimal control approach. Any estimator can be cast in a form that estimates the disturbance inputs, in general, but the estimator must have perfect knowledge of the plant dynamics in order for the zero locations to be estimated accurately. An error in the estimate leads to reductions in the closed-loop attenuation, a situation that can be avoided only through adaptive algorithms. Chapter 8 provides a treatment of adaptive zero placement.

To conclude this section on disturbance rejection, consider the different characteristics of stationary versus transient disturbances. We have just seen that stationary disturbances may be decoupled from the system response by placing zeros at the important frequencies. It is also possible to modify the structure's eigenvectors to achieve this end, although this approach requires a MIMO design architecture to be successful. When the dynamic disturbance is transient in nature, that is, when it consists of pulsed loads or impulsive loads, the structural FRF magnitude is concentrated at the structural resonances. There may be a superimposed forcing frequency when the loads are pulsed, but the dominant spectral energy will be at the structure's damped natural frequencies. In this situation, the adaptive structure design strategy may be changed from a disturbance rejection architecture to one of active damping. This approach requires that the system eigenvalues be changed by increasing the real part of the stable closed-loop poles.

### 7.6.3 Modification of System Eigenvalues

Adaptive structures, with properly selected transducers placed at appropriate locations, can minimize the effects of forces entering the system from any path. The load paths may arise from attached rotating machinery, aerodynamic or hydrodynamic forces, fluid flow through internal piping systems, ambient acoustic pressures, or environmental forces. Many of these forces do not exhibit narrowband autospectra; in fact they are often random in nature, resulting in very broadband excitation of the structure. In those situations, it is efficient to design an active damping compensator that uses some method of pole placement to move the structural poles to regions of higher damping. The damping is increased by some practical amount, limited by the saturation levels of the actuator/power amplification components. The increased real parts of the closed-loop structure's eigenvalues lead to a reduction in the RMS power of the controlled output variable (displacement, velocity, strain, etc.). The details of eigenvalue assignment are discussed next.

***Eigenvalue Assignment*** A specific assignment of eigenvalues to a precise location in the $s$-plane is called pole placement. The effectiveness of pole placement methods is most easily appreciated using the state-space equations

$$\dot{\mathbf{X}}(t) = \mathbf{AX}(t) + \mathbf{B}u(t) \tag{7.132}$$

$$y(t) = \mathbf{CX}(t) \tag{7.133}$$

We keep our focus on SISO adaptive structures and stipulate that the command inputs $r$ or disturbance inputs $d$ be zero in order to simplify the presentation of the pole placement technique. (After the methods are presented, the closed-loop response to some wideband disturbance $d$ will be examined.) It can be shown that the closed-loop poles of the structure may theoretically be placed in *any* desirable locations using a control law which is full state feedback:

$$u(t) = -\mathbf{KX}(t) \tag{7.134}$$

$$= -(K_1x_1(t) + K_2x_2(t) + \cdots + K_Nx_N(t) + K_{N+1}\dot{x}_1(t) + K_{N+2}\dot{x}_2(t) + \cdots + K_{N+N}\dot{x}_N(t)) \tag{7.135}$$

providing the system ($\mathbf{A}$, $\mathbf{B}$) is completely controllable. For output feedback, some reduced number of poles can be moved to nearly arbitrary locations, depending on the dimensionality of the output $y(t)$. For the SISO problem, the use of output feedback $u(t) = -\tilde{K}y(t)$ essentially destroys the ability to place the structure's closed-loop poles at arbitrary locations. When output feedback is required, the modified dual Athens-Levine algorithm for optimal output feedback can be used to compute the feedback gains (see Section 8.6.1).

Pole placement methods are based on equating the characteristic equation of the system matrix for the closed-loop system to some desired polynomial of the same order. For full state feedback, equation (7.132) becomes

$$\dot{\mathbf{X}}(t) = (\mathbf{A} - \mathbf{BK})\mathbf{X}(t) \tag{7.136}$$

so that the characteristic equation for the closed-loop system is written

$$|s\mathbf{I} - \mathbf{A} + \mathbf{BK}| = 0 \tag{7.137}$$

When eigenvalues $(s + \lambda_1)$, $(s + \lambda_2)$, $\ldots$, $(s + \lambda_n)$ of the desired system can be specified, it is possible to write a corresponding *desired* characteristic polynomial of the same order as equation (7.137):

$$\Delta_d = (s + \lambda_1)(s + \lambda_2)\ \cdots\ (s + \lambda_n) = 0 \tag{7.138}$$

Equating these last two expressions, it is necessary to solve for the state-feedback gain vector $\mathbf{K}$ so that the equality is satisfied. This requires solving $n$ equations in the $n$ unknown gains.

A simple example using a SDOF spring-mass-damper system illustrates several key characteristics associated with placing poles in an adaptive structure design. The state equation for the displacement and velocity states is

$$\begin{bmatrix} \dot{\eta}(t) \\ \ddot{\eta}(t) \end{bmatrix} = \begin{bmatrix} 0 & 1 \\ -\omega_n^2 & -2\zeta\omega_n \end{bmatrix} \begin{bmatrix} \eta(t) \\ \dot{\eta}(t) \end{bmatrix} + \begin{bmatrix} 0 \\ 1 \end{bmatrix} [K_1 K_2] \begin{bmatrix} \eta(t) \\ \dot{\eta}(t) \end{bmatrix}$$

The characteristic equation is

$$\begin{vmatrix} s & -1 \\ \omega_n^2 + K_1 & s + 2\zeta\omega_n + K_2 \end{vmatrix} = 0 \tag{7.139}$$

or

$$s^2 + (2\zeta\omega_n + K_2)s + (\omega_n^2 + K_1) = 0 \tag{7.140}$$

As seen earlier, the desired, underdamped structural eigenvalues can be represented as $\lambda = s + (\zeta_d\omega_d \pm \omega_d^2\sqrt{1 - \zeta_d^2})$, so that the desired characteristic equation is

$$s^2 + (2\zeta_d\omega_d)s + (\omega_d^2) = 0 \tag{7.141}$$

Equating coefficients for the previous two characteristic equations, the two state-feedback gains are

$$K_1 = \omega_d^2 - \omega_n^2 \tag{7.142}$$
$$K_2 = 2\zeta_d\omega_d^2 - 2\zeta\omega_n^2 \tag{7.143}$$

These expressions indicate the fundamental limitation of pole placement in the adaptive structure—the gains increase with increasing active damping. As an example, consider the same spring-mass-damper system whose open-loop characteristic equation has coefficients $m = 2$ kg, $c = 3.77$ kg/s, and $k = 17{,}766$ N/m. Suppose we want to increase the damping to $\zeta_d = 0.2$ using state-feedback. From equation (7.143), we write

$$2(0.01)(94.25) + K_2 = 2(0.2)94.25 \tag{7.144}$$

so that $K_2 = 36.1$. If the natural frequency is to remain unchanged, $K_1 = 0$. To check this result, we assume that the velocity of the adaptive structure near resonance is approximately 25 mm/s. Then, the control signal $u = -K_2\dot{\eta}$ amplitude is on the order of 0.1 N/kg, which scales by the mass to arrive at a control force of $F \approx 0.17$ N for a closed-loop damping ratio $\zeta_d = 0.2$. Note that this force will be five times higher if we try to critically damp the structural poles. In reality, all systems are multi-degree-of-freedom (MDOF) systems; therefore, this numerical example shows that controlling the damping for arbitrary numbers of states will likely require forces beyond the capability of the actuators and/or power electronics, which are part of the adaptive structure.

There are numerous algorithms available that determine stable state-feedback gains for pole placement in higher order systems. The calculated feedback gains place the closed-loop structural poles at the desired locations specified by the user. Two numerical approaches that are included in the MATLAB Controls Toolbox are **acker.m** and **place.m**.

## 7.7 SUMMARY

The purpose of Chapter 7 was to review concepts from classical control theory in the context of the disturbance rejection problem, specific to the adaptive structure. Since the disturbance enters the structural plant through some physical path, it is shaped by the dynamics of the closed-loop structure. This must be considered in the design process, as was demonstrated in this chapter. Basic tools such as root-locus, Bode plots, Nyquist plots, and their use in the determination of gain and phase margins were discussed. Series compensators such as lead, lag, lead-lag, and PID controllers were reviewed. Controller design objectives and methods of achieving such objectives for adaptive structures were presented.

## BIBLIOGRAPHY

Bendat, J. S. and A. G. Piersol, 1984. *Random Data Analysis*, Wiley, New York.

Bode, H., 1945. *Network Analysis and Feedback Amplifier Design*, Van Nostrand, New York, 1945.

Cannon, R. H. and D. E. Rosenthal, 1984. "Experiments in Control of Flexible Structures with Noncolocated Sensors and Actuators," *Journal of Guidance and Control*, **7**,(5), 546–553.

Churchill, R. V. and J. W. Ward, 1978. *Fourier Series and Boundary Value Problems*, McGraw-Hill, New York, pp. 66–74.

Clark, R. L., 1996. "Accounting for Out-of-Bandwidth Modes in the Assumed Modes Approach: Implications on Colocated Output Feedback Control," accepted by *Journal of Dynamic Systems, Measurement, and Control.*

Clark, R. L., D. G. Cole, and K. D. Frampton, 1996, "Phase Compensation for Feedback Control of Enclosed Sound Fields," *Journal of Sound and Vibration*, **195**(5), 701–718.

Cremer, L., M. Heckl, and E. Ungar, 1987. *Structure-Borne Sound*, Springer-Verlag, Berlin.

Doyle, J. C., B. A. Francis, and A. R. Tannenbaum, 1992. *Feedback Control Theory*, Macmillan, New York.

Ewins, D. J., 1986. *Modal Testing*, Research Studies Press, Letchworth, England, pp. 87–148.

Franco, S., 1988. *Design with Operational Amplifiers and Analog Integrated Circuits*, McGraw Hill, New York.

Franklin, G. F., J. D. Powell, and A. Emami-Naeini, 1994. *Feedback Control of Dynamic Systems*, 3rd ed., Addison-Wesley Series in Electrical and Computer Engineering, Menlo Park, CA, pp. 601–662.

Goh, C. J. and T. K. Caughey, 1985. "On the Stability Problem Caused by Finite Actuator Dynamics in the Collocated Control of Large Space Structures," *International Journal of Control*, **41**(3), 787–802.

Hale, F. J., 1988. *Introduction to Control System Analysis and Design*, Prentice Hall, Englewood Cliffs, NJ.

Hong, H. and D. S. Bernstein, 1995. "A Comparison of the Fundamental Properties of Feedback and Feedforward Control: Bode Integral Constraints and Spillover," *Proceedings of Active 95*, Newport Beach, CA, pp. 929–940.

Hong, J. and D. S. Bernstein, 1997. "Bode Integral Constraints, Colocation, and Spillover in Active Noise and Vibration Control," *IEEE Transactions on Control Systems and Technology*, to appear.

Inman, D. J., 1994. *Engineering Vibration*, Prentice Hall, Englewood Cliffs, NJ.

Inman, D. J., J. W. Umland, and J. Bellos, 1990. "Controlling Flexible Structures with Second Order Actuator Dynamics," N90-23083, pp. 879–890.

Junkins, J. L. and Y. Kim, 1993. *Introduction to Dynamics and Control of Flexible Structures*, AIAA Education Series.

Kailath, T., 1980. *Linear Systems*, Prentice Hall, Englewood Cliffs, NJ.

Kennedy, C. C. and C. D. Pancu, 1947. "Use of Vectors in Vibration Measurement and Analysis," *Journal of Aeronautical Science*, **14** (11).

Klosterman, A. L., 1971. *On the Experimental Determination and Use of Modal Representations of Dynamic Characteristics*, University Microfilms, Ann Arbor, MI.

Kuo, B. C., 1963. *Analysis and Synthesis of Sampled-Data Control Systems*, Prentice Hall, Englewood Cliffs, NJ, pp. 252–279.

Maciejowski, J. M., 1989. *Multivariable Feedback Design*, Addison-Wesley, Wokingham, England.

Melsa, J. L. and D. G. Schultz, 1969. *Linear Control Systems*, McGraw-Hill, New York, pp. 301–347.

Newland, D. E., 1975. *Random Vibrations and Spectral Analysis*, Longman Group, London.

Ogata, K., 1970. *Modern Control Engineering*, Prentice Hall, Electrical Engineering Series, Englewood Cliffs, NJ.

Santina, M. S., A. R. Stubberud, and G. H. Hostetter, 1994. *Digital Control System Design*, Saunders College Publishing, Fortworth, TX.

Shahian, B. and M. Hassul, 1993. *Control System Design Using MATLAB*, Prentice Hall, Englewood Cliffs, NJ.

Sievers, L. A. and A. H. von Flotow, 1990. "Comparison and Extensions of Control

Methods for Narrowband Disturbance Rejection," Proceedings of 1990 ASME Winter Annual Meeting, Dallas, TX.

Skogestad, S. and I. Postlethwaite, 1996. *Multivariable Feedback Control, Analysis and Design*, Wiley, New York.

Tohyama, M. and R. H. Lyon, 1989. "Zeros of a Transfer Function in a Multi-Degree-of-Freedom Vibrating System," *Journal of Acoust. Soc. of Amer.*, **86**(5), 1854–1863.

Zhou, K., J. C. Doyle, and K. Glover, 1996. *Robust and Optimal Control*, Prentice Hall, Upper Saddle River, NJ.

# 8

# ACTIVE CONTROL: SYSTEM ARCHITECTURES AND ALGORITHMS

## 8.1 INTRODUCTION

Adaptive structures usually require the use of multivariable control system architectures, thereby demanding more complex controller design measures than have been discussed in the previous chapter. Large numbers of sensors and actuators can lead to greater performance and stability robustness because there is an effective increase in the "spatial sampling frequency" in well-designed multi-input, multi-output (MIMO) control architectures. In this chapter, it is assumed that the designer has selected appropriate spatial and temporal signal processing methods for the transducers; hence, the remaining concern is how to design the control system architecture that will be most suited for the application. Thus, it remains only to insert the spatial/temporal filters at appropriate locations in the closed-loop system and design the controller.

We begin by describing a *unified control system architecture* that generalizes the effective actions of any linear control strategy—feedback control, feedforward control, and hybrid combinations of both—possible for adaptive structures. This discussion follows from some of the single-input, single-output (SISO) results explored in earlier chapters, in that the closed-loop structure can still be described by its pole and zero locations in the complex plane. The essential difference for the multivariable plant is that the closed-loop zeros are usually different than the zeros of individual transfer functions. Regardless of that, the effects of different control directions through the plant can still be described as pole placement or zero placement processes, brought about by feedback or feedforward control actions, respectively. Section 8.2 casts the unified approach as the generalized plant, or two-port model, to facilitate specific design procedures.

Once the designer understands how to tailor the closed-loop dynamic response by placing zeros, adding zeros, placing poles, and so on, and how to generate the two-port model description, the only remaining tasks are to decide exactly what the dynamics need to be for the desired performance and the specific design of the compensator. Section 8.3 presents an overview discussion of how the adaptive structure performance objectives ultimately determine which elements of the unified controller are required. Then, we present precise architectural and algorithmic details for the three subsystems identified in the unified architecture. Finite- and infinite-impulse response (FIR and IIR) adaptive digital filter designs for *feedforward control* are reviewed and discussed in the context of adaptive structures. For the *feedback control* forms, the two-port design approach is summarized and used to illuminate differences in static and dynamic compensation methods. These are contrasted to multivariable designs (acceleration feedback and position feedback), which follow more closely the classical control methods discussed previously. Robust control designs are also presented in a heuristic context relying heavily on example problems. Finally, unified *hybrid* controller designs are presented in detail, using several variations that have proven successful for adaptive structure control.

Sampled-data versions of the plant are used throughout this chapter, at appropriate times. Readers who are not familiar with discrete-time representations should review the introductory comments about z-transform methods and properties (Appendix A) before proceeding.

## 8.2 UNIFIED CONTROL SYSTEM ARCHITECTURE

The history of adaptive structure research and development demonstrates an overwhelming preference for the *exclusive* use of feedback *or* adaptive feedforward control systems. There is little technical justification for this exclusivity, as demonstrated throughout the remainder of this book. In most cases, it is likely related to prior training of different designers. Because there are definite advantages for each type of control, and even different advantages for the blend of feedback and adaptive feedforward, it is very helpful to begin by casting each approach in a unified framework and investigating the effects on the closed-loop performance.

### 8.2.1 SISO Considerations

We begin our discussion with a SISO configuration, primarily because the MIMO system leads to a complexity associated with describing the closed-loop zeros. The SISO analysis provides a more direct understanding of the closed-loop response on the zeros of an individual path through the structure and leads to the identical results for the system poles. There are distinctions in the architectures that best represent the most common use of adaptive structures—vibration control. The following discussion clarifies those distinctions.

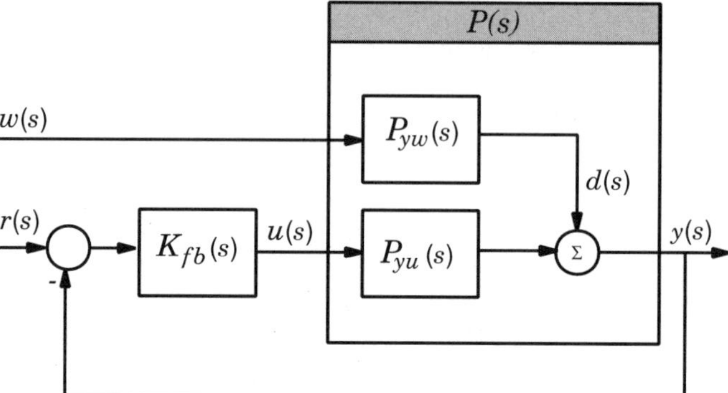


**Figure 8.1:** SISO classical feedback architecture.

The importance of the disturbance rejection problem was noted in the previous chapter; however, the construction of block diagrams that represent the physical manner in which disturbances enter the plant needs further discussion. It is typical in most control theory textbooks to introduce disturbances at the output $y(s)$, as shown in Figure 8.1. If we try to relate this to many adaptive structure designs, the output $y(s)$ would be a measurement variable (i.e., strain, acceleration, etc.), and the addition of a disturbance $d(s)$ at that location would necessarily mean that the units of the disturbance are identical to those of the measurement variable. It is immediately apparent that it will not be possible to isolate the disturbance contribution $d(s)$ to the output signal without some prior knowledge of the structure's dynamics (i.e., $P_{yu}(s)$) and all other inputs to the structural plant, that is, an estimation scheme. For this reason, it will be advantageous to consider the disturbance as the input $w(s)$ in Figure 8.1. In general, the choice is a matter of perspective, with the exception of shape control of course. However, one must recognize that, for the adaptive structure problem, disturbances typically propagate through the structure before contributing to the output. It might be argued that the disturbance $d(s)$ in Figure 8.1 is not actually an exogeneous disturbance input, since it is affected by the closed-loop response via the poles of $P_{yw}(s)$, which are common with the poles of $P_{yu}(s)$. Obtaining the proper perspective on the disturbance path is paramount to understanding the disturbance rejection problem and how it differs from the conventional presentation in most controls-oriented texts. For our discussion, we assume that the disturbance $w(s)$ has some predefined spectral content, and we define its transfer function to the output $y(s)$ as $P_{yw}(s)$. Thus, the disturbance $d(s)$ at the output will be affected through the control actions we take on the propagation of $w(s)$ through the structure. Next, we examine three cases that summarize the effects of feedback control, feedforward control, and combined feedforward/feedback control on disturbances propagating through the structure to the measured output $y(s)$.

**Example 8.1: Feedback Control** Figure 8.1 illustrates a familiar feedback architecture that is motivated by a typical adaptive structure design. The disturbance $w(s)$ excites the structure with resulting dynamic response described by the transfer function $P_{yw}(s)$ to the output $y(s)$. A second reference input $r(s)$ enters the structure at the same location as the output feedback signal $y(s)$, sharing the forward-path transfer function through the structure, given by $K_{fb}(s)P_{yu}(s)$. The disturbance is taken to be $w(s)$, so that the closed-loop response can be written in the following form, specifically isolating the numerator and denominator polynomials of the transfer functions:

$$y(s) = \frac{P_{yu}(s)K_{fb}(s)}{1 + K_{fb}(s)P_{yu}(s)}\, r(s) + \frac{P_{yw}(s)}{1 + K_{fb}(s)P_{yu}(s)}\, w(s) \tag{8.1}$$

$$= S(s)P_{yu}(s)K_{fb}(s)r(s) + S(s)P_{yw}(s)w(s) \tag{8.2}$$

$$= \frac{N_{yu}(s)N_{fb}(s)/D_{yu}(s)D_{fb}(s)}{1 + [N_{yu}(s)/D_{yu}(s)][N_{fb}(s)/D_{fb}(s)]}\, r(s) + \frac{N_{yw}(s)/D_{yw}(s)}{1 + [N_{yu}(s)/D_{yu}(s)][N_{fb}(s)/D_{fb}(s)]}\, w(s) \tag{8.3}$$

Numerator polynomials are represented by $N(s)$ for all transfer functions, and the denominator polynomials are similarly represented by $D(s)$. Subscripts for the different numerator and denominator terms have been chosen with respect to the corresponding transfer function subscripts. For example, the numerator polynomial for the control signal propagation through the structure is written $N_{yu}(s)$. Rewriting,

$$y(s) = \frac{N_{yu}(s)N_{fb}(s)}{D_{yu}(s)D_{fb}(s) + N_{yu}(s)N_{fb}(s)}\, r(s) + \frac{N_{yw}(s)D_{fb}(s)}{D_{yu}(s)D_{fb}(s) + N_{yu}(s)N_{fb}(s)}\, w(s) \tag{8.4}$$

it can be seen that the roots of the denominator, that is, the closed-loop poles, have been modified by the feedback. (Recall that polynomial addition implies that the roots of the polynomial sum will be different from the roots of the two polynomials in the sum. Also, multiplication of two polynomials results in a product polynomial whose roots are composed of the roots of the two polynomials in the product.) These algebraic considerations require that there *must* be a polynomial addition in the numerator (denominator) of the controlled system transfer function if the zeros (poles) are to be freely modified. For this case, we see that the numerator roots, that is, the closed-loop zeros, are equal to the open-loop zeros for the reference input. Stated otherwise, the series feedback compensator cannot be used to *place* the closed-loop zeros in the reference-to-output transfer function. We are equally interested in the closed-loop response to disturbances. Equation 8.4 shows that the zeros of the disturbance path are the combination of the zeros in the disturbance-to-output transfer function and the poles of the compensator. When the compensator is in the feedback path, the closed-loop response $y(s)$ will be nearly identical, except that the closed-loop zeros for the reference input will be composed of the open-loop zeros of the reference-to-output path and the compensator poles, as discussed in Chapter 7. Thus, the conclusion for this first example is that feedback can *modify* only the closed-loop poles, while the zeros of either path are simply combinations of open-loop dynamics. For either input, it is not possible to modify the closed-loop zeros with a feedback architecture only, and as indicated by equation (8.2), disturbance rejection can be achieved only if the sensitivity, $S(s)$, is reduced in the desired bandwidth, a result previously discussed in Chapter 7.

**Example 8.2: Feedforward Control** Figure 8.2 describes a second common architecture seen in adaptive structure design, referred to as feedforward control. Feedforward compensation has traditionally been used in control systems; it is not unique to adaptive structures. However, it is much more prevalent in adaptive structures because of the many situations that call for disturbance rejection as opposed to tracking a reference. The term "feedforward" arises because the strategy is to use a signal generated from some measure of the disturbance, in this case $w(s)$ (or less likely $d(s)$), to counteract the effect of the disturbance as it enters the structure through some alternative path. Consider the block-diagram presented in Figure 8.2. The output, $y(s)$, can be expressed as a function of the disturbance, $w(s)$, as follows:

$$y(s) = P_{yw}(s)w(s) + K_{ff}(s)P_{yu}(s)w(s) \tag{8.5}$$

$$= (P_{yw}(s) + K_{ff}(s)P_{yu}(s))w(s) \tag{8.6}$$

$$= \frac{D_{ff}(s)N_{yw}(s) + N_{ff}(s)N_{yu}(s)}{D_{ff}(s)D_{yu}(s)} w(s). \tag{8.7}$$

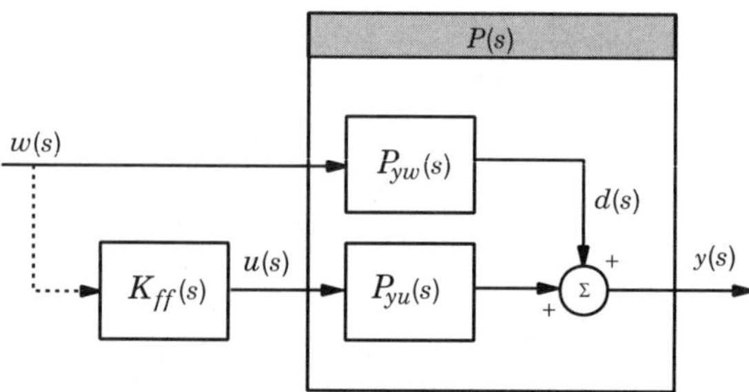


**Figure 8.2:** Common feedforward architecture for known/measurable disturbances.

Two important observations can be made from equation (8.7). First, for the adaptive structure, the poles of both transfer function paths are common and have been factored out of equation (8.7). The remaining poles are the sum of the structural plant poles, $D_{yu}(s)$, and the poles of the feedforward compensator, $D_{ff}(s)$. Second, the zeros of the controlled transfer function from $w(s)$ to the output $y(s)$ are not simply composed of the zeros from the different uncontrolled transfer functions. They can now be modified by the selection of the dynamic feedforward compensator $K_{ff}(s)$. In fact, it is possible to completely eliminate the effect of the disturbance on the plant *at the chosen output*. (The plant response may increase through other structural paths!) Let us consider three possible cases. For the first case, we will assume the typical approach that the disturbance, $w(s)$, enters at the output directly (i.e., $P_{yw}(s) = 1$). For this case, the optimal compensator is $K_{ff}(s) = -1/P_{yu}(s)$. In the second case, we let $P_{yw}(s) = P_{yu}(s)$ (i.e., the disturbance enters through the control path). For this case, the optimal compensator is simply $K_{ff}(s) = -1$ (not a likely application!). Finally, we consider the general case whereby the control path and disturbance path are unique such that the optimal compensator is $K_{ff}(s) = -P_{yw}(s)/P_{yu}(s)$. In general, other than the trivial case ($K_{ff}(s) = -1$), the optimal compensators cannot be realized, because they must be proper transfer functions with minimum phase zeros. However, feedforward control applications are often based upon the rejection of one or more *harmonic* signals that allows us to satisfy the compensator equation $K_{ff}(s) = -P_{yw}(s)/P_{yu}(s)$ without any problems. For such applications, fixed-gain or adaptive filters can be implemented that effectively reject the disturbance at the desired plant output(s) completely.

**Example 8.3: Combined Feedforward and Feedback Control** The final example we consider is shown in Figure 8.3, which is nearly a combination of the two configurations presented in Figures 8.1 and 8.2. The response, $y(s)$, can be expressed as a function of the disturbance, $w(s)$, and the reference, $r(s)$, as follows:

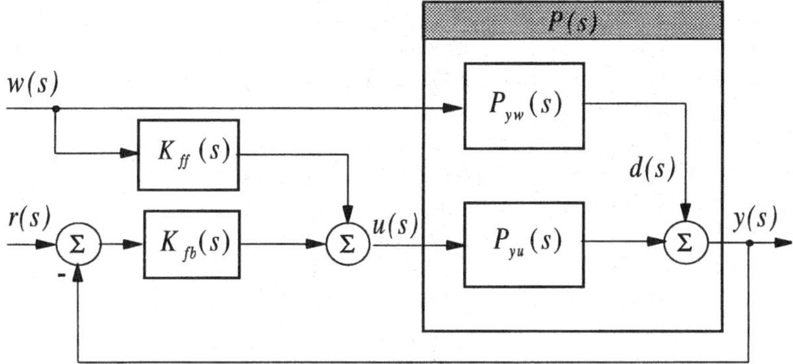


**Figure 8.3:** Combined feedback and feedforward architecture.

$$y(s) = \frac{P_{yu}(s)K_{fb}(s)}{1 + P_{yu}(s)K_{fb}(s)} r(s) + \frac{P_{yw}(s) + P_{yu}(s)K_{ff}(s)}{1 + P_{yu}(s)K_{fb}(s)} w(s) \tag{8.8}$$

$$= S(s)P_{yu}(s)K_{fb}(s)r(s) + S(s)(P_{yw}(s) + P_{yu}(s)K_{ff}(s))w(s) \tag{8.9}$$

As indicated in equation (8.9), the control system designer now has *two* options for disturbance rejection. One option is to decrease the sensitivity in the bandwidth of interest, which is quite effective for stochastic disturbances. However, the second option is to cancel the effect of the disturbance at the chosen output. This is accomplished by choosing a feedforward compensator that places zeros at those frequencies where the signal energy is high in the disturbance-to-output transfer function. This is accomplished through the proper design of $K_{ff}(s)$ in the numerator polynomial $S(s)(P_{yw}(s) + P_{yu}(s)K_{ff}(s))$. As mentioned above, if the designer has access to a periodic reference signal that is correlated with the disturbance, this should become part of the design. These feedforward control strategies can improve upon model-based, feedback disturbance suppression performance, particularly if it is adaptive feedforward, as will be outlined in subsequent sections. It will also be shown that the blended use of feedback and adaptive feedforward leads to better convergence properties of the adaptive algorithms operating on adaptive structures.

### 8.2.2 MIMO Considerations

Generally, we must design for MIMO structures where the mathematical models are described using state variables or transfer matrices. Digital computers have made state-variable models more popular than transfer matrices, but the transfer-matrix models are more closely related to the previous discussion and illuminate two interesting results that are very important. First, it is well known and easy to show that the plant's open-loop poles are changed in the presence of MIMO feedback. Second, the MIMO plant's open-loop and closed-loop zeros must be treated differently than those of the SISO plant. We begin by examining the structure's closed-loop poles. Recalling that the transfer matrix $\mathbf{P}_{yu}(s)$ is related to the state-space description of the structure by

$$\mathbf{P}_{yu}(s) = \mathbf{C}(sI - \mathbf{A})^{-1}\mathbf{B} + \mathbf{D} \tag{8.10}$$

We consider the reference-to-output portion of the feedback system in Figure 8.3. Disturbance inputs can be ignored for this discussion. The inputs, $\mathbf{r}(s)$, are now vectors that represent any number $m$ of distinct inputs to the adaptive structure, propagating through $n$ different structural paths, which results in the output vector $\mathbf{y}(s)$ of the same order as the structural paths. The open-loop transfer function is $\mathbf{P}_{yu}(s)\mathbf{K}(s)$, which loses rank if any two inputs and paths turn out to be identical. The effects of the inputs $\mathbf{r}(s)$ on the closed-loop response $\mathbf{y}(s)$ are directly analogous to those of the SISO description given in Section 7.6.1:

$$\mathbf{y}(s) = (I + \mathbf{P}_{yu}\mathbf{K}_{fb})^{-1}\mathbf{P}_{yu}\mathbf{K}_{fb}\ \mathbf{r}(s) \tag{8.11}$$

where we consider $\mathbf{r}(s) \in R^{m\times 1}$ and $\mathbf{y}(s) \in R^{n\times 1}$ for simplicity. This requires that $\mathbf{P}_{yu}(s) \in C^{n\times m}$, that $\mathbf{K}_{fb}(s) \in C^{m\times n}$, and that $\mathbf{P}_{yu}(s)\mathbf{K}_{fb}(s)$ be always square. After some manipulation, it can be shown that the eigenvalues of the closed-loop system (i.e., closed-loop poles) are given by either of the expressions

$$\Lambda_{cl}(s) = \det(s\mathbf{I} - \mathbf{A} + \mathbf{BC}) \tag{8.12}$$

$$= \det(s\mathbf{I} - \mathbf{A})\det(\mathbf{I} + \mathbf{P}_{yu}(s)) \tag{8.13}$$

Recognizing that $(\mathbf{I} + \mathbf{P}_{yu}(s)) = (\mathbf{I} + \mathbf{C}(s\mathbf{I} - \mathbf{A})^{-1}\mathbf{B})$, equation (8.13) shows that the closed-loop poles are changed from the open-loop poles $\Lambda_{\text{ol}}(s) = (s\mathbf{I} - \mathbf{A})$ in the presence of feedback, a result identical to that of the SISO configuration.

The effects of feedback on the zeros of the transfer matrix $\mathbf{P}(s)$ are less straightforward, unfortunately, and can only be described by looking at the *transmission zeros* of the transfer matrix. Transmission zeros affect the system inputs identically to the zeros of a SISO system; they block or absorb the energy of a sinusoidal input signal with a frequency equal to the zero frequency. However, it is significantly more work to isolate analytical expressions for transmission zeros (MacFarlane and Karcanias, 1975). For simplicity, the generalization of the unified control architecture's effect on transmission zeros is not derived here. Instead, the resulting conclusions are stated without proof:

- The zeros of a transfer matrix are not identical to the zeros of the individual transfer functions of the matrix.
- The zeros of a controllable and observable system are equal to the transmission zeros, defined by the following: *If a system has a transmission zero at $s = z_i$, and if an input $x(t) = x_i \exp z_i t$ is applied to the system, then the output vector $y(t)$ will not contain the exponential* $\exp z_i t$ *in any of its components for $x_i$ in the direction of the transmission zero.*
- If $\mathbf{P}(s)$ is square and nonsingular, the transmission zeros are the roots of the matrix $\mathbf{N_r}(s)$ of the right-coprime factorization $\mathbf{P}(s) = \mathbf{N_r}(s)[\mathbf{D_r}(s)]^{-1}$.
- If $\mathbf{P}(s)$ is nonsquare, the transmission zeros can be identified only by solving for frequencies where the rank of $\mathbf{N_r}(s)$ falls below its normal rank.
- The zeros of the open-loop system are the same as the zeros of the closed-loop system under unity feedback for MIMO systems.

The last item, provided without proof, is a statement that the closed-loop poles for the MIMO system behave similarly to those of the SISO case analyzed earlier. This conclusion is based on a root-locus argument that indicates that all open-loop poles must go to the open-loop zeros. The second-to-last item indicates the general difficulty of showing that the zeros of a MIMO system are changed in the presence of feedforward control. In fact, the transmission zeros are *not* generally modified by feedforward control. The zeros of the individual transfer functions are modified, but these are not guaranteed to be the same, and usually are not the same, as the transmission zeros. This concept is tied to local versus global control for feedforward architectures in Section 8.4.

The previous discussions of a unified control system architecture indicate the common characteristics that are observed for a broad range of compensator architectures. All compensators may incorporate feedback or feedforward designs; therefore, the designer's selection determines whether the controlled system will benefit from changing pole positions or mitigation of disturbances through the structure by modification of the zeros. To complete this section, it is helpful to introduce a more compact representation of the multivariable plant.

The general multivariable description can be represented by the block diagram presented in Figure 8.4. Notice in the block diagram that the *flow* of the inputs and outputs are reversed. This is a format adopted by the controls community, since one can simply read the input-output equations from the general model as follows:

$$\begin{bmatrix} \mathbf{z}(s) \\ \mathbf{y}(s) \end{bmatrix} = \begin{bmatrix} \mathbf{P}_{zw}(s) & \mathbf{P}_{zu}(s) \\ \mathbf{P}_{yw}(s) & \mathbf{P}_{yu}(s) \end{bmatrix} \begin{bmatrix} \mathbf{w}(s) \\ \mathbf{u}(s) \end{bmatrix} \tag{8.14}$$

As indicated in equation (8.14) and illustrated in Figure 8.4, the system matrix $\mathbf{P}(s)$ is partitioned according to the input-output variables. The output variable $\mathbf{z}(s)$ is denoted the performance variable, and for the preceeding SISO examples, $z(s) = y(s)$. Now consider the two-port model illustrated in Figure 8.5. We have chosen to sketch the *flow* of inputs and outputs based upon that of traditional block-diagrams. However, notice that both a feedforward and feedback compensator have been added to the system. From block diagram algebra [or a linear fractional transformation (Zhou et al., 1996)], the response at the performance variable $\mathbf{z}(s)$ can be expressed as a function of the disturbance $\mathbf{w}(s)$ by first letting $\mathbf{u}(s) = \mathbf{K}_{fb}(s)\mathbf{y}(s)$ and substituting to obtain

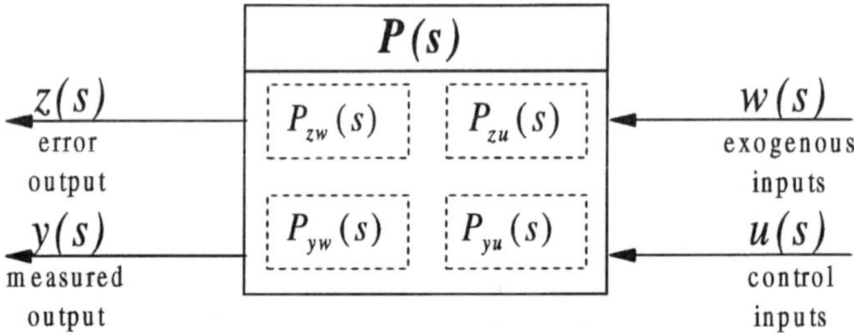


**Figure 8.4:** General block diagram of two-port or two-input, two-output (TITO) system.

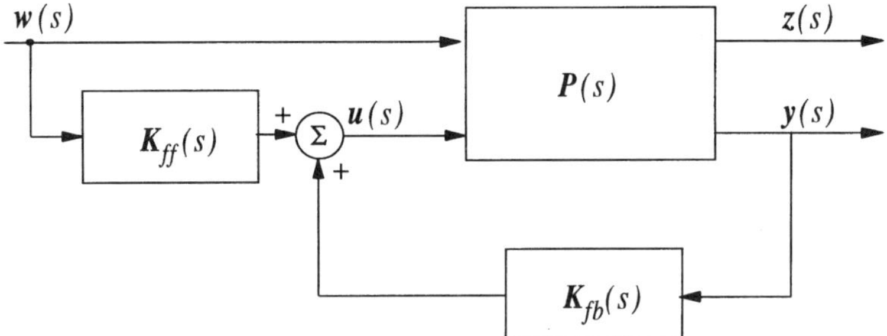


**Figure 8.5:** General MIMO feedback architecture.

$$\mathbf{y}(s) = \mathbf{P}_{yw}(s)\mathbf{w}(s) + \mathbf{P}_{yu}(s)\mathbf{K}_{fb}(s)\mathbf{y}(s) \tag{8.15}$$

Solving for $\mathbf{y}(s)$, one obtains

$$\mathbf{y}(s) = [\mathbf{I} - \mathbf{P}_{yu}(s)\mathbf{K}_{fb}(s)]^{-1}\mathbf{P}_{yw}(s)\mathbf{w}(s) \tag{8.16}$$

Since

$$\mathbf{z}(s) = \mathbf{P}_{zw}(s)\mathbf{w}(s) + \mathbf{P}_{zu}(s)\mathbf{K}_{fb}(s)\mathbf{y}(s) \tag{8.17}$$

one can readily show that

$$\mathbf{z}(s) = [\mathbf{P}_{zw}(s) + \mathbf{P}_{zu}(s)\mathbf{K}_{fb}(s)[\mathbf{I} - \mathbf{P}_{yu}(s)\mathbf{K}_{fb}(s)]^{-1}\mathbf{P}_{yw}(s)]\mathbf{w}(s) \tag{8.18}$$

As explained by Zhou et al. (1996) and Skogestad and Postlethwaite (1996), the bracketed term defines the lower linear fractional transformation (LFT):

$$f_l(\mathbf{P}, \mathbf{K}) \equiv [\mathbf{P}_{zw}(s) + \mathbf{P}_{zu}(s)\mathbf{K}_{fb}(s)[\mathbf{I} - \mathbf{P}_{yu}(s)\mathbf{K}_{fb}(s)]^{-1}\mathbf{P}_{yw}(s)] \tag{8.19}$$

Including the feedforward compensator in the system transfer function, the relationship between the disturbance $\mathbf{w}(s)$ and the performance variable $\mathbf{z}(s)$ can be expressed concisely as follows:

$$\begin{aligned}\mathbf{z}(s) = {} & [\mathbf{P}_{zw}(s) + \mathbf{P}_{zu}(s)\mathbf{K}_{fb}(s)[\mathbf{I} - \mathbf{P}_{yu}(s)\mathbf{K}_{fb}(s)]^{-1}\mathbf{P}_{yw}(s)]\mathbf{w}(s) \\ & + [\mathbf{P}_{zu}(s) + \mathbf{P}_{zu}(s)\mathbf{K}_{fb}(s)[\mathbf{I} - \mathbf{P}_{yu}(s)\mathbf{K}_{fb}(s)]^{-1}\mathbf{P}_{yu}(s)]\mathbf{K}_{ff}(s)\mathbf{w}(s)\end{aligned} \tag{8.20}$$

For the special case when we choose the performance variables and the measured outputs as the same variables [i.e., $\mathbf{z}(s) = \mathbf{y}(s)$], equation (8.20) can be simplified as follows:

$$\mathbf{y}(s) = [\mathbf{P}_{yw}(s) + \mathbf{P}_{yu}(s)\mathbf{K}_{fb}(s)[\mathbf{I} - \mathbf{P}_{yu}(s)\mathbf{K}_{fb}(s)]^{-1}\mathbf{P}_{yw}(s)]\mathbf{w}(s)$$
$$+ [\mathbf{P}_{yu}(s) + \mathbf{P}_{yu}(s)\mathbf{K}_{fb}(s)[\mathbf{I} - \mathbf{P}_{yu}(s)\mathbf{K}_{fb}(s)]^{-1}\mathbf{P}_{yu}(s)]\mathbf{K}_{ff}(s)\mathbf{w}(s) \quad (8.21)$$
$$= \mathbf{S}(s)[\mathbf{P}_{yw}(s) + \mathbf{P}_{yu}(s)\mathbf{K}_{ff}(s)]\mathbf{w}(s) \quad (8.22)$$

Notice the similarity between the SISO expression of equation (8.6) and the MIMO expression of equation (8.22). It is left as an excercise for the reader to verify equation (8.22) *Hint*: $[\mathbf{I} + \mathbf{PK}(\mathbf{I} - \mathbf{PK})^{-1}] = [\mathbf{I} - \mathbf{PK}]^{-1}$. Also note that positive feedback was assumed as illustrated in the block diagram, and $\mathbf{S}(s) = [\mathbf{I} - \mathbf{P}_{yu}(s)\mathbf{K}_{fb}(s)]^{-1}$ is the sensitivity. The key result shown here is that disturbance suppression for the multivariable plant proceeds along the same general guidelines as were evident for the SISO design: minimize the sensitivity or place zeros at appropriate frequencies. However, for the adaptive structure, it is more typical that $\mathbf{z}(s) \neq \mathbf{y}(s)$ (i.e., the performance variables used in the cost function are different from the sensed variables used in feedback). For such cases, the relationship between the disturbance and the performance variables with respect to feedback can be expressed concisely as follows:

$$\mathbf{z}(s) = f_l(\mathbf{P}, \mathbf{K})\mathbf{w}(s) \quad (8.23)$$

Thus, reducing the sensitivity in the desired bandwidth does not necessarily result in a reduction in the performance variables, because $f_l(\mathbf{P}, \mathbf{K})$ is a function of the open-loop response associated with all four transfer function paths. This complexity forces the designer to consider *closed-loop shaping*, which will be the subject of a subsequent discussion on $\mathcal{H}_2$ and $\mathcal{H}_\infty$ control system design. The placement of zeros is more challenging for the multivariable plant as well and prompts the study of local versus global control, as discussed in Section 8.4. Next, we examine in more detail the widely used modeling form for design of the controller.

## 8.3 STRUCTURE OF THE MODEL: THE GENERALIZED PLANT

Before proceeding with a discussion of cost functionals and the general methods of control system design and synthesis, it is helpful to discuss the two-port adaptive structure model used in design. Whether the plant model is obtained experimentally via system identification or through analysis, the following discussion pertains. Herein, we distinguish between the plant $\mathbf{P}(s)$ and the generalized plant $\mathbf{G}(s)$. As presented in Section 8.2 of this chapter, the plant $\mathbf{P}(s)$ is the system on which the active controller is implemented. However, the generalized plant $\mathbf{G}(s)$ is defined as the system including any frequency shaped costs, external noise sources or disturbances, and performance weightings used in the design and synthesis of the controller. Consider the schematic diagram presented in Figure 8.6. There are two vector inputs and two vector outputs to

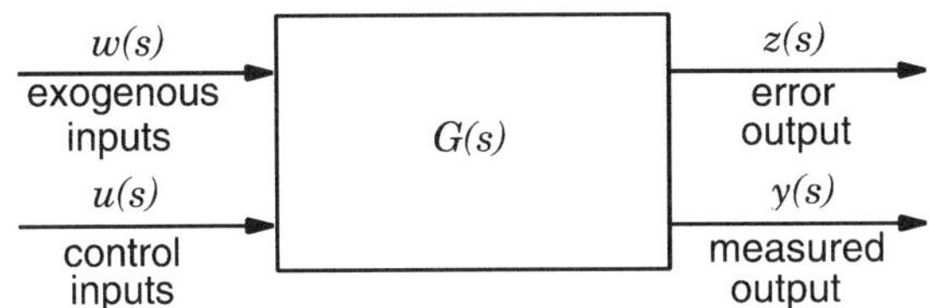


**Figure 8.6:** Schematic diagram of the generalized plant.

the generalized plant. The vector signal $\mathbf{w}(s)$ represents the exogenous inputs and contains all disturbances, sensor noise, and command reference signals. The vector signal $\mathbf{z}(s)$ is termed the error output and contains all signals important in the *design* of the control system for performance (i.e., all of the signals in the cost functional). The vector signal $\mathbf{y}(s)$ contains all of the measured output signals from the plant $\mathbf{P}(s)$ required for *implementation* of the control system, and the vector signal $\mathbf{u}(s)$ contains all of the control inputs to the plant $\mathbf{P}(s)$ required to implement the controller. The model of the generalized plant as presented is frequently termed the two-port model, or the two-input, two-output (TITO) model.

To further illustrate the form of the generalized plant, consider the schematic diagram depicted in Figure 8.7. As illustrated, $\mathbf{P}(s)$, the plant to be controlled, is contained within the system model of $\mathbf{G}(s)$. Additionally, a frequency shaped filter $\mathbf{W}(s)$ is also included in the model. This frequency shaped filter is coupled to selected outputs of the plant model, as well as to the control inputs. Thus, the filter can be used to weight the outputs and inputs to the control system. The outputs of this frequency shaped filter define the error outputs used to evaluate the system performance and generate the cost that will be used in the design process. For example, if structural acoustic control is the objective,

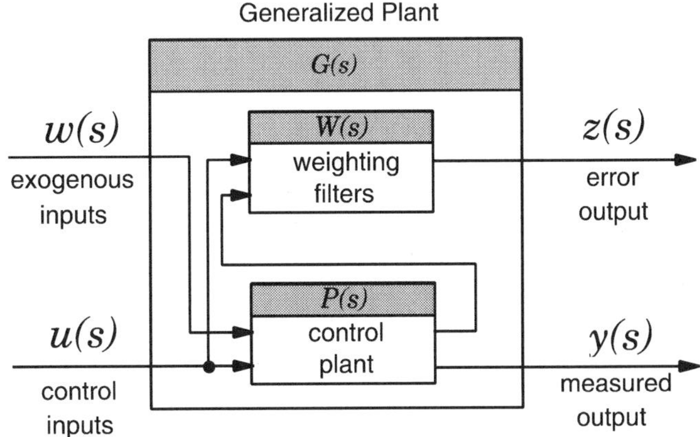


**Figure 8.7:** Schematic diagram of a generalized plant with a frequency shaped filter.

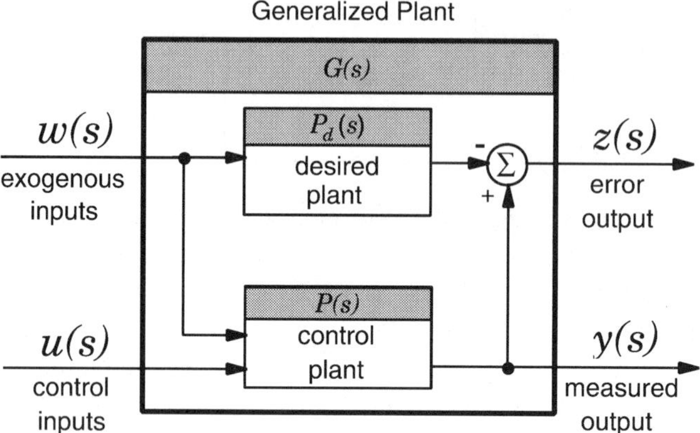


**Figure 8.8:** Schematic diagram of a generalized plant with a frequency shaped filter for servo control system design.

the structural plant $\mathbf{P}(s)$ is coupled to a model of the structural acoustic coupling contained in $\mathbf{W}(s)$ to generate the desired performance measure from the outputs of $\mathbf{W}(s)$, as detailed by Baumann et al. (1991), and later by Clark and Cox (1996).

The format presented is not the only format for creating the generalized plant. In general, one can weight the exogenous inputs for model-reference, servo control system design, as illustrated in Figure 8.8. The desired plant $\mathbf{P}_d(s)$ is modeled and the output of this plant is subtracted from the output of $\mathbf{P}(s)$. The performance measure is thus the difference in the desired response of the plant and the actual plant. The resulting control system is designed to minimize this performance measure. In general, one can incorporate many alternative frequency shaped filters to design controllers for adaptive structures. In fact, the frequency shaped filter used to construct the performance variables used in the cost functional provides a direct measure of the designer's understanding of the particular application physics. However, in the design of dynamic compensators (i.e., a compensator that is a function of frequency), the resulting compensator is frequently of the same order as that of the generalized plant. Thus, the designer must achieve a balance between including the essential physics required to achieve the desired performance objectives while not excessively increasing the order of the generalized plant to the extent that it is impractical to realize the controller (i.e., the order of the compensator is too large). In general, the choice of frequency weighted filters is limited only by the creativity and practical insight of the designer. Once the model is constructed for design, one must choose the type of control to implement—feedback, feedforward, adaptive feedforward, or hybrid.

For feedback control, the general form of the model is presented in Figure 8.9. As illustrated, the measured outputs are coupled to the control inputs through some form of static or dynamic compensator. In general, the cost func-

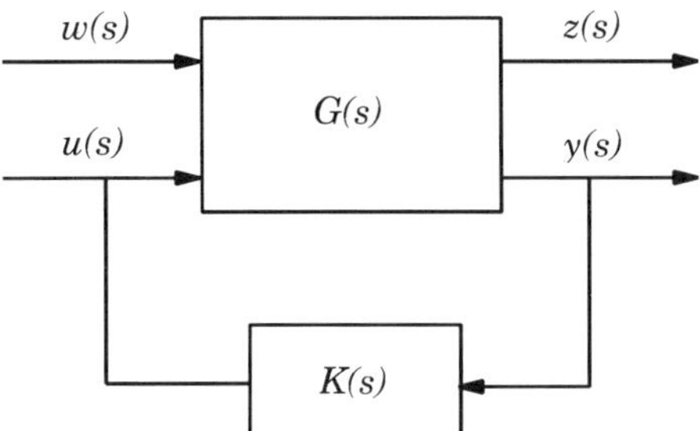


**Figure 8.9:** Schematic diagram of the generalized plant with a feedback compensator.

tional is constructed from some system norm associated with the exogenous inputs $\mathbf{w}(s)$ and the error signal outputs $\mathbf{z}(s)$. For example, one can choose to design a compensator to minimize $\|\mathbf{T}_{zw}(j\omega)\|_2$, which is a measure of the closed-loop, system $\mathcal{H}_2$ norm between $\mathbf{w}(s)$ and $\mathbf{z}(s)$. In the design of adaptive feedforward control systems, one can use the same terminology; however, it is typical to choose the measured outputs $\mathbf{y}(s)$ of the plant in adaptive feedforward control such that they are the same as the error signal outputs $\mathbf{z}(s)$. Since the adaptive feedforward control system is typically designed "on-line," if a frequency shaped output is required, one must construct analog or digital filters to incorporate the desired weighting and the outputs of the generalized plant must be generated "on-line" for control system implementation. As illustrated in Figure 8.10, one assumes that an uncontrollable reference signal (unaffected by the control inputs), correlated with the exogenous inputs, is available for control implementation. However, as illustrated, a measure of the disturbance itself is rarely available. It is usually possible to find a related signal that includes

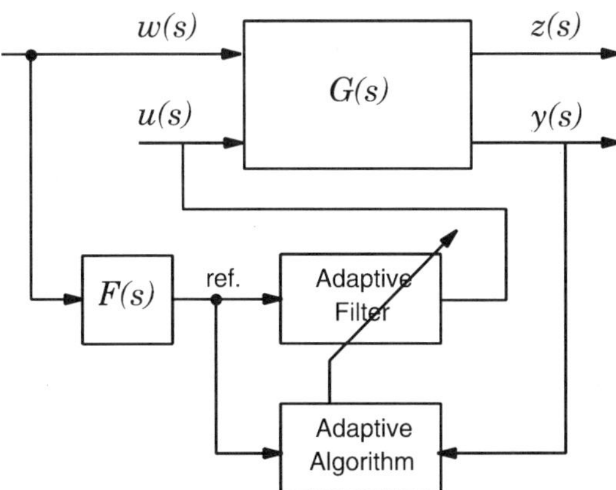


**Figure 8.10:** Schematic diagram of the generalized plant and adaptive feedforward control.

additional dynamics $\mathbf{F}(s)$. For the control of periodic signals, these additional dynamics do not pose a significant problem, as they only affect the relative phase and magnitude of the adaptive filter. Hybrid control (the combination of adaptive feedforward control and feedback control) results from the integration of the control system implementations presented in Figure 8.9 and 8.10 and was previously illustrated in Figure 8.5.

## 8.4 GLOBAL VERSUS LOCAL CONTROL OBJECTIVES: COST SELECTION

Before choosing sensors, actuators, or control system architectures, one must formulate the control system design objectives. These objectives can be the result of performance constraints imposed by a customer or self-imposed constraints based upon the performance requirements of a particular application. Once design constraints are formulated, one must then select transducers that are capable of measuring the appropriate response of the adaptive structure and actuators capable of coupling to the structure to influence its behavior in a manner that will result in the desired performance. Sometimes the sensed variables are identical to the performance variables, which is particularly true when control system design objectives are *local*. However, if the objective is to dissipate energy from the system or modify the input impedance as seen by exogenous disturbances, then a *global* control strategy is employed. Local control can be achieved with feedback, feedforward, or adaptive feedforward control; however, global control requires feedback for global modification of the system dynamics.

To convey the difference between global and local control, consider a reverberant plant; for the purpose of this example, we consider a plate with pinned boundary conditions and configured with three point-force control inputs. The dimensions of the plate, material properties, and location of the transducers are provided in the script file **globloc2.m**, which can be used to produce the results discussed in this section. For this example, a linear quadratic regulator (LQR) controller was designed and implemented based upon two different cost functionals. In the first case, the velocity at the center of the plate and control effort were chosen as the performance metric:

$$\mathbf{z}(t) = [Q^{1/2}\dot{w}(L_x/2, L_y/2, t) \quad R^{1/2}u(t)]^{\mathrm{T}} \tag{8.24}$$

$$= [Q^{1/2}\mathbf{\Phi}^{\mathrm{T}}(L_x/2, L_y/2)\dot{\boldsymbol{\eta}}(t) \quad R^{1/2}u(t)]^{\mathrm{T}} \tag{8.25}$$

where $\dot{w}(x, y, t)$ is the plate response in physical coordinates, $\dot{\boldsymbol{\eta}}(t)$ is the plate response in generalized coordinates, $\mathbf{\Phi}^{\mathrm{T}}(x, y)$ is a vector of the assumed modes shapes evaluated at the appropriate field coordinates, $u(t)$ is the control input, $Q$ is a scalar performance penalty weight, and $R$ is a scalar control effort penalty weight. The controller was designed to minimize the cost that balanced control

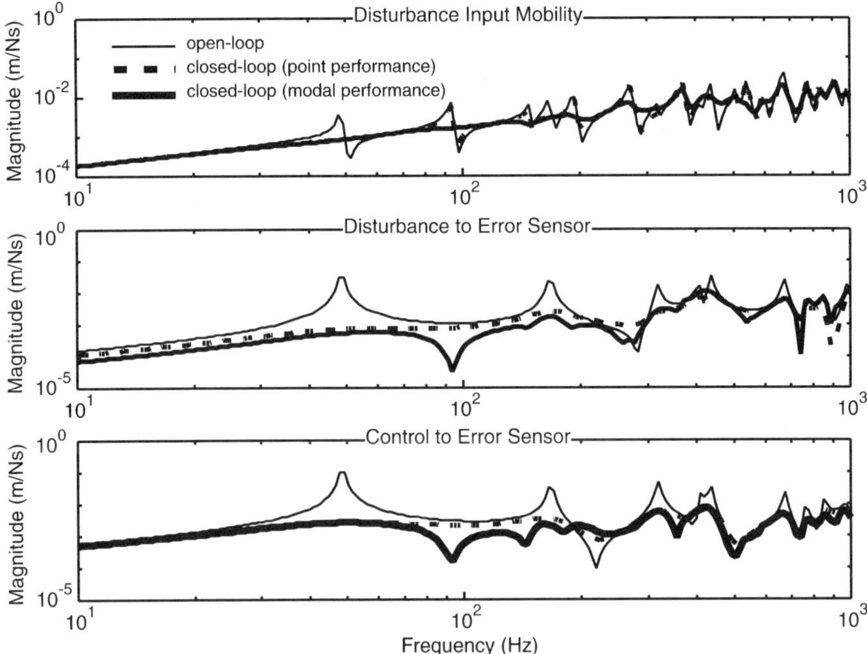


**Figure 8.11:** Comparison of open- and closed-loop frequency responses for a controller designed with a point-velocity performance metric (local control) and a modal-velocity performance metric (global control).

effort against velocity at the chosen structural coordinate. In the second case, the modal velocities of the structure (a measure of the kinetic energy) and control effort were selected as the performance metric:

$$\mathbf{z}(t) = [Q^{1/2}\dot{\boldsymbol{\eta}}^{\mathrm{T}}(t) \quad R^{1/2}u(t)]^{\mathrm{T}}. \tag{8.26}$$

Figure 8.11 shows a comparison of the magnitude of the open- and closed-loop frequency response functions (FRFs) for each test case (for a given choice of $R$ and $Q$ provided in the script file **globloc2.m**); the figure includes the input mobility of a point disturbance force, the mobility between the point disturbance force and a velocity sensor positioned at the center of the plate, and the mobility between the third control input and the velocity sensor at the center of the plate.

As illustrated, the closed-loop responses for both cost functionals are very similar in frequency ranges corresponding to the first, third, fifth, and so on, resonant frequencies of the structure; however, at the second, fourth, sixth, and so on, resonant frequencies, the controller designed using the modal velocities as the performance index provides greater levels of attenuation at the location of the input disturbance (a position other than the center of the plate). However, the closed-loop frequency responses of the disturbance-to-error and control-to-error frequency response functions presented in Figure 8.11 indicate similar levels of attenuation and performance for the two control systems. Why? The error sensor chosen to evaluate the performance is the same error sensor used in the cost functional of the first test case. Because the sensor is centered on the plate, the second, fourth, sixth, and so on, modes of the structure cannot be observed. Thus, the performance of both closed-loop systems appears very similar at the location of this error sensor. However, at locations other than the point of this performance metric, the two closed-loop systems display very different levels of attenuation. Since all of the structural modes are included in the second cost functional, the closed-loop system designed with this cost functional yields a more global effect. In both cases, the poles of the system are modified significantly; however, the objective of the first cost functional is limited to control at a specific point on the structure, whereas the objective of the second cost functional is to control a measure of the structural kinetic energy. Thus, the resulting controller for the second case is capable of affecting the response of structural modes not included in the first cost functional. This does not mean that the first cost functional is a bad choice. If the objective is simply to minimize the response of the structure at its center, then it is pointless to expend energy in structural modes that do not influence the response at this point. However, if the objective is to reduce the structural response over the entire surface of the structure, a more global performance metric is required. Thus, as demonstrated in this example, the performance metric chosen for the cost functional plays a significant role in the spatial influence (global vs. local) of the control system design.

In general, the selection of the cost functional is one of the most important tasks in the design of a control system and typically conveys the designer's understanding of the application physics. In some control approaches, the variables used in the cost functional are the same as the variables measured with the error sensors in the implementation of the controller. However, in many control system implementations, the variables used in the cost functional are not readily accessible through transducers. Hence, such variables, which are associated with *performance*, are used exclusively in the design and synthesis of the controller such that the measured error signals can be appropriately modified through dynamic compensation to generate the control signals necessary to meet performance requirements. In general, the selection of the cost functional provides the designer with a method of:

- Constraining the control system to local or global objectives
- Creating a balance between performance and control effort
- Specifying the bandwidth of desired performance

For example, the cost functional can be chosen such that the displacement or velocity of the structure measured at a finite set of locations is minimized. For typical structures, this results in optimal performance at the chosen locations with no guarantees over the controlled system response at other surface locations of the structure. However, if the modal velocities or displacements are included in the cost functional, then the controller can be designed for global performance.

In the following sections, various cost functionals are formulated for the design of optimal controllers. (*Remember, optimal controllers are only optimal for the specified cost functional.*) Both $\mathcal{H}_2$ and $\mathcal{H}_\infty$ controllers are considered, as well as adaptive, feedforward controllers. Depending upon the chosen performance variables, all of these controllers can be designed for optimal performance locally or globally. The effect of the chosen cost functional on control system performance is demonstrated by example in Chapter 9.

## 8.5 FEEDFORWARD CONTROL ALTERNATIVES AND ADAPTIVE ALGORITHMS

In this section, we discuss the implementation of feedforward control concepts introduced in Chapter 7, including open-loop control (nonadaptive) and two adaptive feedforward control algorithms: the filtered-$x$ LMS algorithm and the time-averaged gradient algorithm. The general two-port formulation for feedforward control was introduced in Section 8.3 and is illustrated again in Figure 8.10. However, the adaptive methods require us to work in the z-domain, as the sampled-data adaptation schemes are almost exclusively implemented using digital signal processing hardware. A schematic for the discrete-time version of the generalized plant with adaptive feedforward control is shown in Figure 8.12. Feedforward control system design and implementation rely upon digital filtering fundamentals and adaptive filter theory presented previously in Chapter 4. In this section, fundamentals of adaptive feedforward control are presented

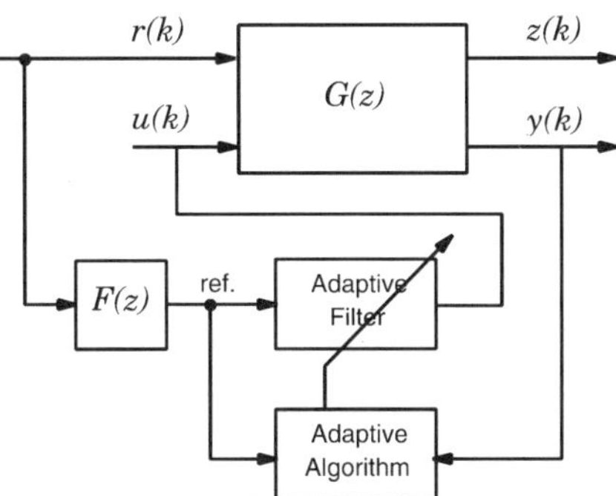


**Figure 8.12:** Schematic diagram of the discrete-time generalized plant for feedforward control.

mostly in the context of adaptive structures, so the presentation is not intended as an exhaustive discussion of feedforward control algorithms. For a more in-depth discussion of adaptive filter theories that underly adaptive feedforward control, the reader is referred to excellent texts by Widrow and Stearns (1985), and Haykin (1991). In addition, the application of adaptive filtering for active noise control systems is also discussed by Kuo and Morgan (1996).

### 8.5.1 Open-Loop Feedforward Control: Fixed-Filter Algorithms

Fixed-filter algorithms refer to a class of feedforward implementations that are not adaptive. They are implemented using a standard feedforward topology, as previously illustrated in Figure 8.12; however, there is no mechanism for adapting the filter coefficients in real-time. The filter coefficients are determined *a priori* using knowledge of an analytical or experimental model of the generalized plant $G(z)$. The fixed-filter algorithms utilize the Wiener filter presented in Section 4.5.1, which is shown for the generalized plant in Figure 8.13. For purposes of this discussion, we limit ourselves to the case where $F(z)$ is simply unity, and therefore the reference input $r(k)$ is identical to the system disturbance. For periodic disturbances, this is a very reasonable assumption if a reference signal correlated to the periodic content of the disturbance is available or can be constructed. A second (obvious) requirement is that the reference signal must not be affected by the control input $u(k)$. We find that this requirement for an *uncontrollable reference* is common for all adaptive structures that utilize true feedforward control.

A detailed version of the fixed-filter feedforward control system is presented in Figure 8.14. The problem is similar to the system identification filter topology for the Wiener filter shown previously in Figure 4.11. In this case, the generalized plant $P(z)$ has dynamics between the control input $u(k)$ and the system output $y(k)$. The error function for the fixed-gain problem is equal to the output $y(k)$. The cost function is selected to be the mean-squared error, defined as follows:

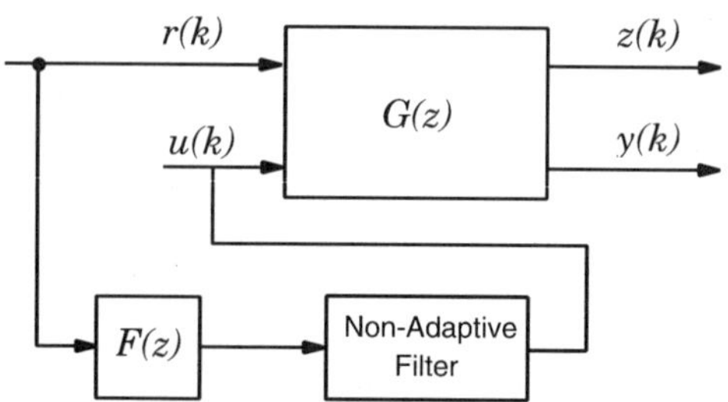


**Figure 8.13:** Schematic diagram of the discrete-time generalized plant for fixed-filter feedforward control.

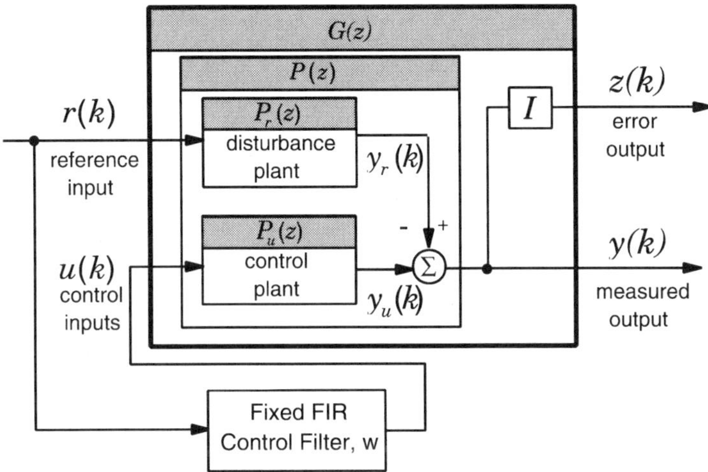


**Figure 8.14:** Detailed schematic diagram of the generalized plant (discrete) for fixed-filter feedforward control.

$$\xi = E[y^2(k)] \tag{8.27}$$

where

$$y(k) = y_r(k) - y_u(k) \tag{8.28}$$

where $y_r(k)$ is the plant response due to the disturbance and $y_u(k)$ is the plant response due to the control input. The determination of the optimal filter weights proceeds in a fashion similar to that shown previously in Section 4.5.1, where the error function is the system output $y(k)$. In vector form, the mean-squared error can be expressed as follows:

$$\xi = \mathbf{b}^{\mathrm{T}}\mathbf{R}\mathbf{b} + \mathbf{w}^{\mathrm{T}}\mathbf{D}\mathbf{w} + 2\mathbf{w}^{\mathrm{T}}\mathbf{C} \tag{8.29}$$

where $\mathbf{R}$ is the input correlation matrix as defined in Section 4.5.1, $\mathbf{b}$ is a vector of the impulse response of the disturbance-to-error path $\mathbf{P_r(z)}$, $\mathbf{D}$ represents the correlation matrix of the reference signals filtered through the control path $\mathbf{P_u(z)}$, $\mathbf{w}$ is the vector of filter coefficients, and $\mathbf{C}$ is the cross-correlation vector between $y_u(k)$ (with the control filter set to 1) and the desired response at the error $y_r(k)$.

The optimal solution is obtained by taking the derivative of the mean-squared error with respect to the weight vector and setting the vector equations equal to zero:

$$\frac{\partial \xi}{\partial \mathbf{w}} = 2[\mathbf{D}]\mathbf{w} + 2\mathbf{C} = 0 \tag{8.30}$$

Solving for the optimal solution, $\mathbf{w_{opt}}$, one obtains

$$\mathbf{w_{opt}} = -\mathbf{D}^{-1}\mathbf{C} \tag{8.31}$$

Thus, the minimum mean-squared error is found by substitution of the resultant $\mathbf{w_{opt}}$ into equation (8.29), which is expressed as follows:

$$\xi_{\min} = \mathbf{b}^{\mathrm{T}}\mathbf{R}\mathbf{b} + \mathbf{C}^{\mathrm{T}}\mathbf{D}^{-1}\mathbf{C} \tag{8.32}$$

As can be seen, the minimum mean-squared error is controlled by the inverse of $\mathbf{D}$. The system is implemented by using the optimal control filter $\mathbf{w_{opt}}$ in the topology shown previously in Figure 8.14, which produces the minimum possible $E[y^2(k)]$.

### 8.5.2 Adaptive Feedforward Control

The increase in computational capabilities of modern digital signal processing systems has permitted the real-time implementation (at high bandwidths) of adaptive filter algorithms developed in the 1960s and 1970s. These algorithms provide a control system that can adapt to changes in the plant $P(z)$ and provide outstanding disturbance rejection for certain classes of inputs. Algorithms presented in this section include the filtered-$x$ LMS algorithm and the TAG algorithm. The fundamental topology of these algorithms is presented in the following sections. Further discussion of adaptive filters and feedforward adaptive noise control systems can be found in the texts by Widrow and Stearns (1985), Haykin (1991), and Kuo and Morgan (1996).

***Filtered-x LMS Algorithm—Single Channel*** The filtered-$x$ LMS algorithm is the most widely accepted feedforward control algorithm because of its ease of implementation and remarkable performance. Consider the schematic of the feedforward control system shown previously in Figure 8.12. For the purpose of this discussion, we will limit ourselves to cases where $F(z)$ equals unity (no delay) and where the control filter is an FIR filter for reasons outlined previously. A detailed schematic of this control system is shown in Figure 8.15. In a fashion similar to that discussed in Chapter 4, the output of the control filter can be expressed in terms of the filter coefficients $w_n$ and the reference signal $r(k-n)$:

$$u(k) = \sum_{n=0}^{N} w_n r(k-n) \tag{8.33}$$

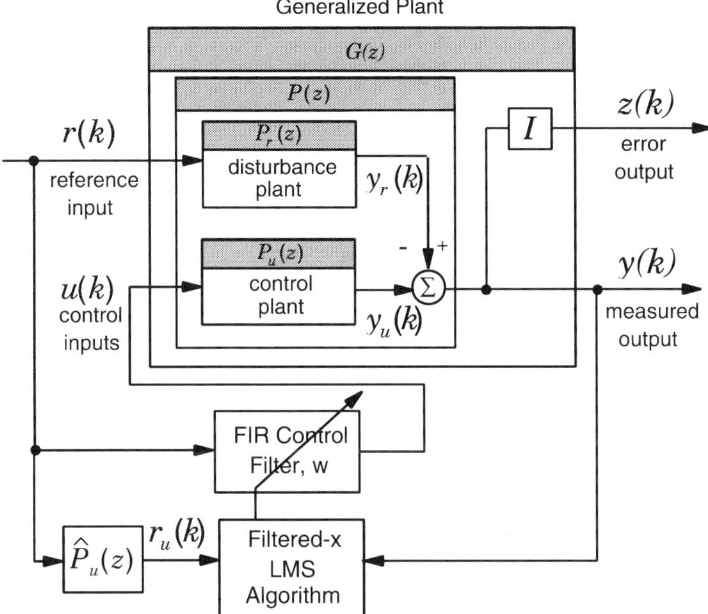


**Figure 8.15:** Detailed schematic diagram of the discrete-time generalized plant for filtered-$x$ LMS feedforward control.

The output of the plant due to the control inputs only can be computed as follows:

$$y_u(k) = \sum_{l=0}^{L} h_l \sum_{n=0}^{N} w_n r(k-n-l) \tag{8.34}$$

where $h_l$ is the discrete impulse response of the control input-to-output path, which is assumed to be of order $L$. The net output of the system can be written as follows:

$$y(k) = y_r(k) - \sum_{l=0}^{L} h_l \sum_{n=0}^{N} w_n r(k-n-l) \tag{8.35}$$

The order of convolution can be interchanged without changing the result, as follows:

$$y(k) = y_r(k) - \sum_{n=0}^{N} w_n \sum_{l=0}^{L} h_l r(k-n-l) \tag{8.36}$$

which can be further rewritten as:

$$y(k) = y_r(k) - \sum_{n=0}^{N} w_n r_u(k-n) \tag{8.37}$$

where:

$$r_u(k-n) = \sum_{l=0}^{L} h_l r(k-n-l) \tag{8.38}$$

By rearranging the convolution, a signal $r_u$ is created, which is to be estimated by the filtered-$x$ operation. (Note that the term "filtered-$x$" originates from adaptive filter theory, where $x$ is assumed to be a symbol for the reference signal. Due to conflicts in terminology, $r$ is used as the reference herein, but the algorithm name was left as in the original presentation.) If the true impulse response of the plant $P_u(z)$ is estimated by an FIR filter of coefficients $\hat{h}_l$, then an estimate to the resultant $r_u(k)$ is given as $\hat{r}_u(k)$ in Figure 8.15. The cost function can be expressed as:

$$\xi = E[y^2(k)] \tag{8.39}$$

For the LMS algorithm (Widrow and Stearns, 1985), the instantaneous value of $y^2(k)$ is used as an estimate of the expected value of the $\xi$. We can thus write an estimate to the gradient of $\xi$ as

$$\hat{\nabla}(\xi(k)) = -2r_u(k)y(k) \tag{8.40}$$

Now we write the controller weights in vector form as

$$\mathbf{w} = [w_0, w_1, \ldots, w_{(N-1)}]^T \tag{8.41}$$

The controller weights can be adapted as previously presented in Section 4.5.2, which produces the filtered-$x$ LMS algorithm as follows:

$$w(k+1) = w(k) + \mu r_u(k)y(k) \tag{8.42}$$

Bounds for the value of the convergence parameter $\mu$ of the LMS algorithm have been presented in Section 4.5.4. The bounds are controlled by two factors. First, the maximum value is limited such that the filter weights and mean-squared error converge. Second, the value of the convergence parameter $\mu$ controls the misadjustment $M$ of the mean-squared error. The bounds set on the convergence parameter $\mu$ are dependent on properties of the input correlation matrix, $R$. For the filtered-$x$ LMS algorithm, the bounds of $\mu$ are controlled by the reference input *filtered* through the estimate $\hat{P}_u(z)$ of the control-to-error plant $P_u(z)$. This filtered input correlation matrix is defined to be $D$. Thus, the bounds of the convergence parameter $\mu$ are given by

$$0 < \mu < \frac{2}{\lambda_{D,\max}} \tag{8.43}$$

where $\lambda_{D,\max}$ is the maximum eigenvalue of the filtered input correlation matrix $D$. Equation (8.43) provides the necessary and sufficient condition for the convergence of the weight vector. Additionally, the necessary and sufficient condition for the convergence of the mean-squared error is given by

$$\sum_{n=1}^{N} \frac{\mu\lambda_{D,n}}{2 - \mu\lambda_{D,n}} < 1 \tag{8.44}$$

where $N$ is the order of the adaptive filter and $\lambda_{D,n}$ is the $n$th eigenvalue of the filtered input correlation matrix $D$. Equations (8.43) and (8.44) constitute the necessary and sufficient conditions for the convergence of the mean-squared error and the control filter weights. In a fashion similar to the derivation presented in Section 4.5.4, the misadjustment can be approximated by the filtered input correlation with zero time lag, as follows:

$$M \approx \frac{\mu I r_{\hat{r}_u \hat{r}_u}}{2} \tag{8.45}$$

The time constant of the filtered-$x$ LMS algorithm can be approximated as follows:

$$\tau_{av,\xi} \approx \frac{1}{2\mu I r_{\hat{r}_u \hat{r}_u}} \tag{8.46}$$

The convergence rate of the filtered-$x$ LMS algorithm is governed by the reference input filtered through the estimate of the control-to-error plant $\hat{P}_u(z)$. As discussed later in this chapter, the resultant bounds on $\mu$ dictate the convergence of the feedforward portion of a hybrid control system. Feedback control can be used to modify the system dynamics (through matrix $D$) such that the feedforward system converges more rapidly.

***Filtered-x LMS Algorithm—Multichannel*** In this section, the filtered-$x$ LMS algorithm is expanded to encompass the configuration of multiple control inputs and multiple error outputs. The algorithm was first presented by Elliott et al. (1987) as an expansion of earlier work by Widrow and Sterns (1985). The discussion is presented in matrix form following the work by Kuo and Morgan (1996) for a single-channel reference case. The algorithm can be readily expanded to multiple references as detailed in Kuo and Morgan (1996).

In the schematic diagram presented in Figure 8.15, the error output sensors now consist of M outputs arranged in a vector, $\mathbf{y}(k)$, of length M as follows:

$$\mathbf{y}(k) = [y_1(k), y_2(k), \ \ldots, \ y_M(k)]^{\mathrm{T}} \tag{8.47}$$

The control filters consist of an array of $P$ actuators (or control inputs), each of order $N$. These aggregate control weights are arranged in one vector, as follows:

$$\mathbf{w}(k) = [\mathbf{w}_1^{\mathrm{T}}(k), \mathbf{w}_2^{\mathrm{T}}(k), \ \ldots, \ \mathbf{w}_\mathbf{P}^{\mathrm{T}}(k)]^{\mathrm{T}} \tag{8.48}$$

where

$$\mathbf{w}_p(k) = [w_{p,1}(k), w_{p,2}(k), \ \ldots, \ w_{p,N}(k)]^{\mathrm{T}}, \qquad p = 1, 2, \ \ldots, \ P \tag{8.49}$$

The outputs of the control filters are the control inputs to the generalized plant, and they are presented as follows:

$$\mathbf{u}(k) = [u_1(k), u_2(k), \ \ldots, \ u_P(k)]^{\mathrm{T}} \tag{8.50}$$

The output of the signal $u_p(k)$ is found by filtering the reference signal through the corresponding FIR control filter $\mathbf{w}_p(k)$ as follows:

$$u_p(k) = \mathbf{w}_p^{\mathrm{T}}(k)\mathbf{r}(k) \tag{8.51}$$

where the single reference signal is

$$\mathbf{r}(k) = [r(k), r(k-1), \ \ldots, \ r(k-N+1)]^{\mathrm{T}} \tag{8.52}$$

The vector of the current and past reference signal values can be arranged to form a $PN \times P$ matrix as follows:

$$\hat{\mathbf{R}}(k) = \begin{bmatrix} \mathbf{r}(k) & \mathbf{0} & \cdots & \mathbf{0} \\ \mathbf{0} & \mathbf{r}(k) & \mathbf{0} & \vdots \\ \vdots & \mathbf{0} & \ddots & \mathbf{0} \\ \mathbf{0} & \cdots & \mathbf{0} & \mathbf{r}(k) \end{bmatrix} \tag{8.53}$$

Note that this matrix is block diagonal with column vectors $\mathbf{r}(k)$ along the diagonal. Note also that this matrix should not be confused with the input (reference) correlation matrix $\mathbf{R}$, defined previously. Thus the outputs of the control filters can be written in matrix form as

$$\mathbf{u}(k) = \hat{\mathbf{R}}^{\mathbf{T}}(k) * \mathbf{w}(k) \tag{8.54}$$

where $*$ denotes convolution. The response at the error sensors due to the control inputs is contained in the vector $\mathbf{y}_{\mathbf{u}}(k)$, which is found by the convolution of the appropriate control-to-output transfer functions as follows:

$$\mathbf{y}_u(k) = \mathbf{P}_u(k) * \mathbf{u}(k) \tag{8.55}$$

$\mathbf{P}_u(k)$ is an $M \times P$ matrix of the control to output impulse response functions. The matrix $\mathbf{P}_u(k)$ is of the form

$$\mathbf{P}_u(k) = \begin{bmatrix} \mathbf{p}_{u11}(k) & \mathbf{p}_{u12}(k) & \cdots & \mathbf{p}_{u1P}(k) \\ \mathbf{p}_{u21}(k) & \mathbf{p}_{u22}(k) & \cdots & \mathbf{p}_{u2P}(k) \\ \vdots & \vdots & \ddots & \vdots \\ \mathbf{p}_{uM1}(k) & \mathbf{p}_{uM2}(k) & \cdots & \mathbf{p}_{uMP}(k) \end{bmatrix} \tag{8.56}$$

The error signal (system output) is expressed as the difference between the response due to the primary disturbance $\mathbf{y}_{\mathbf{r}}(k)$ and that due to the secondary control inputs $\mathbf{y}_{\mathbf{u}}(k)$:

$$\mathbf{y}(k) = \mathbf{y}_r(k) - \mathbf{y}_u(k) \tag{8.57}$$

Equation (8.55) can be substituted into equation (8.57), producing the output relation

$$\mathbf{y}(k) = \mathbf{y}_r(k) - [\mathbf{P}_u(k) * \mathbf{u}(k)] \tag{8.58}$$

After substituting equation (8.54), the output becomes

$$\mathbf{y}(k) = \mathbf{y}_r(k) - \mathbf{P}_u(k) * [\hat{\mathbf{R}}^{\mathrm{T}}(k) * \mathbf{w}(k)] \tag{8.59}$$

For multiple output signals, the cost function is defined to be the sum of the expected values of the squared error signals:

$$\xi = \sum_{m=0}^{M} \mathrm{E}[y_m^2(k)] \tag{8.60}$$

In a fashion similar to that used in the single-channel filtered-$x$ LMS algorithm, the expected value of the mean-square error cost function is approximated by the instantaneous value, and thus equation (8.60) is approximated as

$$\hat{\xi} = \sum_{m=0}^{M} y_m^2(k) \tag{8.61}$$

The control filters are updated in a fashion similar to that for the SISO case such that the weights are adapted in the direction of the negative of the instantaneous gradient:

$$\mathbf{w}(k+1) = \mathbf{w}(k) - \frac{\mu}{2}\,\hat{\nabla}(\xi(k)) \tag{8.62}$$

The gradient is estimated, again in a fashion similar to that for the SISO case, producing the following relation:

$$\hat{\nabla}(\xi(k)) = -2 \begin{bmatrix} \hat{p}_{u11}(k) * \mathbf{r}(k) & \hat{p}_{u21}(k) * \mathbf{r}(k) & \cdots & \hat{p}_{uM1}(k) * \mathbf{r}(k) \\ \hat{p}_{u12}(k) * \mathbf{r}(k) & \hat{p}_{u22}(k) * \mathbf{r}(k) & \cdots & \hat{p}_{uM2}(k) * \mathbf{r}(k) \\ \vdots & \vdots & \ddots & \vdots \\ \hat{p}_{u1K}(k) * \mathbf{r}(k) & \hat{p}_{u2K}(k) * \mathbf{r}(k) & \cdots & \hat{p}_{uMK}(k) * \mathbf{r}(k) \end{bmatrix} \mathbf{y}(k) \tag{8.63}$$

$$= -2\,[\hat{\mathbf{P}}_{\mathbf{u}}^{\mathbf{T}}(k) \otimes \mathbf{r}(k)]\,\mathbf{y}(k) \tag{8.64}$$

where $\hat{p}_{uij}$ are the FIR estimates of the impulse response functions between the $i$th error sensor from the $j$th actuator and $\otimes$ denotes a Kronecker product convolution obtained by convolving each element $\hat{p}_{uij}$ with $\mathbf{r}(k)$. The response function of the filtered reference signal is given by

$$\hat{r}_{uij}(k) = \hat{p}_{uij}(k) * \mathbf{r}(k) \tag{8.65}$$

The matrix of filtered reference signals can be written separately as

$$\hat{\mathbf{r}}_{\mathbf{u}} = -2 \begin{bmatrix} \hat{r}_{u11}(k) & \hat{r}_{u21}(k) & \cdots & \hat{r}_{uM1}(k) \\ \hat{r}_{u12}(k) & \hat{r}_{u22}(k) & \cdots & \hat{r}_{uM2}(k) \\ \vdots & \vdots & \ddots & \vdots \\ \hat{r}_{u1K}(k) & \hat{r}_{u2K}(k) & \cdots & \hat{r}_{uMK}(k) \end{bmatrix} \tag{8.66}$$

Thus, the gradient of equation (8.64) can be written as follows:

$$\hat{\nabla}(\xi(k)) = -2 \begin{bmatrix} \hat{r}_{u11}(k) & \hat{r}_{u21}(k) & \cdots & \hat{r}_{uM1}(k) \\ \hat{r}_{u12}(k) & \hat{r}_{u22}(k) & \cdots & \hat{r}_{uM2}(k) \\ \vdots & \vdots & \ddots & \vdots \\ \hat{r}_{u1K}(k) & \hat{r}_{u2K}(k) & \cdots & \hat{r}_{uMK}(k) \end{bmatrix} \mathbf{y}(k) \tag{8.67}$$

Combining equations (8.66) and (8.67) with equation (8.62) produces the MIMO filtered-$x$ LMS algorithm:

$$\mathbf{w}(k+1) = \mathbf{w}(k) + \mu \hat{\mathbf{r}}_u(k)\mathbf{y}(k) \tag{8.68}$$

As indicated by equation (8.68), the update for the filter coefficients is obtained from current filter coefficients $\mathbf{w}(k)$, a *filtered* reference signal $\hat{\mathbf{r}}_u(k)$, and the output error signal $\mathbf{y}(k)$. Once the filters representing the control-to-error transfer function path are obtained, the implementation of this algorithm is quite efficient.

***TAG Descent Algorithm—Single Channel*** The filtered-$x$ LMS algorithm and its derivatives have been used extensively in a wide variety of active control applications. The algorithm has an inherent drawback because of the required control-to-error transfer function model. This model is typically obtained off-line and thus is not adaptive to changes in the control-to-error plant $P_u(z)$. An alternative algorithm, based on the gradient search using Newton's method (Widrow and Stearns, 1985) and termed the time-averaged gradient (TAG) algorithm, is presented in this section. The schematic of the TAG algorithm is shown in Figure 8.16. As can be seen from the figure, the updated algorithm only requires input from the error signal $y(k)$. This algorithm is adaptive and does not require an off-line model of the plant $P(z)$. The TAG algorithm follows closely the work of Gibbs and Clark (1993) and Kewley et al. (1995). In their work, a modified version, termed the harmonic TAG algorithm, was studied. This algorithm allows for the control of multiple tones that are harmonics of a fundamental tone. The algorithm only requires a reference signal from the fundamental tone and the harmonics are generated internally using trigonometric relations. This section introduces the TAG algorithm for SISO cases. The algorithm is then expanded to the MIMO and harmonic reference cases.

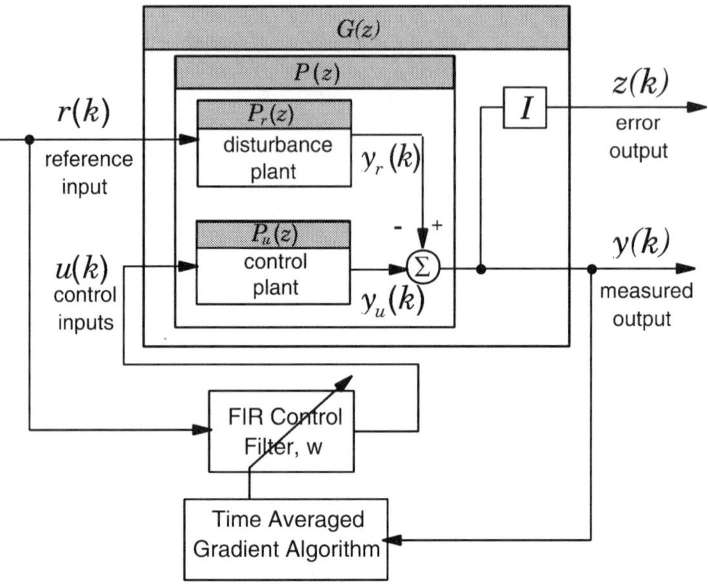


**Figure 8.16:** Detailed schematic diagram of the discrete-time generalized plant for TAG feedforward control.

First, Newton's method as applied to control filter updates is reviewed. With the use of Newton's method (Widrow and Stearns, 1985), the control filter weights are adapted as follows:

$$w_n(k+1) = w_n(k) - \frac{\xi'(w_n(k))}{\xi''(w_n(k))} \tag{8.69}$$

where $\xi'(w_n(k))$ and $\xi''(w_n(k))$ are the first and second derivatives of the cost function $\xi(w_n(k))$. Theoretically, if the cost function is quadratic, then the weight vector converges to the solution in a single iteration. Furthermore, if the weights are orthogonal, then the update of one weight is independent of all other weights. Under these conditions, the optimal solution can be achieved by updating the entire weight vector in one iteration.

Implementation of this algorithm requires an estimate of the first and second derivatives of the cost function. One technique previously presented by Gibbs and Clark (1993) uses central finite difference relations to estimate these derivatives:

$$\xi'(w_n) \approx \frac{\xi(w_n + \Delta w_n) - \xi(w_n - \Delta w_n)}{2\Delta w_n} \tag{8.70}$$

and

$$\xi''(w_n) \approx \frac{\xi(w_n - \Delta w_n) - 2\xi(w_n) + \xi(w_n + \Delta w_n)}{(\Delta w_n)^2} \tag{8.71}$$

The mean-squared error for this technique is estimated by time averaging as follows:

$$\xi = E[y^2(k)] = \frac{1}{N_s N_p} \sum_{k=1}^{N_s N_p} y^2(k) \tag{8.72}$$

where $N_s$ is the number of samples per cycle for a periodic signal and $N_p$ is the number of periods of a periodic signal. In essence, the error function is averaged over a sufficient length of time such that the result is a sufficiently accurate estimate of the expected value of the signal. In the absence of noise, the cost function can be estimated in one period for a periodic signal that is synchronously sampled. A higher sampling rate and larger number of cycles are necessary for systems with increased noise levels. Approximating the mean-squared error by averaging a large data set minimizes the effect of noise on the estimation. Substitution of equations (8.70), (8.71), and (8.72) into equation (8.69) produces the following update equation:

$$w_n(k+1) = w_n(k) - \frac{\Delta w(\Sigma_k[y_k^2(w_k + \Delta w_k)] - \Sigma_k[y_k^2(w_k - \Delta w_k)])}{2(\Sigma_k[y_k^2(w_k + \Delta w_k)] - 2\Sigma_k[y_k^2(w_k)] + \Sigma_k[y_k^2(w_k - \Delta w_k)])} \tag{8.73}$$

Under certain circumstances, it may be desirable to reduce the amount by which the weight is adapted based upon Newton's method. This can be accomplished by incorporating an adaptation parameter $\mu$, as follows:

$$w_n(k+1) = w_n(k) - \mu \frac{\Delta w(\Sigma_k[y_k^2(w_k + \Delta w_k)] - \Sigma_k[y_k^2(w_k - \Delta w_k)])}{2(\Sigma_k[y_k^2(w_k + \Delta w_k)] - 2\Sigma_k[y_k^2(w_k)] + \Sigma_k[y_k^2(w_k - \Delta w_k)])} \tag{8.74}$$

where $\mu$ can range from zero to one. With $\mu$ equal to one, the system updates the control filter weights based on the standard TAG method.

The adaptation process can be briefly outlined as follows: First, the system is initiated with an initial guess for the optimal control filter (usually zero), and the cost function is estimated through time averaging. Second, a single filter coefficient is perturbed an amount $\Delta w$, and the cost function is estimated through time averaging. The weight is then perturbed an amount $-\Delta w$ from the original state, and the cost function is estimated through time averaging. Then the new weight value is determined from equation (8.73), and the procedure is repeated for each additional weight. Finally, the entire process is repeated to provide convergence and maintain adaptability.

***TAG Descent Algorithm—Multichannel*** The conventional TAG algorithm updates each coefficient independently and is thus easily adapted to a multichannel configuration where each control filter coefficient for each channel is updated independently. The major difference is in the estimate of the mean-squared error cost function. The cost function for the MIMO case is given by

$$\xi = \mathrm{E}\left[\sum_{m=1}^{M} y_m^2(k)\right] = \frac{1}{N_s N_p} \sum_{k=1}^{N_s N_p} \sum_{m=1}^{M} y_m^2(k) \tag{8.75}$$

The update equation (8.74) is thus modified to include the multiple error sensors, becoming

$$w_n(k+1) = w_n(k) - \mu \frac{\Delta w(\Sigma_m \Sigma_k [y_k^2(w_k + \Delta w_k)] - \Sigma_m \Sigma_k [y_k^2(w_k - \Delta w_k)])}{2(\Sigma_m \Sigma_k [y_k^2(w_k + \Delta w_k)] - 2\Sigma_m \Sigma_k [y_k^2(w_k)] + \Sigma_m \Sigma_k [y_k^2(w_k - \Delta w_k)])} \tag{8.76}$$

where $\Sigma_m$ is included to allow for the summation across the error sensors. As discussed in Chapter 4, if the sample rate is chosen such that the number of samples per period, $N_s$, for a sinusoidal reference is four, then the principal axes of the resultant mean-squared error surface $\xi$ are along the weight axes, which corresponds to a decoupled system. Thus, each filter coefficient's optimal value can be determined independently of the other filter coefficients. The TAG algorithm benefits from the lack of system modeling requirements but usually suffers from slower convergence rates compared to the filtered-$x$ LMS algorithm.

## 8.6 FEEDBACK CONTROL ALTERNATIVES: DESIGN AND SYNTHESIS

In Section 8.3, the two-port model was introduced in terms of the generalized plant $\mathbf{G}(s)$, which incorporates both the plant $\mathbf{P}(s)$ and any additional static or dynamic weights required in the design process to produce a compensator that yields the desired closed-loop performance. In this section, we discuss methods of designing static and dynamic compensators for the realization of feedback control systems. Additionally, the linear quadratic regulator (LQR) and the linear quadratic Gaussian (LQG) problems originally presented in Chapter 3, will be cast from the two-port model and are shown to be subsets of the more general $\mathcal{H}_2$ controller design problem.

### 8.6.1 Static Compensation

Static compensation is the feedback control law resulting when a constant-gain matrix is used to couple the system outputs to the system inputs for the general MIMO control case. Both output-feedback control and state-feedback control fall under this categorization. However, it is generally recognized that one rarely has access to a measure of all system states in control, and thus dynamic compensation is typically required due to the implementation of a state estimator. The discussion in this section is limited to static compensation for both direct, output-feedback control and state-feedback control. In general, both problems can be cast in terms of an optimal LQR problem that seeks to minimize a linear quadratic cost function balancing performance with control effort.

***The $\mathcal{H}_2$ Controller Design for State-Feedback Control*** The LQR problem has been introduced earlier in Chapter 3. In this section, following the work of Boyd and Barratt (1991), we demonstrate that the LQR problem can be cast from the more general $\mathcal{H}_2$ control problem by appropriately defining our generalized plant model. In general, the system is described in terms of the state equations

$$\dot{\mathbf{x}}(t) = \mathbf{A}\mathbf{x}(t) + \mathbf{B}_u\mathbf{u}(t) + \mathbf{w}(t) \tag{8.77}$$

where $\mathbf{w}(t)$ is a zero-mean, white-noise process such that the spectral density matrix $\mathbf{S}_{ww}(j\omega) = \mathbf{I}$ for all $\omega$. We further assume that all states are observable and that $\mathbf{u}(t) = \mathbf{K}\mathbf{x}(t)$. Given the dynamic constraint of equation (8.77), we seek to find the control law that minimizes the following cost functional:

$$J = \lim_{t \to \infty} \mathbf{E}(\mathbf{x}^{\mathrm{T}}(t)\mathbf{Q}\mathbf{x}(t) + \mathbf{u}^{\mathrm{T}}(t)\mathbf{R}\mathbf{u}(t)) \tag{8.78}$$

where $\mathbf{Q}$ is a positive semidefinite weighting matrix on performance and $\mathbf{R}$ is a positive definite weighting matrix on control effort. As outlined by Boyd and Barrat (1991), this LQR problem can be cast from the more general $\mathcal{H}_2$ control design by appropriate choice of the output error signals. Let

$$\mathbf{z}(t) = \begin{bmatrix} \mathbf{R}^{1/2}\mathbf{u}(t) \\ \mathbf{Q}^{1/2}\mathbf{x}(t) \end{bmatrix} \tag{8.79}$$

where $\mathbf{Q}^{1/2}$ is understood to be the square root of the entries of the diagonal matrix $\mathbf{Q}$, and similarly for $\mathbf{R}$. Given this newly defined output error signal, the cost can be expressed concisely as

$$J = \lim_{t \to \infty} \mathbf{E}(\mathbf{z}^{\mathrm{T}}(t)\mathbf{z}(t)) \tag{8.80}$$

This cost function is simply the square of the $\mathcal{H}_2$ norm when $\mathbf{w}$ is a random process:

$$J = \|\mathbf{T}_{zw}\|_2^2 \tag{8.81}$$

where $\mathbf{T}_{zw}$ represents the closed-loop transfer function between $\mathbf{w}$ and $\mathbf{z}$.

Consider the schematic diagram of the generalized plant as illustrated in Figure 8.17. The control plant $\mathbf{P}(s)$ can be expressed concisely as demonstrated in equation (8.77) and is represented in the block diagram of Figure 8.17. Notice that the white-noise process $\mathbf{w}(s)$ and the control signal $\mathbf{u}(s)$ are both inputs to $\mathbf{P}(s)$. The weighting filter $\mathbf{W}(s)$ for the generalized plant $\mathbf{G}(s)$ is simply a block diagonal matrix expressed in terms of the static weights $\mathbf{R}^{1/2}$ and $\mathbf{Q}^{1/2}$, weighting the control effort and performance, respectively, to construct the error signal $\mathbf{z}(s)$. For the case presented (state-feedback), one assumes that the measured outputs, $\mathbf{y}(s)$, are the system states $x(s)$.

For a realizable system with a bounded $\|\mathbf{T}_{zw}\|_2$ (i.e., $\mathbf{D}_{zw} = 0$), a controller can be synthesized that minimizes the desired norm and results in constant state-feedback:

$$\mathbf{u}(t) = \mathbf{K}_{\mathrm{opt}}\mathbf{x}(t) \tag{8.82}$$

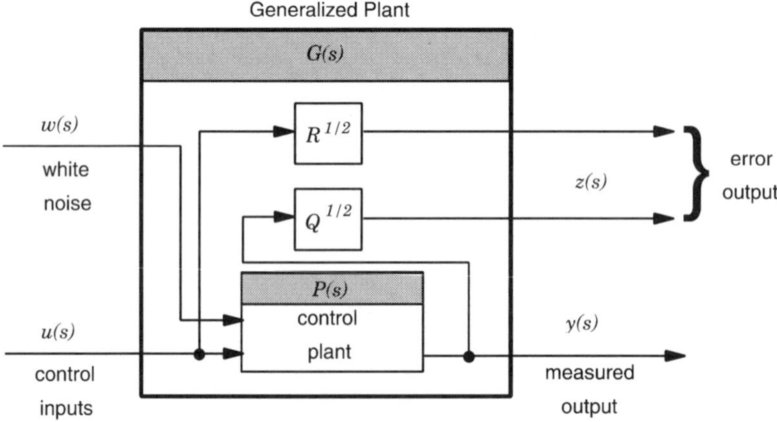


**Figure 8.17:** Schematic diagram of the generalized plant as formed for the LQR problem.

Note also that a realizable system requires that there be some controller **K** such that

$$\mathbf{T}_{zw} = \mathbf{P}_{zw} + \mathbf{P}_{zu}\mathbf{K}(\mathbf{I} - \mathbf{P}_{yw}\mathbf{K})^{-1}\mathbf{P}_{yw} \tag{8.83}$$

The solution to the general optimization problem for dynamic compensation was originally derived by Doyle et al. (1989) and will be discussed in a subsequent section. As detailed by Boyd and Barratt (1991), the objective is to find the solution to the associated algebraic Riccati equation (ARE):

$$\mathbf{A}^{\mathrm{T}}\mathbf{X}^{*} + \mathbf{X}^{*}\mathbf{A} - \mathbf{X}^{*}\mathbf{B}\mathbf{R}^{-1}\mathbf{B}^{\mathrm{T}}\mathbf{X}^{*} + \mathbf{Q} = \mathbf{0} \tag{8.84}$$

where $\mathbf{X}^{*}$ is the unique, positive-definite solution. The solution is typically obtained as follows. One forms the associated Hamiltonian matrix:

$$\mathbf{H} = \begin{bmatrix} \mathbf{A} & -\mathbf{B}\mathbf{R}^{-1}\mathbf{B}^{\mathrm{T}} \\ -\mathbf{Q} & -\mathbf{A}^{\mathrm{T}} \end{bmatrix} \tag{8.85}$$

Once this matrix is formed, a matrix **T** is sought such that

$$\mathbf{T}^{-1}\mathbf{H}\mathbf{T} = \begin{bmatrix} \tilde{\mathbf{A}}_{11} & \tilde{\mathbf{A}}_{12} \\ \mathbf{0} & \tilde{\mathbf{A}}_{22} \end{bmatrix} \tag{8.86}$$

and $\tilde{\mathbf{A}}_{11}$ is stable. The resulting solution can then be obtained from the matrix **T**:

$$\mathbf{X}^{*} = \mathbf{T}_{21}\mathbf{T}_{11}^{-1} \tag{8.87}$$

where

$$\mathbf{T} = \begin{bmatrix} \mathbf{T}_{11} & \mathbf{T}_{12} \\ \mathbf{T}_{21} & \mathbf{T}_{22} \end{bmatrix} \tag{8.88}$$

Given $\mathbf{X}^{*}$, the optimal state-feedback gain matrix can be computed:

$$\mathbf{K}_{\mathrm{opt}} = \mathbf{R}^{-1}\mathbf{B}^{\mathrm{T}}\mathbf{X}^{*} \tag{8.89}$$

For an expanded discussion of the solution to this problem, see Boyd and Barratt (1991). The application of this solution is presented through example in Chapter 9.

***The Dual Levine-Athans Algorithm for Direct, Output-Feedback Control System Design*** The minimization of equation (8.80) can also be carried out under the assumption that $\mathbf{u}(t) = \mathbf{G}\mathbf{y}(t)$, where $\mathbf{y}(t)$ is the output signal from available sensor measurements. This implies that the control signal is generated as a constant-gain matrix $\mathbf{G}$ times the measured output of an array of sensors (not states). This choice of control input forms a parametric linear quadratic problem, and several techniques for obtaining the optimal controller have been investigated (Makila and Toivonen, 1987). However, unlike the state-feedback case, there is still no guaranteed unique solution. The approach taken here is a variation on the method proposed by Levine and Athans (1970), and its development follows Toivonen (1985). It has the advantage of being globally convergent to a solution from a stabilizing initial guess, and the solution satisfies the necessary conditions for optimality.

Consider the cost function in equation (8.80). By letting $\mathbf{u}(t) = \mathbf{G}\mathbf{C}_y\mathbf{x}(t)$, we can write

$$J = \int_0^\infty \mathbf{x}^{\mathrm{T}}(t)(\mathbf{Q} + \mathbf{C}_y^{\mathrm{T}}\mathbf{G}^{\mathrm{T}}\mathbf{R}\mathbf{G}\mathbf{C}_y)\mathbf{x}(t)\, dt \tag{8.90}$$

where $\mathbf{C}_y$ is the matrix of coefficients that weight each state of $\mathbf{P}(s)$ to produce the measured or sensed outputs (i.e., the participation of each generalized coordinate resulting from the operation and spatial aperture of each sensor). As illustrated in Figure 8.18, the states of the control plant $\mathbf{P}(s)$ are passed to a weighting filter $\mathbf{W}(s)$, and the output of this filter is weighted by the performance penalty $\mathbf{Q}^{1/2}$. This peformance penalty is applied to the states of $\mathbf{W}(s)$ as well as $\mathbf{P}(s)$. Thus, for the cost function presented in equation (8.90), $\mathbf{x}(t)$ combines states of both $\mathbf{P}(s)$ and $\mathbf{W}(s)$. (Note that the system presented in Figure 8.17 can be modified similarly to include a frequency weighted cost functional.) The cost functional of equation (8.90) depends upon the initial conditions of the state. Given random initial conditions with $\mathbf{E}(\mathbf{x}(0)^{\mathrm{T}}\mathbf{x}(0)) = \mathbf{F}$, the cost can then be expressed, using the trace operator, as

$$J = \mathrm{Tr}((\mathbf{Q} + \mathbf{C}_y^{\mathrm{T}}\mathbf{G}^{\mathrm{T}}\mathbf{R}\mathbf{G}\mathbf{C}_y)\mathbf{L}) = \mathrm{Tr}(\mathbf{K})\mathbf{F} \tag{8.91}$$

where $\mathbf{K}$ and $\mathbf{L}$ satisfy the equations

$$0 = \mathbf{K}\mathbf{A}_c + \mathbf{A}_c^{\mathrm{T}}\mathbf{K} + \mathbf{C}_y^{\mathrm{T}}\mathbf{G}^{\mathrm{T}}\mathbf{R}\mathbf{G}\mathbf{C}_y + \mathbf{Q} \tag{8.92}$$

$$0 = \mathbf{L}\mathbf{A}_c + \mathbf{A}_c^{\mathrm{T}}\mathbf{L} + \mathbf{F} \tag{8.93}$$

with $\mathbf{A}_c = \mathbf{A} - \mathbf{B}_u\mathbf{G}\mathbf{C}_y$. The gain

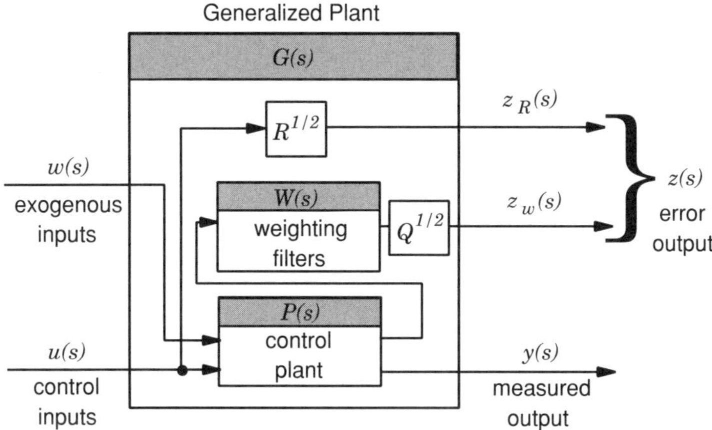


**Figure 8.18:** Schematic diagram of the generalized plant used to develop the output-feedback control problem.

$$\mathbf{G}_{\text{opt}} = \mathbf{R}^{-1}\mathbf{B}_u^{\mathrm{T}}\mathbf{K}\mathbf{L}\mathbf{C}_y^{\mathrm{T}}(\mathbf{C}_y\mathbf{L}\mathbf{C}_y^{\mathrm{T}})^{-1} \tag{8.94}$$

is a stationary point on the cost function; that is, $dJ/d\mathbf{G} = 0$. Therefore, $\mathbf{G}_{\text{opt}}$ satisfies the necessary conditions for optimality.

The procedure for calculating $\mathbf{G}$ involves first determining an initial gain matrix $\mathbf{G}_o$, which stabilizes the system. Equation (8.93) is then solved for $\mathbf{L}$, and the nonlinear equations

$$0 = \mathbf{K}\mathbf{A}_c + \mathbf{A}_c^{\mathrm{T}}\mathbf{K} + \mathbf{C}_y^{\mathrm{T}}\mathbf{G}^{\mathrm{T}}\mathbf{R}\mathbf{G}\mathbf{C}_y + \mathbf{Q} \tag{8.95}$$

$$\mathbf{G} = \mathbf{G}_o + \theta(\mathbf{R}^{-1}\mathbf{B}_u^{\mathrm{T}}\mathbf{K}\mathbf{L}\mathbf{C}_y^{\mathrm{T}}(\mathbf{C}_y\mathbf{L}\mathbf{C}_y^{\mathrm{T}})^{-1} - \mathbf{G}_o), \tag{8.96}$$

must be solved for $\mathbf{G}$ and $\mathbf{K}$. The scalar parameter $\theta$ is constrained to $0 < \theta < 1$ and can be adjusted downward to ensure that $J(\mathbf{G}) < J(\mathbf{G}_o)$. The numerical algorithm proceeds by iterating on equations (8.93), (8.95), and (8.96), and updating $\mathbf{G}$ until $|\mathbf{G}_{i+1} - \mathbf{G}_i| < \epsilon$. In practice, $\mathbf{G}_o = \mathbf{0}$ can be used as the initial stabilizing gain matrix, and $\mathbf{F}$ is typically chosen to be $\mathbf{B}_w^{\mathrm{T}}\mathbf{B}_w$, where $\mathbf{B}_w$ describes a disturbance input influence matrix.

To check equations (8.95) and (8.96), let us consider what happens when the measured output is modified to contain all of the system states [i.e., $\mathbf{C}_y = \mathbf{I}$ such that the states of $\mathbf{P}(s)$ and $\mathbf{W}(s)$ are all observed]. For this case, if the scalar parameter $\theta$ is set to unity, then the standard AREs result, and the optimal solution to the LQR problem is obtained. Thus, as expected, the approach yields the same solution as that obtained in the previous section.

### 8.6.2 Dynamic Compensation

Dynamic compensation involves a dynamic model of the feedback control law that can be SISO or MIMO. In LQG control theory, the dynamic compensator results from the combination of the state-estimator and feedback control law, as detailed in Chapter 3. The more general approach to the design of dynamic compensators was presented by Doyle et al. (1989) for general $\mathcal{H}_2$ and $\mathcal{H}_\infty$ control problems. The small-gain theorem and $\mathcal{H}_\infty$ control problems are used extensively in the development of robust control theory. The general theory for developing $\mathcal{H}_2$ and $\mathcal{H}_\infty$ controllers is presented and follows from that of Doyle et al. (1989). Upon completing this development, the LQG problem will be cast as an $\mathcal{H}_2$ controller design to demonstrate that it can be viewed as a subset of this more general approach to control system design.

To begin, we consider the generalized plant illustrated in Figure 8.7 and outline the standard assumptions made for $\mathcal{H}_2$ and $\mathcal{H}_\infty$ controller synthesis. One can define the realization of the transfer matrix $\mathbf{G}(s)$

$$\mathbf{G}(s) = \left[\begin{array}{c|cc} \mathbf{A} & \mathbf{B}_w & \mathbf{B}_u \\ \hline \mathbf{C}_z & \mathbf{0} & \mathbf{D}_{zu} \\ \mathbf{C}_y & \mathbf{D}_{yw} & \mathbf{0} \end{array}\right] \tag{8.97}$$

where the state-space realization has been partitioned to separate the two vector inputs, disturbance ($\mathbf{w}$) and control ($\mathbf{u}$), and the two vector outputs, error ($\mathbf{z}$) and sensor ($\mathbf{y}$). Consistent with Zhou et al. (1996) and Skogestad and Postlethwaite (1996), we assume:

1. $(\mathbf{A}, \mathbf{B}_u)$ is stabilizable and $(\mathbf{C}_y, \mathbf{A})$ is detectable.
2. $\mathbf{D}_{zu}$ and $\mathbf{D}_{yw}$ have full rank.
3. $\begin{bmatrix} \mathbf{A} - j\omega\mathbf{I} & \mathbf{B}_u \\ \mathbf{C}_z & \mathbf{D}_{zu} \end{bmatrix}$ has full column rank for all $\omega$.
4. $\begin{bmatrix} \mathbf{A} - j\omega\mathbf{I} & \mathbf{B}_w \\ \mathbf{C}_y & \mathbf{D}_{yw} \end{bmatrix}$ has full row rank for all $\omega$.
5. $\mathbf{D}_{zw} = \mathbf{0}$ and $\mathbf{D}_{yu} = \mathbf{0}$.

Assumptions (3) and (4) guard against the cancellation of poles and zeros on the imaginary axis in the design of the optimal controller. Assumption (1) is required for the existence of a controller $\mathbf{K}$. Assumption (2) guarantees that the controller is realizable, and assumption (5) eliminates direct feedthrough from the disturbance to the error and from the control to the error, which is typical of most control problems that are properly cast. We now proceed with the development of the $\mathcal{H}_2$ and $\mathcal{H}_\infty$ optimal controller designs.

***$\mathcal{H}_2$ Controller Design*** In $\mathcal{H}_2$ control, particularly for LQG, one frequently imposes an additional assumption:

6. $\mathbf{D}_{zu}^{\mathrm{T}}\mathbf{C}_z = 0$ and $\mathbf{B}_w\mathbf{D}_{yw}^{\mathrm{T}} = \mathbf{0}$,

which states that there are no cross terms in the cost functional between the control and the error output or between the disturbance and the sensed output (process noise and measurement noise are statistically uncorrelated). The $\mathcal{H}_2$ controller design proceeds by defining the following two Hamiltonian matrices:

$$\mathbf{M}_2 = \begin{bmatrix} \mathbf{A} & -\mathbf{B}_u\mathbf{B}_u^{\mathrm{T}} \\ -\mathbf{C}_z^{\mathrm{T}}\mathbf{C}_z & -\mathbf{A}^{\mathrm{T}} \end{bmatrix} \tag{8.98}$$

and

$$\mathbf{N}_2 = \begin{bmatrix} \mathbf{A}^{\mathrm{T}} & \mathbf{C}_y^{\mathrm{T}}\mathbf{C}_y \\ -\mathbf{B}_w\mathbf{B}_w^{\mathrm{T}} & -\mathbf{A} \end{bmatrix} \tag{8.99}$$

which must belong to the domain of Ric (*dom(Ric)*), as outlined by Doyle et al. (1989). If a Hamiltonian matrix $\mathbf{H}$ belongs to the domain of Ric, then $\mathbf{H}$ has no eigenvalues on the imaginary axis, and the two subspaces

$$\mathrm{Im}\begin{bmatrix} \mathbf{X}_1 \\ \mathbf{X}_2 \end{bmatrix}, \qquad \mathrm{Im}\begin{bmatrix} \mathbf{0} \\ \mathbf{I} \end{bmatrix} \tag{8.100}$$

are complementary. The matrix $[\mathbf{X}_1^{\mathrm{T}}\ \mathbf{X}_2^{\mathrm{T}}]^{\mathrm{T}}$ corresponds to an invariant, $n$-dimensional, spectral subspace of the Hamiltonian matrix in which the eigenvalues in Re $s < 0$, and $X_1, X_2 \in \Re^{n\times n}$. If the two matrices of equation (8.100) are complementary, then $\mathbf{X} := \mathrm{Ric}(\mathbf{H}) := \mathbf{X}_2\mathbf{X}_1^{-1}$.

Letting $\mathbf{m}_2 := \mathrm{Ric}(\mathbf{M}_2)$ and $\mathbf{n}_2 := \mathrm{Ric}(\mathbf{N}_2)$, one can define $\mathbf{F}_2 := -\mathbf{B}_u^{\mathrm{T}}\mathbf{m}_2$ and $\mathbf{L}_2 := -\mathbf{n}_2\mathbf{C}_y^{\mathrm{T}}$. As proven by Doyle et al. (1989), the unique optimal controller can be expressed as follows:

$$\mathbf{K}_{\mathrm{opt}}(s) := \left[\begin{array}{c|c} \mathbf{A} + \mathbf{B}_u\mathbf{F}_2 + \mathbf{L}_y\mathbf{C}_y & -\mathbf{L}_2 \\ \hline \mathbf{F}_2 & \mathbf{0} \end{array}\right] \tag{8.101}$$

For details of the proof, the interested reader is referred to Doyle et al. (1989) or Zhou et al. (1996). In passing, we also note that, for the Hamiltonian matrix $\mathbf{M}_2$ of equation (8.98), $\mathbf{M}_2 \in$ *dom*(*Ric*) if and only if $(\mathbf{A}, \mathbf{B}_u)$ is stabilizable and $(\mathbf{C}_z, \mathbf{A})$ has no unobservable modes on the imaginary axis. This is physically satisfying for the control system realization (Zhou et al., 1996). Basic design approaches and examples of $\mathcal{H}_2$ controllers are presented in Chapter 9, and numerical routines for computing the optimal compensator are readily available in commercial software packages such as the $\mu$-Analysis & Synthesis Toolbox for MATLAB.

***$\mathcal{H}_\infty$ Controller Design*** As in the description of the $\mathcal{H}_2$ controller design, the generalized plant illustrated in Figure 8.7 is used as the model. The suboptimal $\mathcal{H}_\infty$ control problem is considered herein and is consistent with the presentation of Doyle et al. (1989). The objective is to find all admissible controllers $\mathbf{K}(s)$ such that $\|\mathbf{T}_{zw}\|_\infty < \gamma$ (i.e., the $\mathcal{H}_\infty$ norm of the closed-loop transfer function between $w$ and $z$ is bounded by $\gamma$). Thus, $\gamma$ must be greater than the $\mathcal{H}_\infty$ optimal level. Given the assumptions outlined at the beginning of this section, including a requirement that $(\mathbf{A}, \mathbf{B}_w)$ be stabilizable and $(\mathbf{A}, \mathbf{C}_z)$ be detectable, a stabilizing controller $\mathbf{K}(s)$ can be determined such that the lower linear fractional transformation is bounded by $\gamma$ (i.e., $\|\mathbf{T}_{zw}\|_\infty = \|f_l(\mathbf{P}, \mathbf{K})\|_\infty < \gamma$) if and only if the following three conditions hold (Skogestad and Postlethwaite, 1996):

1. $\mathbf{M}_\infty \geq 0$ is a solution to the ARE

$$\mathbf{A}^{\mathrm{T}}\mathbf{M}_\infty + \mathbf{M}_\infty\mathbf{A} + \mathbf{C}_z^{\mathrm{T}}\mathbf{C}_z + \mathbf{M}_\infty(\gamma^{-2}\mathbf{B}_w\mathbf{B}_w^{\mathrm{T}} - \mathbf{B}_u\mathbf{B}_u^{\mathrm{T}})\mathbf{M}_\infty = \mathbf{0} \tag{8.102}$$

   such that

$$\mathrm{Re}\lambda_i[\mathbf{A} + (\gamma^{-2}\mathbf{B}_w\mathbf{B}_w^{\mathrm{T}} - \mathbf{B}_u\mathbf{B}_u^{\mathrm{T}})\mathbf{M}_\infty] < 0 \tag{8.103}$$

   for all $i$. This means that the iteration on $\gamma$ proceeds until an eigenvalue $\lambda_i$ crosses the imaginary axis. Additionally,
2. $\mathbf{N}_\infty \geq 0$ is a solution to the ARE

$$\mathbf{A}\mathbf{N}_\infty + \mathbf{N}_\infty\mathbf{A}^{\mathrm{T}} + \mathbf{B}_w\mathbf{B}_w^{\mathrm{T}} + \mathbf{N}_\infty(\gamma^{-2}\mathbf{C}_z^{\mathrm{T}}\mathbf{C}_z - \mathbf{C}_y^{\mathrm{T}}\mathbf{C}_y)\mathbf{N}_\infty = \mathbf{0} \tag{8.104}$$

   such that

$$\mathrm{Re}\lambda_i[\mathbf{A} + \mathbf{N}_\infty(\gamma^{-2}\mathbf{C}_z^{\mathrm{T}}\mathbf{C}_z - \mathbf{C}_y^{\mathrm{T}}\mathbf{C}_y)] < 0 \tag{8.105}$$

   for all $i$. Again, this means that the iteration on $\gamma$ proceeds until an eigenvalue $\lambda_i$ crosses the imaginary axis; finally,
3. $\rho(\mathbf{M}_\infty\mathbf{N}_\infty) < \gamma^2$, where $\rho(\mathbf{V})$ is defined as the spectral radius of the matrix $\mathbf{V}$ and is the magnitude of the largest eigenvalue of the matrix.

Given these conditions, the suboptimal controller can be expressed as follows:

$$\mathbf{K}_{\text{sub}}(s) := \left[\begin{array}{c|cc} \hat{\mathbf{A}}_\infty & -\mathbf{Z}_\infty\mathbf{L}_\infty & \mathbf{Z}_\infty\mathbf{B}_u \\ \hline \mathbf{F}_\infty & \mathbf{0} & \mathbf{I} \\ -\mathbf{C}_y & \mathbf{I} & \mathbf{0} \end{array}\right] \tag{8.106}$$

where

$$\hat{\mathbf{A}}_\infty = \mathbf{A} + \gamma^{-2}\mathbf{B}_w\mathbf{B}_w^{\mathrm{T}}\mathbf{M}_\infty + \mathbf{B}_u\mathbf{F}_\infty + \mathbf{Z}_\infty\mathbf{L}_\infty\mathbf{C}_y \tag{8.107}$$

$$\mathbf{F}_\infty := -\ \mathbf{B}_u^{\mathrm{T}}\mathbf{M}_\infty \tag{8.108}$$

$$\mathbf{L}_\infty := -\ \mathbf{N}_\infty\mathbf{C}_y^{\mathrm{T}} \tag{8.109}$$

and

$$\mathbf{Z}_\infty := (\mathbf{I} - \gamma^{-2}\mathbf{N}_\infty\mathbf{M}_\infty)^{-1} \tag{8.110}$$

Details of the proof can be found in the reference by Doyle et al. (1989) or Zhou et al. (1996). Design approaches and examples are considered in Chapter 9, and the reader is expected to have the $\mu$-Synthesis and Analysis Toolbox and MATLAB for the design examples.

***The LQG Problem Cast from the $\mathcal{H}_2$ Controller Design*** As discussed earlier in this section, the LQG problem can be developed as one example of an $\mathcal{H}_2$ controller design. For completeness, the development is summarized here. Consider the schematic diagram of the generalized plant illustrated in Figure 8.19. The control effort penalty $\mathbf{R}$, performance penalty $\mathbf{Q}$, process noise weight $\mathbf{W}$, and the measurement noise weight $\mathbf{V}$ are all incorporated into the generalized plant. The dynamic response of the generalized plant $\mathbf{G}(s)$ can be cast in state-variable form as follows:

$$\dot{\mathbf{x}}(t) = \mathbf{A}\mathbf{x}(t)\mathbf{B}\mathbf{u}(t) + \mathbf{w}_{\text{process}}(t) \tag{8.111}$$

$$\mathbf{y}(t) = \mathbf{C}\mathbf{x}(t) + \mathbf{v}_{\text{sensor}}(t) \tag{8.112}$$

Following the discussion presented by Boyd and Barratt (1991), the output error can be expressed as follows:

$$\mathbf{z}(t) = \begin{bmatrix} \mathbf{R}^{1/2} & \mathbf{0} \\ \mathbf{0} & \mathbf{Q}^{1/2} \end{bmatrix} \begin{bmatrix} \mathbf{u}(t) \\ \mathbf{x}(t) \end{bmatrix} \tag{8.113}$$

Additionally, the exogenous input to the system can be expressed as follows:

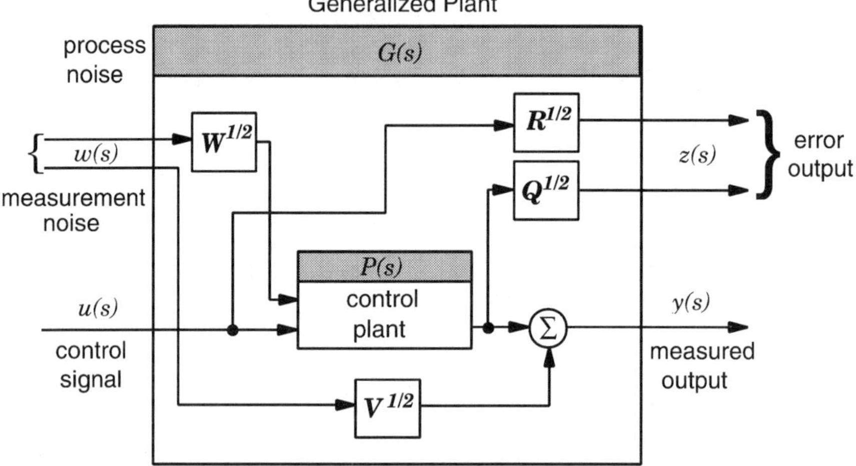


**Figure 8.19:** Schematic diagram of the generalized plant used to develop the LQG problem.

$$\begin{bmatrix} \mathbf{w}_{\text{process}}(t) \\ \mathbf{v}_{\text{sensor}}(t) \end{bmatrix} = \begin{bmatrix} \mathbf{W}^{1/2} \\ \mathbf{V}^{1/2} \end{bmatrix} \mathbf{w}(t) \tag{8.114}$$

where $\mathbf{w}(t)$ is a white process with $\mathbf{S}_{ww} = \mathbf{I}$. The cost functional is again formed from the square of the $\mathcal{H}_2$ norm:

$$J = \|\mathbf{T}_{zw}\|_2^2 \tag{8.115}$$

For the LQG problem, one assumes that the plant is controllable and observable from the inputs and outputs, respectively, and $\mathbf{R} > \mathbf{0}$ and $\mathbf{V} > \mathbf{0}$. Given this formulation, the optimal dynamic compensator can be obtained from the general approach outlined in this section.

### 8.6.3 Singular Values and MIMO Loop Shaping

In Chapter 7, we found expressions for the sensitivity function and the complementary sensitivity function, and used Bode plots to consider stability of SISO systems. For MIMO control system design and synthesis, we have an analogous tool, the singular value decomposition (SVD) (see Appendix B). However, before proceeding, the reader should be warned that the discussion is targeted primarily to systems where the disturbance is not shaped by the plant dynamics. In other words, $\mathbf{d}(s) = \mathbf{w}(s)$ in Figure 8.2 and is thus unaffected by the plant dynamics before entering at the output. For such cases, the closed-loop transfer matrix between the sensed output and the disturbance can be expressed concisely as follows:

$$\mathbf{y}(s) = [\mathbf{I} - \mathbf{P}_{yu}(s)\mathbf{K}_{fb}(s)]^{-1}\mathbf{d}(s) \tag{8.116}$$
$$= \mathbf{S}(s)\mathbf{d}(s) \tag{8.117}$$

Thus, any reduction in the sensitivity leads directly to a reduction in the disturbance. Even if the disturbance is influenced by the plant dynamics through some alternative plant transfer matrix $\mathbf{P}_{yw}(s)$, the output can be expressed as a function of $\mathbf{w}(s)$ as follows:

$$\mathbf{y}(s) = [\mathbf{I} - \mathbf{P}_{yu}(s)\mathbf{K}_{fb}(s)]^{-1}\mathbf{P}_{yw}(s)\mathbf{w}(s) \tag{8.118}$$
$$= \mathbf{S}(s)\mathbf{P}_{yw}(s)\mathbf{w}(s) \tag{8.119}$$

Again, any reduction in the sensitivity serves to reduce the response *at the sensed output* due to the disturbance, since the sensitivity simply weights the open-loop transfer matrix between the disturbance $\mathbf{w}(s)$ and the sensed output $\mathbf{y}(s)$. However, if the error output $\mathbf{z}(s)$ (i.e., the performance variables used in the cost functional) do not coincide with the sensed variables $\mathbf{y}(s)$, the closed-loop response is described by a linear fractional transformation:

$$\mathbf{z}(s) = [\mathbf{P}_{zw}(s) + \mathbf{P}_{zu}(s)\mathbf{K}_{fb}(s)[\mathbf{I} - \mathbf{P}_{yu}(s)\mathbf{K}_{fb}(s)]^{-1}\mathbf{P}_{yw}(s)]\mathbf{w}(s) \tag{8.120}$$
$$= [\mathbf{P}_{zw}(s) + \mathbf{P}_{zu}(s)\mathbf{K}_{fb}(s)\mathbf{S}(s)\mathbf{P}_{yw}(s)]\mathbf{w}(s) \tag{8.121}$$

As demonstrated in equation (8.121), a reduction in the sensitivity matrix $\mathbf{S}(s)$ does not guarantee that the response is decreased at the error output, since $\mathbf{P}_{zw}(s)$ and $\mathbf{P}_{zu}(s)\mathbf{K}_{fb}(s)$ both influence the closed-loop response at the error $\mathbf{z}(s)$ used in the cost functional. This motivates the use of closed-loop design procedures for the synthesis of compensators. However, open-loop shaping based upon the singular values of the MIMO transfer matrix provides the designer with significant insight concerning robust stability and rejection of sensor noise. For the special case in which the sensed variables and error variables are the same, open-loop shaping can be used to evaluate performance as well. However, this is typically not the case in the application of adaptive structures. We proceed here with a discussion of MIMO loop shaping for such special cases, because it is frequently encountered in the controls literature and is quite important to the understanding of robust stability.

The topics of MIMO loop shaping and loop transfer recovery are addressed in the prior work by Doyle and Stein (1981), and much of the qualitative discussion herein follows from that work. In Chapter 3, we discussed norms of signals and norms of transfer functions. For a MIMO system described by

$$\mathbf{y}(s) = \mathbf{P}(s)\mathbf{u}(s) \tag{8.122}$$

one can show that

$$\frac{\|\mathbf{y}(s)\|_2}{\|\mathbf{u}(s)\|_2} = \frac{\|\mathbf{P}(s)\mathbf{u}(s)\|_2}{\|\mathbf{u}(s)\|_2} \tag{8.123}$$

This equation is analogous to the SISO expression:

$$\left|\frac{y(s)}{u(s)}\right| = |P(s)| \tag{8.124}$$

The largest magnitude of equation (8.123) can be computed from the maximum singular value of $\mathbf{P}(s)$: $\sigma_{\max}(\mathbf{P}(s))$. Similarly, the minimum response can be calculated from the minimum singular value of $\mathbf{P}(s)$: $\sigma_{\min}(\mathbf{P}(s))$. The bounds set by the singular values of the system transfer function provide a means of evaluating the performance and stability of the MIMO control system.

Analogous to the SISO control system, we can define the *loop transfer matrix*: $\mathbf{P}(s)\mathbf{K}(s)$, where $\mathbf{P}(s)$ is the plant as outlined previously and $\mathbf{K}(s)$ is the dynamic compensator. The *return difference transfer matrix* is defined as $\mathbf{I} + \mathbf{P}(s)\mathbf{K}(s)$, the *sensitivity matrix* is defined as $[\mathbf{I} + \mathbf{P}(s)\mathbf{K}(s)]^{-1}$, and the *complementary sensitivity matrix* is defined as $[\mathbf{I} + \mathbf{P}(s)\mathbf{K}(s)]^{-1}\mathbf{P}(s)\mathbf{K}(s)$. As in the design of SISO systems, reference tracking, disturbance rejection, response to noise, and sensitivity can all be related to the previously defined MIMO matrices. To track a reference, the loop transfer matrix should be large over the frequency range of interest. For disturbance rejection, both the loop transfer matrix and the return difference matrix must be large over the frequency range of the disturbances. To minimize the sensitivity to variations in the plant, one desires that the sensitivity matrix be small over the frequency range of expected variations in plant dynamics (for example at frequencies where the model is known to be poor). To minimize the effects of sensor noise, one desires the complementary sensitivity matrix to be small over the applicable bandwidth. As in SISO design, these performance metrics guide the design and selection of dynamic compensators.

For good reference tracking and disturbance rejection, it is necessary that the return difference transfer matrix be large or that the sensitivity matrix be small. One can use SVD to assess these conditions by computing the maximum singular value of the sensitivity matrix or the minimum singular value of the return difference transfer matrix:

$$\sigma_{\max}([\mathbf{I} + \mathbf{P}(s)\mathbf{K}(s)]^{-1}) << 1 \tag{8.125}$$

or

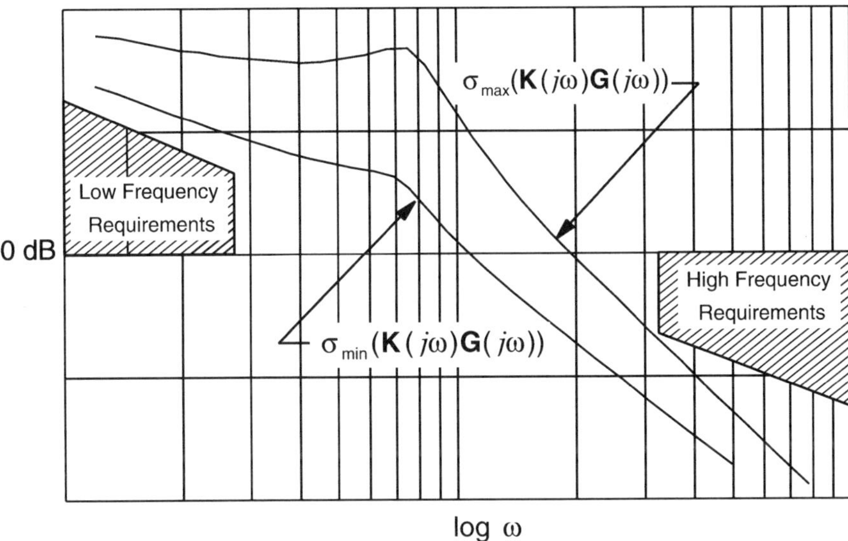


**Figure 8.20:** Schematic diagram of performance and stability boundaries.

$$\sigma_{\min}([\mathbf{I} + \mathbf{P}(s)\mathbf{K}(s)]) >> 1 \tag{8.126}$$

Hence, for good reference tracking and disturbance rejection, the minimum singular value of the return difference transfer matrix should be large over the bandwidth containing the disturbances and reference inputs. One can also show that this condition is met when the minimum singular value of the loop transfer matrix is large over the bandwidth of interest:

$$\sigma_{\min}[\mathbf{G}(s)\mathbf{K}(s)] >> 1 \tag{8.127}$$

As illustrated in Figure 8.20, a low-frequency bound is typically set by design constraints to assure appropriate disturbance rejection and command reference tracking. As depicted, the minimum singular values must exceed this bound.

For sensor noise, which is introduced on the output, the objective is to minimize the complementary sensitivity transfer matrix. Thus, the maximum singular value of this matrix should be small:

$$\sigma_{\max}[(\mathbf{I} + \mathbf{G}(s)\mathbf{K}(s))^{-1}\mathbf{G}(s)\mathbf{K}(s)] << 1 \tag{8.128}$$

One can also show that this corresponds to requiring that the maximum singular value of the loop transfer matrix is small over the bandwidth of the sensor noise:

$$\sigma_{\max}[\mathbf{G}(s)\mathbf{K}(s)] << 1 \tag{8.129}$$

Thus, equations (8.127) and (8.129) serve as design guides (particularly when the sensed output and the error output are the same) for loop shaping in MIMO control system design, as illustrated in the schematic diagram presented in Figure 8.20. If one sets boundaries on low-frequency performance for disturbance rejection and reference tracking and high-frequency rolloff for unmodeled dynamics and sensor noise, then the maximum and minimum singular values of the loop transfer function must be within these bounds to achieve the desired stability and performance requirements. Once a control system design is obtained, the maximum and minimum singular values of the MIMO loop transfer matrix are essential to studying the performance and stability of the control system. In general, one should always evaluate the maximum and minimum singular values of the loop transfer matrix to assess stability and performance through the control path; however, a formal structured design approach is recommended for MIMO loop shaping. Stein and Athans (1987) outlined a MIMO, LQG/LTR design procedure, and in general, the use of singular value loop shaping is at the heart of robust control system design and synthesis. Robust performance and robust stability are both important topics in the design and implementation of adaptive structures; however, a detailed presentation of this material is beyond the scope of this text. The interested reader is referred to Maciejowski (1989), Zhou et al. (1996), and Skogestad and Postlethwaite (1996), for greater detail on this subject.

## 8.7 HYBRID CONTROL

Hybrid control, in this text, is defined as the combination of both feedback and adaptive feedforward control, as previously outlined by Saunders et al. (1993). The primary purpose of hybrid control systems is to design the disturbance control problem with the architecture best suited for overall performance. If the reverberant plant is exposed to exogenous inputs consisting of *both* stochastic and harmonic disturbances, for which a reference correlated with the harmonic disturbance is available, it is likely that feedback control strategies are better suited for the stochastic disturbances, while adaptive feedforward control strategies are better suited for the harmonic disturbances. Hybrid control utilizes both forms of control to maximize the performance of the adaptive structure to mixed-signal disturbances. It will also be shown that the operation of adaptive feedforward control in collaboration with feedback control improves certain performance metrics of the adaptive feedforward signals. For example, earlier work by Sievers and von Flotow (1990) demonstrated that the convergence of the MIMO, LMS algorithm was limited by the "slowest and fastest time scales of the system." Thus, by modifying the timescales of the system with feedback (the timescales are controlled by the poles of the reverberant plant), one can decrease the convergence time of the feedforward controller.

Another motivation for hybrid control architectures is that performance robustness for narrowband disturbance rejection is often unacceptable for fixed-gain control strategies. Controllers based upon output-feedback control typically require large gains that are quite sensitive to variations in the plant model, and fixed-gain feedforward controllers depend highly upon model fidelity for optimal performance. Due to the stringent requirements placed upon model fidelity for robust performance, adaptive feedforward control strategies have become increasingly popular for narrow-band disturbance rejection. This is because of the advantages of adaptive zero placement discussed earlier. Unfortunately, this approach requires the existence of an *uncontrollable* reference signal. Uncontrollable reference refers to a signal that is correlated with the disturbance signal but cannot be affected, that is, it is uncontrollable, by the control on the plant. For impulsive or stochastic disturbance inputs, an uncontrollable reference signal correlated with the disturbance is typically not available, which leaves only feedback control alternatives. However, it is usually possible to measure or synthesize uncontrollable references for narrow-band disturbances directly. Therefore, the hybrid combination of feedback and adaptive feedforward control provides a method of rejecting each disturbance type with the best control approach. The essential features of hybrid control for adaptive structures can be summarized as follows:

1. Robust disturbance rejection of narrowband, periodic signals
2. Simultaneous transient suppression of impulsive and stochastic disturbances
3. Improved performance and convergence of the adaptive, feedforward controller

Consider the reverberant plant, which can be described as follows:

$$\mathbf{y}(s) = \mathbf{P}(s)\mathbf{u}(s) \tag{8.130}$$

where $\mathbf{y}(s)$ is the Laplace transform of the $L \times 1$ vector of outputs, $\mathbf{u}(s)$ is the $K \times 1$ vector of inputs, and

$$\mathbf{P}(s) = \begin{bmatrix} P_{11}(s) & P_{12}(s) & \bullet & \bullet & P_{1K}(s) \\ P_{21}(s) & P_{22}(s) & \bullet & \bullet & P_{2K}(s) \\ \bullet & \bullet & \bullet & \bullet & \bullet \\ P_{L1}(s) & P_{L2}(s) & \bullet & \bullet & P_{LK}(s) \end{bmatrix}_{L \times K} \tag{8.131}$$

where $P_{ni}(s)$ is the transfer function between the $i$th input and the $n$th output. For a reverberant structure, the individual transfer functions' poles and zeros can be expressed as a function of the modal coordinates, as discussed in Chapter 7:

$$P_{ni}(s) = \frac{\sum_{m=1}^{M} \Gamma_{mn}(s)\Gamma_{mi} \prod_{j=1, m \neq j}^{M} (s^2 + 2\xi_j\omega_j s + \omega_j^2)}{\prod_{j=1}^{M} (s^2 + 2\xi_j\omega_j s + \omega_j^2)} \tag{8.132}$$

where

$$\Gamma_{mn}(s) = \int_{\mathbf{L}} k_n(s)\beta_n(\mathbf{x})M_x^n\,[\Phi_m(\mathbf{x})]\,d\mathbf{x} \tag{8.133}$$

$$\Gamma_{mi} = \int_{\mathbf{L}} \Phi_m(x)L_x^i[\gamma_i(x)]\,dx \tag{8.134}$$

which is consistent with the terminology developed in Chapter 6 to describe the aperture and function of the transducers used. Note that a finite number of modes, $M$, are used to compute the transfer functions and that $\xi_j$ and $\omega_j$ represent the damping ratio and circular natural frequency of the $j$th mode, respectively. Now, the individual transfer functions all share a common denominator, so the plant transfer matrix can be expressed as follows:

$$\mathbf{P}(s) = \frac{\begin{bmatrix} N_{11}(s) & N_{12}(s) & \bullet & \bullet & N_{1K}(s) \\ N_{21}(s) & N_{22}(s) & \bullet & \bullet & N_{2K}(s) \\ \bullet & \bullet & \bullet & \bullet & \bullet \\ N_{L1}(s) & N_{L2}(s) & \bullet & \bullet & N_{LK}(s) \end{bmatrix}_{L \times K}}{\prod_{j=1}^{M} (s^2 + 2\xi_j\omega_j s + \omega_j^2)} \tag{8.135}$$

where $N_{ni}(s)$ is the polynomial numerator of the transfer function $P_{ni}(s)$. The frequency response for any input-output path can be obtained by letting $s = j\omega$. In adaptive feedforward control, a cost function is typically constructed from measureable outputs such that

$$J = \mathrm{E}\left[\sum_{n=1}^{N} y_n^2(t)\right] \tag{8.136}$$

where $E[*]$ is the expectation, $y_n(t)$ is the output of the $n$th error sensor, and $J$ is the cost functional to be minimized. For adaptive feedforward control, the output of the $n$th error sensor can be expressed in terms of the transfer function between the disturbance and the error sensor $P_{nd}(s)$, and the transfer functions between each control channel and error sensor $P_{ni}(s)$ such that

$$y_n(s) = r(s)\left[P_{nd}(s) + \sum_{i=1}^{K} P_{ni}(s)K_i(s)\right] \tag{8.137}$$

where $r(s)$ is the disturbance input (also used as the reference here) and $K_i(s)$ is the $i$th feedforward compensator. Since the denominator is common, one can express the output as follows:

$$y_n(s) = \frac{r(s)}{\prod_{j=1}^{M}(s^2 + 2\xi_j\omega_j s + \omega_j^2)}\left[N_{nd}(s) + \sum_{i=1}^{K} N_{ni}(s)K_i(s)\right] \tag{8.138}$$

where $N_{nd}(s)$ is the numerator of the transfer function between the $n$th sensor and the disturbance and $N_{ni}(s)$ is the numerator of the transfer function between the $n$th sensor and the $i$th control input. As long as the number of sensors exceeds the number of control inputs, an LMS approach can be formulated to solve for the optimal compensator. More generally, the optimal compensator can be described from the following expression:

$$\mathbf{Nc}(s)\mathbf{K}(s) = -\mathbf{Nd}(s) \tag{8.139}$$

where $\mathbf{Nc}(s)$ is an $L \times K$ matrix of polynomial numerators of the transfer functions between each control input and error sensor, $\mathbf{K}(s)$ is a $K \times 1$ vector of compensators, and $\mathbf{Nd}(s)$ is a $L \times 1$ vector of polynomial numerators of the transfer function between the disturbance and each error sensor. If $L = K$,

$$\mathbf{K}(s) = -\frac{\text{adjoint}\{\mathbf{Nc}(s)\}}{|\mathbf{Nc}(s)|}\mathbf{Nd}(s) \tag{8.140}$$

However, for $L > K$,

$$\mathbf{K}(s) = -[\mathbf{Nc}^{\mathrm{T}}(s)\mathbf{Nc}(s)]^{-1}\mathbf{Nc}^{\mathrm{T}}(s)\mathbf{Nd}(s) \tag{8.141}$$

The similarity of these compensator equations to the SISO compensators given by equations (7.130) and (7.131) should be noted. Thus, the optimal compensator is described in terms of an autoregressive-moving average model; that is, the compensator has both numerator and denominator dynamics. For broadband reference signals, the compensator can only be realized if all of the poles of the denominator, defined by the numerator polynomials of the control-to-error transfer functions, are in the left-half $s$-plane. This will not be possible whenever there are right-half $s$-plane zeros, of course. This is a marked limitation of actually implementing these optimal compensators. Furthermore, equations (8.140) and (8.141) serve to demonstrate that model fidelity is of ultimate importance in the design of fixed compensators. If the dynamics of the plant change such that the zeros are affected, the compensator is no longer optimal. It is well-known that the zeros of structural transfer functions are, in fact, very sensitive to slight changes in boundary conditions and ambient conditions, such as temperature and pressure.

It is informative to examine the optimal compensator for periodic signals. The phase and magnitude of the compensator can be obtained by letting $s = j\omega_o$ so that

$$\mathbf{K}(j\omega_o) = -\frac{\text{adjoint}\{\mathbf{Nc}(j\omega_o)\}}{|\mathbf{Nc}(j\omega_o)|}\mathbf{Nd}(j\omega_o), \qquad \text{for } L = K \quad (8.142)$$

$$\mathbf{K}(j\omega_o) = -[\mathbf{Nc}^{\mathrm{T}}(j\omega_o)\mathbf{Nc}(j\omega_o)]^{-1}\mathbf{Nc}^{\mathrm{T}}(j\omega_o)\mathbf{Nd}(j\omega_o), \qquad \text{for } L > K \quad (8.143)$$

which can be transformed directly into a discrete-time filter once the sampling rate is chosen. One observes again that the performance of feedforward control systems depends upon the placement of zeros that coincide with the existing narrowband disturbance frequencies (i.e., for select choices of $\omega$). As indicated by equations (8.142) and (8.143), if the design frequency $\omega_o$ differs from that of the disturbance frequency, performance will be less than optimal.

An alternative to fixed-gain, feedforward control is fixed-gain, feedback control for the rejection of harmonic disturbances. Gupta (1980) first proposed a method for modifying the standard LQR design to address the narrowband disturbance rejection problem. The technique involved a frequency weighted performance penalty, which, as discussed earlier in this chapter, can be achieved through the standard $\mathcal{H}_2$ control system design. As illustrated in Figure 8.18, one can augment the plant to include a frequency weighted filter on the output. For MIMO, narrowband, disturbance rejection, the frequency weighted filter $\mathbf{W}(s)$, depicted in Figure 8.18, is typically chosen such that the inputs to the filter are the measured outputs of the plant $\mathbf{P}(s)$, and the filter is a diagonal matrix such that

$$\mathbf{W}(s) = \frac{\omega_o^2}{s^2 + \omega_o^2}\mathbf{I} \qquad (8.144)$$

where $\omega_o$ is the frequency of the disturbance and $\mathbf{I}$ is an identity matrix of dimension equal to the number of measured outputs. As detailed by Kwakernaak and Sivan (1972), as well as by Sievers and von Flotow (1990), the location of the closed-loop regulator poles can be determined by the root-square locus method and are the stable roots of

$$\Phi(-s)\Phi(s)\det\left(\mathbf{I} + \frac{1}{R}\,[\mathbf{W}(-s)\mathbf{P}(-s)]^{\mathrm{T}}][\mathbf{W}(s)\mathbf{P}(s)]\right) = 0 \tag{8.145}$$

where

$$\Phi(s) = \Phi_P(s)\Phi_W(s) \tag{8.146}$$

$$= |s\mathbf{I} - \mathbf{A}_P||s\mathbf{I} - \mathbf{A}_W| \tag{8.147}$$

$R$ is the scalar control effort penalty weighting and $\mathbf{A}_P$ and $\mathbf{A}_W$ are the system matrices for $\mathbf{P}(s)$ and $\mathbf{W}(s)$, respectively. As outlined by Sievers and von Flotow (1990), the polynomial of equation (8.145) can be expressed as follows:

$$\Phi_P(-s)\Phi_P(s)(s^2 + \omega_o^2)^{2L} \det\left(\mathbf{I} + \frac{\omega_o^2}{R(s^2 + \omega_o^2)^2}\,[\mathbf{P}^{\mathrm{T}}(-s)\mathbf{P}(s)]\right) = 0 \tag{8.148}$$

For large values of $R$, the closed-loop pole locations near $s = j\omega_o$ can be obtained from the following expression:

$$(s - j\omega_o)^{2L}(2j\omega_o^2)^{2L}\det\left(\mathbf{I} + \frac{\omega_o^2}{4R(s - j\omega_o)^2}\,[\mathbf{P}^H(j\omega_o)\mathbf{P}(j\omega_o)]\right) = 0 \tag{8.149}$$

where we assume that the plant dynamics are relatively far away from $j\omega_o$. This expression can be reduced to the following:

$$\det\left[(s - j\omega_o)^2\mathbf{I} + \frac{\omega_o^2}{4R}\,[\mathbf{P}^H(j\omega_o)\mathbf{P}(j\omega_o)]\right] = 0 \tag{8.150}$$

with poles defined such that

$$s_i = j\omega_o - \frac{\omega_o}{2\sqrt{R}}\,\sigma_i(\mathbf{P}(j\omega_o)) \tag{8.151}$$

where $\sigma_i(\mathbf{P}(j\omega))$ is the $i$th singular value of $\mathbf{P}(j\omega)$, as discussed previously. The time constants are controlled by the real part of the poles:

$$\tau_i = \frac{2\sqrt{R}}{\omega_o\sigma_i(\mathbf{P}(j\omega_o))} \tag{8.152}$$

Recall that this point was reached through the LQR formulation, which assumes that all of the system states are available. In practice, as we have discussed, this is seldom the case, and state estimation is required. Even so, the development demonstrates the sensitivity of this MIMO internal model approach to the modeled plant dynamics. As indicated in equation (8.152), the time constants of the closed-loop system are controlled by the dynamics of the system evaluated at $\omega_o$. If the dynamics of the plant "drift," or if the frequency of the disturbance $\omega_o$ drifts during the operation, then the closed-loop performance will be sacrificed and the notch filter made possible by equation (8.144) will no longer be aligned with the new frequency of the disturbance $\hat{\omega_o}$.

Both of the control approaches presented for the rejection of a periodic disturbance motivate us to consider an alternative approach: adaptive feedforward control, which has been discussed in detail earlier in this chapter. The implementation of an adaptive controller for the rejection of periodic disturbances provides a means of adapting the notch filter such that it is always aligned with the period of the disturbance presented to the system. The feedback control system can be used in parallel to suppress those exogenous disturbance inputs for which a reference is not available, typically broadband signals that are less subject to the robustness problems observed for narrowband disturbances.

The feedback control system provides an additional benefit when implemented simultaneously with adaptive, feedforward control. That benefit is improved convergence properties of the adaptive filters. Recall that a filtered reference signal is required to implement the filtered-$x$ LMS algorithm for adaptive feedforward control. The filtered reference is created by modeling the control-to-error dynamics over the bandwidth of the disturbance signals that are to be controlled with the feedforward control. It can be shown that the feedback controller reduces the eigenvalue spread of the input correlation matrix by reducing the spectral norm of those control-to-error dynamics. The power spectral density of the filtered reference signal can be obtained from the power spectral density of the reference and the plant dynamics, as follows (Saunders et al., 1993):

$$\mathbf{S}_{r_P r_P}(j\omega) = |\mathbf{P}(j\omega)|^2 \mathbf{S}_{rr}(j\omega) \tag{8.153}$$

$$= \mathbf{P}(j\omega)\mathbf{S}_{rr}(j\omega)\mathbf{P}^H(j\omega) \tag{8.154}$$

where $\mathbf{S}_{r_P r_P}(j\omega)$ is the power spectral density of the filtered signal and $\mathbf{S}_{rr}(j\omega)$ is the power spectral density of the reference signal. If we let $\mathbf{T}(j\omega)$ represent the stable, closed-loop plant, then the power spectral density of the filtered reference in the presence of feedback can be expressed as follows:

$$\mathbf{S}_{r_T r_T}(j\omega) = |\mathbf{T}(j\omega)|^2 \mathbf{S}_{rr}(j\omega) \tag{8.155}$$

$$= \mathbf{T}(j\omega)\mathbf{S}_{rr}(j\omega)\mathbf{T}^H(j\omega) \tag{8.156}$$

where $\mathbf{S}_{r_T r_T}(j\omega)$ is the power spectral density of the signal filtered through the stable, closed-loop plant. The total power of this filtered reference can be obtained from the trace of this matrix:

$$\mathrm{Tr}[\mathbf{S}_{r_T r_T}(j\omega)] = \mathrm{Tr}[\mathbf{T}(j\omega)\mathbf{S}_{rr}(j\omega)\mathbf{T}^H(j\omega)] \tag{8.157}$$

If the feedback control system is designed to reject impulsive and stochastic disturbance inputs, then the signal power of the closed-loop filtered reference will be smaller than that of the open-loop filtered reference. Thus, in the presence of feedback, the trace of the filtered, input correlation matrix decreases. The spectral norm of this matrix is controlled by its largest eigenvalue, which is reduced, since the feedback control system serves to reduce the principal gains of $\mathbf{T}(j\omega)r(j\omega)$. This reduction in the spectral norm results in a corresponding reduction in the spread of eigenvalues of the input correlation matrix. Thus, the disparity in the time constants also decreases. As pointed out in Chapter 4, the stability of the LMS algorithm is controlled by the convergence parameter, which is bounded by the reciprocal of the trace of the input correlation matrix. Hence, the stability bounds are increased whenever the trace of the input correlation matrix is decreased.

The hybrid configuration also benefits other adaptive feedforward control strategies. For example, if the TAG descent algorithm is implemented after the feedback portion of the adaptive structure is designed, the time required for convergence is decreased also. As has been discussed in Section 8.5.2 of this chapter, the time required for convergence of the TAG descent algorithm depends upon the timescales of the plant, because each perturbation of the filter weight produces an impulsive disturbance that must be allowed to decay adequately before averaging the mean-squared error (MSE) to minimize the bias on the estimate of the MSE. However, the closed-loop structure usually exhibits faster timescales, which serves to decrease the time required for decay of system transient response when the filter coefficients are perturbed. Thus, the hybrid configuration provides for faster, real-time convergence of the TAG descent algorithm. The use of hybrid control algorithms is not widespread at this time and remains the subject of ongoing research.

## 8.8 SUMMARY

Chapter 8 was devoted to multivariable control system design. As in Chapter 7, the disturbance rejection problem specific to the adaptive structure was developed. As outlined, one must consider the linear fractional transformation that yields the disturbance-to-error path in the design of adaptive structures, and as such, the reduction of the closed-loop response typically cannot be guaranteed by a simple reduction in the sensitivity transfer matrix over the bandwidth of interest. This observation results from the fact that the disturbance inputs and control inputs to an adaptive structure are rarely the same, and the error outputs and sensed or measured outputs are rarely the same. The two-port model was developed, and the structure of this model with respect to the essential physics of the control problem was identified. The development of the two-port model led to a discussion of the cost functional and its importance in the design of adaptive structures. Methods of design—both static (constant gain) and dynamic (frequency dependent) compensators—were developed based upon optimal control theory. Both $\mathcal{H}_2$ and $\mathcal{H}_\infty$ design approaches were presented. Methods of implementing multivariable, adaptive feedforward controllers were introduced, and the combination of adaptive feedforward and feedback control (i.e., hybrid control) was outlined. The complementary nature of these two control strategies was identified, and the importance of concurrent consideration in the design process was emphasized.

## BIBLIOGRAPHY


Baumann, W. T., W. R. Saunders, and H. H. Robertshaw, 1991. "Active Suppression of Acoustic Radiation from Impulsively Excited Structures," *Journal of the Acoustical Society of America*, **90**(6), 3202–3208.

Boyd, S. and C. Barratt, 1991. *Linear Controller Design: Limits of Performance*, Prentice-Hall, Englewood Cliffs, NJ.

Clark, R. L. and D. E. Cox, 1996. "Multi-Variable Structural Acoustic Control with Static Compensation," submitted to the *Journal of the Acoustical Society of America*.

Doyle, J. C. and G. Stein, 1981. "Multivariable Feedback Design: Concepts for a Classical/Modern Synthesis," *IEEE Transactions on Automatic Control*, **AC-26**(1), 4–16.

Doyle, J. C., K. Glover, P. P. Khargonekar, and B. A. Francis, 1989. "State-Space Solutions to Standard $H_2$ and $H_\infty$ Control Problems," *IEEE Transactions on Automatic Control*, **34**(8), 831–847.

Elliot, S. J., I. M Stothers, and P. A. Nelson, 1987. "A Multiple Error LMS Algorithm and Its Application to the Active Control of Sound and Vibration," *IEEE Transactions on Acoustics, Speech, and Signal Processing*, **ASSP-35**(10), 1423–1434.

Gibbs, G. P. and Clark, R. L., 1993. "Feedforward Higher Harmonic Control Using the H-TAG Algorithm," *Proceedings of Noise-Con 1993*, Williamsburg, VA.

Gupta, N. K., 1980. "Frequency Shaped Cost Functionals: Extension of Linear-Quadratic-Gaussian Design Methods," *Journal of Guidance and Control*, **3**(6), 529–535.

Haykin, S., 1991. *Adaptive Filter Theory*, Prentice-Hall, Inc., Englewood Cliffs, NJ.

Kewley, D. L., R. L. Clark, and S. C. Southward, 1995. "Feedforward Control Using the Higher-Harmonic, Time-Averaged Gradient Descent Algorithm," *Journal of the Acoustical Society of America*, **97**(5), 2892–2905.

Kuo, S. M. and D. R. Morgan, 1996. *Active Noise Control Systems*, Wiley, New York.

Kwakernaak, H. and R. Sivan, 1972. *Linear Optimal Control Systems*, Wiley, New York.

Levine, W. S. and M. Athans, 1970. "On the Determination of the Optimal Constant

Output Feedback Gains for Linear Multivariable Systems," *IEEE Transactions on Automatic Control*, **AC-15**(1), 44–48.

Maciejowski, J. M., 1989. *Multivariable Feedback Design*, Addison-Wesley, New York.

Makila, P. M. and H. T. Toivonen, 1987. "Computational Methods for Parametric LQ problems—A Survey," *IEEE Transactions on Automatic Control*, **AC-32**(8), 658–671.

MacFarlane, A. G. J. and N. Karcanias, 1976. "Poles and zeros of linear multivariable systems: A survey of the algebraic, geometric and complex variable theory." *International Journal of Control*, **24**, 33–74.

Saunders, W. R., H. H. Robertshaw, and R. A. Burdisso, 1993. "An Evaluation of Feedback, Adaptive Feedforward, and Hybrid Controller Designs for Active Structural Control of Lightly-Damped Structures," *Proceedings of the Conference on Recent Advances in Active Control*, Blacksburg, VA, Technomic Press, pp. 339–354.

Sievers, L. A. and A. H. von Flotow, 1990. "Comparison and Extensions of Control Methods for Narrowband Disturbance Rejection," *Proceedings of the ASME Winter Annual Meeting*, Dallas, TX.

Skogestad, S. and I. Postlethwaite, 1996. *Multivariable Feedback Control, Analysis and Design*, Wiley, New York.

Stein, G. and M. Athans, 1987. "The LQG/LTR Procedure for Multivariable Feedback Control Design," *IEEE Transactions on Automatic Control*, **AC-32**(2), 105–114.

Toivonen, H. T., 1985. "A Globally Convergent Algorithm for the Optimal Constant Output Feedback Problem," *International Journal of Control*, **41**, 1589–1599.

Widrow, B. and S. D. Stearns, 1985. *Adaptive Signal Processing*, Prentice-Hall, Englewood Cliffs, NJ.

Zhou, K., J. C. Doyle, and K. Glover, 1996. *Robust and Optimal Control*, Prentice Hall, Upper Saddle River, NJ.

# 9

# ADAPTIVE STRUCTURES: DYNAMICS AND CONTROL

The objective of this chapter is to illustrate the use of a general approach that can be used for prototype design. At this point, we have reviewed in detail a number of areas of importance for the design and synthesis of adaptive structures. The tutorial examples presented in this chapter serve to integrate the topics previously presented in a manner that aids sequential design of the dynamically controlled system. For each example provided, the following format has been adopted for presentation:

1. Problem Overview
2. Description of the Plant
3. Transducer Selection, Design, and Placement
4. Development of the Augmented Plant Model
5. Control Strategy
6. Simulation or Experimental Results
7. Summary

This basic outline forms a *recipe* for adaptive structures and guides the reader in the basic thought process required in the development of the problem. Before initiating the design of an adaptive structure, one should review the general performance requirements, physics associated with the system dynamics, transducers available for implementation, bandwidth requirements, types of external disturbances present, and realizable control approaches. Organizing the initial design accordingly ensures that each topic of importance has been addressed.

## 9.1 ACTIVE CONTROL OF VIBRATION

### 9.1.1 Problem Overview

The primary purpose of this example is to investigate the design of feedback, adaptive feedforward, and hybrid vibration suppression in a panel structure configured with piezoelectric wafer transducers. The structural response is assumed to derive from the presence of either periodic or stochastic point-force disturbances. This constrained adaptive structure design relies on the existence of a dynamic structural model and/or the availability of an actual test structure. In developing a dynamic model, the importance of including a sufficient number of structural modes for analysis and design of any adaptive structure is demonstrated. Next, both $\mathcal{H}_2$ and $\mathcal{H}_\infty$ controllers are designed to mitigate the effects of a stochastic disturbance source. An adaptive feedforward controller is then implemented for a periodic source, and the final example combines both the feedback and adaptive feedforward elements to realize the hybrid controller. After completing the active vibration control example, the reader is given an opportunity to execute the Matlab script files that were used to create the accompanying results. The following discussion will lead to a better appreciation of the varied design methods and effects brought about by the different control system architectures. However, the reader is strongly encouraged to experiment with the corresponding script files in order to investigate countless other variations of adaptive structure designs for panel piezostructures.

### 9.1.2 Description of the Plant

A schematic diagram of a simply supported plate is presented in Figure 9.1. The plate is augmented with piezoelectric transducers that were used as both sensors and actuators (sensoriactuators). The model development is thus consistent with that of the piezostructure presented in Section 5.3.1. The dimensions of the plate and transducers are illustrated in Figure 9.1 and have been used in the script file **p_model.m**. For the example presented, the material properties of the plate are based on aluminum, using an assumption of proportional damping (damping ratio = 5%) to provide a dissipative mechanism. Both stochastic and periodic exogenous disturbances are applied to the structure as concentrated, transverse forces. The modeling is presented using a generalized coordinate system, with the disturbance forces exciting each mode of the system uniformly.

### 9.1.3 Transducer Selection, Design, and Placement

As illustrated in Figure 9.1, the plate is configured with three piezoelectric transducers, which are used as colocated transducer pairs (sensoriactuators). The locations of the transducers are not optimized for controllability or observability as part of the formal design process. They are simply selected in an arbitrary fashion. (Transducer placement is a very important consideration in the

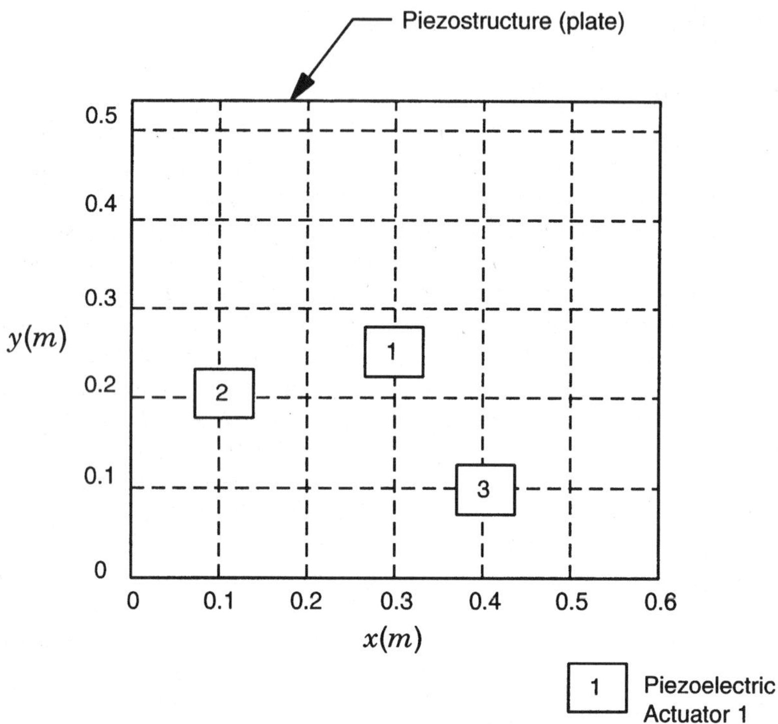


**Figure 9.1:** Schematic diagram of plate configured with piezoelectric transducers.

design process; numerous schemes exist in the literature based on theoretical and/or experimental dynamic descriptions of the structure. The reader is encouraged to pursue this advanced topic after mastering the modeling and controller design issues described here.) The dimensions of the transducers are based upon realistic dimensions of commercially available piezoceramic elements. In this example, it is assumed that perfect compensation of the feedthrough dynamics associated with the electrically induced charge has been obtained (i.e., perfect resolution of the mechanical response).

Executing the script file **p_model.m** produces three different state-variable models of the piezostructure, including $N = 10$, 30, and 40 structural modes, respectively. Assuming the modal parameters are known, the out-of-plane displacement can be represented using the modal expansion theorem

$$w(\mathbf{x}, t) = \sum_{n=1}^{N} \phi_n(\mathbf{x})\eta_n(t) \tag{9.1}$$

For this example, the measured outputs $y(t)$ are considered to be strain rate proportional signals (i.e., current for the sensoriactuator), and the control inputs $u(t)$

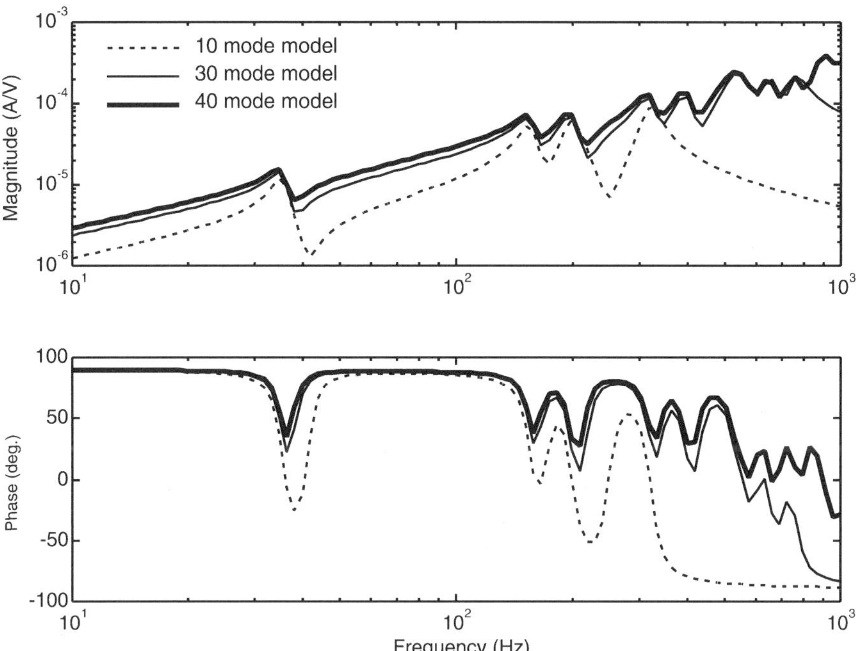


**Figure 9.2:** Comparison of plate frequency responses with increasing number of modes included in the dynamic model.

correspond to the applied piezoelectric voltage. The model also includes inputs and outputs in generalized coordinates, $\eta_n(t)$ for the structural system. As an example, for the 30-mode model, the first 30 inputs correspond to the generalized control forces and the first 30 outputs correspond to the structural velocities in generalized coordinates. The last three inputs and outputs correspond to the physical signals of applied voltages and mechanically induced currents for the piezoelectric sensoriactuator transducers. A typical frequency response of the piezostructure is presented in Figure 9.2 for the colocated transducer pair labeled 1 in Figure 9.1. For comparison, the same frequency responses, computed with 10, 30, and 40 mode models, are illustrated to demonstrate convergence of the zeros over the bandwidth of interest (0–400 Hz). This convergence is critical if the objective is to predict realistic closed-loop performance as outlined by Clark (1997). For the discussions within the remainder of this section, the 40-mode model of the plate is used to evaluate the control system performance over a bandwidth extending from 0 to 400 Hz.

### 9.1.4 Development of the Augmented Plant Model

The objective of the control system design for this example involves minimizing the vibration response of the structure subject to periodic or stochastic distur-

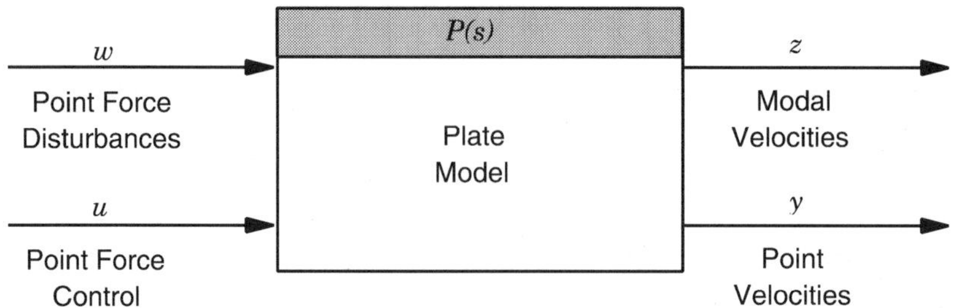


**Figure 9.3:** Block diagram of plate model used in the example illustrating the appropriate inputs and outputs.

bances. To accomplish this objective, one must choose exogenous inputs and a measure of the vibration to establish the *performance path* in the general two-port model of the system. Consider the block diagram presented in Figure 9.3. As illustrated, the performance path requires generalized force inputs and generalized velocity outputs. Thus, each mode of the structure can be driven with a stochastically independent disturbance. The outputs allow for a means of minimizing the vibration of each structural mode. Of course, the lower control path contains the inputs and outputs to the piezoelectric sensoriactuators. Hence, for a 40-mode model, the plate has 43 inputs and 43 outputs (40 for the generalized velocities and 3 for the piezoelectric transducers). The generalized plant includes standard LQG weightings, as illustrated in Figure 8.19.

### 9.1.5 Control Strategy

Selection of the control strategy depends on many factors, which have been discussed in earlier chapters: model accuracy, global versus local control, convergence requirements for adaptive disturbance rejection, availability of colocated transducers, and many more. The active vibration control system for this example may be accomplished using either feedback or feedforward methods. Because the problem is cast using generalized coordinates, the control performance will be global, but the design is essentially local in its architecture; that is, local minimization of each coordinate results in global minimization of the structural vibration related to that coordinate (mode).

The applicability of both feedforward and feedback control for this problem allows us to investigate the design of two different feedback compensators, an adaptive feedforward compensator, and a hybrid controller. For the augmented plant discussed in the previous section, dynamic compensators are designed using an $\mathcal{H}_2$ norm and $\mathcal{H}_\infty$ norm. Open-loop and closed-loop response is compared. As noted previously, in the development of the generalized plant $\mathbf{G}(s)$, the plant $\mathbf{P}(s)$ is augmented such that the design approach follows the standard LQG weightings. Upon completing the feedback examples, adaptive feedforward control is demonstrated using the filtered-$x$ LMS algorithm. Both the system identification and control system implementation are demonstrated through the use of MATLAB script files. The final example includes the demonstration of a hybrid control system that combines the $\mathcal{H}_2$ compensator and the adaptive, feedforward controller.

### 9.1.6 Results from Analysis and Simulation

***An $\mathcal{H}_2$ Vibration Controller for a Stochastic Disturbance*** Consider the block diagram of the generalized plant $\mathbf{G}(s)$, presented in Figure 8.19, with a dynamic plant $\mathbf{P}(s)$, illustrated in Figure 9.3. A dynamic compensator is designed to control the kinetic energy of the structure such that the generalized velocities of the structural modes are minimized. The script file **pv_cont.m** has been used to develop the dynamic compensator. The magnitudes of the control effort penalty, performance penalty, and so on, are documented in the script file. The open-loop and closed-loop responses are compared in Figure 9.4. The results illustrate the $\mathcal{H}_2$ norm of the matrix transfer function at each frequency between the disturbance input and the generalized velocities, $\|\mathbf{G}_{vw}(j\omega)\|_2$, where the subscript $v$ is used to indicate the velocity output and the subscript $w$ is used to represent the disturbance input. It is assumed that the stochastic processes drive each generalized force independently and the data plotted have been computed from the square root of the sum of the squared singular values at each frequency. As

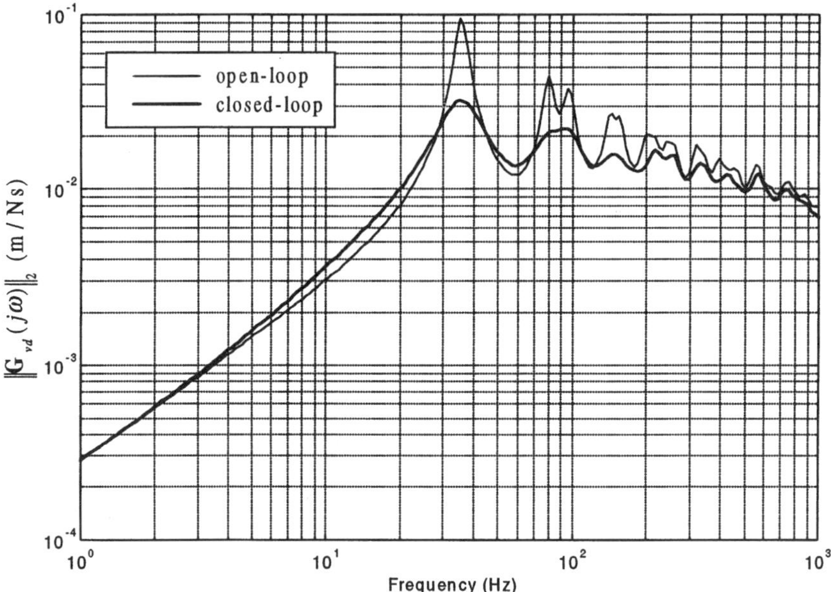


**Figure 9.4:** Comparison of open-loop and closed-loop RMS power of plate for $\mathcal{H}_2$ controller design.

indicated, for the chosen parametric values, the dynamic compensator provides significant levels of attenuation over the bandwidth.

For comparison, an additional script file is provided: **lqg_comp.m**. Within this file, the same structural model is created, and a typical LQR regulator and LQG estimator design are completed. The resulting open-loop and closed-loop structural responses generated with this script file are identical to that obtained with the $\mathcal{H}_2$ controller design approach. As discussed in the previous chapter and by Doyle et al. (1989), the LQG problem can be cast from the more general $\mathcal{H}_2$ controller design, and this example serves to substantiate this conclusion.

***An $\mathcal{H}_\infty$ Vibration Controller for Stochastic Disturbance*** As an alternative, an $\mathcal{H}_\infty$ controller has been designed using the identical piezostructure panel. The same performance weightings are used in both cases. To design the controller, the script file **pv_cont.m** can be used again. For the $\mathcal{H}_\infty$ design, the variable *Norm* should be changed to *Norm='inf';* in the script file. Choosing between *'inf'* and *'two'* determines which controller is implemented. The results from this control system design are presented in Figure 9.5. As in the previous example, the figure shows the $\mathcal{H}_2$ norm of the transfer function matrix between the disturbance and the output velocities for the open-loop and closed-loop responses. Upon comparing the closed-loop response of the $\mathcal{H}_\infty$ controller to that of the $\mathcal{H}_2$ controller, Figure 9.4, one observes very little difference in

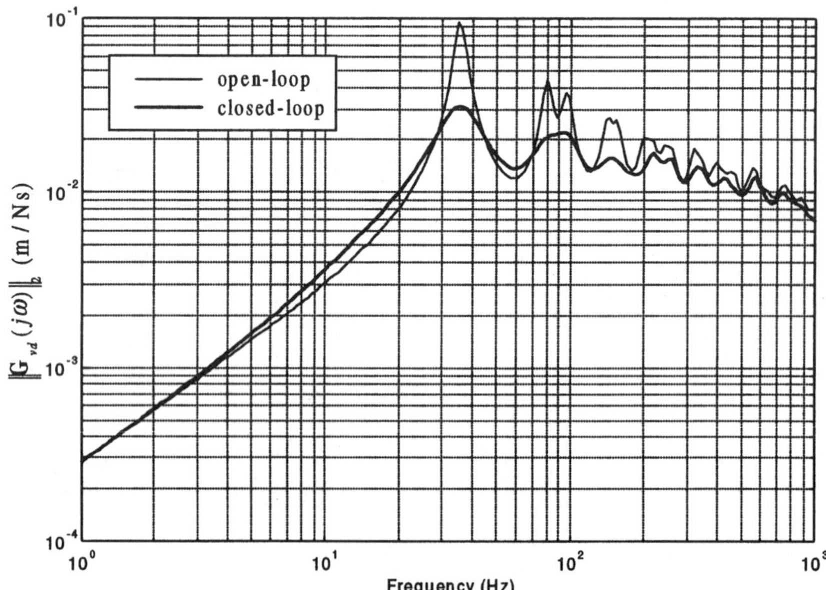


**Figure 9.5:** Comparison of open-loop and closed-loop RMS power of plate for $\mathcal{H}_\infty$ controller design.

the control effectiveness for this particular example. However, this observation is specific to the design example. The reader should recognize that, if the plant displayed a significant peak in the response at a particular frequency or group of frequencies (e.g., for lightly damped structures, $\xi < 1$) then the resulting controllers would show less similarity, as the $\mathcal{H}_\infty$ design is more aggressive on the peak gain of the system. Additionally, the weighting functions chosen for the design example presented are not intended as a general choice for weighting functions; weighting functions are design specific. Rather, the weightings have been selected primarily for comparison with the $\mathcal{H}_2$ design. Recall that the weightings for the $\mathcal{H}_\infty$ controller are the same as the $\mathcal{H}_2$ controller that is responsible for the strikingly similar results. The reader is encouraged to excercise **pv_cont.m** with different damping ratios to observe the differences in the $\mathcal{H}_2$ and the $\mathcal{H}_\infty$ controllers.

***Simulation of Adaptive Feedforward Control*** In the following example, a discrete-time model of the piezostructure panel has been generated from the continuous-time model used in the previous examples. The discrete-time model is required to simulate the digital implementation of the filtered-$x$, adaptive feedforward control system. For the purpose of this example, the strain rate "measured by" each of the sensoriactuators is used to formulate the cost function for the least-mean-squares (LMS) algorithm. Thus, a local cost function is used in the convergence of the adaptive control system. Note that, by defining a local cost function for the feedforward control, the minimization of response is guaranteed only at the outputs of the sensors that are used to construct the error. If generalized coordinates had been selected to construct the error, the effect would have been global, as seen in the previous examples. For this local control method, the plant response can actually increase at other locations and the global dynamics (i.e., poles) are not modified. Before implementing the adaptive control system, the control system path must be identified at the frequency or frequencies to be controlled. In this example, we wish to suppress a sinusoidal, point-force disturbance force entering at a physical coordinate corresponding to $(0.6L_x, 0.6L_y)$ and oscillating at a frequency of 600 Hz. Thus, only a two-coefficient filter is required to create the filtered reference signal. Since the control system incorporates three inputs and three outputs, a total of nine filters with two coefficients each are required to create the filtered reference. The script file **lmsidmim.m** can be used to build the discrete-time model of the structure, and the LMS algorithm is used to identify the plant at the disturbance frequency. These filters are stored in the file **wid.mat** for use in the filtered-$x$ implementation.

After identifying the plant along the control path to the error sensors, one can implement the filtered-$x$ LMS algorithm. For the test structure used in this study, the script file **filtx.m** is used to simulate the control system response. Upon executing the files, the results presented in Figures 9.6 and 9.7 are generated. In general, all control-to-error transfer function paths must be identified. In the script file, this is accomplished by sequentially driving each control channel

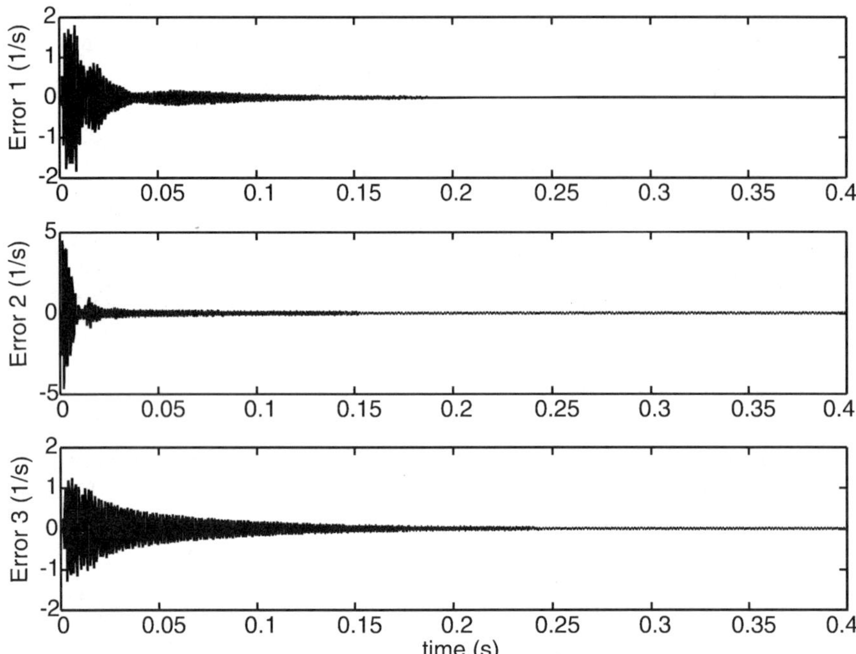


**Figure 9.6:** Error signals during convergence of the filtered-$x$ LMS algorithm for adaptive feedforward control of a harmonic disturbance.

and converging the filters for the respective control-to-error transfer functions. The results presented in Figures 9.6 and 9.7 represent the convergence of the algorithm. The convergence of the adaptive filters is apparent in Figure 9.6, since the error signal at each of the sensors is monitored as a function of time. For the chosen convergence parameter, the error converges in approximately 0.25 seconds. The adaptation of the filter weights associated with the third control channel is depicted in Figure 9.7. Again, one observes that the filter weights adapt to their steady-state values in approximately 0.25 seconds. These results are typical for all of the control channels required to obtain the $3 \times 3 = 9$ transfer function paths. Recall that the convergence parameter can be used to increase or decrease the convergence time of the algorithm, constrained by the stability requirements placed upon the adaptive algorithm (discussed in Chapter 4).

***Simulation of Hybrid Control*** While the filtered-$x$ LMS algorithm can be used to successfully control a periodic disturbance, as discussed in Chapter 8, feedback control is often more appropriate for control of stochastic disturbances. This is particularly true when an uncontrollable reference signal cannot be identified. In this example, we consider the general case when the structure is subject to both periodic and stochastic disturbances. To demonstrate the complimentary performance of the two control approaches for this situation, the hybrid

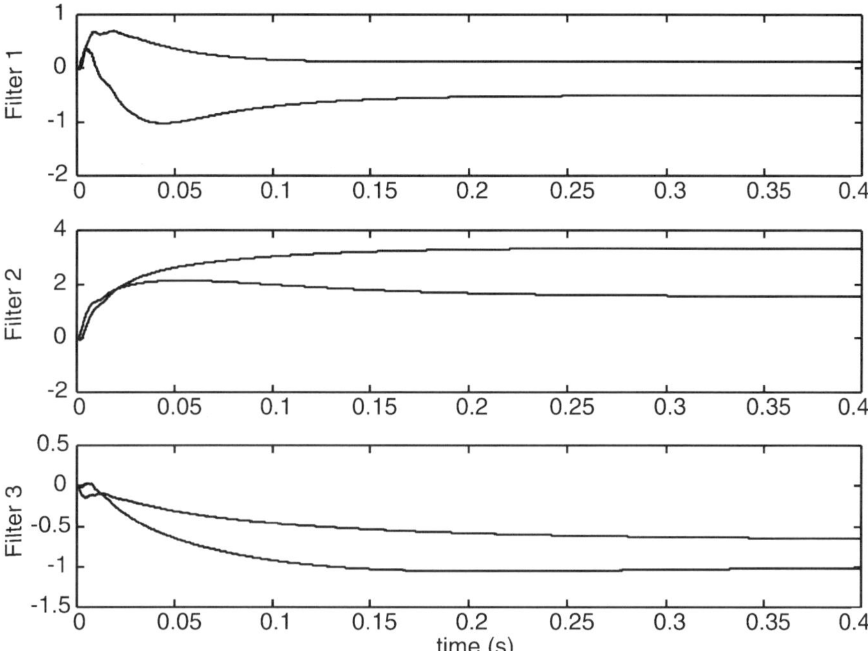


**Figure 9.7:** Adaptation of filter weights associated with the third control channel during convergence of the filtered-$x$ LMS algorithm for feedforward control of the plated presented with a harmonic disturbance.

control approach described in Chapter 8 is implemented on the piezostructure panel. The same three-input, three-output system used in the previous example is used here; however, the dynamic compensator designed earlier from the $\mathcal{H}_2$ synthesis for the stochastic disturbances is also incorporated into the control path. This modifies the plant that is acted on by the adaptive feedforward controller. Because the feedback path is closed by the dynamic compensator, the control inputs and sensed outputs of the closed-loop plant are used to implement the adaptive feedforward controller. The script file **hyblmsid.m** can be used to identify the modified control-to-error path filters required to generate the filtered reference signal, and the script file **hybrid.m** can be used to generate the controlled responses presented in Figures 9.8 and 9.9.

When the hybrid control system is engaged, linearly combining both feedback and adaptive feedforward control, each of the three error signals converge within approximately 0.1 seconds for the chosen convergence parameter, as illustrated in Figure 9.8. The results are plotted on the same scale that was used in Figure 9.6 to display the improved convergence characteristics and transient response performance. Note that there is an initial transient response of the structure when the sinusoidal disturbance is introduced at $t = 0$. This transient response decays much faster, and the initial magnitude of the response is

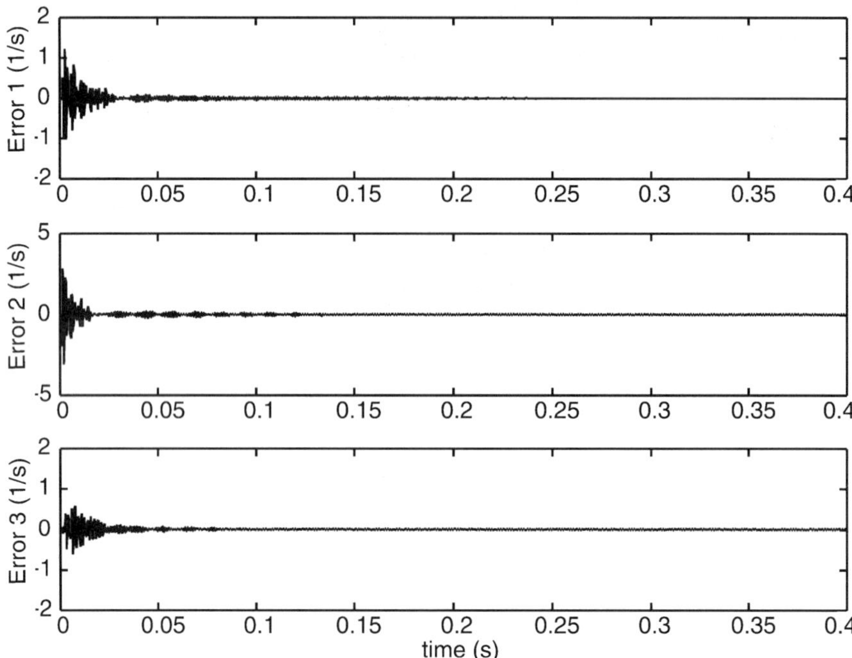


**Figure 9.8:** Error signals during convergence of the filtered-$x$ LMS algorithm for hybrid control of the plated presented with a harmonic disturbance.

much less when the hybrid control system is implemented due to the increased input impedance associated with the feedback control system. Additionally, the convergence parameter can be increased even further for the hybrid control scenario, resulting in convergence gains that would cause instabilities for the open-loop implementation of the filtered-$x$ LMS algorithm. Recall that the rate of convergence increases as a result of less disparity between eigenvalues of the autocorrelation matrix of the filtered reference signal, as noted previously in Section 8.7. While not displayed in the results presented, the hybrid controller also results in improved performance associated with stochastic disturbances. The performance of the dynamic compensator over the bandwidth of interest was demonstrated earlier in Section 9.1.6.

It is appropriate at this point to address the relative importance of the system identification process required for implementation of the filtered-$x$ LMS algorithm in the hybrid control system configuration. As in the prior example for adaptive feedforward control, the frequency responses of the structure from the control inputs to the sensed outputs are measured at the frequency of the disturbance input. This information is used to ensure that the relative phase of the reference signal is accurately modeled for convergence of the algorithm. As noted in equation (8.68), the convergence parameter $\mu$ is multiplied by the "filtered" reference signal and the error signal. As long as the feedback controller does not

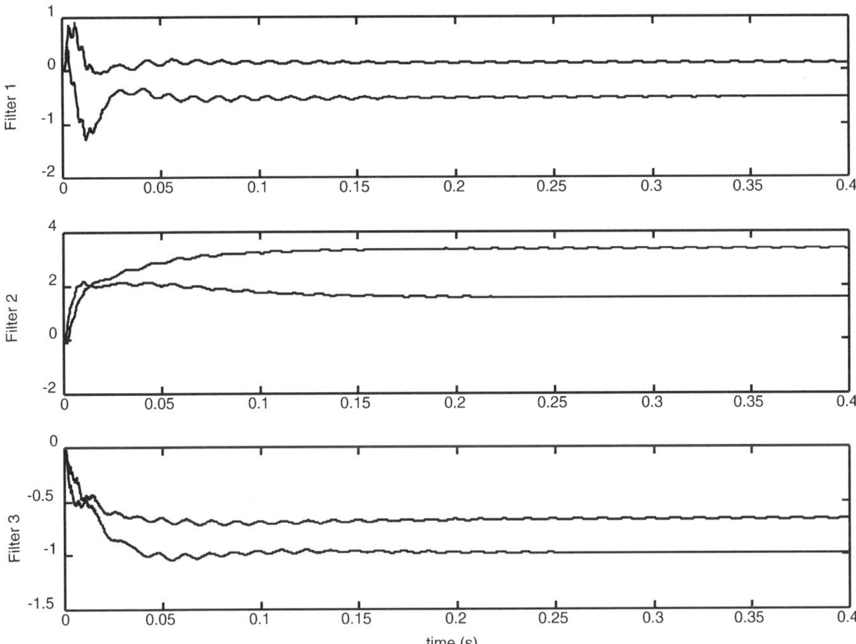


**Figure 9.9:** Adaptation of filter weights associated with the third control channel during convergence of the filtered-$x$ LMS algorithm for hybrid control of the plated presented with a harmonic disturbance.

significantly alter the phase associated with the plant frequency response at a frequency corresponding to that of the disturbance, the algorithm will converge using the system identification obtained for the open-loop plant. However, if the frequency or frequencies of the input disturbance are in close proximity to resonances of the open-loop system, the addition of feedback control may shift the resonances above or below the frequency of the harmonic disturbance. The relative phase of the structural output can change theoretically by as much as 180 degrees (undamped response only) at frequencies above and below resonances of the reverberant plant. Thus, a phase inversion may occur between the open-loop and closed-loop system identification process. This inversion is equivalent to a change in sign on the convergence parameter, which will obviously lead to instability of the LMS algorithm. Thus, the use of open-loop system identification results for hybrid control implementation should be carefully considered. As discussed, it is most warranted when designing hybrid controllers for structures with significant open-loop damping levels.

To conclude this discussion, we refer to the stability bounds on the filtered-$x$ LMS algorithm. As shown by Morgan (1980) for the single-channel case, the phase difference between the identified model and the plant must be bounded by $\pm 90$ degrees for convergence. The extension to multivariable concepts is a bit more involved, but the stated bound provides the reader with a reasonable stability requirement. Ideally, one would want to model the closed-loop response of the system at the desired frequencies; however, the filtered-$x$ LMS algorithm is relatively robust, and, unless the disturbance is in close proximity to a resonant frequency or the plant dynamics are significantly modified (i.e., both magnitude and phase over the entire bandwidth), the open-loop system identification can be used in most cases for implementation of the filtered-$x$ LMS algorithm. To demonstrate this point, the reader is encouraged to use the system identification obtained in the prior example (from script file **lmsidmim.m**) with the hybrid control system example.

### 9.1.7 Summary of Vibration Control Example

In the preceeding tutorial examples, feedback, adaptive feedforward, and hybrid control were demonstrated. The significance of including a sufficient number of structural modes in the dynamic model to provide convergence of the open-loop zeros of the transfer function was demonstrated. If an insufficient number of modes are included in the modeling effort, the structure will appear much stiffer than it is in reality. For the feedback applications, controllers were designed based upon $\mathcal{H}_2$ synthesis, $\mathcal{H}_\infty$ synthesis, and standard LQG synthesis. LQG style weighting functions were used in the generalized plant, and for such cases, the compensator resulting from the $\mathcal{H}_2$ design was identical to that achieved with the LQG design, as was expected. The filtered-$x$ LMS algorithm was simulated for control of a single-frequency disturbance input. Both system identification and control were demonstrated. The adaptive feedforward controller was implemented parallel with the feedback compensator resulting from the $\mathcal{H}_2$ design, and hybrid control was demonstrated. The transient response of the adaptive control system was observed to converge on a much faster timescale due to the modification of the system dynamics resulting from the feedback control. The rate of convergence for the adaptive controller is thus increased in the hybrid configuration.

## 9.2 ACTIVE STRUCTURAL ACOUSTIC CONTROL (FREQUENCY SHAPED COST)

### 9.2.1 Problem Overview

The purpose of this example is to demonstrate methods of including a frequency shaped cost functional and the relative importance of the cost functional in the design of feedback compensation (Gupta, 1980). There are many reasons for including a frequency shaped compensator, which include the need for band-limited performance, the need to limit control effort to a specified number of states, the need to specify the interaction of structural modes in the response, and so on. Thus, whether the frequency shaped cost functional is "flat" over the bandwidth or "weighted," the designer is forced to identify the desired performance variables. If the performance variables cannot be measured directly with transducers placed on the structure, the resulting controller is said to be "model based." In structural acoustic control applications, the designer is concerned primarily with the sound power radiated from the structure. Since this cannot be measured directly, a model of the essential physics must be developed to synthesize the controller. As will be demonstrated in the example provided, the standard two-port formulation of the plant naturally accommodates the additional dynamics required to create the performance variables (frequency shaped cost). In this tutorial, active structural acoustic control (ASAC) is compared to active vibration control (AVC), and the importance of the frequency shaped cost (radiation filters) required to develop a performance metric that relates the structural response to the radiated acoustic energy is demonstrated. To compare the controller resulting from the ASAC design and AVC design in a consistent manner, the control system design parameters (i.e., ratio of control effort penalty to performance penalty) are adjusted until the control energy resulting from each design is equivalent. Since the cost functionals are different, this common performance parameter can be used to compare the results on a "level playing field." Additionally, static compensation with direct output-feedback control is utilized and compared to the optimal (with respect to the stated cost functional) LQR control.

### 9.2.2 Description of the Plant

As in the prior example, an analytical model of a plate with simply supported boundary conditions has been developed for the purpose of this tutorial. A schematic diagram of the plate is presented in Figure 9.10. The plate is configured with three colocated point-force and point-velocity transducers at the locations illustrated. In addition to the inputs used for control, five point-force inputs are used to generate the disturbances and are also illustrated in Figure 9.10. All of the disturbances are driven with statistically uncorrelated stochastic processes. The dimensions of the plate and positions of the inputs and outputs illustrated are consistent with those contained in the script file **p_model.m**. The plate is made from aluminum, and a proportional damping ratio of 5% is assumed.

### 9.2.3 Transducer Selection, Design, and Placement

The emphasis of this tutorial is the realization of frequency shaped cost functionals and the subsequent design of the augmented plant. Thus, simple point-force and point-velocity inputs and outputs have been used. Distributed transducer designs are considered in the next tutorial. For this problem, the placement of the control inputs and disturbance inputs have been chosen arbitrarily, using a random number generator. The objective is to produce inputs that are capable of coupling to all of the structural modes and minimize the designer's bias in preferential selection. This process of "transducer placement" is not sug-

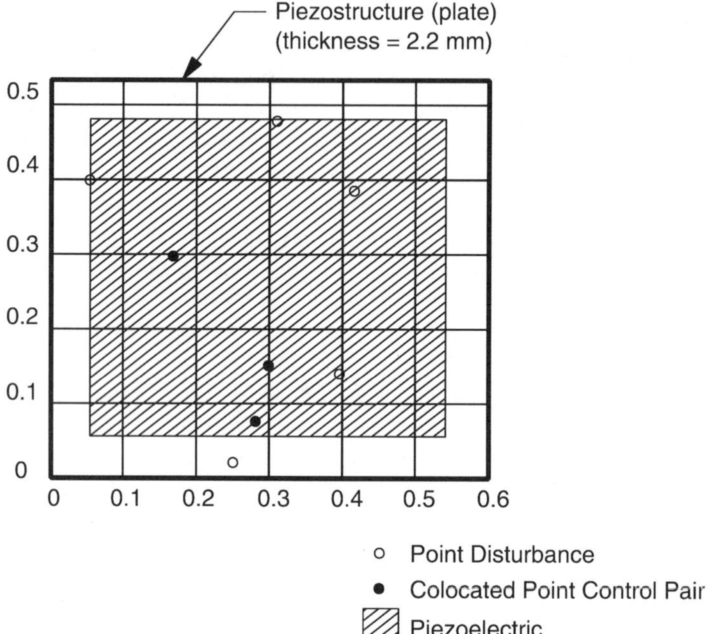


**Figure 9.10:** Schematic diagram of plate with pinned boundaries and configured with a single piezoelectric transducer and three colocated, point-force, point-velocity transducers.

gested for general design purposes, but the objectives in this example are aimed at demonstrating the importance of selecting an appropriate cost function, not at selecting or placing a transducer.

### 9.2.4 Development of the Augmented Plant Model

Consistent with the presentation described in Chapter 8, the augmented plant model follows the general two-port form illustrated in Figure 8.6. The modal velocities and the physical velocities colocated with the control forces are the outputs of the physical plant $\mathbf{P}(s)$, which is internal to the augmented plant $\mathbf{G}(s)$. For the purpose of this example, a 30-mode model of the panel is used for the design and analysis of the closed-loop system. As discussed previously, a sufficient number of modes must be used in the structural model to accurately capture the compliance of the system. It is important to note that the number of modes used to construct the cost function has no effect on the structural compliance of the plant. The number of modes used in the cost function are an essential variable in designing the performance path and are used to "shape" the closed-loop performance of the system. In constructing the augmented plant, the designer must identify whether there are physically motivated frequency dependencies of the states and outputs that should be considered. For example, a structure's modal displacements or modal velocities can be weighted as a function of frequency, and for structural acoustic control, it is necessary to weight the modal velocities.

In order to assemble the augmented plant, a model of the desired frequency weighted cost functional must be constructed. Since the objective is to design a controller that reduces the structural acoustic radiation, we must examine the essential physics that underly the dynamic coupling between the structure and the fluid (see Appendix C). This leads to the development of a frequency dependent model of the structural acoustic coupling, which becomes the distinguishing feature of the augmented plant $\mathbf{G}(s)$. For the purpose of this example, we assume that the panel is bounded by an infinite baffle, as illustrated in Figure C.1 of Appendix C. (Practically, a baffle appears "infinite" to the acoustic source if its dimensions are much greater than the acoustic wavelength. The face of a loudspeaker can be approximated as an "infinite" baffle at high frequencies.) This simplifying assumption leads to the easier Rayleigh integral formulation of the structural acoustic interaction. Details of computing the sound power radiating from a vibrating structure follow directly from solutions of the Rayleigh integral. As detailed in Appendix C, one can develop an expression that relates the modal velocity to the sound power radiated:

$$\overline{P}(j\omega) = \frac{k^2 \rho_o c}{4\pi^2} \mathbf{v}(j\omega)\mathbf{M}(j\omega)\mathbf{v}^*(j\omega) \tag{9.2}$$

The matrix $\mathbf{M}$ is frequency dependent and defines the interactive exchange of energy between the structural modes and the acoustic waves in the surrounding medium; $\mathbf{v}$ is a vector of the modal velocities of the structure, which are also functions of frequency. The objective is to generate a cost function that quantifies the sound power radiated, so we wish to design filters that exhibit frequency-response functions identical to the frequency-dependent characteristics of equation (9.2). One can lump the constants into a modified matrix $\tilde{\mathbf{M}}$ and express the sound power radiated as follows:

$$\overline{P}(j\omega) = \mathbf{v}(j\omega)\tilde{\mathbf{M}}(j\omega)\mathbf{v}^*(j\omega) \tag{9.3}$$

If the Laplace domain equivalent of the matrix $\tilde{\mathbf{M}}(j\omega)$ is expressed as $\tilde{\mathbf{M}}(s)$, then one can consider the spectral factorization of the matrix

$$\mathbf{M}(s) = \mathbf{W}^{\mathrm{T}}(-s)\mathbf{W}(s) \tag{9.4}$$

where $\mathbf{W}^{\mathrm{T}}(-s)$ is the parahermitian conjugate to $\mathbf{W}(s)$ and is simply the Laplace domain analog to the Hermitian conjugate of a matrix in the frequency domain. Then, using the frequency-response function notation, we can write equation (9.3) as

$$\overline{P}(j\omega) = \mathbf{v}(j\omega)\mathbf{W}^{\mathrm{T}}(-j\omega)[\mathbf{W}(j\omega)\mathbf{v}^{*}(j\omega)] \tag{9.5}$$

where the bracketed term represents a filtering operation by the matrix $\mathbf{W}(j\omega)$ on the modal velocities $\mathbf{v}^{*}(j\omega)$. The quantity $\mathbf{W}(s)$ is called a radiation filter, as detailed by Baumann et al. (1991). This filter can be obtained through a spectral factorization, as outlined by Francis (1986), and a state-variable model for the filter can be used to augment the system dynamics. The time-averaged sound power radiated from the structure when driven by a white noise excitation can thus be computed from the output of these augmented states.

For this example, the radiation filters have been designed from the self and mutual radiation efficiencies detailed in Appendix C, providing a normalized measure of the RMS acoustic power. This normalization only affects the choice of **Q** (i.e., the performance penalty) in the design optimization process. For comparison, the squared magnitude of several radiation filters are compared to respective plots of self and mutual radiation efficiencies from which they were derived in Figure 9.11. (The results presented in Figure 9.11 can be generated with the script file **plotflt.m**.) The squared magnitudes of the filters are represented by bold lines, while the computed radiation efficiencies are represented by thin lines. The self and mutual radiation efficiencies of the selected modes are indicated in the parentheses next to the appropriate curves, with the first two indices representing a particular mode and the second two indices representing the interacting mode. As

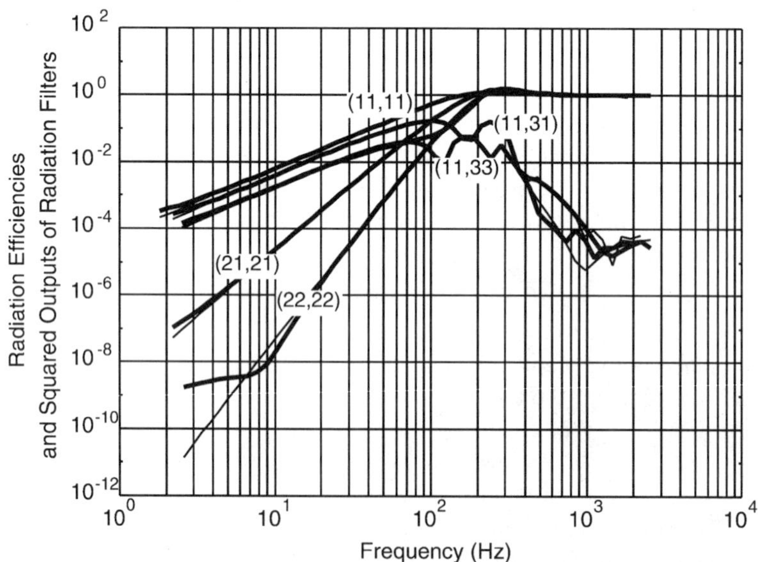


**Figure 9.11:** Comparison of radiation efficiencies of structural modes to the squared outputs of the radiation filters used to provide a measure of the power radiated.

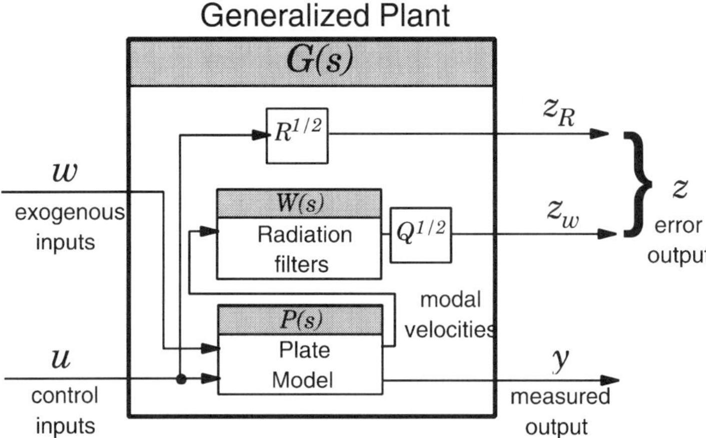


**Figure 9.12:** Block diagram of the generalized plant using the control system design for ASAC.

illustrated, the dynamics of the radiation filters accurately capture the frequency response of the self and mutual radiation efficiencies.

Given the radiation filter, one can readily form the generalized plant. As illustrated in Figure 9.12, the modal velocities from the plate model $\mathbf{P}(s)$ are the inputs to the radiation filters $\mathbf{W}(s)$. The outputs of the radiation filters are weighted by a performance penalty $\mathbf{Q}^{1/2} = \mathbf{I}$. The control inputs are fed to the plate model and also directly to a control effort penalty: $\mathbf{R}^{1/2} = \sqrt{R}\ \mathbf{I}$. The outputs from the control effort penalty and the performance penalty result in the error output $\mathbf{z}(s)$, which is used to form the cost function. The scalar $R$ allows the designer to trade-off attenuation in radiated sound power against the control energy required. Since the units of the control signal are not consistent with those of the radiated sound power, one must "experiment" with the scalar until reasonable values of control energy are obtained. One method of determining if an unrealistic level of control energy is required involves evaluating the singular values of the transfer matrix between the input disturbance and the resulting control signals. Evaluating the sum of the squared singular values at each frequency provides a measure of the ratio of RMS signal energy at each frequency. One would not want to implement a controller that required more control energy than disturbance energy, and, ideally, an order of magnitude less is a good design target. The disturbance inputs and measured outputs of the augmented model $\mathbf{G}(s)$ are obtained directly from the respective input and output paths for the plant model $\mathbf{P}(s)$, as illustrated in Figure 8.6.

### 9.2.5 Control Strategy

Two static compensators (i.e., constant gain compensators) are examined for this control system design example. Selection of the static versus dynamic compensator is motivated by the inclusion of the appropriate structural acoustic dynamics via the radiation filters/augmented plant. In other words, we wish to reduce the radiated power according to the frequency distribution determined by $\mathbf{W}(\mathbf{s})$. This leads to a static compensation strategy. For the first example, a static compensator is designed based upon the LQR approach for direct output feedback control. Details of the control system synthesis have been presented in Section 8.6.1. The measured outputs are the velocities of the structure, colocated with the three point-force inputs. Of course, these "ideal" transducers cannot be physically realized, but the purpose of this example is to convey the importance of frequency weighted cost functions. For the second example, the LQR approach has been used to design the optimal state-feedback controller. The controller is optimal with respect to the selected cost function. All of the states of the generalized plant—those of the radiation filter and those of the structural model—have been used to design the controller. Thus, the LQR controller represents the best possible control system performance for the chosen control input paths. The output feedback and state feedback control system designs serve as lower and upper bounds for the performance that can be expected when implementing a dynamic compensator based upon an $\mathcal{H}_2$ or LQG design.

To demonstrate the importance of the frequency shaped cost function, both controllers are designed using three different cost functions. Reductions in the panel's radiated sound power and estimates of the expended control energy are compared for each of the design cases. The first cost function is based upon the radiation filters and is designated as the ASAC cost. The second cost function is constructed directly from the modal velocities (i.e., flat weighting) and is designated as the AVC cost. This cost function provides a measure of structural energy in the panel. The third cost function is based upon a measure of the structural velocities over an equal grid of 20 points on the structure and is designated as the AVPTS cost. It serves to provide an alternative measure or formulation for the structural energy. Since the engineering units associated with each of the different cost functions vary, the scalar control effort penalty $R$ must be adjusted in each case such that the same amount of control energy is expended in the implementation of the controller. In making this adjustment, one assures that one controller does not out-perform the other because more control energy is available. This requires a little iteration on the part of the designer in terms of choosing $R$ and evaluating the $\mathcal{H}_2$ norm of the closed-loop system between the disturbance input and the control input. A more advanced approach based upon sequential quadratic programming (Gill et al., 1981) and nonlinear optimization can be used to choose the appropriate $R$ if desired.

### 9.2.6 Results from Analysis

***Static Compensation with Direct, Output-Feedback Control*** The initial design is performed assuming direct, rate-feedback control between the MIMO colocated point forces and point-velocity transducers. The controller design and simulation using the three different cost functions are demonstrated by example in the script file **perfilt.m**. The script file **perfilt.m** can be used to generate the results illustrated in Figures 9.13 and 9.14; however, if the reader desires to replicate the actual figures, the closed-loop results produced with the script file must be saved and plotted on a common graph.

To demonstrate the role of the frequency shaped cost functional, the performance results of the three controllers designed with the three different cost functionals are compared. Since the objective is structural acoustic control, the radiated sound power is chosen as the first basis for comparison. In Figure 9.13, the sum of the squared singular values of the transfer matrix between the disturbance path and the output of the radiation filters is plotted as a function of frequency. This is given by the square of the $\mathcal{H}_2$ norm of the transfer function matrix evaluated at each frequency: $\|\mathbf{G}_{ad}(j\omega)\|_2^2$. For this transfer matrix, the subscripts $a$ and $d$ represent the output of the radiation filters (acoustic power) due to the disturbance path, respectively. Both open-loop and closed-loop results are presented. As illustrated, the ASAC controller outperforms the AVC and AVPTS controller at low frequency, particularly in proximity to the fundamental mode of the panel. Notice also that the radiated power at resonances corresponding to the (2,1) and (1,2) modes (the first two resonances following the fundamental) is minimal, owing to their relative inefficiency with respect

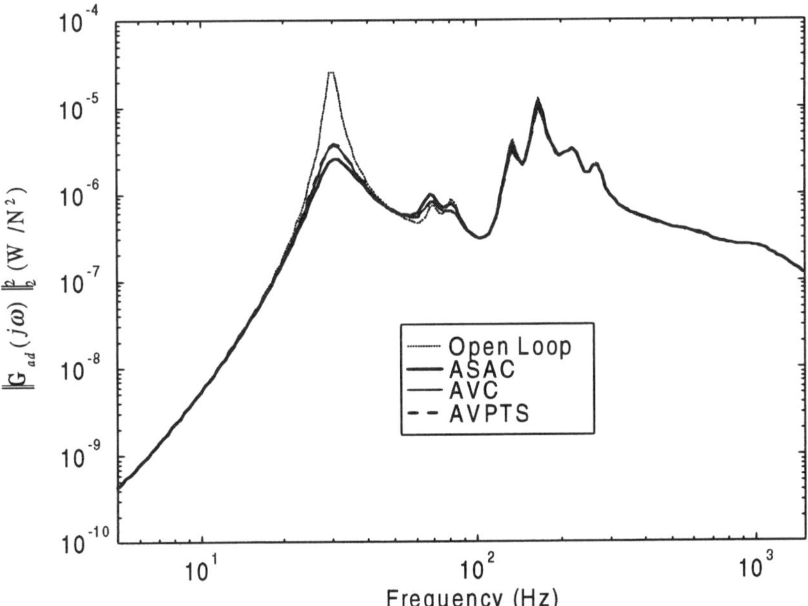


**Figure 9.13:** Comparison of the open-loop and closed-loop performance with direct, rate-feedback control for the three alternative cost functionals: ASAC, AVC, and AVPTS, and colocated point-force, point-velocity transducers.

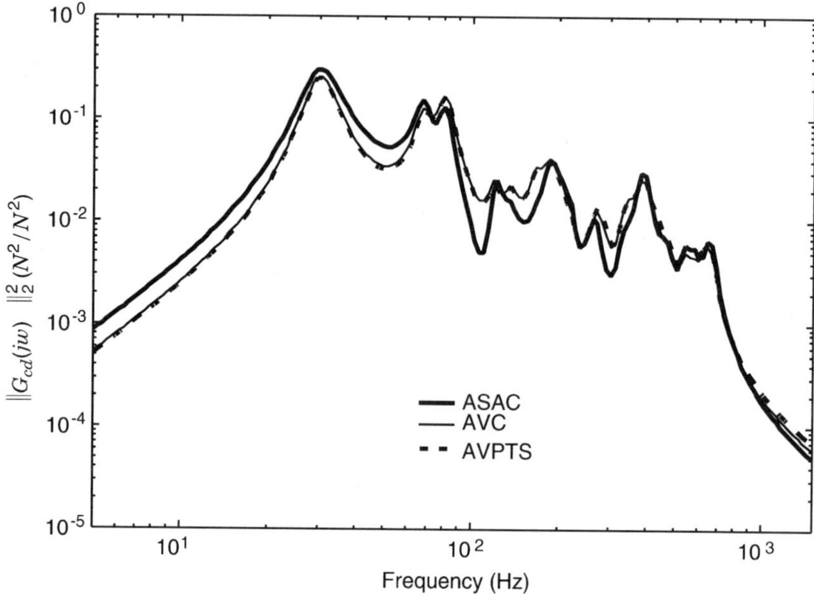


**Figure 9.14:** Comparison of the ratio of control signal energy to disturbance signal energy with direct, rate-feedback control designed with the three alternative cost functionals: ASAC, AVC, and AVPTS, and colocated point-force, point-velocity transducers.

to acoustic radiation. Since the first, or fundamental, resonance is an efficient acoustic radiator, the emphasis of the ASAC control system over this bandwidth makes physical sense. However, there are marginal differences in the resulting designs, because the controller gains are limited by the low-authority control approach chosen: static-gain, velocity-feedback.

Comparative analyses of the expended control energy for the same design cases are presented in Figure 9.14. The figure shows the sum of the squared singular values for the transfer matrix between the disturbance and control, which is the square of the $\mathcal{H}_2$ norm of the transfer function matrix evaluated at each frequency, $\|\mathbf{G}_{cd}(j\omega)\|_2^2$. The subscripts $c$ and $d$ indicate the control and disturbance paths, respectively. The control energy required for implementation of the three designs resulting from the three alternative cost functionals is illustrated. For vibration control, the frequency shaped control effort required is essentially the same as identified by the legends for AVC ($R = 2.10 \times 10^{-4}$, modal coordinates) and AVPTS ($R = 6.99^{-4}$, physical coordinates). However, the control system designed with the structural acoustic cost functional indicated by the legend for ASAC ($R = 3 \times 10^{-11}$) yields more control effort at low frequency and less control effort at high frequency. The bandwidth over which the control energy is expended is directly influenced by the frequency shaped cost functional. Since the first, or fundamental, resonance of the panel is an efficient acoustic radiator, the static compensator resulting from the ASAC cost puts more control energy in this bandwidth. Again, the differences are minimal, but this is related to the rate-feedback design. In the following example, a static-gain, full state-feedback controller is realized for the same structural acoustic system.

***Static Compensation with State-Feedback Control*** When the system states can be measured (or estimated) directly (both for the structure and the radiation filters), LQR theory is used to design an optimal state-feedback controller. As in the prior example, the control effort penalty must be adjusted for each of the different cost functions to assure that the expended control energy is equal for each case. The same three cost functions used in the previous rate-feedback controller design have been used to design the control system for this example. As illustrated in Figure 9.15, the acoustic response due to the (1,2) and (2,1) mode is negligible in proximity to the corresponding resonances. The results presented are obtained from the sum of the squared singular values from the transfer matrix of the disturbance path and the output of the radiation filters as a function of frequency. The closed-loop performance of the state-feedback controller (ASAC) differs primarily at low frequency. The ASAC control system design demonstrates greater levels of sound power reduction in proximity to the

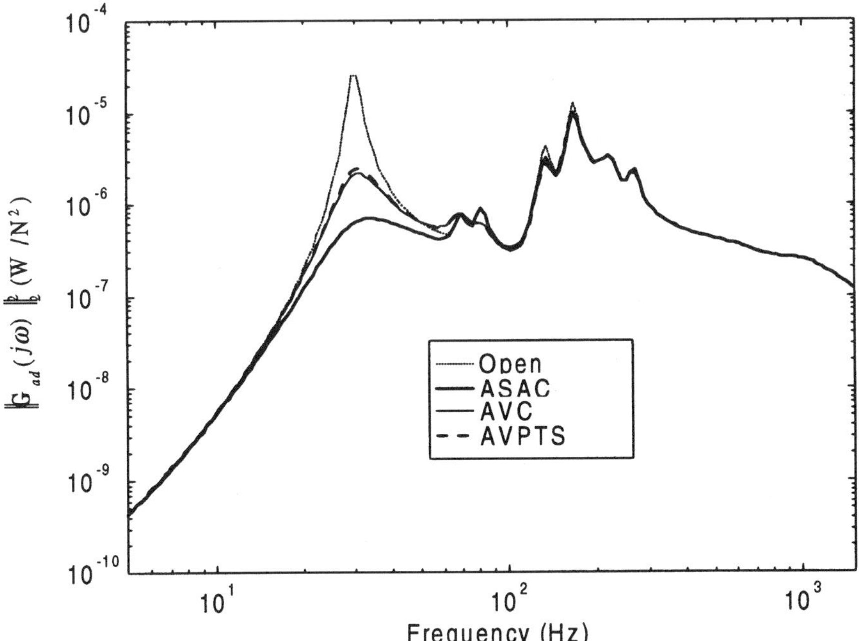


**Figure 9.15:** Comparison of the open-loop and closed-loop performance with state-feedback control for the three alternative cost functionals: ASAC, AVC, and AVPTS, and point-force transducers.

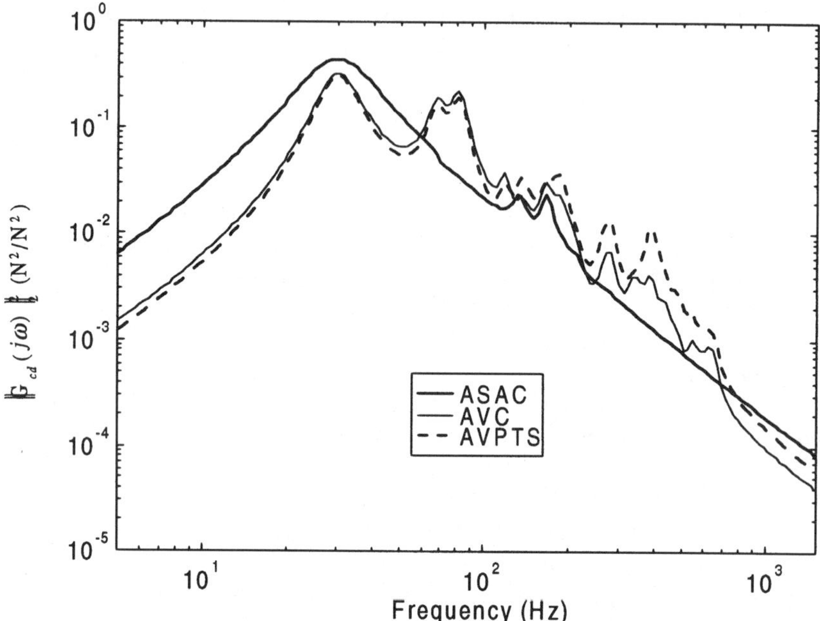


**Figure 9.16:** Comparison of the ratio of control signal energy to disturbance signal energy with state-feedback control designed with the three alternative cost functionals: ASAC, AVC, and AVPTS, and point-force transducers.

fundamental mode when compared to that of the AVC and AVPTS control system designs. This result was anticipated, considering the frequency-dependent variation in the control energy presented in Figure 9.16. The frequency shaped cost function associated with the ASAC system provides a direct measure of the desired performance variable, acoustic power, by appropriately weighting the most efficient acoustic radiators. As a result, more control effort and performance are observed at low frequency, which corresponds to the bandwidth of the fundamental mode of the panel.

The sums of the squared singular values for the transfer matrix between the disturbance path and control path are illustrated as a function of frequency in Figure 9.16. The sum total of the control energy over the bandwidth is the same for all three designs; however, the ASAC ($R = 2.77 \times 10^{-11}$) control system design expends more energy in the low-frequency regime near the fundamental mode of the panel. The AVC ($R = 2.11 \times 10^{-4}$) and AVPTS ($R = 7.26 \times 10^{-4}$) control system designs demonstrate very similar trends in terms of the frequency-dependent variation in the control energy, and the control effort expended at higher frequencies exceeds that of the ASAC control system design. Specifically, notice that a significant level of control effort is expended by the AVC and AVPTS controllers in the proximity of the second and third resonances [the (1,2) and (2,1) modes] of the structures. Since these modes are inefficient radiators, the ASAC controller essentially ignores these dynamics.

### 9.2.7 Summary of Active Structural Acoustic Control Example

The importance of the frequency shaped cost function was demonstrated in the preceding examples using both output-feedback and state-feedback control system designs. For both types of control, greater reduction in low-frequency sound radiation resulted from inclusion of the ASAC frequency shaped cost, which was based upon the essential physics of the structural acoustic coupling. In general, the impact of the different cost functions on the resulting closed-loop performance was more readily observed in the state feedback control approach. This is simply related to the improved controllability of the system when all of the states are available for static feedback. For our purposes, the control energy can be "shaped" more appropriately to minimize the intended cost function. In practice, one can very rarely implement a state-feedback controller. For adaptive structures, it is virtually impossible. State estimators must be designed for such applications, leading to the well-known LQG control. The use of state estimation does not change the fundamental influence of the frequency shaped cost functional on the controlled system performance. The emphasis still remains on the development of appropriate models of the "essential physics," the generatation of performance variables, and inclusion of these dynamic models in the generalized plant.

## 9.3 ASAC WITH A DISTRIBUTED PIEZOELECTRIC TRANSDUCER

### 9.3.1 Problem Overview

The purpose of this example is to demonstrate the importance of spatial signal processing in structural control. For this example, a SISO, ASAC controller is implemented with a single piezoceramic transducer operated in a sensoriactuator configuration. The transducer aperture covers 90% of the plate surface area, thereby providing a method to spatially filter the response of the structural mode with the highest radiation efficiency. The distributed transducer leads to improved control performance when compared to more localized estimations of the structural acoustic response.

### 9.3.2 Description of the Plant

The plant consists of a plate with simply supported boundary conditions. The model is identical to the one discussed in Section 9.2.2. However, the control input and measured output are obtained from a single piezoceramic transducer, as illustrated in Figure 9.10. The transducers are incorporated into the dynamic model of the plant using the methods presented in Section 5.3.1.

### 9.3.3 Transducer Selection, Design, and Placement

In the previous examples of ASAC applications, point-force actuators have been used with output-feedback and state-feedback static compensator designs. The apertures of most commercially available piezoceramic transducers are relatively small (e.g., 1 in. × 1 in. × 0.01 in.) in comparison to the wavelengths on typical test structures. Thus, static compensation typically does not provide a viable alternative in the control of low-frequency structural modes, since the transducer naturally couples to the higher-frequency modes with greater efficiency (see Chapter 6). To alleviate this problem for structural acoustic control, the transducer dimension is "stretched" to extend over 90% of the plate surface area and is centered on the plate, as illustrated in Figure 9.10. This is a very large actuator, but its advantage is that the spatial aperture of this dimension efficiently couples the induced strain with the low-frequency, long-wavelength structural modes. The transducer can then be made much thinner, and for this example, it is constrained to have the same volume as that of a typical transducer. For this large aperture, the number of modes to which the transducer couples is limited through spatial signal processing and the amount of voltage required to control the structure are reduced. At the writing of this manuscript, the transducer described herein has not been realized in hardware; however, the results of this example affirm that the possibilities merit investigation.

Before proceeding with a discussion of the remaining design requirements, consider the frequency-response function illustrated in Figure 9.17. The FRF is

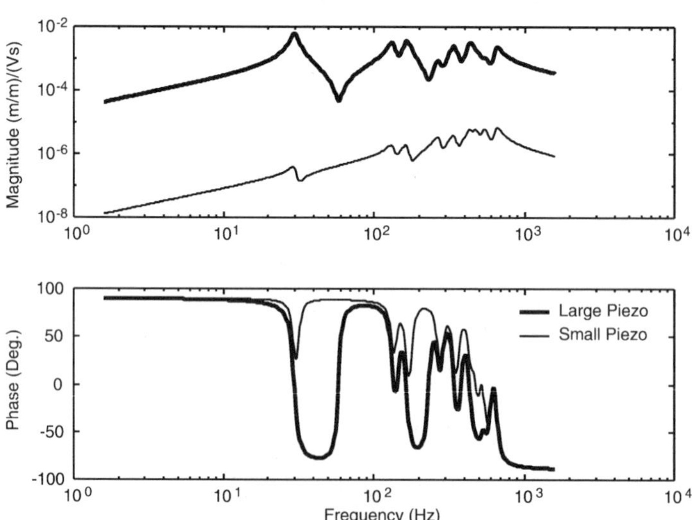


**Figure 9.17:** Comparison of colocated frequency response functions (FRFs) for a piezoelectric sensoriactuator covering 0.25% of the plate surface area and one covering 90% of the plate surface area.

from a simply supported plate configured with a single, centered, pieozoceramic transducer, covering 0.25% of the surface and operated as a sensoriactuator. The script file **lspiezo.m** can be used to reproduce the result. Using magnitude and phase of the FRFs, the "small" transducer is compared to that of a "large" transducer covering 90% of the plate surface area. We select strain-rate (i.e., current) as the output and applied voltage as the input. As illustrated, the magnitude for the large transducer aperture is much greater, by orders of magnitude at low frequency, compared to that of the small aperture transducer. This increase in magnitude results from the spatial signal processing that occurs. Since the large colocated transducer can only couple efficiently to the low-order modes, the structure appears more compliant as seen through the transducer aperture. Additionally, the pole-zero separation is much greater for the transducer with the large aperture, since the higher-order modes are essentially filtered through spatial signal processing. Recall that the volumes of the transducers were constrained to be the same. Thus, the larger transducer is much thinner (a necessity for practical constraints on weight) and will depole at a much lower voltage. However, this effect is offset by the gain in the magnitude of the response at low frequency due to the spatial filtering. The large transducer aperture is thus more practical for the realization of structural acoustic control systems designed to control low-frequency sound radiation.

### 9.3.4 Development of the Augmented Plant Model

The augmented plant development follows directly from Section 9.2.4. The primary difference for the generalized plant used in this example is that a single sensoriactuator is used to provide the measured output (current $\propto$ strain-rate) and the control input (applied voltage).

### 9.3.5 Control Strategy

The control strategy is consistent with that presented in Section 9.2.5, that is, static compensation designs for both output and state feedback. The only difference is that a SISO controller is realized with a sensoriactuator as opposed to a MIMO controller with point-force and velocity inputs and outputs, respectively.

### 9.3.6 Results from Analysis

***Static Compensation with Direct, Output-Feedback Control*** For the piezoceramic transducer covering 90% of the plate surface area, the optimal feedback gain is computed using the algorithm described in Section 8.6.1. To generate the results, the script file **perfilt2.m** can be used. The control effort penalty is adjusted to assure that the total energy expended by each of the different control systems is the same. The sum of the squared singular values for the transfer matrix between the disturbance path and the control path are plotted as a function of frequency in Figure 9.18. As illustrated, the results are identi-

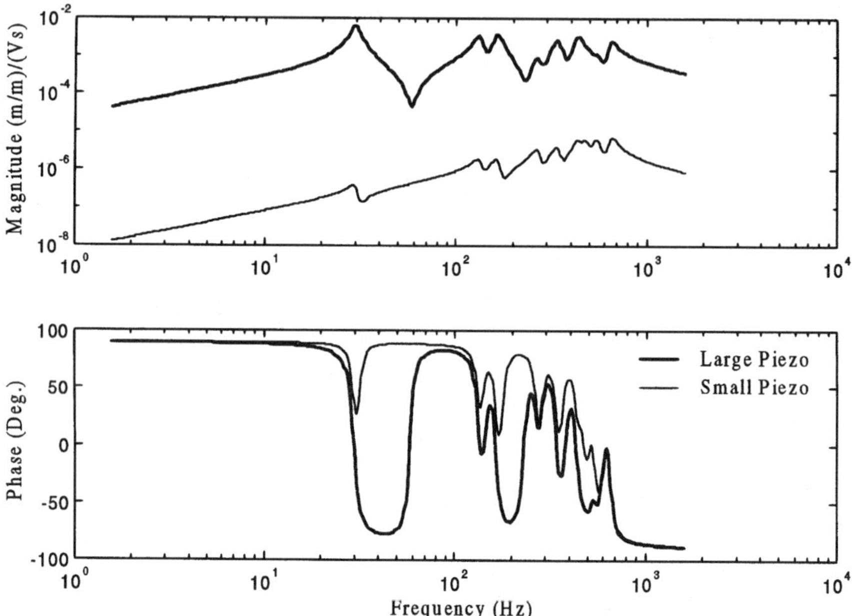


**Figure 9.18:** Comparison of the ratio of control signal energy to disturbance signal energy with direct, rate-feedback control designed with the three alternative cost functionals: ASAC, AVC, and AVPTS, and a single-input, single-output (SISO), pieozoelectric sensoriactuator.

cal for the ASAC ($R = 7.25^{-12}$), AVC ($R = 6.39^{-6}$), and AVPTS ($R = 2.11^{-5}$) cost functions. This seems intuitive, since for this SISO, direct rate-feedback, control system design, only a single feedback gain constant can be modified. The equal performance of the three alternative designs is demonstrated in Figure 9.19, which illustrates the sum of the squared singular values of the transfer matrix between the disturbance path and the output of the radiation filters. Since the design flexibility is constrained by a single coefficient, the frequency shaped cost function is of little value in this case.

***Static Compensation with State-Feedback Control*** The full state-feedback control produces a more interesting result. Figure 9.20 shows the sum of the squared singular values for the transfer matrix between the disturbance path and the control path as a function of frequency. These results can be reproduced with the script file **perfilt2.m**. Note that the ASAC ($R = 6.56 \times 10^{-13}$) controller emphasizes the control effort at low frequency, while the AVC ($R = 5.16 \times 10^{-6}$) and AVPTS ($R = 1.95 \times 10^{-5}$) designs provide more control effort at high frequencies and less at low frequencies. Again, this makes physical sense, because the fundamental mode of the panel is an efficient acoustic radiator at low frequency for a structural acoustic cost function, while for the

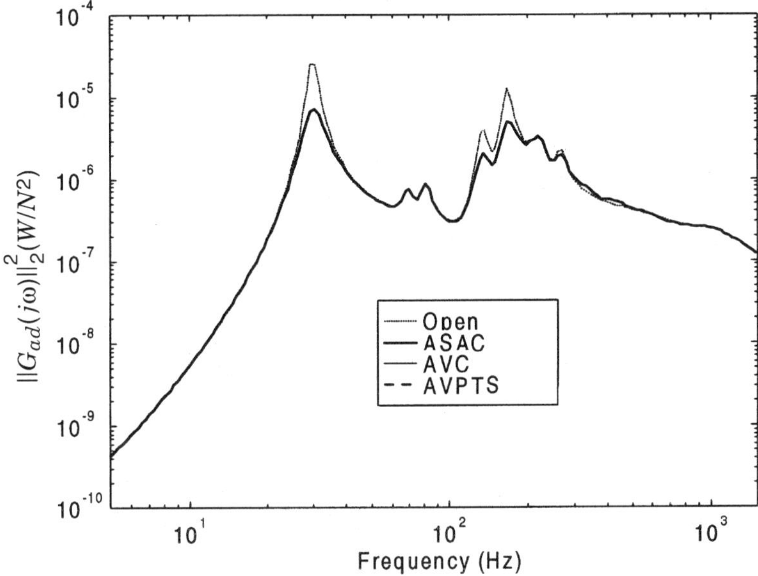


**Figure 9.19:** Comparison of the open-loop and closed-loop performance with direct, rate-feedback control for the three alternative cost functionals: ASAC, AVC, and AVPTS, and a SISO, piezoelectric sensoriactuator.

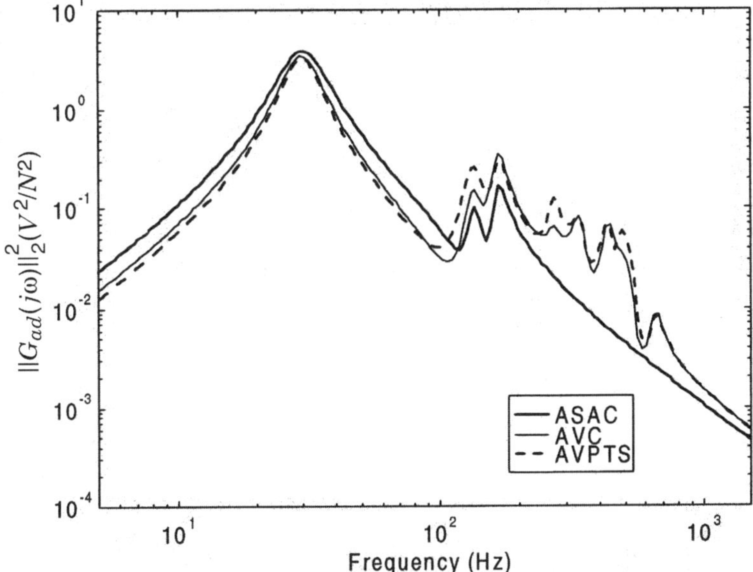


**Figure 9.20:** Comparison of the ratio of control signal energy to disturbance signal energy with state-feedback control designed with the three alternative cost functionals: ASAC, AVC, and AVPTS, and a single piezoelectric transducer.

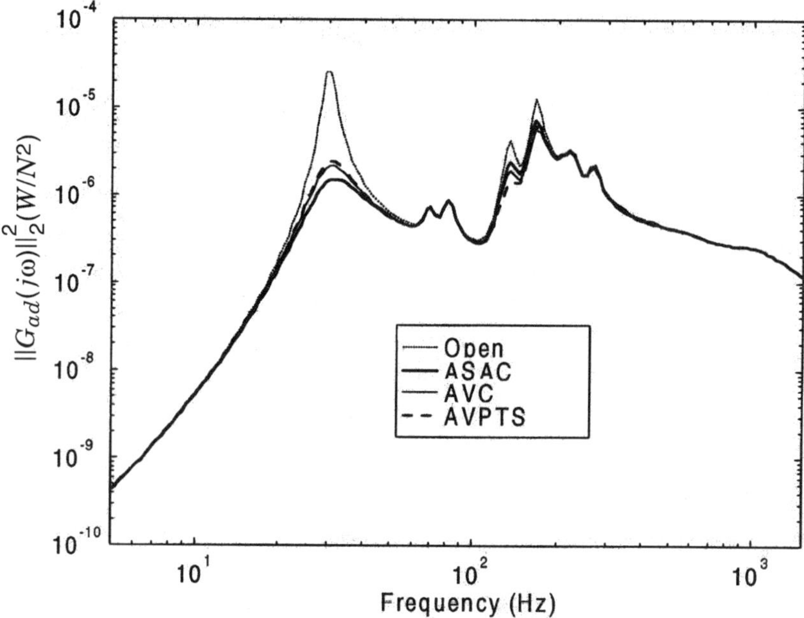


**Figure 9.21:** Comparison of the open-loop and closed-loop performance with state-feedback control for the three alternative cost functinoals: ASAC, AVC, and AVPTS, and a single piezoelectric transducer.

vibration cost functions, all of the structural modes are equally important. The performance of the feedback control system designs are compared in Figure 9.21. As illustrated, the ASAC controller yields greater performance at low frequency, in proximity to the fundamental mode, with very little performance at higher frequencies. However, the AVC and AVPTS control system designs provide moderate performance at low and high-frequency. As indicated in Figure 9.21, at some resonant frequencies of the panel, the control system has no effect at all. Lack of performance at these system resonances results from the symmetry of the transducer relative to the structure (i.e., the spatial aperture). Structural modes with even modal indices cannot be controlled with the SISO piezoelectric transducer used in this example. However, the importance of a frequency shaped cost is demonstrated for this simple example as one observes that a greater level of control energy and resulting performance is achieved in proximity to the fundamental resonance for active structural acoustic control.

### 9.3.7 Summary of ASAC with a Distributed Transducer

In this design example, selection of transducer aperture significantly changes the adaptive structure's performance. Upon comparing the frequency response of a sensoriactuator covering 0.25% of the plate to that of a sensoriactuator covering

90% of the plate, one observes that much more control effort can be expended at low frequency due to the spatial aperture of the large transducer. For structural acoustic control, the low-frequency, fundamental mode of a panel is the most efficient radiator. Shaping the transducer to maximize the control system's coupling to this mode enhances the closed-loop performance. Additionally, for the example presented, modes with even indices do not couple efficiently to the sound field at low frequency. The spatially centered transducer is coupled to the panel at nodal lines of such modes, effectively ignoring both response and actuation of all even modes. The closed-loop response demonstrated for the SISO controllers, even with simple output feedback control, serves to demonstrate that significant levels of reduction in sound power can be achieved with distributed transducers. It is currently difficult to realize such transducers, but this example serves to demonstrate new possibilities as active material design continues to evolve.

## 9.4 AN EXPERIMENTAL MODEL FOR ACTIVE VIBRATION CONTROL

### 9.4.1 Problem Overview

The purpose of this tutorial is to demonstrate adaptive structure design, beginning with a model developed through experimental system identification. Active vibration control is demonstrated on a plate with "simply-supported" boundary conditions and configured with an array of four adaptive piezoceramic sensoriactuators. The design objective is to minimize the vibrational energy in the plate. Both predicted and measured closed-loop responses are compared. The control system has been implemented using a TMS320C31 digital signal processor. In contrast to the prior examples, the model provided for review in this example has been obtained from a "curve-fit" of measured frequency response functions as opposed to an analytical development.

### 9.4.2 Description of the Plant

A picture of the test structure is provided in Figure 9.22. As illustrated, the structure consists of a plate supported by thin shim stock at the boundaries, which are rigidly anchored to an optical bench. The thin shim stock plate supports are very compliant to rotation but are very stiff to out-of-plane displacement. Thus, the resulting boundary conditions emulate "ideal" simply supported conditions (Clark and Fuller, 1992). The plate is constructed from steel, with dimensions of 0.6 m $\times$ 0.525 m $\times$ 1.62 m. As illustrated, the structure is configured with four piezoelectric transducers operated as adaptive sensoriactuators. Additionally, a Ling dynamic shaker is attached to the structure and serves as a disturbance force. An accelerometer is also attached to the structure to provide an independent measure (independent from the sensoriactuators) of the closed-loop

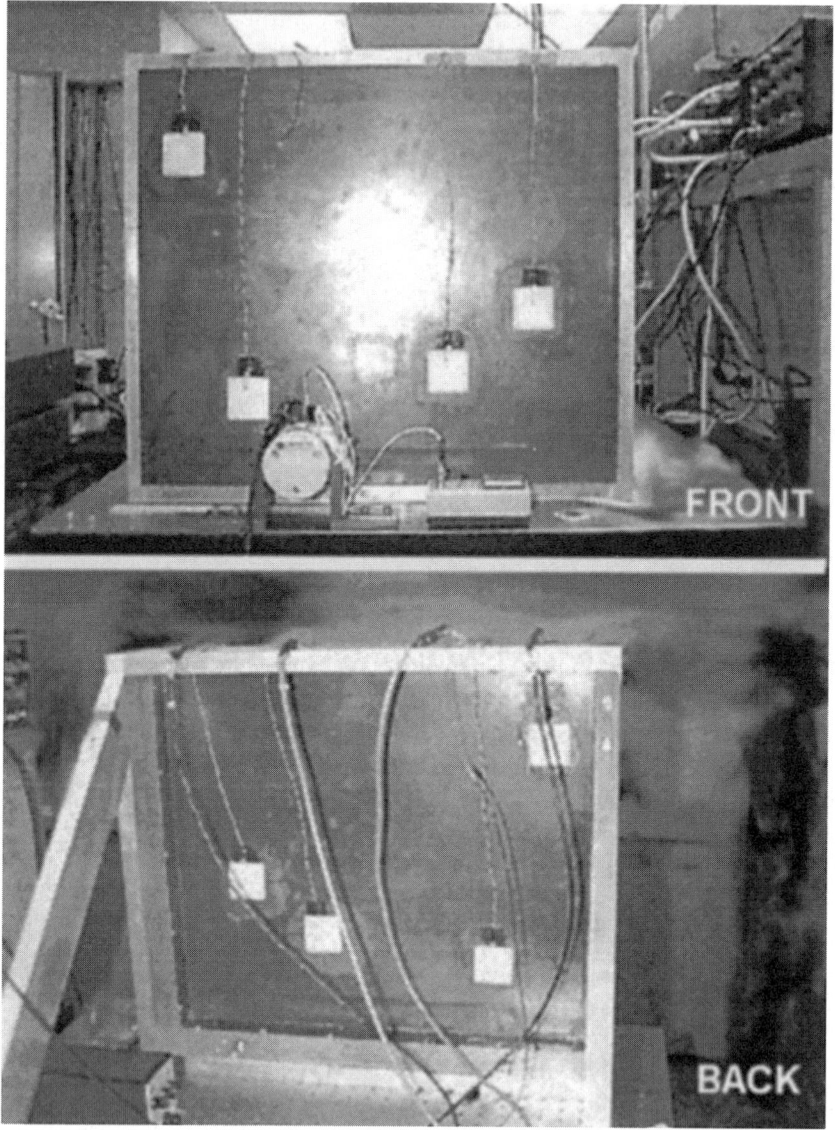


**Figure 9.22:** Picture of experimental simply supported plate testbed.

performance. The positions of the sensoriactuators, shaker, and accelerometer are identified in the schematic diagram of Figure 9.23.

The frequency responses of the four-input, four-output adaptive piezoelectric sensoriactuator array, measured with a SigLab spectrum analyzer, is stored for the purpose of system identification over a bandwidth extending from DC to 800 Hz. The software used for system identification is called SmartID (AcX, 1996).

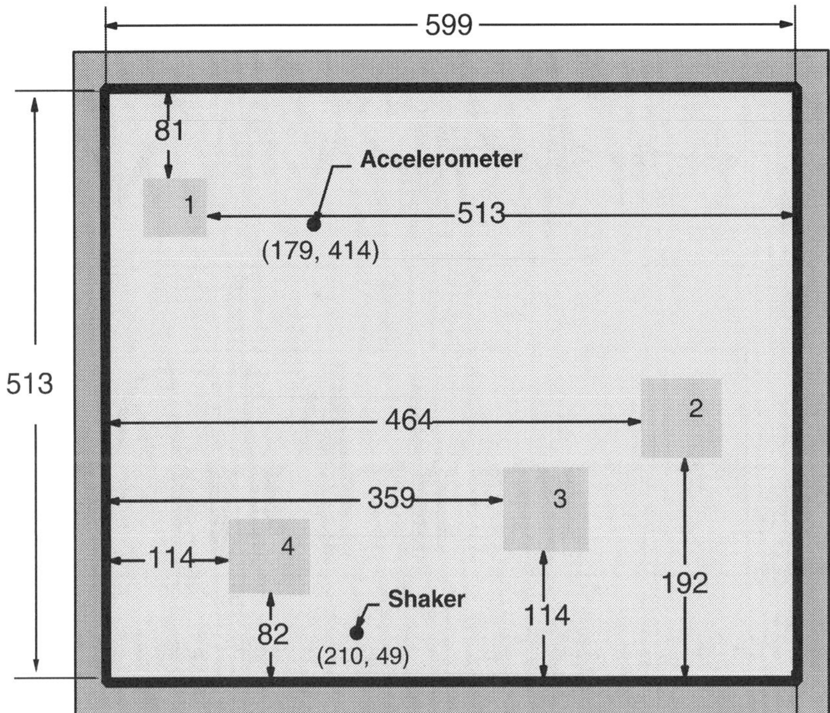


**Figure 9.23:** Schematic diagram of the plate configured with an array of piezoelectric sensoriactuators (dimensions are in millimeters).

The basic method of operation involves a frequency domain observability range space extraction (FORSE), as detailed by Liu et al. (1996). The algorithm provides a state-space model directly from frequency domain data, using proper state-space parameterization and singular value decomposition (SVD) for multivariable and high-order systems. The advantage of this curve-fitting routine is that it is capable of accurately fitting the dynamics associated with closely spaced poles and zeros, a characteristic typical of piezostructure FRFs. Referring back to Figure 9.23 for naming conventions, typical results of the curve fits for a colocated FRF (transducer 1) and a noncolocated FRF (transducer 2 to transducer 4) are presented in Figures 9.24 and 9.25, respectively. As illustrated, both poles and zeros are accurately identified, leading to very good estimates of both magnitude and phase of the FRFs. The resulting test-based model is stored in the file **psmd5s6a.mat**. The representations presented in Figures 9.24 and 9.25 can be generated with the script file **plotsid.m**. Using this script file, the reader can also generate the FRFs corresponding to all of the other input-output paths. Once the system identification has been completed, one can address the control system design problem.

The disturbance for the control experiment was generated with the Ling

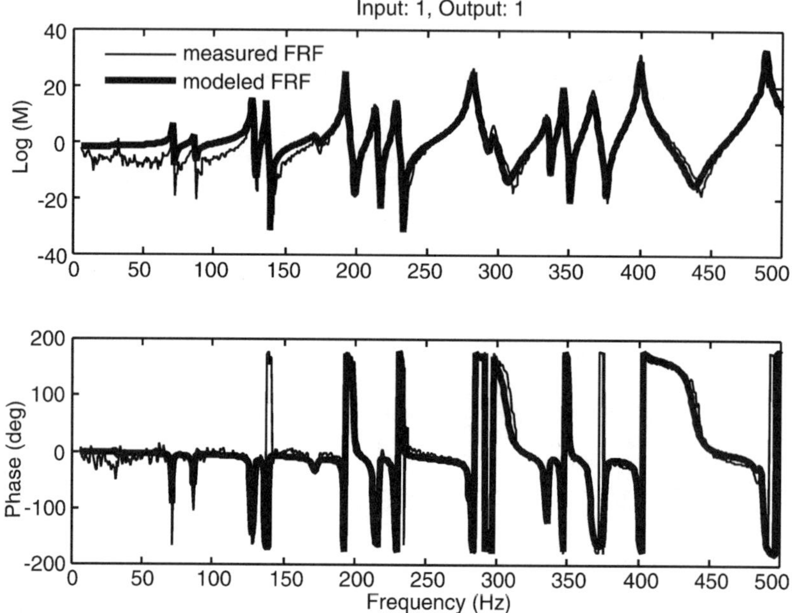


**Figure 9.24:** Measured frequency response of adaptive sensoriactuator 1 and frequency response of model of the same resulting from the system identification.

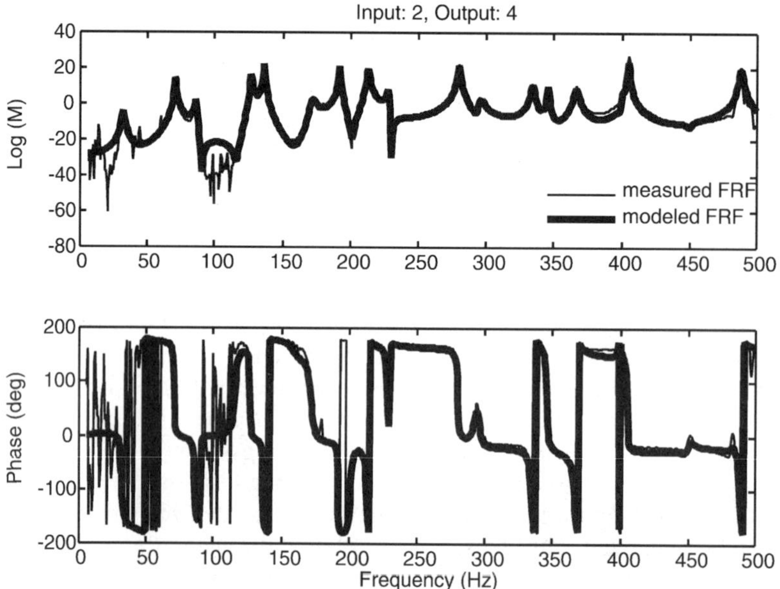


**Figure 9.25:** Measured frequency response between adaptive sensoriactuator 2 and adaptive sensoriactuator 4 (noncolocated), and frequency response of model of the same resulting from the system identification.

dynamic shaker. Using a band-limited (DC to 800 Hz) random signal to drive the structure, a model of the disturbance path was obtained. Note that this path must include the shaker dynamics also.

### 9.4.3 Transducer Selection, Design, and Placement

The experimental system used for this example serves primarily as a test bed for demonstrating the adaptive piezoelectric sensoriactuator. The locations of the transducers on the structure are determined from a group of randomly selected locations. Since this test bed is intended to support both vibration and structural acoustic control experiments, the transducers are not oriented to emphasize any particular performance variables. The dimensions of the transducers are constrained by the dimensions of commercially available piezoceramic materials. Larger apertures would actually be preferred, as a significant percentage of the vibrational energy resides in the fundamental structural mode whose wavelength is twice that of the diagonal dimension of the plate structure.

### 9.4.4 Development of the Augmented Plant Model

Given the test-based model of the plant, performance objectives must be defined before designing a controller. Thus, the augmented plant model must be developed. For this example, the objective is vibration control. The control problem is cast as illustrated previously in the block diagram of Figure 8.19; however, one must select a disturbance input path and a performance path for the augmented (two-port) model. For an experiment conducted in the laboratory, it is very tempting to simply "measure the disturbance" and the related structural response, thereby providing a reasonable estimate of the disturbance-to-performance path. In general applications, the disturbance and/or the path to error cannot be measured. As a result, for the work presented in this example, a controller is designed assuming that independent, stochastic noise sources are presented to all inputs of the state equations obtained for this model, and the performance penalty weights all states of the modeled plant equally. Consistent with the LQG model, sensor noise and control effort penalties are selected in addition to the previously defined process noise and performance penalty. All of the weightings—process noise, sensor noise, performance penalty, and control effort penalty—are applied uniformly over the bandwidth. Details of numerical values used in the design process are included in the script file **contdes.m**, which can be used to formulate the augmented plant and design the controller.

### 9.4.5 Control Strategy

The $\mathcal{H}_2$ controller synthesis outlined in Chapter 8 is used to design a dynamic compensator. According to the discussion of the augmented plant model, the weightings are consistent with those of the LQG control system design.

Before implementing the controller on the test model, the controller is evalu-

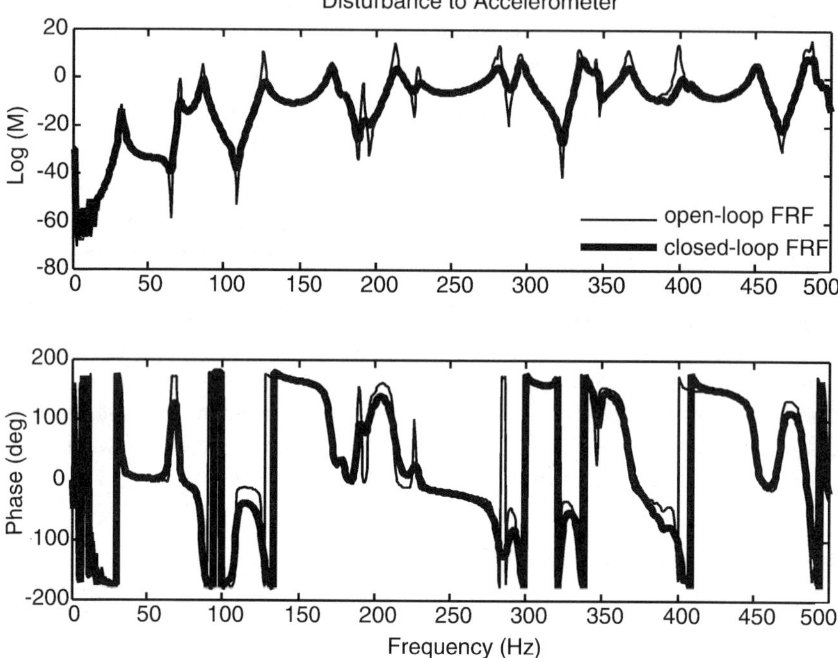


**Figure 9.26:** Predicted open- and closed-loop FRFs from disturbance shaker input to accelerometer output.

ated using the numerical model obtained from the system identification, and the closed-loop response is predicted. Executing the script file **contdes.m** produces the results illustrated in Figure 9.26. Since the states of the structure cannot be measured directly to evaluate performance, the frequency response between the attached disturbance shaker and an accelerometer positioned on the structure, as illustrated in Figure 9.23, is predicted. Both the magnitude and phase of the open-loop and closed-loop response for the modeled system are illustrated. As can be seen in Figure 9.26, the prototype controller is expected to provide approximately 6 dB of reduction in response at the resonant frequencies of the structure over the bandwidth illustrated. Next, the compensator is realized in DSP hardware and tested on the actual structure.

### 9.4.6 Results from the Experimental Implementation

The design and analysis of the previous section is for control of a continuous system, so the resulting dynamic compensator must be converted to discrete time before implementing it on a TMS320C31 digital signal processor. A sample rate of 4 kHz is selected, and a discrete-time model is realized using a Tustin transform (**c2dm** in MATLAB). In general, one should always sample as fast as possible to minimize the effects of time-delay in the bandwidth. However, the computational power of the digital signal processor, the number of statements that must be executed, and the efficiency of the program implemented on the DSP will eventually place an upper limit on sample rate. Since the actual compensator design begins with a continuous-time model, it is extremely important to check the stability of the resulting closed-loop discrete time model (i.e., all of the poles in the unit circle) before proceeding with the implementation. Depending upon the level of time delay, it is possible that a stable continuous-time realization will be unstable in the discrete time. However, one can incorporate this into the design procedure by accounting for the time delay in the continuous-time model or by simply designing in discrete-time compensator from the onset.

After stability is confirmed for the sampled-data system, the compensator can be cautiously applied. For this case, the measured open-loop and closed-loop FRFs of the structure from the input of the disturbance shaker to the output of an accelerometer are presented in Figure 9.27. Just as expected, the closed-loop response is attenuated by approximately 6 dB or more at the resonant frequencies of the structure over the bandwidth presented. Upon comparing the predicted and measured closed-loop performances of the system in Figures 9.26

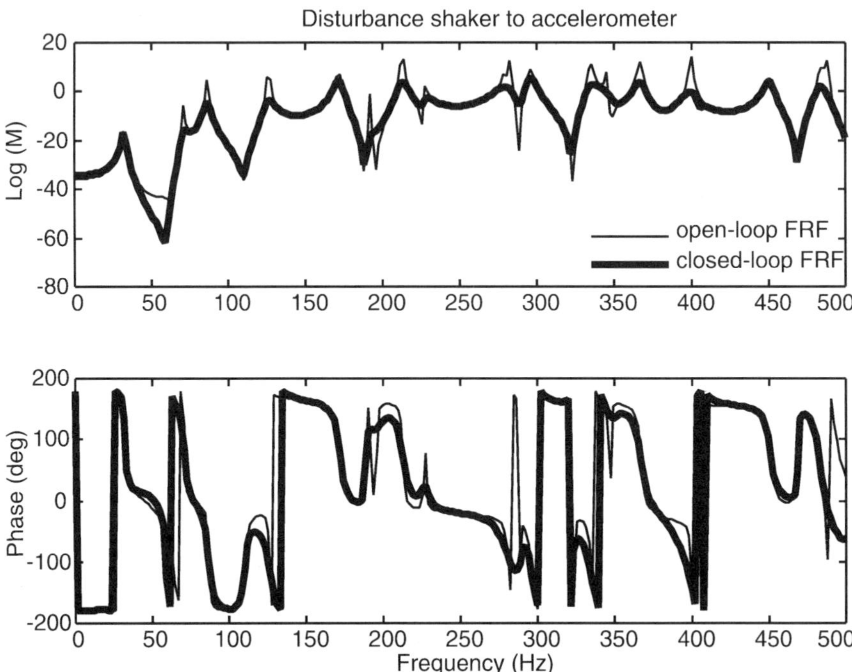


**Figure 9.27:** Measured open- and closed-loop FRFs from disturbance shaker input to accelerometer output.

and 9.27, one observes that the results are consistent at most frequencies. Significant levels of energy are dissipated in the proximity of the resonant frequencies. As expected, reverberant energy within this lightly damped structure can be dissipated using feedback control. The input impedance at the resonances of the structure has increased significantly due to the added damping.

### 9.4.7 Summary of Experimental Vibration Control

In this example, we obtained a frequency-domain model of a steel plate with nearly simply supported boundary conditions using system identification software. The model was used to develop an augmented plant for control system analysis and synthesis. Adaptive, piezoelectric sensoriactuators were implemented in hardware on the steel plate test bed, and the reverberant response of the structure was attenuated by 6 dB or more in the proximity of resonant frequencies over the desired bandwidth (DC to 500 Hz). The measured closed-loop response served to reflect the predicted closed-loop response of the structure. The reader is encouraged to "experiment" with the model provided and develop alternative controllers for the system. Evaluate your design and compare the predicted performance to that of the compensator provided.

## 9.5 AN EXPERIMENTAL MODEL FOR ASAC

### 9.5.1 Problem Overview

In the previous example of structural acoustic control, a plate with pinned boundaries was controlled using analytical models of the structural acoustic response. The radiation filters were developed in modal coordinates, since the mode shapes were simple sinusoidal shape functions. In this example, we examine the experimental implementation of a controller designed to minimize structure-borne sound radiation from a panel with clamped boundary conditions. The primary difficulty in designing adaptive structures with complex geometry and/or complex boundary conditions is that the modes must usually be developed from a series expansion (see the discussion on assumed modes in Section 2.3.2). This can be tedious for certain boundary conditions or high bandwidth applications. In this example, we explore an alternative—the discretization of Rayleigh's integral (see Appendix C) to predict the radiated power. The approach presented follows that proposed by Elliott and Johnson (1993).

A test structure mounted in the Transmission Loss Test Facility at the Structural Acoustics Branch of the NASA-Langley Research Center is illustrated in Figure 9.28. The plate is configured with four piezoelectric transducers. Again, each transducer is configured to operate as an adaptive sensoriactuator, as detailed in Chapter 5. The control system is implemented on a TMS320C40 digital signal processor, and the identified test-based model and frequency shaped

**Figure 9.28:** Picture of the Transmission Loss Test Facility at the Structural Acoustics Branch of the NASA–Langley Research Center.

cost functional required for structural acoustic control are detailed in the following discussion.

The primary purpose of this example is to identify the level of complexity introduced to the frequency shaped cost in an attempt to accurately model the essential physics of the structural acoustic radiation. Radiation filters are designed based upon a spatial discretization of the structure into elemental radiators. Singular value decomposition of this frequency-dependent radiation matrix is used to identify the dominant radiation modes and the dominant singular values. Model reduction/truncation is then applied to minimize the number of states required in the radiation filters. Finally, effects of the model truncation on closed-loop performance are studied.

### 9.5.2 Description of the Plant

The test structure illustrated in Figure 9.28 measures 355.6 mm × 254 mm × 1 mm, and the positions of the sensoriactuators are identified in the schematic diagram presented in Figure 9.29. In addition to the strain measurements obtained from the sensoriactuators, velocity measurements have been obtained at 20 locations on the plate using an Ometron VPI sensor (scanning laser Doppler velocimeter). The laser provides a noninvasive means of measuring the open-loop and closed-loop velocities of the structure at the discrete points indicated.

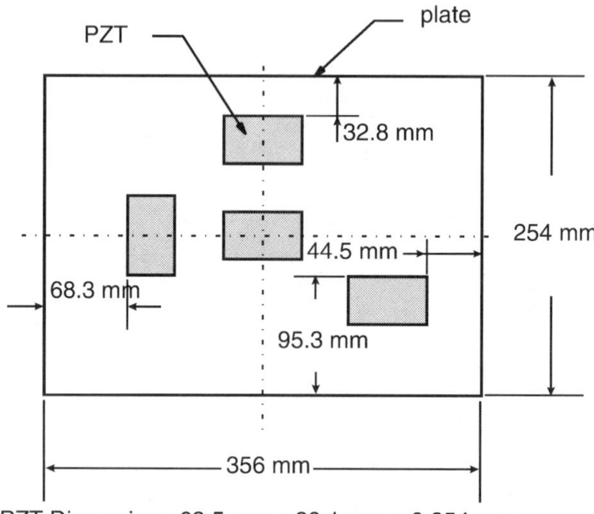


**Figure 9.29:** Schematic diagram of test panel used in structural acoustic control experiments.

Additionally, as will be outlined in a subsequent section, discrete velocity measurements are required to generate the performance objective, since the structural acoustic radiation is estimated from a discretized version of the Rayleigh integral (Elliott and Johnson, 1993; Maillard and Fuller, 1994).

The frequency responses of the 4-input, 24-output system are determined using a Bruel & Kjaer 2032 spectrum analyzer over a bandwidth extending from DC to 1600 Hz. The SmartID software package is used to identify the state-variable model of the system from the measured FRFs. The FRF of adaptive piezoelectric sensoriactuator (APSA) 1 is presented in Figure 9.30. As illustrated, the measured response and predicted response with the identified system are approximately the same up to a frequency of 700 Hz. The desired control bandwidth is between 20 Hz and 500 Hz, so the structural model is truncated at 800 Hz. As discussed in the section devoted to control synthesis, this rolloff due to the unmodeled dynamics can be used to assist in the loop shaping, since increasing the level of sensor noise in proportion to the plant response provides a means of indicating the bandwidth over which noise dominates the measured response.

The frequency responses between all 4 transducers and the 20 velocity measurements are all used in the identification of the model. The model of the structure with the 4-input, 4-output array of sensoriactuators is required to implement the controller, and the model of the structure from the 4 piezoelectric inputs to the 20 velocity outputs is required to generate the performance metric. An acoustic loudspeaker is used to generate the acoustic disturbance on one side of the panel in the transmission loss facility, and a model of the disturbance

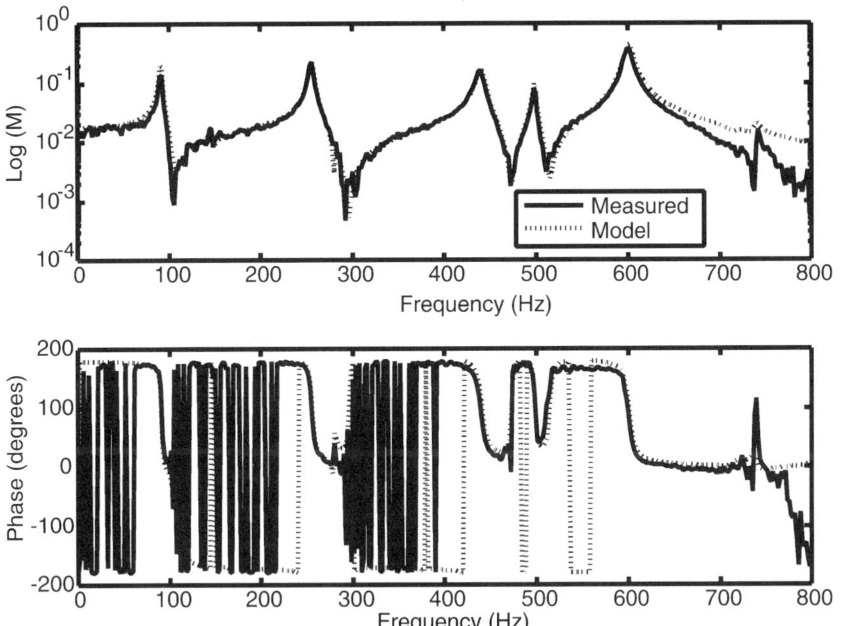


**Figure 9.30:** Exemplary frequency response of sensoriactuator and curve fit of the dynamic response.

source is also included to create the performance path associated with the typical two-port model (disturbance to error). The radiation filter design, required to implement a cost functional emphasizing reduction of radiated sound power, will be presented in Section 9.5.4.

### 9.5.3 Transducer Selection, Design, and Placement

Piezoceramic elements are selected to meet the need for compact transducers that might ultimately be integrated into the host structure for the ASAC application. Additionally, the piezoceramic elements are readily surface mounted and thus require no reaction force from an additional support structure. The transducers are operated as adaptive piezoelectric sensoriactuators.

### 9.5.4 Development of the Augmented Plant Model

Before assembling the augmented plant model, a model that incorporates the physics associated with the structural acoustic coupling must be developed. This dynamic model is used to "filter" the discrete velocity measurements such that the RMS sound power radiated from the structure can be estimated. The radiated sound power and the control energy generate the performance metric for the structural acoustic control problem. Thus, once the frequency-dependent filter modeling the structural acoustic coupling is developed, the augmented plant can be assembled.

***Approximation of Radiation Filters*** In this example, the modes of the test structure will not be explicitly identified; therefore, a different method is required to implement the radiation filters. Following the work of Elliott and Johnson (1993), the sound power radiating from a vibrating planar surface mounted in an infinite rigid baffle can be estimated using a discretization of Rayleigh's integral. The formulation is omitted from this text for brevity but can be found in the reference (Elliott and Johnson, 1993). The radiation of a planar surface is discussed in Appendix C on structural acoustics. The Rayleigh integral presented in equation (C.2) describes the radiation from a continuous vibrating surface in an infinite baffle in terms of continuous functions. For the test panel of this example, analytic functions describing the spatial vibration characteristics are unknown. Using a laser doppler velocimeter (laser vibrometer), the out-of-plane velocities at a finite number of discrete locations are determined. This discretized measurement approach allows the surface to be considered as differential elements vibrating with known velocity. The pressure at some location off the surface can then be found through numerical integration of equation (C.2).

The radiated power from a vibrating surface described by a discrete number of velocity measurements can be found by the following relation:

$$\overline{P} = \mathbf{v}(j\omega)^H \mathbf{R}(j\omega)\mathbf{v}(j\omega) \tag{9.6}$$

where $\mathbf{v}(j\omega)$ is an $m \times 1$ vector of panel velocity measurements and $\mathbf{R}(j\omega)$ is an $m \times m$ radiation matrix. $\mathbf{R}(j\omega)$ is real, symmetric, and positive definite. Thus, the total radiated power of a vibrating surface (estimated by a discrete number of points) can be determined with only local panel vibration measurements. $\mathbf{R}(j\omega)$ can be calculated for a distribution of velocity points as follows [dropping the $(j\omega)$ terminology for convenience]:

$$\mathbf{R} = \frac{\omega^2 \rho S^2}{4\pi c} \begin{bmatrix} 1 & \dfrac{\sin(kr_{12})}{kr_{12}} & \cdots & \dfrac{\sin(kr_{1m})}{kr_{1m}} \\ \dfrac{\sin(kr_{21})}{kr_{21}} & 1 & \dfrac{\sin(kr_{23})}{kr_{23}} & \vdots \\ \vdots & \vdots & \ddots & \\ \dfrac{\sin(kr_{m1})}{kr_{m1}} & \cdots & \dfrac{\sin(kr_{mm-1})}{kr_{mm-1}} & 1 \end{bmatrix} \tag{9.7}$$

where $\omega$ is the circular frequency in rad/sec, $\rho$ is the density of air, $S$ is the area associated with the discretized radiator, $c$ is the speed of sound in air, $k$ is the wavenumber $(\omega/c)$, and $r_{ij}$ is the distance between the center of the $i$th and $j$th velocity locations. Since $r_{ij} = r_{ji}$, the radiation matrix is symmetric. Since the radiation matrix is also a function of frequency, a singular value decomposition can be performed to determine the dominant radiation modes of the system at each frequency. Singular value decomposition is performed at frequency $\omega_i$ as follows:

$$\mathbf{R}(\omega_i) = \mathbf{u}\boldsymbol{\Sigma}\mathbf{u}^{\mathbf{H}} \tag{9.8}$$

where $\mathbf{u}$ is an $m \times m$ matrix whose columns are the normalized radiation modes of the matrix at frequency $\omega_i$ and $\Sigma$ is an $m \times m$ diagonal matrix whose elements are the singular values, decreasing monotonically along the diagonal. The magnitude of the $j$th diagonal element of $\Sigma$ determines the relative importance of the $j$th radiation mode compared to other radiation modes.

For the aluminum panel used in this experiment (355.6 × 254 × 1 mm), a plot of the first six radiation modes at 5 Hz and 500 Hz are shown in Figures 9.31 and 9.32. Note how the shape of the first radiation mode transitions from a uniform "piston" mode to one with rounded edges ("dome" shaped). A plot of

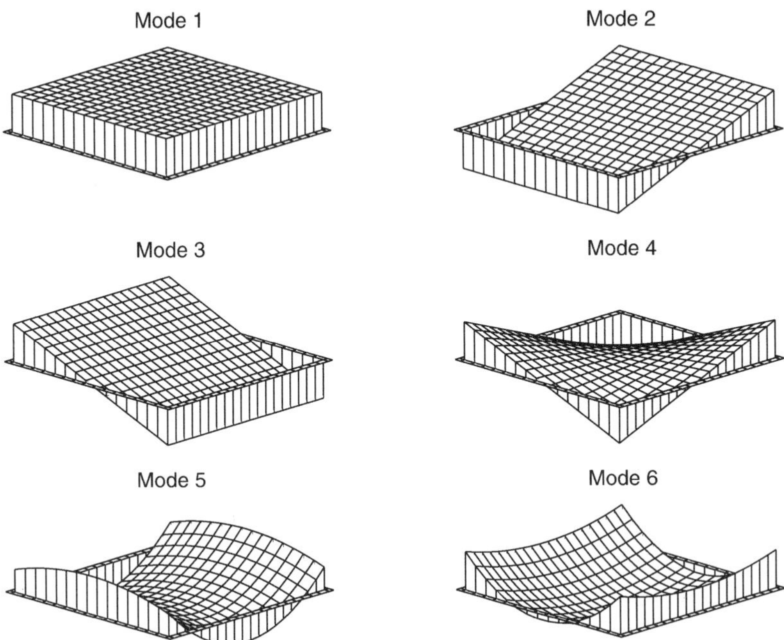


**Figure 9.31:** Radiation modes 1–6 at 5 Hz.

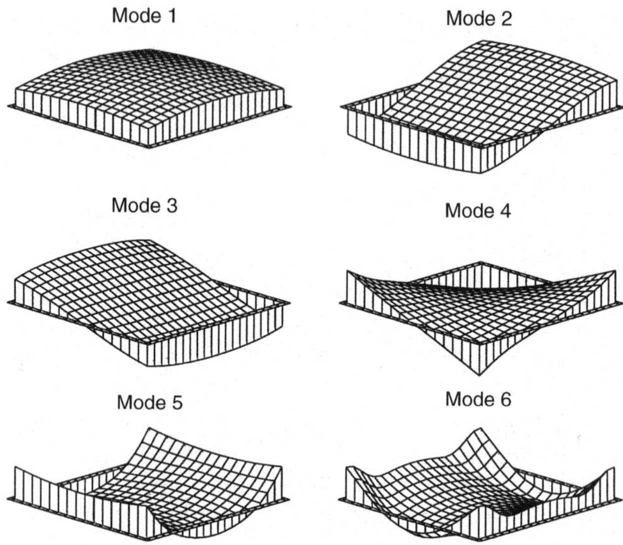


**Figure 9.32:** Radiation modes 1–6 at 500 Hz.

the first six singular values as a function of frequency are presented in Figure 9.33. Note that the radiation in the low-frequency range is dominated by the first mode, and, for frequencies between 300 and 500 Hz, the radiation is dominated by at most six modes.

In order to predict the total power radiated over some bandwidth, the char-

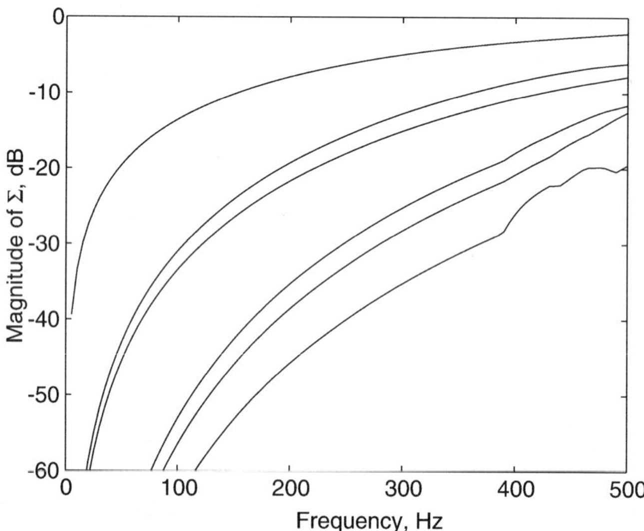


**Figure 9.33:** Six dominant singular values of the radiation matrix **R**.

acteristics of the **R** matrix must be modeled over that bandwidth. Thus, for an $m \times m$ matrix, a total of $m^2$ transfer functions must be modeled (curve fit). For this experiment, a $5 \times 4$ grid of velocity points is used, which corresponds to 400 transfer functions or approximately 1000 states. This level of complexity is impractical for control system design and synthesis, not to mention real-time implementation, since the order of the dynamic compensator will be *at least* the order of the augmented plant. However, a special curve-fitting technique is presented that provides a reasonable approximation to the dynamics of the **R** matrix using a small fraction of the original number of states.

Work by Borgiotti and Jones (1994) describes the "nesting" property of the radiation modes. In their work, they assert that the space spanned by the *significant* radiating modes at frequencies below some arbitrary maximum frequency of the bandwidth $\omega_{\max}$ is a subspace of the space spanned by the radiating modes at the frequency $\omega_{\max}$. Thus, the set of singular vectors at $\omega_{\max}$ corresponding to the significant radiating modes can be used as a basis to describe the radiation at any frequency below $\omega_{\max}$.

The key to incorporating the essential physics of structural acoustic coupling rests in curve fitting the dominant radiating modes over the bandwidth, which is performed using a technique termed radiation modal expansion (RME) (Gibbs et al., 1997). In this technique, the significant radiation modes at the upper frequency of the bandwidth of interest are used as a basis to curve fit the properties of the **R** matrix over the entire bandwidth. The amplitude weighting coefficients are determined by the radiated power of each respective normalized radiation mode shape at each frequency $\omega$ over the bandwidth as follows:

$$\sigma_i(\omega)^2 = \mathbf{u}_{i,\max}^H \mathbf{R}(j\omega)\mathbf{u}_{i,\max} \tag{9.9}$$

where $\sigma_i^2(\omega)$ is the radiated power of the $i$th radiating mode (shape determined at $\omega_{\max}$) and $\mathbf{u}_{i,\max}$ is the $i$th radiating mode shape determined at $\omega_{\max}$. A plot of the amplitude weighting coefficients $\sigma_i(\omega)$ for the first six radiation modes is shown in Figure 9.34. Note that the magnitudes of the RME coefficients agree reasonably well with the actual singular values depicted previously in Figure 9.33. In order to create a model of the RME system, it is only necessary to curve fit the dynamics described by the RME coefficients shown in Figure 9.34. Thus, the model transforms the velocity measurements into radiation mode coordinates with a set of constant transformation matrices ($20 \times 6$ in this example), and then each mode passes through a model of the RME coefficients. The complexity is reduced from modeling 400 transfer functions to that of modeling 6 transfer functions and a simple transformation matrix.

### 9.5.5 Assembling the Augmented Plant Model

From the system identification process, we obtain a model of the structural plant, including 1 disturbance input, 4 piezoelectric inputs, 4 piezoelectric out-

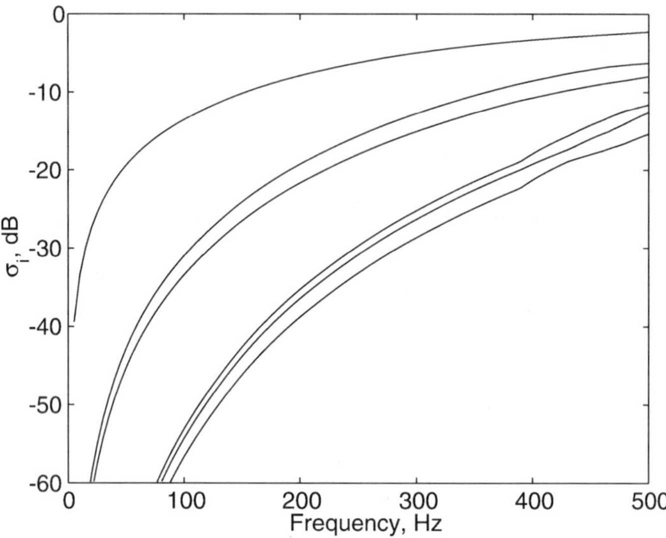


**Figure 9.34:** Radiation modal expansion coefficients for first six radiation modes.

puts, and 20 velocity outputs. To incorporate the appropriate performance metric, radiation filters are designed to develop a frequency weighted cost functional that incorporates a physical model of the structural acoustic coupling. For this example, three different performance metrics are explored, and closed-loop performance resulting from each design metric is evaluated. In developing the augmented plant, a state-variable model is generated that includes all three performance metrics, and the **sel** function of the $\mu$-analysis and synthesis toolbox is used to choose appropriate outputs for the error incorporated in the different cost functionals.

Consider the block diagram of the augmented plant presented in Figure 9.35. As illustrated, there is a state-variable model of the structural plant (*Ap*, *Bp*, *Cp*, *Dp*); a state-variable model of the radiation filters, including six radiation modes (*A*6, *B*6, *C*6, *D*6); a state-variable model of the radiation filter, including one radiation mode (*A*1, *B*1, *C*1, *D*1); and a state-variable model that weights the structural velocities (*Av*, *Bv*, *Cv*, *Dv*). All of these filters are defined in the script file **bld_mod.m**, which can be used to build the composite state-variable model of the augmented system. The inputs to each of the performance filters are obtained from the velocity outputs of the structural model. Thus, for the radiation filters, the volumetric velocity of each elemental radiator of the panel is assumed to be that of the velocity measured at the center point of the elemental radiator. The remaining ouputs of the augmented plant are constructed from the vector sum of the piezoelectric sensoriactuators and the assumed sensor noise, which is filtered with the state-variable model (*Aw*, *Bw*, *Cw*, *Dw*), as illustrated in Figure 9.35. The inputs to the system include the model of the disturbance path, the four piezoelectric sensoriactuators, and the sensor noise.

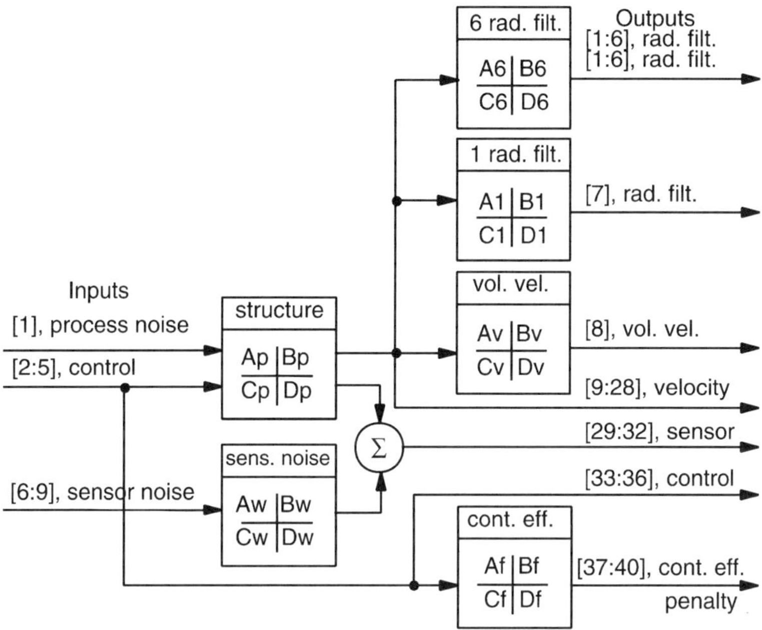


**Figure 9.35:** Block diagram of the structural acoustic system used in the design process.

Notice that the control effort penalty is weighted by a filter ($Af$, $Bf$, $Cf$, $Df$) also. Filters for control effort penalty and sensor noise are included for future flexibility as both of these vector inputs can be frequency weighted; however, for the purpose of this example, a uniform weighting over the bandwidth is assumed.

Once the augmented plant model is developed, there are a total of 9 inputs and 40 outputs. Now, at this point, you might question why the system architecture includes three different performance paths. As outlined earlier, the objective is to try each of the three different error paths, but not all three concurrently. Using the $\mu$-analysis and synthesis toolbox, it turns out that it is easiest to simply include all dynamics desired in a single model and then select the desired inputs and outputs when you are ready to design the control system. The **sel** function can be used for this purpose and is incorporated into the script file that is used to design the control system: **tlc_desn.m**. The vector array of inputs and outputs are grouped, and when "selecting" the design system, the appropriate inputs and outputs are identified for the desired plant.

### 9.5.6 Control Strategy

So far, we have identified a model of the structural plant, constructed a model that incorporates the physics of structural acoustic coupling, and coupled the system models as illustrated in Figure 9.35. We are now ready to proceed with the compensator design. The objective is to compare the performance (with respect to control of sound radiation) of each control system design based upon the three different cost functionals: six radiation filters, one radiation filter, and no radiation filters (pure velocity). Executing the script file **tlc_desn.m** will design a controller for the six radiation filter case. The reader is encouraged to design controllers for the other two cases as well, and the only modification to the script file required is to place comment characters (%) in front of the lines of code describing the six radiation filter design and remove the characters from the lines preceding the other design alternatives. The design parameters that are utilized in this excercise are the control effort penalty weighting and the sensor noise weighting. Due to the chosen weighting functions, the frequency weighted $\mathcal{H}_2$ synthesis is structured analogous to the standard LQG problem. By increasing the control effort penalty, the closed-loop performance can be limited. Additionally, due to the rolloff in the modeled system (only a finite number of modes are included in the system identification), increasing the sensor noise tends to decrease the performance and the effective bandwidth of operation for the closed-loop system. Conceptually, one recognizes that the frequency at which the sensor noise is of the same order of magnitude as the plant output corresponds to the limit of the bandwidth over which the control system should perform. Thus, considering the rolloff in the plant model, as the level of sensor noise is increased, the frequency at which the sensed variables and the sensor noise are of the same order of magnitude decreases. By considering the maximum and minimum singular values of the loop transfer function, including the dynamic compensator, one can assess the gain crossover frequency. For a fixed control effort penalty, increasing the sensor noise relative to the plant response tends to decrease the gain crossover frequency.

The script file **tlc_desn.m** produces three figures, two of which are used extensively in the design process: a comparison of the open-loop and closed-loop radiated sound power and a plot of the singular values corresponding to the loop transfer matrix. As detailed in Chapter 8, the maximum and minimum singular values of the MIMO loop transfer matrix have an interpretation that is analogous to that of the Bode plot of a SISO system. For this example, the objective is to maximize performance over the bandwidth of operation while minimizing the control effort expended at higher frequencies where the dynamics of the structure are not modeled. To accomplish this "rolloff," the sensor noise is adjusted to effectively lower the corner frequency of the dynamic compensator. As illustrated in Figure 9.36, the maximum singular values never exceed unity above a frequency of approximately 600 Hz. This is below the bandwidth of the modeled dynamics but exceeds the bandwidth targeted for performance (20–500 Hz). Weightings for sensor noise and process noise are set at $10^{-3}$ and $2 \times 10^{-2}$, respectively, as indicated in the script file **tlc_desn.m** for the six radiation filter design case.

The $\mathcal{H}_2$ norm at each frequency of the closed-loop transfer matrix is computed between the disturbance input and the radiation filter output. For the given compensator design, the result is presented in Figure 9.37. Note that the inputs

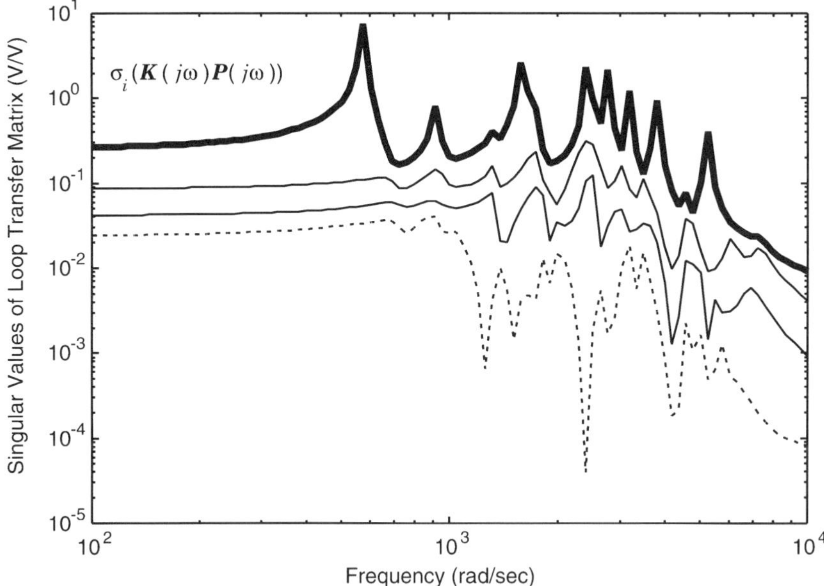


**Figure 9.36:** Magnitude of the singular values of the loop transfer matrix.

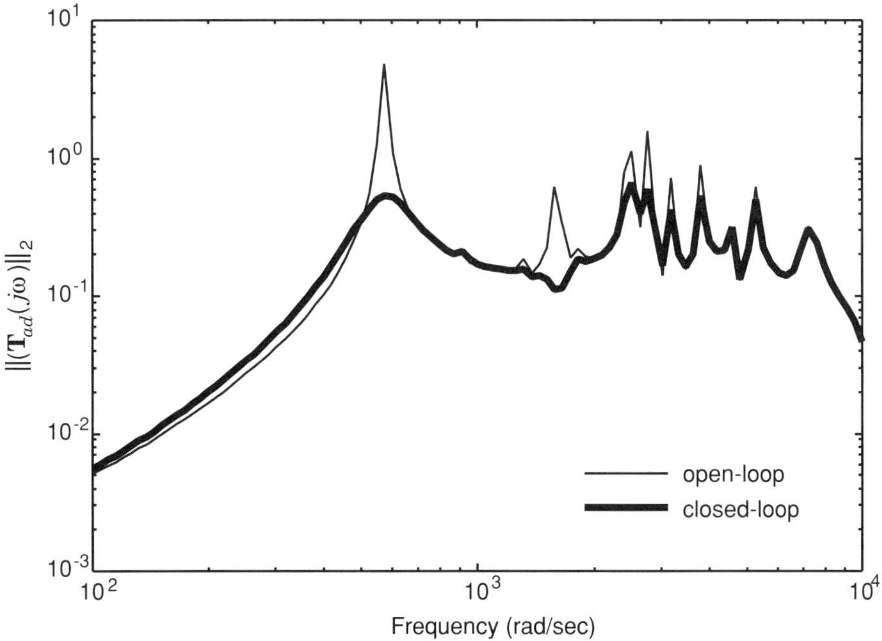


**Figure 9.37:** Predicted open- and closed-loop performance of test panel based upon dynamics obtained through modeling and system identification.

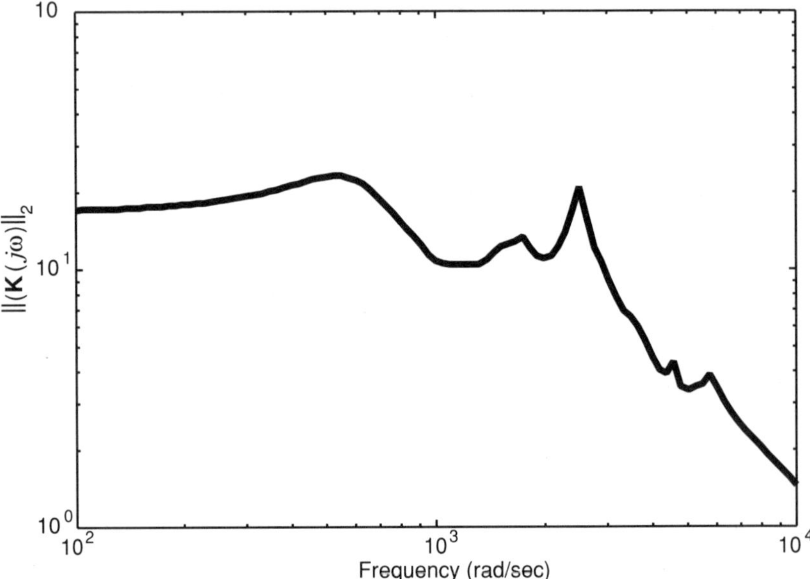


**Figure 9.38:** Root mean-square of squared singular values of the dynamic compensator plotted as a function of frequency.

and outputs are not converted to engineering units; however, this does not affect the relative performance of the system, as the results presented in Figure 9.37 are representative of the relative power for the open- and closed-loop system. As illustrated, the predicted closed-loop performance is greatest at low frequency, in the bandwidth of the resonant frequency corrresponding to the fundamental mode with approximately 20 dB of attenuation. The closed-loop performance extends to approximately 600 Hz, which is the frequency at which the maximum singular values of the loop transfer matrix fall below unity.

In Figure 9.38, the $\mathcal{H}_2$ norm at each frequency of the dynamic compensator is presented. As illustrated, the bandwidth of the compensator is limited to approximately 500 Hz. Beyond this bandwidth, the dynamic response decreases with increasing frequency, which is also reflected in the singular values of the loop transfer matrix presented in Figure 9.36. This rolloff is necessary to minimize the control energy expended at higher frequencies where model uncertainty is greatest. The measured performance of the closed loop system with the given compensator is presented in the following section.

### 9.5.7 Results from Analysis

As detailed at the beginning of this example, the objective is to design a compensator for three different cost functions. Using the script file **tlc_desn.m**, this can be accomplished; however, the discussion to this point has been limited to a single design. The reader is encouraged to experiment with the script file and study the alternative designs. However, in this section, the closed-loop performances for compensators designed from each alternative cost function are presented.

As illustrated in Figure 9.39, the open-loop and closed-loop sound power radiating from the structure is presented. In the legend, there are four cases identified: no control (No control), feedback control based upon a six radiation filter design (FB Control, 6 rad), feedback control based upon a single radiation filter design (FB Control, 1 rad), and feedback control based upon minimizing the velocity over an array of points on the structure (FB Control, vib). As indicated, approximately 15 dB of reduction in sound power is observed in proximity to the resonance corresponding to the fundamental panel mode for both the six radiation filter design and the single radiation filter design. The vibration control system design results in approximately 12 dB of attenuation. Also notice that the acoustic response at some of the resonant frequencies (between 100 Hz and 200 Hz) is insignificant compared to that corresponding to the resonance of the fundamental mode. This results from the radiation efficiency of the modes that characterize the structural acoustic control problem, since structural modes with even indices are inefficient radiators at low frequency (i.e., when the structural wavelength is shorter than the acoustic wavelength). To see this effect, consider the plot of the structural power illustrated

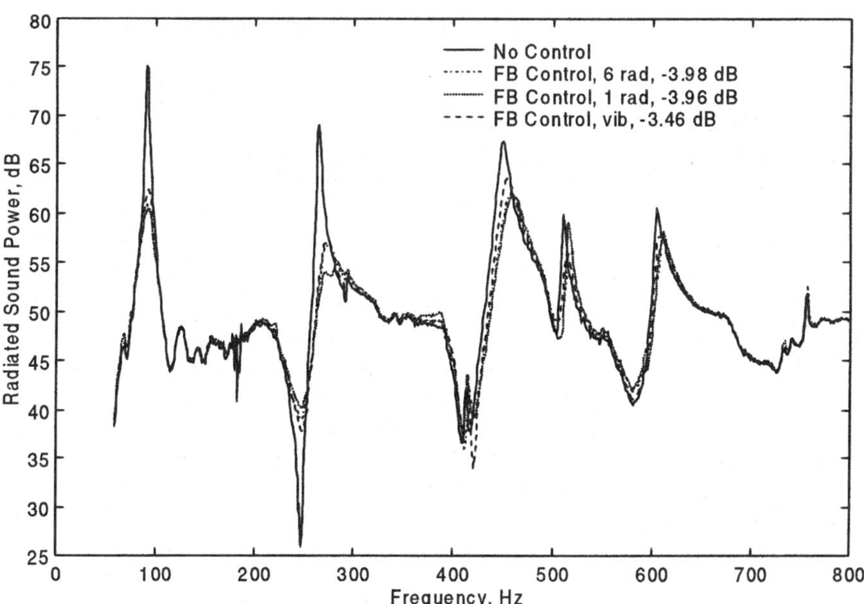


**Figure 9.39:** Measured open- and closed-loop acoustic power radiated.

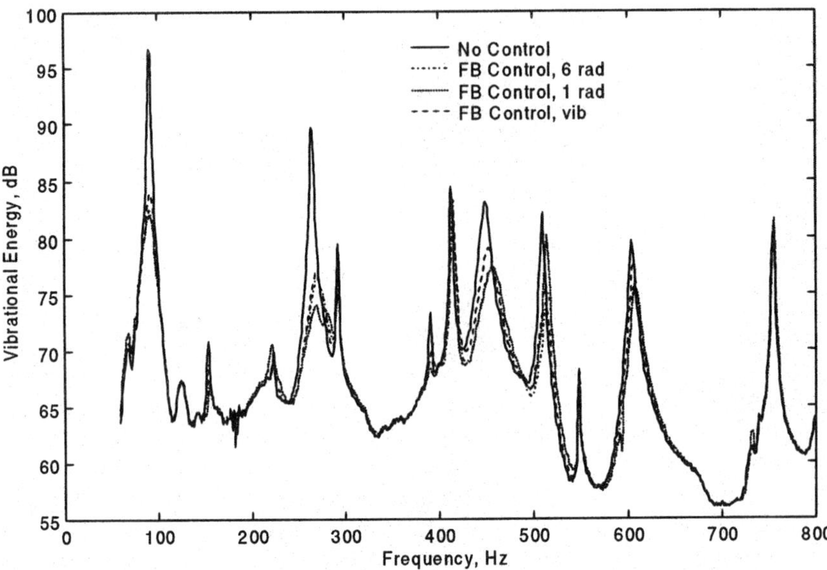


**Figure 9.40:** Measured open- and closed-loop structural power.

in Figure 9.40 and obtained from an RMS average of the velocity measurements. As indicated in the results, the resonant response of structural modes that are not significant radiators now appears in the plot of the vibrational energy. However, the modes corresponding to these resonances do not play a significant role in the frequency shaped cost functional, since their relative contribution to the radiated sound power is insignificant when the acoustic wavelength is much greater than the structural wavelength. Thus, the frequency shaped cost functional, made possible by the implementation of the radiation filters, provides a means of incorporating this physical mechanism into the control system design. The overall performance of the two control system designs based upon radiation filters as opposed to vibration control resulted in an additional 0.5 dB of sound power attenuation over the bandwidth. Considering that the total reduction for the vibration controller was 3.5 dB, the increase in performance was on the order of 15%.

### 9.5.8 Summary of Structural Acoustic Control Experiments

A method of developing a frequency shaped cost functional incorporating the structural acoustic coupling for an experimental test structure was presented. The order of the state-variable model of the structural acoustic coupling was reduced through singular value decomposition, and a model incorporating six radiation filters and a single radiation filter was used to design a dynamic compensator for control of a structure clamped at all boundaries and configured with four piezoelectric sensoriactuators. The relative performance of the compensator designs from the two different choices of radiation filters was indistinguishable; however, a marginal improvement was observed when compared to the implementation of a controller based solely on a measure of the structural vibration. The importance of the frequency shaped cost was demonstrated, as was the need to provide a practical method of rendering such model-based control system designs.

## 9.6 ACTIVE CONTROL OF POWER FLOW IN BEAMS

### 9.6.1 Problem Overview

For this example, we experimentally demonstrate the control of vibrational energy from a power flow (wave control) perspective. Vibrations may travel through beam-like structures in the form of flexural, extensional, and torsional motion, and often couple to a fluid and radiate sound. As previously shown in Example 2.6 of Chapter 2, the power flow (positive in the $+\,x$-direction) due to flexural motion in beams is proportional to the difference of the square of the positive and the square of the negative traveling wave amplitudes. This result was shown previously in Equation (2.197) and is presented again here for the purpose of discussion.

$$P_f(\omega) = EIk^3\omega(|A_1|^2 - |A_2|^2 - 2\mathrm{Imag}(A_3A_4^*)) \tag{2.197}$$

The traveling waves can be considered "carriers" of energy whereby the positive traveling wave carries energy in the positive direction and the negative traveling wave carries energy in the negative direction. Thus, any reduction in the amplitude of the traveling waves provides a corresponding reduction in power flow.

Work by Fuller et al. (1990), Scheuren (1990), and Gibbs (1993) has demonstrated the efficiency of controlling power flow leaving finite beams terminated by an impedance (representative of a structural joint). In their work, power flow could be controlled with one actuator per wave type (e.g., flexural). Wave based control is in direct contrast to the usual modal approaches, which attempt to reduce the total vibrational energy in a beam section. Implementation of control using a modal approach typically requires the application of one actuator and sensor per mode of interest to achieve efficient attenuation (Meirovitch and Norris, 1984). In the following sections, the experimental results for control of flexural and extensional motion in a beam from a wave perspective are presented. Since active control using power flow methods is related to sensing and control of traveling waves, a technique for the real-time sensing of traveling flexural and extensional waves in beams is required. This technique has been previously presented in Sections 6.5.1 and 6.5.2 and is applied in the following example to determine the net traveling flexural and extensional waves in real time. The wave estimates are used as "error" signals in a feedforward

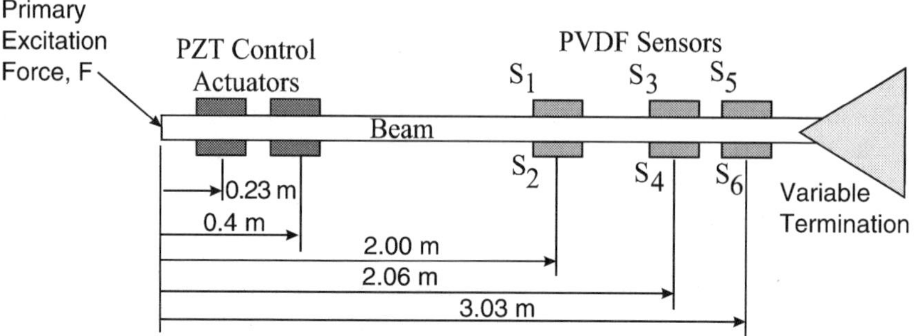


**Figure 9.41:** Schematic of beam power flow experimental arrangement.

active control scheme, and the controller is configured to minimize traveling waves in the beam (and corresponding power flow). Independent control performance is experimentally determined using experimental spatial dynamics modeling (ESDM). The ESDM technique uses velocity measurements made with a scanning laser velocimeter, and postprocessing techniques are applied to determine the net three-dimensional motion of the structure.

### 9.6.2 Description of the Plant—Experimental Setup

A schematic for the test apparatus used in this experiment is shown in Figure 9.41. The test beam is made of aluminum and has the dimensions 3.5 m × 76.2 mm × 3.175 mm. The beam has a free boundary condition at one end and is configured with a variable termination at the other. The termination consists of a diverging pyramid-shaped box filled with sand that is "anechoic" for frequencies above 150 Hz, having a flexural power reflection coefficient less than 0.1 for this frequency range. The termination is highly reflective for extensional motion (due to the extremely long wavelengths). For part of the experiment, the termination is intentionally fitted with aluminum blocks, causing the termination to be highly reflective. The primary noise source is provided by a 25-lb shaker attached with a stinger at the free end of the beam and oriented at an angle of 20 degrees relative to the neutral axis of the beam. This arrangement produces a vibrational response in both flexure and extension.

### 9.6.3 Transducer Selection, Design, and Placement

Piezoelectric devices are used exclusively for sensing and actuation in this experiment because of their light weight, compact size, efficient operation, and relative ease in specifying the desired dimensions of the aperture. The control inputs (secondary sources) consist of a set of two pairs of piezoceramic actuators mounted, as shown in Figure 9.41. The actuators are constructed from PZT-5A, having a dimension of 38.1 mm × 63.5 mm × 0.254 mm. These actuators can be configured to excite pure flexure (symmetric pairs driven out-of-phase), pure extension (symmetric pairs driven in-phase), or any combination of the two wave motions (arbitrary independent input). A detailed discussion of the possible excitation scenarios and resulting beam responses can be found in the reference by Gibbs (1993).

In Sections 6.5.1 and 6.5.2, real-time wave filtering has been presented that utilizes a combination of spatial and temporal (using digital filters) signal processing to estimate the traveling waves in both dispersive media (flexural waves in beams) and nondispersive media (extensional waves in beams). In this technique, the beam is configured with two like sensors (e.g., accelerometers, polyvinylidene fluoride (PVDF) strips sensors, etc.) per wave type, whose outputs are sampled and passed through a relatively simple array of digital filters to produce estimates of the net positive and net negative traveling waves in the beam.

For this experiment, the beam is also fitted with a total of six PVDF sensors, as shown in Figure 9.41. Sensor pairs 1 and 2, as well as sensor pairs 3 and 4, are differenced to provide a response due to flexure only at their respective locations. Sensor pairs 3 and 4, as well as sensor pairs 5 and 6, are summed to provide a response due to extensional motion only at their respective locations. The outputs derived from the three pairs of transducers provide the necessary signals to estimate the traveling waves in that region of the beam.

### 9.6.4 Control Strategy

A block diagram of the feedforward adaptive control system used in these experiments is shown in Figure 9.42. The controller is synthesized from a combination of the MIMO filtered-$x$ LMS algorithm previously discussed in Chapter 8 and the wave vector filter scheme presented in Chapter 6. Thus, the performance variables must be estimated in real time with the wave vector filter such that the adaptive algorithm can adapt the feedforward compensator to minimize the desired cost functional (the power flow in flexure and extension). For the given application, the outputs of the wave vector filters provide estimates of the traveling flexural and extensional waves in the beam (region of the sensors), forming the "error" signals for the adaptive algorithm. For this experiment, the control system is configured to minimize the flexural and extensional wave traveling *toward* the termination (the reflected waves are minimized as a result).

### 9.6.5 Experimental Work

***Full 3-D Velocity Measurement using a Laser Vibrometer and ESDM Techniques*** The control system performance is independently verified by experimentally scanning the beam with a laser Doppler velocimeter both before and after control at over 10,000 data points. The velocity is obtained from a variety of directions and at an arbitrary array of noncoincident data locations. Because of the arbitrary placement and scanning of the laser, each measurement is unique

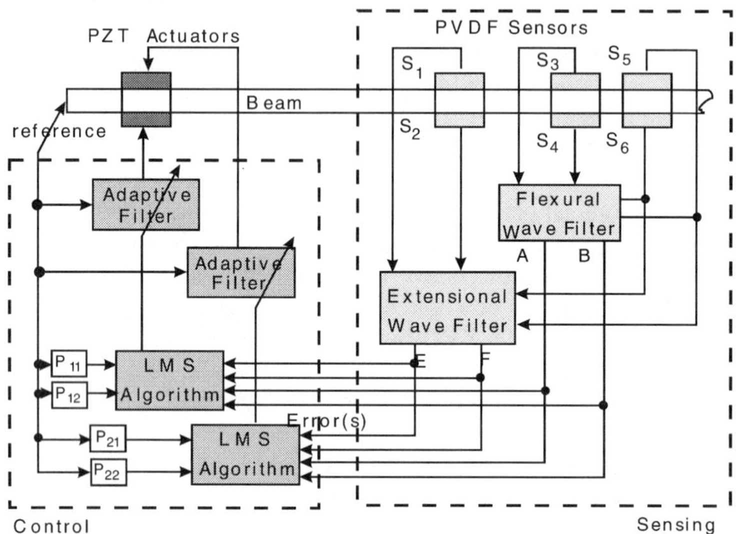


**Figure 9.42:** Block diagram of feedforward active control system with wave vector filters.

in location and direction. The location of the beam, the scanning laser, and the direction of the scan are well known and tightly controlled such that the corresponding vector of measurements is well defined. A technique termed experimental spatial dynamics modeling (ESDM) is used to process the velocity data (arbitrary in location and direction), by employing a specially formulated finite element technique. The results of the ESDM technique provide continuous velocity functions describing the full three-dimensional motion of the structure. For a detailed presentation of the ESDM technique, the interested reader should consult the work of Montgomery and West (1996).

***Experimental Results*** The beam is driven with a pure tone disturbance at 550 Hz using the primary shaker. Active control is applied using the aforementioned techniques. The magnitude of the resulting velocity distribution for the uncontrolled beam is shown with gray scale contour in Figure 9.43. In the figure, the response for the $x$, $y$, and $z$ directions are shown separately as a function of beam width and length. The $x$ direction corresponds to motion along the beam axis, and the response is dominated by an extensional standing wave with a node at approximately $x = 1.5$ m. The antiresonances appear to be located at $x = 0$ m and $x = 3.0$ m. The $y$ direction corresponds to in-plane motion perpendicular to the neutral axis of the beam, and the response illustrated appears to be dominated by a flexural wave in the $y$ direction. The $z$ direction corresponds to out-of-plane motion, and the response appears to be dominated by a combination of flexural and torsional motion. Note that the PVDF strip sensors are designed to be insensitive to asymmetric motions about the neutral axis, and

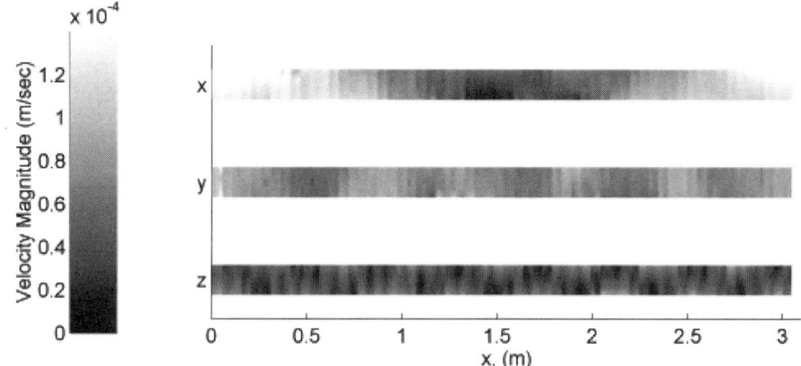


**Figure 9.43:** Total beam response, 550 Hz, no control.

thus the torsional motion and flexural motion in the $y$ direction are unobservable by the control system, and thus are not controlled during the experiment. Waves of this type can be controlled with the same wave filtering technique, assuming that the proper arrangement of actuators and sensors is developed.

In order to determine the performance of the active control system, each wave type is analyzed independently. The symmetric portion of the $z$-direction velocity response corresponds to flexural motion of the system (torsional motion is asymmetric). The symmetric $z$-direction velocity magnitude for the control and no control cases is presented in Figure 9.44. The no control case shows a significant velocity field that does not contain nodal regions; however, it does contain regions of constructive and destructive interference due to wave reflections at the termination. Note the significant control of waves in the region between the control actuators and the termination, which corresponds to a reduction of the velocity field on the order of 30 dB.

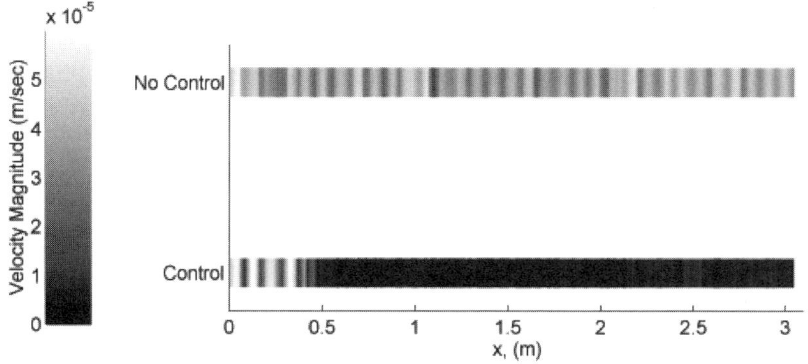


**Figure 9.44:** Flexural response, 550 Hz, before and after control.

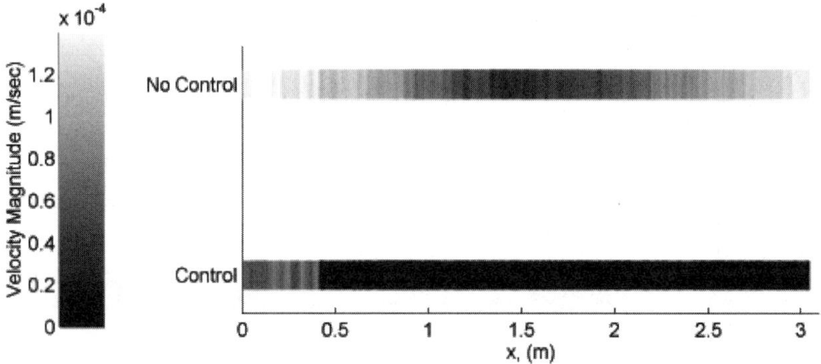


**Figure 9.45:** Extensional response, 550 Hz, before and after control.

The extensional response ($x$ direction) of the beam for the control and no control cases is presented in Figure 9.45. Before applying control, the beam response is dominated by a standing wave with a nodal region near $x = 1.5$ m. A single traveling wave (e.g., toward the termination) would have a uniform velocity magnitude distribution with only a linear phase change. A standing wave consists of two equal (or near equal) and opposite traveling waves that are superimposed. This combination produces a standing wave that has one or more equally spaced nodal regions. After applying control, the beam response between the control actuators and the termination is significantly attenuated. The attenuation of the velocity field in this region is on the order of 20 dB. Note that in Figures 9.44 and 9.45, the response between the primary shaker and control actuators is not necessarily reduced. These experimental results correlate well with the analytical results found in the reference by Gibbs (1993).

The total beam response upon applying control is presented in Figure 9.46. The remaining uncontrolled velocity field appears to consist of a torsional wave

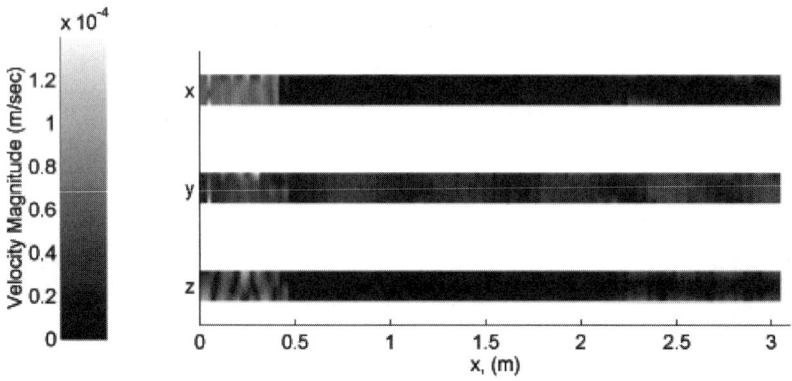


**Figure 9.46:** Total response, 550 Hz, control.

about the neutral axis and a flexural wave in the $y$ direction that are not targeted by the control system.

In the second series of tests, the beam is driven with band limited pink noise (500–1500 Hz), corresponding to the frequency range where the flexural wave filters are designed to operate (see Chapter 6 and Gibbs, 1993). The control system is again configured to simultaneously control both flexural and extensional waves traveling *toward* the termination. The configuration of the laser scanning system used in these experiments takes approximately 18 hours to obtain the response of the system at a single frequency and is thus not suitable for broadband measurements. The outputs of the wave vector filters (error functions) of the control system are available for analysis. The frequency responses of the incident flexural wave ($A_1$ wave) and the reflected flexural wave ($A_2$ wave) are presented in Figure 9.47. The signals corresponding to the incident and reflected wave estimates have been reduced by 0 to 15 dB in the frequency range of 400–1000 Hz. The response of the incident extensional wave filter ($E$ wave) and the reflected wave filter ($F$ wave) are shown in Figure 9.48 before and after control, and these waves have been reduced between 0 and 15 dB in the frequency range of 400–1200 Hz.

### 9.6.6 Summary

In summary, the wave vector filtering-based control system has been demonstrated to control flexural and extensional motion simultaneously in beams by at least 20 dB for pure tones and by 0–15 dB for band-limited disturbances. The scanning laser vibrometer and ESDM techniques have proved to be extremely useful in providing independent verification of control performance with detail far superior to previous measurement schemes.

## 9.7 SUMMARY

The purpose of this chapter is to draw from the topics presented in the previous chapters and implement some of the various modeling and control strategies described. All of the models can be developed from the tools presented in earlier chapters, and script files were included to facilitate the material. A basic procedure was identified for the design of adaptive structures that included:

1. Problem Overview
2. Description of the Plant
3. Transducer Selection, Design, and Placement
4. Development of the Augmented Plant Model
5. Control Strategy
6. Simulation or Experimental Results
7. Summary

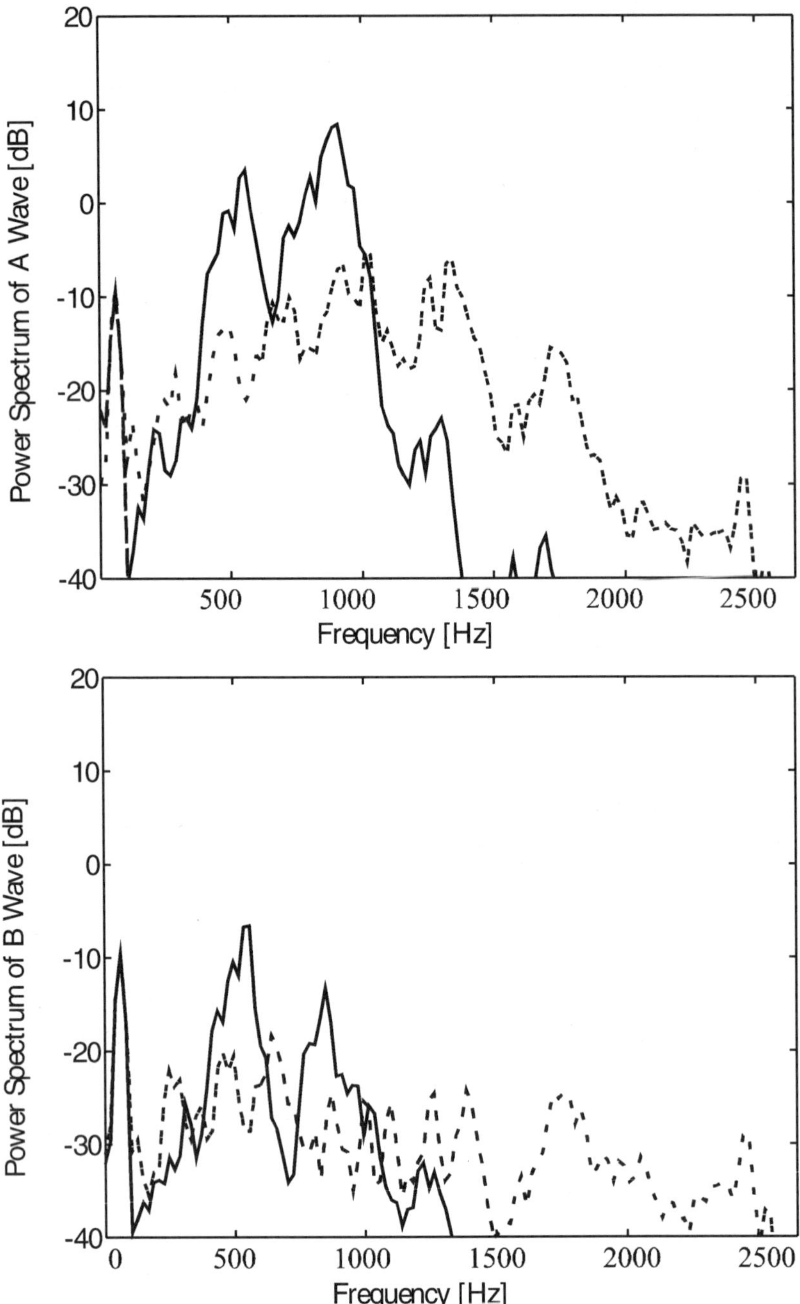


**Figure 9.47:** Frequency response, flexural wave vector filters, before (solid) and after (dashed) broadband control.

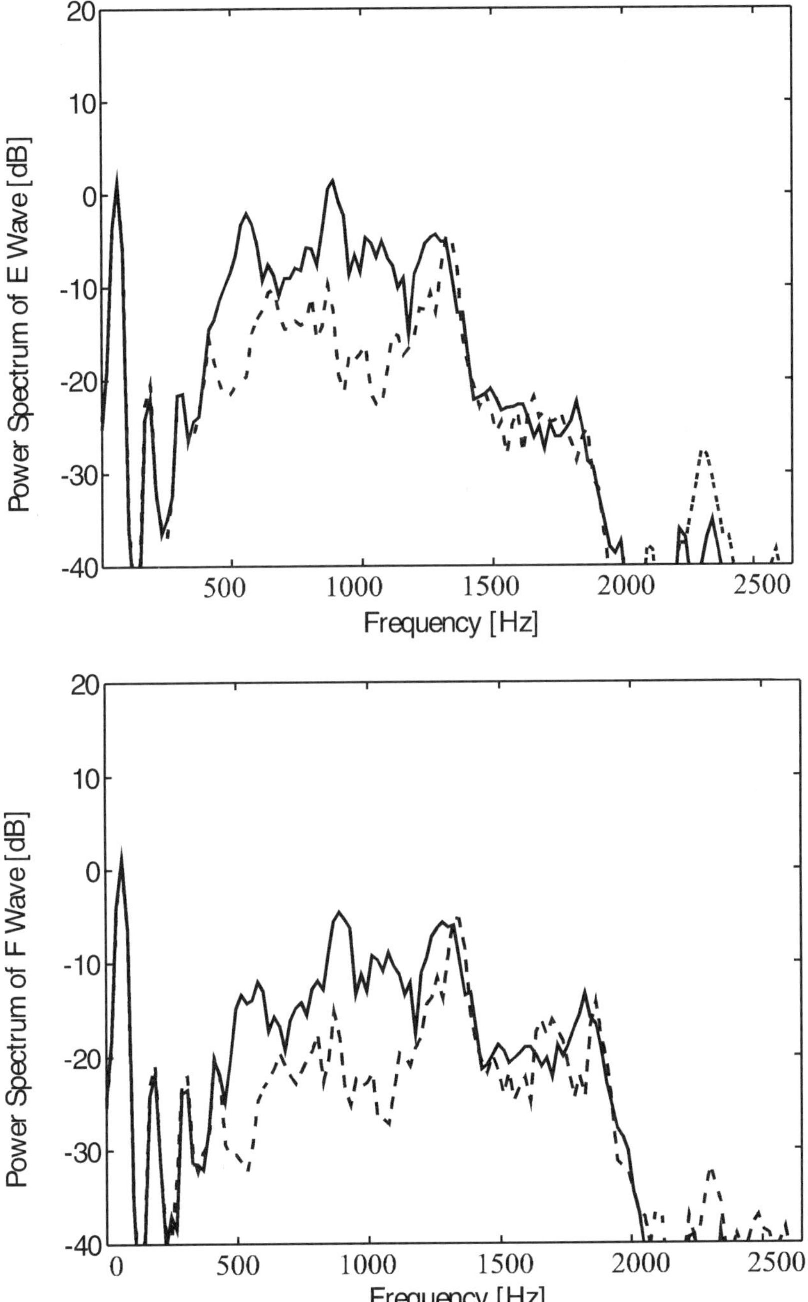


**Figure 9.48:** Frequency response, extensional wave vector filters, before (solid) and after (dashed) broadband control.

This format can be considered as a "recipe" for analysis and design of adaptive structures. All of the examples presented in this chapter followed this format, and the reader is encouraged to break the design problem into such components.

Upon completing this manuscript, the reader should be capable of generating all of the script files included within this work independently. Much of the work in digital signal processing was "taken for granted" in this manuscript; however, a number of good texts in digital control are available (Franklin et al. 1992; Phillips and Nagle, 1990), and the interested reader is encouraged to explore this topic further. The authors sincerely hope that this work is *enabling* for the reader and assists in creating future directions yet to be determined.

## BIBLIOGRAPHY


Active Control eXperts, Inc., 215 First Street, Cambridge, MA, 12142, *Smart ID System Identification Software*.

Baumann, W. T., W. R. Saunders, and H. H. Robertshaw, 1991. "Active Suppression of Acoustic Radiation from Impulsively Excited Structures," *Journal of the Acoustical Society of America*, **90**(6), 3202–3208.

Borgiotti, G. V. and K. E. Jones, 1994. "Frequency Independence Property of Radiation Spatial Filters," *Journal of the Acoustical Society of America*, **96**(6), 3516–3524.

Clark, R. L., 1997. "Accounting for Out-of-Bandwidth Modes in the Assumed Modes Approach: Implications on Colocated Output Feedback Control," accepted for publication in the *Journal of Dynamic Systems, Measurement, and Control.*

Clark, R. L. and C. R. Fuller, 1992. "Experiments on Active Control of Structurally Radiated Sound Using Multiple Piezoceramic Actuators," *Journal of the Acoustical Society of America*, **91**(6), 3313–3320.

Elliott, S. J. and M. E. Johnson, 1993. "Radiation Modes and the Active Control of Sound Power," *Journal of the Acoustical Society of America*, **94**(4), 2194–2204.

Francis, B., 1986. *A Course in $\mathcal{H}_\infty$ Control Theory*, Springer-Verlag, New York.

Franklin, G. F., J. D. Powell, and M. L. Workman, 1992. *Digital Control of Dynamic Systems*, Addison-Wesley, New York.

Fuller, C. R., G. P. Gibbs, and R. J. Silcox, 1990. "Simultaneous Active Control of Flexural and Extensional Waves in Beams," *Journal of Intelligent Material Systems and Structures*, **1**(2), 235–247.

Gibbs, G. P., 1993. "Simultaneous Active Control of Flexural and Extensional Power Flow in Thin Beams," Ph.D. Dissertation, Virgina Polytechnic Institute and State University, Blacksburg, VA.

Gibbs, G. P., R. L. Clark, D. E. Cox, and J. S. Vipperman, 1997. "Structural Acoustic Control and the Importance of Radiation Filters in the Experimental Design," submitted to the *Journal of the Acoustical Society of America.*

Gill, P. E., W. Murray, and M. H. Wright, 1981. *Practical Optimization.* Academic Press, San Diego, CA.

Gupta, N. K., 1980. "Frequency Shaped Cost Functionals: Extension of Linear-

Quadratic-Gaussian Design Methods," *Journal of Guidance and Control*, **3**(6), 529–535.

Liu, K., R. N. Jacques, and D. W. Miller, 1996. "Frequency Domain Structural System Identification by Observability Range Space Extraction," *Journal of Dynamic Systems, Measurement, and Control*, **118**, 211–220.

Maillard, J. and C. R. Fuller, 1994. "Advanced Time Domain Wave-Number Sensing for Structural Acoustic Systems. I. Theory and Design," *Journal of the Acoustical Society of America*, **95**(6), 3252–3261.

Meirovitch, L. and M. A. Norris, 1984. "Vibration Control," *Proceedings of Internoise 84*, pp. 477–482.

Montgomery, D. E. and R. L. West, 1996. "A Least-Squares Discrete Finite Element Method for Modeling Three Dimensional Velocity Response Fields Measured by a Scanning LDV," to be presented in *The Journal of Mechanical Systems and Signal Processing*.

Morgan, 1980. "An Analysis of Multiple Correlation Cancellation Loops with a Filter in the Auxiliary Path," *IEEE Transactions on Acoustics, Speech, and Signal Processing*, **ASSP-28**, 454–467.

Phillips, C. L. and H. T. Nagle, 1990. *Digital Control System Analysis and Design*, Prentice Hall, Englewood Cliffs, NJ.

Scheuren, J., 1990. *Acoustica*, **72**, 33–46.

# APPENDIX A

# TRANSFORMS FOR CONTINUOUS AND DISCRETE-TIME SYSTEMS

## A.1 INTEGRAL TRANSFORMS FOR LINEAR SYSTEMS

The advantage of dealing with linear systems is that we have an arsenal of mathematical tools available for design and analysis. In this section, we deal with integral transforms, namely the Fourier transform and the Laplace transform. We begin with the general concept of an integral transform and discuss the terminology from the engineer's perspective in view of applications for adaptive structures.

As was the case in Chapter 2, the objective is to find a solution to the homogeneous differential equation presented in equation (3.1). In operator notation, this reduces to:

$$L[w(\mathbf{x}, t)] = 0 \tag{A.1}$$

where $\mathbf{x}$ is a position vector in a general $n$-dimensional field. With integral transformations, one essentially maps the partial differential equation into a domain that hopefully facilitates finding the solution. For example, one can typically find a solution to equation (A.1), $w(\mathbf{x}, t; \kappa)$, in terms of an arbitrary constant $\kappa$ taken from a set $A$. For linear differential operators, it is likely that

$$g(\mathbf{x}, t) = \int_A F(\kappa) w(\mathbf{x}, t; \kappa) d\kappa \tag{A.2}$$

is also a solution of the homogeneous partial differential equation as long as the integral converges for all $\mathbf{x} \in D$. As an example, consider a position vector that is a function of two spatial coordinates $x$ and $y$ (as is typical for plate-like structures) such that

$$g(x, y, t) = \int_A F(\kappa)w(x, y, t; \kappa)d\kappa \tag{A.3}$$

Now, if we let $f(x) = g(x, 0, 0)$, then

$$f(x) = \int_A F(\kappa)w(x, 0, 0; \kappa)d\kappa \tag{A.4}$$

where $D_x$ is the domain of the variable $x$. The *kernel* is defined by the equation

$$K(x; \kappa) = w(x, 0, 0; \kappa) \tag{A.5}$$

Substituting equation (A.5) into equation (A.4), one obtains a general expression for an integral transform in terms of the kernel:

$$f(x) = \int_A F(\kappa)K(x; \kappa)d\kappa, \qquad x \in D_x \tag{A.6}$$

By definition, $f(x)$ is the *integral transform* of $F(\kappa)$ by the *kernel* $K(x; \kappa)$.

One might wonder at this point why so much detail has been provided concerning the integral transform. The concept of a kernel is integral (pardon the pun) to distributed sensing and actuation, as conveyed in Chapter 6. Each time a sensor or actuator is placed on a structure, one can describe the transduction device in terms of some spatial aperture and location with respect to the structure. In terms of the distributed strain sensors and actuators, the spatial aperture can be shaped in terms of specific functions. In the event that the aperture is shaped in the form of a solution to the homogenous partial differential equation describing the structural response, the distributed sensor can be interpreted as a kernel. For example, each eigenfunction of the simply supported beam constitutes a kernel, and a sensor shaped as a function of that mode will integrate the structural response and yield a signal that is *transformed* into modal coordinates.

### A.1.1 The Fourier Transform

The Fourier transform is a special kind of integral transform, and it is uniquely defined by its kernel. In general, Fourier transforms can be used to represent any arbitrary nonperiodic function. While Fourier transforms can be performed on any coordinate, the following discussion uses the variables $\omega$ and $t$, since the reader is most likely familiar with the temporal Fourier transform. For the Fourier transform,

$$f(t) = \int_{-\infty}^{\infty} F(\omega)K(t;\omega)d\omega \tag{A.7}$$

where $K(t;\omega) = (2\pi)^{-1/2}\exp(j\omega t)$. Substituting for the kernel,

$$f(t) = \frac{1}{\sqrt{2\pi}} \int_{-\infty}^{\infty} F(\omega)\exp(j\omega t)d\omega \tag{A.8}$$

where $f(t)$ is piecewise continuously differentiable and absolutely integrable in $\Re$. The inversion is obtained as follows:

$$F(\omega) = \frac{1}{\sqrt{2\pi}} \int_{-\infty}^{\infty} f(t)\exp(-j\omega t)dt \tag{A.9}$$

For the Fourier transform to exist,

$$\int_{-\infty}^{\infty} |f(t)|dt \tag{A.10}$$

must be bounded.

Thus, the Fourier integral transform can be used to map a function from the time domain to the frequency domain. In terms of spatial variables, the transform is typically denoted the wavenumber transform, which maps the solution from the spatial domain to the wavenumber domain. This method of analysis is particularly useful in structural acoustics.

### A.1.2 The Laplace Transform

The Laplace transform is an essential tool in the analysis and design of linear systems and adaptive structures. While the Laplace transform is associated with the analysis of continuous-time signals, one can readily map a system from the Laplace domain to the $z$-domain. The $z$-domain results from taking the $z$-transform of a discrete-time signal, as outlined in Section A.2. A word of warning is noted at this point because, while it is possible to map directly from the Laplace domain to the $z$-domain, the direct inversion is not possible since each pole in the $z$-domain will in general result in an infinite set of poles in the Laplace domain. Regardless, the Laplace transform provides a convenient method of studying causal systems and the transient and steady-state response of such systems.

The Laplace transform can be derived from the Fourier transform if we let $\omega = js$ and substitute into equation (A.8):

$$f(t) = \frac{1}{\sqrt{2\pi}} \int_{+j\infty}^{-j\infty} F(s)\exp(st)d(js) \tag{A.11}$$

which can be expressed as:

$$f(t) = \frac{1}{j\sqrt{2\pi}} \int_{-j\infty}^{j\infty} F(s)\exp(st)ds \tag{A.12}$$

By similar substitution into equation (A.9)

$$F(s) = \frac{1}{\sqrt{2\pi}} \int_{-\infty}^{\infty} f(t)\exp(-st)dt \tag{A.13}$$

which defines the *bilateral Laplace transform.* One should recognize that not every function that has a Fourier transform will have a Laplace transform, since

$$\int_{-\infty}^{\infty} f(t)\exp(-st)dt \tag{A.14}$$

may very well diverge because the kernel $\exp(-st)$ dominates the growth of the integrand in the inversion process. At this point it is valid to question why we went to the trouble of deriving the Laplace transform at all, since it appears to afford little advantage in terms of solving problems that could not be solved with the Fourier transform. However, in the analysis of *causal functions* (i.e., functions such that $f(t) = 0$ for $t < 0$), the primary function of interest in control system design, the Laplace transform is of considerable use. In fact, the Laplace transform can be used to find solutions of causal functions that are damped, that oscillate, or even that grow at an exponential rate.

To proceed, we derive the *unilateral Laplace transform*, which is associated with causal functions. In the study of linear systems, we are often interested in the transient response of the system, since the transient response yields information about system stability. Causal functions are tied to the concept of transient response, since we assume that at some point in time an input is applied to the system, and before that instant in time, the system response was zero. Because a number of functions exist for which

$$\int_{-\infty}^{\infty} |f(t)|dt \tag{A.15}$$

is not convergent (i.e., like $\sin(\omega t)$ for real values of $\omega$), we consider an alternative function:

$$f_\sigma(t) = \exp(-\sigma t) f(t) H(t) \tag{A.16}$$

where $H(t)$ is the Heaviside function and $\sigma$ is a positive constant such that

$$\int_{0}^{\infty} \exp(-\sigma t) f(t)dt \tag{A.17}$$

is convergent.

From the Fourier transform,

$$\begin{aligned} F_\sigma(\omega) &= \frac{1}{\sqrt{2\pi}} \int_{-\infty}^{\infty} f(t)H(t)\exp(-(\sigma + j\omega)t)dt \\ &= \frac{1}{\sqrt{2\pi}} \int_{0}^{\infty} f(t)\exp(-(\sigma + j\omega)t)dt \end{aligned} \tag{A.18}$$

Let $s = \sigma + j\omega$ and substitute into equation (A.18):

$$F_\sigma(s) = \frac{1}{\sqrt{2\pi}} \int_{0}^{\infty} f(t)\exp(-st)dt \tag{A.19}$$

Now we define

$$F(s) \equiv \int_{0}^{\infty} f(t)\exp(-st)dt \tag{A.20}$$

$$= \sqrt{2\pi} F_\sigma(s) \tag{A.21}$$

The inversion of $F_\sigma(\omega)$ is obtained as follows:

$$f_\sigma(t) = \frac{1}{\sqrt{2\pi}} \int_{-\infty}^{\infty} F_\sigma(\omega)\exp(-j\omega t)d\omega \tag{A.22}$$

Recall that $\omega = -j(s - \sigma)$ such that the limits of integration in equation (A.22) can be obtained as follows:

$$\omega = -\infty, \qquad s = \sigma + j\omega$$
$$\omega = +\infty, \qquad s = \sigma - j\omega$$

Thus, equation (A.22) can be expressed in terms of $s$:

$$f_\sigma(t) = \frac{1}{\sqrt{2\pi}} \int_{\sigma + j\infty}^{\sigma - j\infty} F_\sigma(s) \exp(-(\sigma - s)t) d(js) \tag{A.23}$$

Multiplying both sides of equation (A.23) by $\exp(\sigma t)$,

$$f_\sigma(t) \exp(\sigma t) = \frac{-j}{\sqrt{2\pi}} \int_{\sigma - j\infty}^{\sigma + j\infty} F_\sigma(s) \exp(st) ds \tag{A.24}$$

But $F_\sigma(s) = (2\pi)^{-1/2} F(s)$ and $f(t) = f_\sigma(t) \exp(\sigma t)$; hence

$$f(t) = \frac{1}{2\pi j} \int_{\sigma - j\infty}^{\sigma + \infty} F(s) \exp(st) ds \tag{A.25}$$

Equation (A.25) constitutes the inverse Laplace transform, which must be computed by contour integration. The evaluation of this integral is the subject of a number of texts (references) and is not expanded here other than to mention that the application of Cauchy's theorem greatly facilitates the process. The majority of the Laplace transforms required in this text can be found in practically any math handbook.

## A.2 DISCRETE-TIME TRANSFORMS FOR LINEAR SYSTEMS

Integral transforms are applicable to the analysis of continuous-time systems; however, for discrete-time systems, we implement discrete-time transforms, which can be derived from the integral transforms as illustrated in this section. In the derivation of both the discrete Fourier transform (DFT) and the $z$-transform, we assume that a perfect, continuous-time to discrete-time converter is available. If a continuous-time signal $x(t)$ is sampled periodically with sample period $T$, then the sequence can be expressed as a multiplication of the continuous-time signal with an infinite series of Dirac functions:

$$\{x(kT)\} = x(t) \sum_{k=-\infty}^{\infty} \delta(t - kT) \tag{A.26}$$

where $\{x(kT)\}$ is the discretized sequence, mapped from the continuous-time domain to the discrete-time domain. To develop the DFT or the $z$-transform, we proceed by applying the integral transforms developed in the previous section.

### A.2.1 The Discrete Fourier Transform

Taking the Fourier transform of the sequence represented in equation (A.26), one obtains

$$X(\omega) = \frac{1}{\sqrt{2\pi}} \int_{-\infty}^{\infty} x(t) \sum_{k=-\infty}^{\infty} \delta(t - kT) \exp(-j\omega t) dt \tag{A.27}$$

$$= \frac{1}{\sqrt{2\pi}} \sum_{k=-\infty}^{\infty} x(kT) \exp(-j\omega kT) \tag{A.28}$$

which is the discrete Fourier transform. The inverse transform is obtained as follows:

$$x(kT) = \frac{1}{\sqrt{2\pi}} \int_{-T/2}^{T/2} X(\omega) \exp(j\omega kT) d\omega \tag{A.29}$$

By convention, the DFT is typically defined as follows:

$$X(\omega) = \sum_{k=-\infty}^{\infty} x(kT) \exp(-j\omega kT) \tag{A.30}$$

with the corresponding inverse transform

$$x(kT) = \frac{1}{2\pi} \int_{-T/2}^{T/2} X(\omega) \exp(j\omega kT) d\omega \tag{A.31}$$

The only difference bctween the conventional definition and that derived here is a constant of proportionality $(1\sqrt{2\pi})$ used in defining the kernel of the original Fourier integral transform of equation (A.8). If we let $z = \exp(j\omega T)$ in equation (A.30), then the *bilateral z-transform* is obtained:

$$X(z) = \sum_{k=-\infty}^{\infty} x(kT)z^{-k} \tag{A.32}$$

Thus $z$ is, in general, a complex number with real and imaginary parts:

$$z = \cos(\omega T) + j\sin(\omega T) \tag{A.33}$$

As can be seen in equation (A.33), the magnitude of $z$ is one and the phase is described as a function of $\omega T$. Hence, the linear frequency axis of the discrete Fourier transform wraps around the unit circle in the $z$-plane, as illustrated in Figure A.1. The DFT is observed to repeat itself every $2\pi$ radians, which clearly illustrates the periodicity of the DFT. For $\omega T = \pi/2$, from equation (A.33) we obtain $z = j$, and the result is plotted in Figure A.1. Similarly, if $\omega T = \pi$, $z = -1$ and so on.

### A.2.2 The $z$-Transform

By definition, the *bilateral z-transform* of a sequence $x(k)$ is defined as

$$X(z) = \mathcal{Z}\{x(k)\} = \sum_{k=-\infty}^{\infty} x(kT)z^{-k} \tag{A.34}$$

where $\mathcal{Z}$ is the $z$-transform operator. The bilateral $z$-transform can be obtained from the bilateral Laplace transform of the sequence represented in equation (A.26):

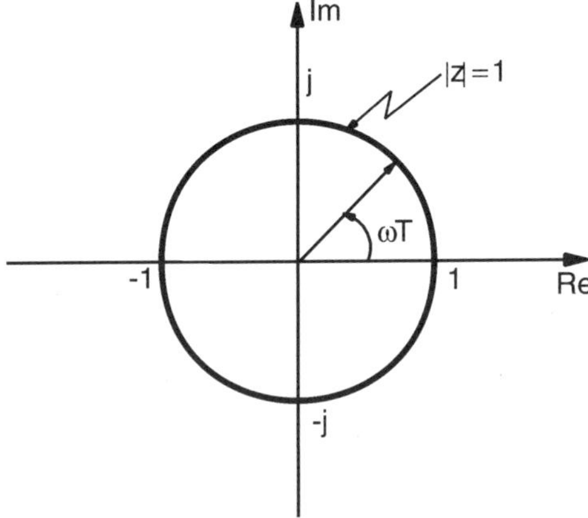


**Figure A.1:** Schematic of unit circle in the $z$-plane.

$$X(s) = \frac{1}{\sqrt{2\pi}} \int_{-\infty}^{\infty} x(t) \sum_{k=-\infty}^{\infty} \delta(t - kT) \exp(-st) dt \tag{A.35}$$

$$= \frac{1}{\sqrt{2\pi}} \sum_{k=-\infty}^{\infty} x(kT) \exp(-skT) \tag{A.36}$$

In this case, we let $z = \exp(sT)$ and substitute into the previous equation:

$$X(z) = \frac{1}{\sqrt{2\pi}} \sum_{k=-\infty}^{\infty} x(kT) z^{-k} \tag{A.37}$$

which is the bilateral $z$-transform scaled by the factor $1/\sqrt{2\pi}$ due to the definition of the kernel in the original Fourier integral transform of equation (A.9). In terms of control systems, we are typically concerned with finite sequences such that $x(t) = 0$ for $t < 0$ (i.e., causal events). We can thus proceed by applying the unilateral Laplace transform to the sequence to obtain:

$$X(s) = \int_{0}^{\infty} x(t) \sum_{k=-\infty}^{\infty} \delta(t - kT) \exp(-st) dt \tag{A.38}$$

$$= \sum_{k=0}^{\infty} x(kT) \exp(-sKT) \tag{A.39}$$

Again, substituting $z$ for $\exp(sT)$:

$$X(z) = \sum_{k=0}^{\infty} x(kT) z^{-k} \tag{A.40}$$

which is by definition the *unilateral z-transform*. Due to the fact that we typically deal with causal sequences in the analysis and design of adaptive structures, the unilateral $z$-transform is used exclusively in the analysis of discrete-time systems for this text. Recall that $s = \sigma + j\omega$, so

$$z = \exp(\sigma T)(\cos(\omega T) - j \sin(\omega T)) \tag{A.41}$$

which is, in general, complex. However, the magnitude of $z$ will depend upon the magnitude of $\exp(\sigma T)$. Thus, one can map the roots from the Laplace domain directly to the $z$-domain, as discussed earlier. For example, consider a pair of complex conjugate roots of an underdamped second-order system illus-

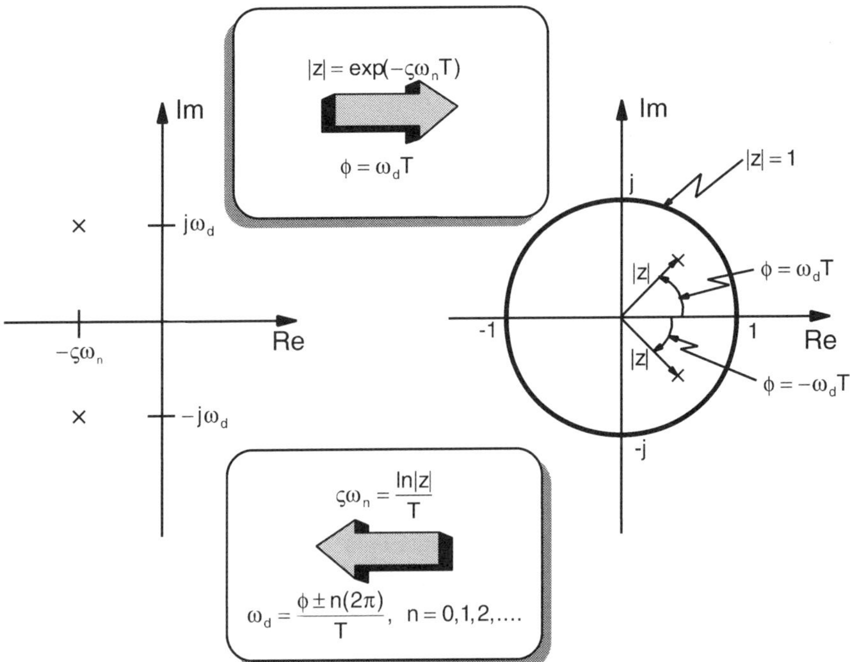


**Figure A.2:** Mapping of complex conjugate pair poles from (*a*) the *s*-plane to (*b*) the *z*-plane.

trated in the Laplace domain of Figure A.2*a*. Assuming that the real part of the roots resides in the left-half plane, which is a necessary condition for stability, one finds the magnitude of the roots in the $z$-plane to be

$$|z| = \exp(-\xi\omega_n T) = \frac{1}{\exp(\xi\omega_n T)} \tag{A.42}$$

where $\xi$ is the damping ratio and $\omega_n$ is the natural frequency. The phase in the $z$-plane is determined from the product of the damped natural frequency and the sample period ($\omega_d T$), where

$$\omega_d = \omega_n\sqrt{1-\xi^2} \tag{A.43}$$

The mapping of the set of complex conjugate poles from the $s$-plane to the $z$-plane is illustrated in Figure A.2*b*. For asymptotic stability, $|z| \le 1$, as discussed in Section A.2.3. Thus, one should recognize that the stability of a discrete-time system is dependent upon the stability of the continuous-time system and the chosen sample rate. In addition, while the mapping of the system poles from the $s$-domain to the $z$-domain is unique with respect to the sample rate, the reverse mapping will result in an infinite number of roots from which the real part and imaginary part can be computed, respectively, as follows:

$$\sigma = \frac{\ln|z|}{T} \tag{A.44}$$

and

$$\omega T = \cos^{-1}(\phi) \tag{A.45}$$

Equation (A.45) will yield an infinite number of poles in the $s$-plane separated by $2\pi/T$.

A detailed mapping of the lines of constant damping ratio and natural frequency from the $s$-plane to the upper half of the $z$-plane is presented in Figure A.3. As illustrated, 10 unique lines of constant damping ratio (the light lines) are presented, ranging in even increments from 0 to 1, and 20 unique lines of constant natural frequency (the dark lines) are displayed. The $z$-plane provides response information normalized to the sample rate, as opposed to time as in the $s$-plane.

One should also note that, due to the circular nature of the mapping from the $s$-plane to the $z$-plane, frequencies greater than $\pi/T$ map on top of the corresponding lower frequencies. This is typically denoted *aliasing* and gives rise

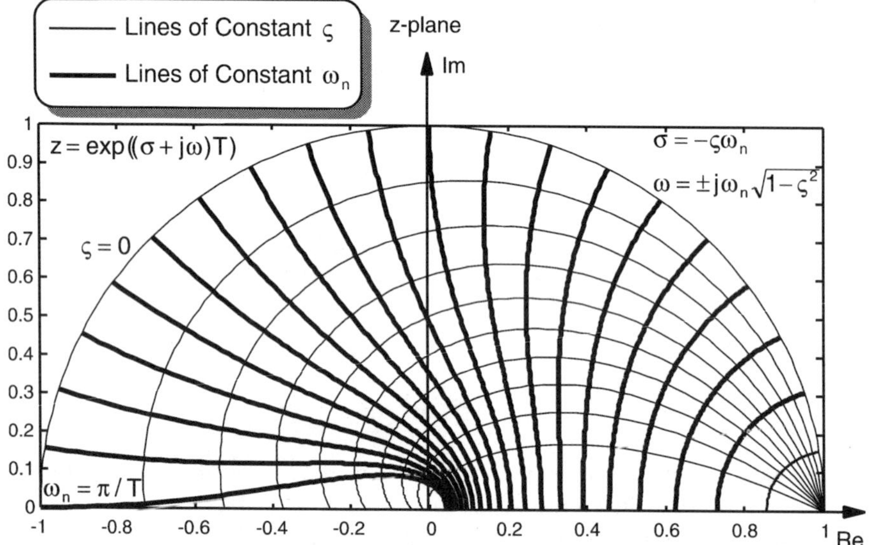


**Figure A.3:** Lines of constant damping and constant natural frequency in the $z$-plane.

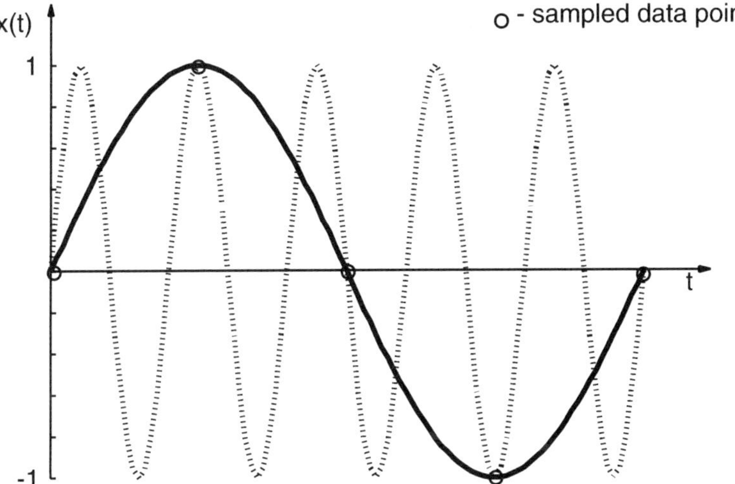


**Figure A.4:** Example of aliasing.

to the *Nyquist-Shannon sampling theorem*, which basically states that it is necessary to sample a time sequence at least two times faster than the highest frequency present in the waveform to uniquely resolve that frequency from the lower frequencies. The concept of aliasing is illustrated in Figure A.4, where the sampling rate is four times faster than the lower frequency waveform. As illustrated, the higher frequency waveform intersects the lower frequency waveform at each of the chosen sampled data points. Thus, once the data is acquired, the contribution from the higher frequency waveform cannot be resolved from that of the lower frequency waveform. In practice an *anti-aliasing filter* is employed, which is nothing more than a low-pass filter with a corner frequency set at half the sampling rate. Due to the finite rolloff of the filter, one typically selects a corner frequency somewhat less than half the sample rate.

Since the $z$-transform is often used to compute transfer functions of sampled data systems, one must consider the inversion process if the goal is to obtain an expression for the impulse response of the system (i.e., the response to a unit impulse). The inverse $z$-transform is obtained through the use of contour integration as follows:

$$x(k) = \frac{1}{2\pi j} \oint X(z) z^{k-1}\, dz \qquad \text{(A.46)}$$

where the path of integration is taken counterclockwise with respect to a circle centered at the origin in the $z$-plane where $Z\{x(k)\}$ converges. One can readily prove the inversion process by substituting for $X(z)$ in equation (A.46):

$$x(k) = \frac{1}{2\pi j} \oint \sum_{n=-\infty}^{\infty} x(n) z^{k-n-1}\, dz$$

$$= \sum_{n=-\infty}^{\infty} x(n) \frac{1}{2\pi j} \oint z^{k-n-1}\, dz$$

$$= \sum_{n=-\infty}^{\infty} x(n) \frac{1}{2\pi j} \oint z^{k-n-1}\, dz$$

But from Cauchy's theorem,

$$\frac{1}{2\pi j} \oint z^{k-1}\, dz = \begin{cases} 0 & m \neq -1 \\ 1, & m = -1 \end{cases}$$

Letting $m = k - n - 1$, one obtains

$$x(k) = x(k) \tag{A.47}$$

which completes the proof. Thus, in the event that $X(z)$ is a polynomial in $z$, as is the case for finite impulse response (FIR) filters discussed in Chapter 4, $x(k)$ is simply one of the coefficients of the polynomial.

**Example A.1** Given the following polynomial in $z^{-1}$:

$$X(z) = 1 + 2z^{-1} + 3z^{-2} \tag{A.48}$$

find $x(k)$:

$$x(k) = \frac{1}{2\pi j} \oint (1 + 2z^{-1} + 3z^{-2}) z^{k-1}\, dz \tag{A.49}$$

Separating the integrals,

$$x(k) = \frac{1}{2\pi j} \oint 1 z^{k-1}\, dz + 2\left( \frac{1}{2\pi j} \oint z^{k-2} \right) dz + 3\left( \frac{1}{2\pi j} \oint z^{k-3} \right) dz \tag{A.50}$$

From Cauchy's theorem,

$$x(k) = \begin{cases} 1, & k = 0 \\ 2, & k = 1 \\ 3, & k = 2 \end{cases}$$

If $X(z)$ is a ratio of polynomials, the Cauchy's residue theorem can be used to evaluate the contour integral:

$$\begin{aligned} x(k) &= \frac{1}{2\pi j} \oint X(z) z^{k-1}\, dz \\ &= \sum_p [\text{residues of } X(z)z^{k-1} \text{ at the pole } z_p] \end{aligned} \tag{A.51}$$

If $X(z)z^{k-1}$ is a rational function of $z$, then one can express the polynomial fraction as follows:

$$X(z)z^{k-1} = \frac{\Phi(z)}{(z - z_p)^r} \tag{A.52}$$

where $X(z)z^{k-1}$ has $r$ poles at $z = z_p$, and $\Phi(z)$ has no poles at $z = z_p$. The residue at $z_p$ is

$$\text{Res}[X(z)z^{k-1}]_{z=z_p} = \frac{1}{(r-1)!} \left[ \frac{d^{r-1}\Phi(z)}{dz^{r-1}} \right]_{z=z_p} \tag{A.53}$$

In the event that there is a single first-order pole, the expression can be simplified as follows:

$$\text{Res}[X(z)z^{k-1}]_{z=z_p} = \Phi(z_p) \tag{A.54}$$

**Example A.2** Consider the following polynomial fraction in $z$:

$$\begin{aligned} X(z) &= \frac{1}{1 - z^{-1} \exp(-a)} \\ x(k) &= \frac{1}{2\pi j} \oint \frac{z^{k-1}}{1 - z^{-1}\exp(-a)}\, dz \\ &= \frac{1}{2\pi j} \oint \frac{z^k}{z - \exp(-a)}\, dz \end{aligned}$$

Thus,

$$\Phi(z) = z^k$$

and

$$\begin{aligned} x(k) &= \Phi(z_p) \\ &= \Phi(e^{-a}) \\ &= (\exp(-a))^n \\ &= \exp(-an), \qquad n \geq 0 \end{aligned}$$

For sequences such that $n \leq 0$ (left-handed sequences), the evaluation of the contour integral is more complicated. The emphasis in this text is placed upon causal sequences for which $n \geq 0$, and thus the inversion of the "left-handed" sequences is not addressed. The interested reader can refer to the text by Oppenheim and Schafer (1989).

### A.2.3 Relationship Between Continuous and Discrete Systems

The realization and implementation of digital controllers is an important topic that was omitted from this text due to contraints on length. However, a brief overview of the relationship between continuous- and discrete-time systems is presented, since a number of the controllers discussed in Chapter 9 were implemented on a digital signal processor. Many references on digital control are available by authors such as Ackermann (1985), Phillips and Nagle (1990), and Franklin et al. (1992), to name a few. The interested reader is encouraged to pursue this topic in greater detail.

The objective of this brief section is to describe the transformation of a continuous-time system into a discrete-time system and detail the stability requirements. The discussion is limited to a zero-order hold, which is common for digital to analog converters. The transfer function of a zero-order hold, including that of the plant $G(s)$, can be expressed as follows:

$$H(s) = \left( \frac{1 - \exp(-Ts)}{s} \right) G(s) \tag{A.55}$$

where $T$ is the sample period and $s$ is the Laplace transform variable. The impulse response of the system can be obtained from the inverse Laplace transform:

$$h(t) = \mathcal{L}^{-1}\left\{\left(\frac{1 - \exp(-Ts)}{s}\right) G(s)\right\} \tag{A.56}$$

$$= \mathcal{L}^{-1}\left\{\frac{G(s)}{s}\right\} - \mathcal{L}^{-1}\left\{\left(\frac{G(s)\exp(-Ts)}{s}\right)\right\} \tag{A.57}$$

If we let $q(t) = \mathcal{L}^{-1}\{G(s)/s\}$, then

$$h(t) = q(t) - q(t - T) \tag{A.58}$$

Taking the $z$-transform,

$$H(z) = \mathcal{Z}\{q(kT) - q(kT - T)\} \tag{A.59}$$

$$= (1 - z^{-1})\mathcal{Z}\{q(kT)\} \tag{A.60}$$

$$= \frac{z - 1}{z} \mathcal{Z}\left\{\mathcal{L}^{-1}\left\{\frac{G(s)}{s}\right\}_{t - kT}\right\} \tag{A.61}$$

Thus, the relationship between the $z$-transform and the Laplace transform depends upon the order of the hold. First- and second-order holds can be considered, but are omitted here for brevity.

As detailed by Franklin et al. (1992) for a zero-order hold with no delay, the discrete-time response of a continuous system can be expressed as follows:

$$\mathbf{x}(kT + T) = \exp(\mathbf{A}T)\mathbf{x}(kT) + \int_0^T \exp(\mathbf{A}\eta)d\eta\ \mathbf{B}\mathbf{u}(kT) \tag{A.62}$$

Let

$$\mathbf{\Phi} = \exp(\mathbf{A}T) \tag{A.63}$$

and

$$\mathbf{\Gamma} = \int_0^T \exp(\mathbf{A}\eta)d\eta\ \mathbf{B} \tag{A.64}$$

Then,

$$\mathbf{x}(k + 1) = \mathbf{\Phi}\mathbf{x}(k) + \mathbf{\Gamma}\mathbf{u}(k) \tag{A.65}$$

$$\mathbf{y}(k) = \mathbf{C}\mathbf{x}(k) + \mathbf{D}\mathbf{u}(k) \tag{A.66}$$

Taking the $z$-transform and assuming zero initial conditions, one obtains the $z$-transfer function:

$$z\mathbf{X}(z) = \mathbf{\Phi X}(z) + \mathbf{\Gamma U}(z) \tag{A.67}$$

With basic linear algebra, one can show that

$$\mathbf{X}(z) = [z\mathbf{I} - \mathbf{\Phi}]^{-1}\mathbf{\Gamma U}(z) \tag{A.68}$$

Since

$$\mathbf{Y}(z) = \mathbf{CX}(z) + \mathbf{DU}(z) \tag{A.69}$$

the $z$-transform of the output can be expressed as follows:

$$\mathbf{Y}(z) = [\mathbf{C}[z\mathbf{I} - \mathbf{\Phi}]^{-1}\mathbf{\Gamma} + \mathbf{D}]\mathbf{U}(z) \tag{A.70}$$

The $z$-transfer function of the multi-input, multi-output (MIMO) system can thus be expressed as follows:

$$\mathbf{H}(z) = \mathbf{C}[z\mathbf{I} - \mathbf{\Phi}]^{-1}\mathbf{\Gamma} + \mathbf{D} \tag{A.71}$$

$$= \frac{\mathbf{C}\,\mathrm{adj}[z\mathbf{I} - \mathbf{\Phi}]\mathbf{\Gamma} + \mathbf{D}|z\mathbf{I} - \mathbf{\Phi}|}{|z\mathbf{I} - \mathbf{\Phi}|} \tag{A.72}$$

Performing a partial fraction expansion, one can express the $z$-transfer function as follows:

$$\mathbf{H}(z) = \sum_{n=1}^{N} \frac{k_n z}{z - p_n} + D \tag{A.73}$$

where $N$ is the number of poles $p_n$ of the polynomial $|z\mathbf{I} - \mathbf{\Phi}|$. Now,

$$Z^{-1}\left[\frac{k_n z}{z - p_n}\right] = k_n (p_n)^k \tag{A.74}$$

From the previous inverse $z$-transform, it is obvious that the only way that the series will converge as $k \rightarrow \infty$ is if $|p_n| < 1$. Thus, for stability, all of the poles must reside within the unit circle in the $z$-plane. Poles within this unit circle map into the left-half plane of the $s$-plane.

## BIBLIOGRAPHY

Ackermann, J., 1985. *Sampled-Data Control Systems: Analysis and Synthesis, Robust System Design*, Springer-Verlag, New York.

Doetsch, G., 1971. *Guide to the Applications of the Laplace and Z-Transforms*, Van Nostrand Reinhold, New York.

Franklin, G. F., J. D. Powell, and M. L. Workman, 1992. *Digital Control of Dynamic Systems*, Addison-Wesley, New York.

Oppenheim, A. V. and R. W. Schafer, 1989. *Discrete-Time Signal Processing*, Prentice Hall, Englewood Clifs, NJ.

Phillips, C. L. and H. T. Nagle, 1990. *Digital Control System Analysis and Design*, Prentice Hall, Englewood Cliffs, NJ.

Sneddon, I. A., 1972. *The Use of Integral Transforms*, McGraw-Hill, New York.

Watson, E. J., 1981. *Laplace Transforms and Applications*, Van Nostrand Reinhold, New York.

Wolf, K. B., 1979. *Integral Transforms in Science and Engineering*, Plenum Press, New York.

# APPENDIX B

# SINGULAR VALUE DECOMPOSITION

Singular value decomposition (SVD) is commonly used in the fields of control and signal processing. It has found widespread application in engineering and science, as detailed by Klema and Laub (1980). SVD is at the heart of developing reduced-order models of dynamic systems as well as loop-shaping for multi-input, multi-output (MIMO) control system design and synthesis. The maximum singular value of a system defines the $\mathcal{H}_\infty$ norm as was detailed in Chapter 3. A brief summary of SVD is provided herein. For greater details of computational algorithms used in SVD, the reference by Haykin (1991) is suggested.

Let $\mathbf{A}$ be a real or complex valued, $n \times m$ matrix. The singular values of $\mathbf{A}$ are the strictly positive square roots of the nonzero eigenvalues of $\mathbf{A}^H\mathbf{A}$ or $\mathbf{A}\mathbf{A}^H$:

$$\sigma_j(\mathbf{A}) = [\lambda_j(\mathbf{A}^H\mathbf{A})]^{1/2} = [\lambda_j(\mathbf{A}\mathbf{A}^H)]^{1/2} > 0, \qquad j = 1, 2, \ldots, p \tag{B.1}$$

where $\text{rank}(\mathbf{A}) = p$. The singular values are frequently ordered as follows:

$$\sigma_1(\mathbf{A}) \geq \sigma_1(\mathbf{A}) \geq \cdots \geq \sigma_p(\mathbf{A}) > 0 \tag{B.2}$$

and the maximum and minimum singular values are defined as

$$\sigma_{\max}(\mathbf{A}) = \sigma_1(\mathbf{A}) \tag{B.3}$$

$$\sigma_{\min}(\mathbf{A}) = \sigma_p(\mathbf{A}) \tag{B.4}$$

Now, for a linear system of equations such that

$$\mathbf{y} = \mathbf{A}\mathbf{x} \tag{B.5}$$

one can represent the matrix $\mathbf{A}$ as a product of an $n \times n$ unimodal matrix $\mathbf{U}$, an $m \times m$ unimodal matrix $\mathbf{V}$, and an $n \times m$ matrix of singular values:

$$\boldsymbol{\Sigma} = \begin{bmatrix} \operatorname{diag}(\sigma_1, \sigma_2, \ldots, \sigma_p) & \mathbf{0} \\ \mathbf{0} & \mathbf{0} \end{bmatrix} \tag{B.6}$$

such that

$$\mathbf{y} = \mathbf{U}\boldsymbol{\Sigma}\mathbf{V}^H\mathbf{x} \tag{B.7}$$

Since $\mathbf{U}\mathbf{U}^H = \mathbf{I}$ and $\mathbf{V}^H\mathbf{V} = \mathbf{I}$, one can show that

$$\mathbf{A}\mathbf{V} = \mathbf{U}\boldsymbol{\Sigma} \tag{B.8}$$

and thus,

$$\mathbf{A}\mathbf{V}_i = \sigma_i(\mathbf{A})\mathbf{u}_i, \quad i = 1, \ldots, p \tag{B.9}$$

where $\mathbf{v}_i$ and $\mathbf{u}_i$ are the $i$th columns of the unimodal matrices $\mathbf{V}$ and $\mathbf{U}$, respectively.

If the inverse transform is desired, one can obtain a useful expression for the pseudoinverse of $\mathbf{A}$:

$$\left(\mathbf{A}^H\mathbf{A}\right)^{-1}\mathbf{A}^H = \mathbf{V}\boldsymbol{\Sigma}^{-1}\mathbf{U}^H \tag{B.10}$$

where

$$\boldsymbol{\Sigma}^{-1} == \begin{bmatrix} \operatorname{diag}(\sigma_1^{-1}, \sigma_2^{-1}, \ldots, \sigma_p^{-1}) & \mathbf{0} \\ \mathbf{0} & \mathbf{0} \end{bmatrix} \tag{B.11}$$

Thus, once the singular values and singular vectors are obtained, one can readily formulate the pseudoinverse and solve for the inputs as a function of the outputs by simply taking the reciprocal of the individual singular values and pre- and postmultiplying this matrix by the unimodal matrices:

$$\mathbf{x} = \mathbf{V}\boldsymbol{\Sigma}^{-1}\mathbf{U}^H\mathbf{y} \tag{B.12}$$

## BIBLIOGRAPHY

Haykin, S., 1991. *Adaptive Filter Theory*, Prentice Hall, Englewood Cliffs, NJ.

Klema, V. C. and A. J. Laub, 1980. "The Singular Value Decomposition: Its Computation and Some Applications," *IEEE Transactions on Automatic Control*, **AC-25**, 164–176.

# APPENDIX C

# STRUCTURAL ACOUSTICS

The study of sound radiation from vibrating bodies begins with the description of the Kirchhoff-Helmholtz integral equation. The derivation of this integral equation can be found in the reference by Pierce (1981), and for simple harmonic motion, the acoustic pressure is expressed as follows:

$$p(\mathbf{r})\exp(j\omega t) = \frac{1}{4\pi}\exp(j\omega t)\int_S \left\{ p(\mathbf{r}_s)\frac{\partial}{\partial n}\left(\frac{\exp(-jkR)}{R}\right) + j\omega\rho_o v_n(\mathbf{r}_s)\left(\frac{\exp(-jkR)}{R}\right)\right\} dS \tag{C.1}$$

where $p(\mathbf{r})$ is the acoustic pressure at the vector position $\mathbf{r}$, $p(\mathbf{r}_s)$ is the pressure on the vibrating surface at vector position $\mathbf{r}_s$, $R$ is the magnitude of the difference between the vector to the vibrating surface $\mathbf{r}_s$ and the vector to the field point $\mathbf{r}$: $|\mathbf{r} - \mathbf{r}_s|$, $v_n(\mathbf{r}_s)$ is the normal velocity of the vibrating surface at vector position $\mathbf{r}_s$, $\rho_o$ is the density of the fluid, and $\omega$ is the circular frequency. This integral equation shows that the acoustic radiation is coupled to both the velocity at the surface and the pressure at the surface. For arbitrary surfaces in free space, the solution of this integral equation is quite involved and is typically performed through numerical techniques. However, for a vibrating surface extending over an infinite half-space, the complexity of the integral equation is greatly reduced, as discussed in Section C.1.

## C.1 THE RAYLEIGH INTEGRAL

If a planar surface extends over an infinite half-space, then one can demonstrate that the acoustic pressure at any field point can be described as follows (Fahy, 1985):

$$p(\mathbf{r}) \exp(j\omega t) = \frac{j\omega\rho_o}{4\pi} \exp(j\omega t) \int_S \left\{ 2v_n(\mathbf{r}_s) \left( \frac{\exp(-jkR)}{R} \right) \right\} dS \qquad \text{(C.2)}$$

Thus, one can view each differential element of the vibrating surface as a point source with a source strength of $2v_n\, \delta S$, as outlined in a number of texts (Pierce, 1981; Fahy, 1985). While the theory applies strictly to a vibrating surface of infinite extent, the formulation can be used to predict sound radiation from vibrating surfaces for which the surface dimension is much greater than the acoustic wavelength of interest. The accuracy of the solution under this assumption is known to deteriorate at the boundaries of the surface. From an academic perspective, the study of vibrating structures residing within an infinite half-space with rigid boundaries makes for a physically motivating structure to study structural acoustic coupling and various control strategies.

## C.2 PLANAR-BAFFLED RADIATORS

Assuming that the structural response of a baffled panel can be expressed in terms of modal coordinates, the acoustic response of the panel can be obtained from Rayleigh's integral as follows:

$$p(\mathbf{r}) \exp(j\omega t) = \frac{j\omega\rho_o}{2\pi} \exp(j\omega t) \int_0^{L_x} \int_0^{L_y} \cdot \left\{ \sum_{m=1}^{\infty} v_m \phi_m(x, y) \left( \frac{\exp(-jkR)}{R} \right) \right\} dy\, dx \qquad \text{(C.3)}$$

where $v_m$ is the velocity of the $m$th mode, $L_x$ and $L_y$ are the dimensions of the structure in the $x$- and $y$-coordinates, respectively, $\phi_m(x, y)$ is the $m$th mode shape function, and $R = \sqrt{(x - x')^2 + (y - y')^2 + (z')^2}$. A schematic diagram of the panel is presented in Figure C.1. If, as outlined by Bies (1976),

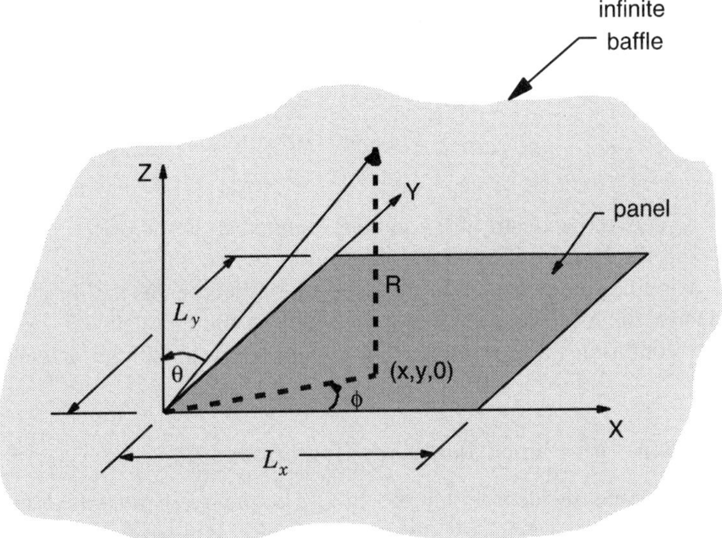


**Figure C.1:** Coordinate system for baffled panel.

$$R \gg \frac{\lambda}{2\pi} \tag{C.4}$$

$$R \gg l \tag{C.5}$$

$$R \gg \frac{\pi l^2}{2\lambda} \tag{C.6}$$

where $\lambda$ is the acoustic wavelength and $l$ is the maximum source dimension, then the acoustic field point is considered to be in the farfield. For far-field assumptions,

$$R \approx r - x\sin(\theta)\cos(\phi) - y\sin(\theta)\cos(\phi) \tag{C.7}$$

Under these assumptions, equation (C.3) can be reduced to the following expression:

$$p(\mathbf{r})\exp(j\omega t) = \frac{j\omega\rho_o \exp(-jkr)\exp(j\omega t)}{2\pi r} \int_0^{L_x}\int_0^{L_y} \cdot \left\{ \sum_{m=1}^{\infty} v_m\phi_m(x,y)\exp\left[ j\,\frac{\alpha x}{L_x} + j\,\frac{\beta y}{L_y} \right] \right\} dy\,dx \tag{C.8}$$

where

$$\alpha = kL_x \sin(\theta)\cos(\phi) \tag{C.9}$$

$$\beta = kL_y \sin(\theta)\sin(\phi) \tag{C.10}$$

and $k = \omega/c$ is the acoustic wavenumber, defined in terms of the circular frequency and the speed of sound $c$ in the fluid.

Two examples are provided in this section: baffled beams and baffled plates. In each case, the beam or plate is assumed to have finite dimensions and pinned boundary conditions, and reside within a *rigid* baffle of infinite extent.

### C.2.1 Beam with Pinned Boundary Conditions

Consider the panel resident within the infinite boundary as illustrated with coordinate system in Figure C.1. If the panel response can be described by structural modes that are functions of a single dimension, such as a beam with pinned boundary conditions (i.e., $\phi_m(x, y) = \sin(m_x \pi x/L_x)$), then one can integrate equation (C.8) and express the acoustic response of the system as follows (Wallace, 1972a):

$$p(\mathbf{r})\exp(j\omega t) = \rho_o ckL_y \frac{\exp(-jkr)\exp(j\omega t)}{2\pi r} \sum_{m_x=1}^{\infty} \frac{v_{m_x} L_x}{m_x \pi} \cdot \left[\frac{1-(-1)^{m_x}\exp(j\alpha)}{1-(\alpha/m_x\pi)^2}\right]\left(\frac{1-\exp(j\beta)}{\beta}\right) \tag{C.11}$$

### C.2.2 Plate with Pinned Boundary Conditions

For a panel, as illustrated in Figure C.1, if the response can be described by structural modes that are a function of both $x$- and $y$-coordinates, such as a plate with pinned boundary conditions (i.e., $\phi_m(x, y) = \sin(m_x \pi x/L_x)\sin(m_y \pi y/L_y)$), then one can integrate equation (C.8) and express the acoustic response of the system as follows (Wallace, 1972b):

$$p(\mathbf{r})\exp(j\omega t) = j\rho_o ck \frac{\exp(-jkr)\exp(j\omega t)}{2\pi r} \sum_{m_x=1}^{\infty}\sum_{m_y=1}^{\infty} \frac{v_{m_x m_y} L_x L_y}{m_x m_y \pi^2} \cdot \left[\frac{1-(-1)^{m_x}\exp(j\alpha)}{1-(\alpha/m_x\pi)^2}\right]\left[\frac{1-(-1)^{m_y}\exp(j\beta)}{1-(\beta/m_y\pi)^2}\right] \tag{C.12}$$

## C.3 RADIATION EFFICIENCY

The radiation efficiency is defined as the ratio of the average acoustic power radiated by a vibrating surface to that which would be radiated by a piston of equal surface area vibrating with the same average normal mean-square velocity (Fahy, 1985). The radiation efficiency can be expressed as follows:

$$\sigma = \frac{\overline{P}}{\rho_o c S \langle \overline{v}_n^2 \rangle} \tag{C.13}$$

where $\overline{P}$ is the time-averaged power of the vibrating surface, $S$ is the surface area of the vibrating structure, and $\langle \overline{v}_n^2 \rangle$ is the average mean-square normal velocity. There are a number of methods used in computing the power radiated into a half-space, one of which includes integrating the acoustic intensity resulting from the vibrating surface over a hemisphere in the farfield as illustrated in Figure C.2. The time-averaged acoustic intensity can be expressed as a function of the pressure and velocity:

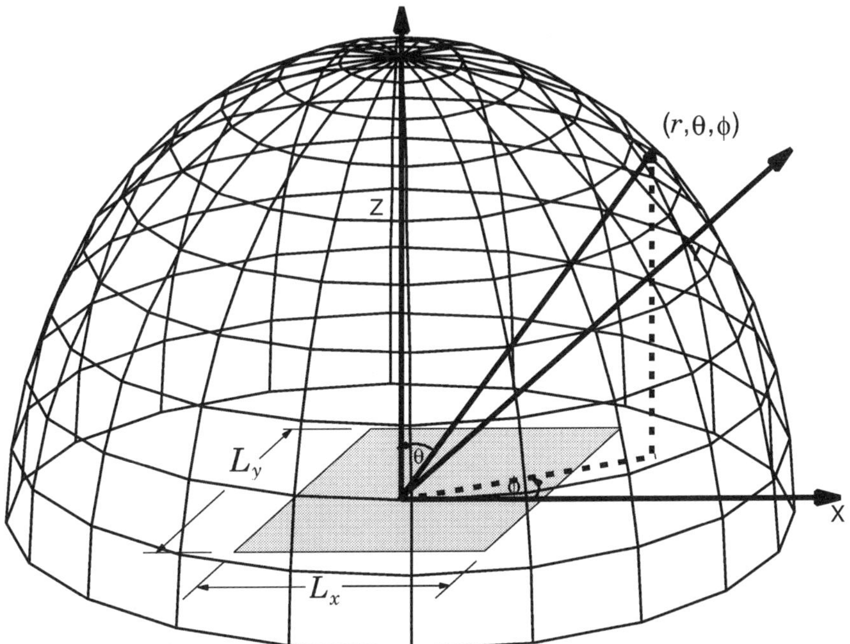


**Figure C.2:** Coordinate system for computation of power over a hemisphere of a baffled vibrating surface.

$$I(r,\theta,\phi) = \langle p(r,\theta,\phi,t)v(r,\theta,\phi,t)\rangle = \frac{1}{T}\int_0^T p(r,\theta,\phi,t)v(r,\theta,\phi,t)dt \quad \text{(C.14)}$$

Far from the source, the acoustic field is characterized by plane waves traveling away from the source, and in this field, the particle velocity and pressure are related by the fluid density $\rho_o$ and speed of sound $c$:

$$p(r,\theta,\phi,t) = \rho_o c v(r,\theta,\phi,t) \quad \text{(C.15)}$$

Hence, if one chooses to compute the intensity in the farfield, the expression for the power reduces to the integral of the intensity over the hemisphere, which can be expressed in terms of an integral over the surface of the average mean-squared pressure:

$$\overline{P} = \int_0^{\pi}\int_0^{2\pi} I(r,\theta,\phi)r^2\sin(\theta)d\theta d\phi \quad \text{(C.16)}$$

$$= \int_0^{\pi}\int_0^{2\pi}\left[\frac{1}{T}\int_0^T \frac{p^2(r,\theta,\phi,t)}{\rho_o c}dt\right] r^2\sin(\theta)d\theta\, d\phi \quad \text{(C.17)}$$

The average mean-squared normal velocity of the vibrating surface can be expressed as follows:

$$\langle \overline{v}_n^2\rangle = \frac{1}{L_x L_y}\int_0^{L_x}\int_0^{L_y}\left[\frac{1}{T}\int_0^T v_n^2(x,y,t)dt\right] dy\, dx \quad \text{(C.18)}$$

Given equation (C.8) and assuming harmonic motion, one can develop an expression for the squared pressure in terms of the velocity of the structural modes:

$$|p(\mathbf{r})|^2 = \frac{\omega^2\rho_0^2}{4\pi^2 r^2}\left\{\int_0^{L_x}\int_0^{L_y}\left\{\sum_{r=1}^{\infty} v_r\phi_r(x,y)\exp\left[j\frac{\alpha x}{L_x} + j\frac{\beta y}{L_y}\right]\right\} dy\, dx\right\}$$

$$\times\left\{\int_0^{L_x}\int_0^{L_y}\left\{\sum_{s=1}^{\infty} v_s\phi_s(x,y)\exp\left[j\frac{\alpha x}{L_x} + j\frac{\beta y}{L_y}\right]\right\} dy\, dx\right\} \quad \text{(C.19)}$$

$$= \frac{\omega^2\rho_0^2}{4\pi^2 r^2}\sum_{r=1}^{\infty} v_r H_r(\theta,\phi,\omega)\sum_{s=1}^{\infty} v_s H_s(\theta,\phi,\omega) \quad \text{(C.20)}$$

where the function $H_r(\theta, \phi, \omega)$ is used to replace the integral term. Substituting equation (C.20) into equation (C.17), one obtains

$$\overline{P}(j\omega) = \frac{k^2 \rho_o c}{4\pi^2} \int_0^{\pi} \int_0^{2\pi} \sum_{r=1}^{\infty} v_r H_r(\theta, \phi, \omega) \cdot \sum_{s=1}^{\infty} v_s H_s(\theta, \phi, \omega) \sin(\theta) d\theta\, d\phi \tag{C.21}$$

Notice upon substitution that the $r^2$ term is no longer part of the expression, and thus as expected, the power is not a function of the distance from the source. One can express equation (C.21) in vector form:

$$\overline{P}(j\omega) = \frac{k^2 \rho_o c}{4\pi^2} \int_0^{\pi} \int_0^{2\pi} \mathbf{vHH} * \mathbf{v} * \sin(\theta) d\theta\, d\phi \tag{C.22}$$

$$= \frac{k^2 \rho_o c}{4\pi^2} \mathbf{v} \left\{ \int_0^{\pi} \int_0^{2\pi} \mathbf{HH} * \sin(\theta) d\theta\, d\phi \right\} \mathbf{v}* \tag{C.23}$$

$$= \frac{k^2 \rho_o c}{4\pi^2} \mathbf{vMv}* \tag{C.24}$$

The frequency-dependent matrix **M** defines the structural acoustic interaction of modes, which leads to power radiated. The diagonal entries of this matrix are proportional to the self radiation efficiencies and the off-diagonal entries are proportional to the mutual radiation efficiencies of structural modes (Lomas and Hayek, 1977; Baumann et al., 1991; Cunefare, 1991). To obtain the radiation efficiencies of select modes as defined by Wallace (1972a; b), one must normalize the diagonal entries by the mean-squared velocity of the corresponding mode. At low frequency (below coincidence for the low-order structural modes), the matrix contains off-diagonal terms; however, at high frequency (above coincidence), the off-diagonal terms become negligible. The physical result of this observation is that the structural modes interact in the acoustic field at low frequency and radiate independently at high frequency.

To provide an example, the script file **pradeff.m** can be used to generate the self and mutual radiation efficiencies of a panel with pinned boundary conditions. The ratio of the panel dimensions can be defined within the script file as well as the number of modes for which the radiation efficiences are computed. For the purpose of this example, the panel dimensions were chosen such that $L_y/L_x = 0.875$ and the number of modes was set to 5. The modes are sorted in the program with a script file **pfreq.m** such that the lowest order modes (in frequency) of a panel with pinned boundary conditions are selected. Upon

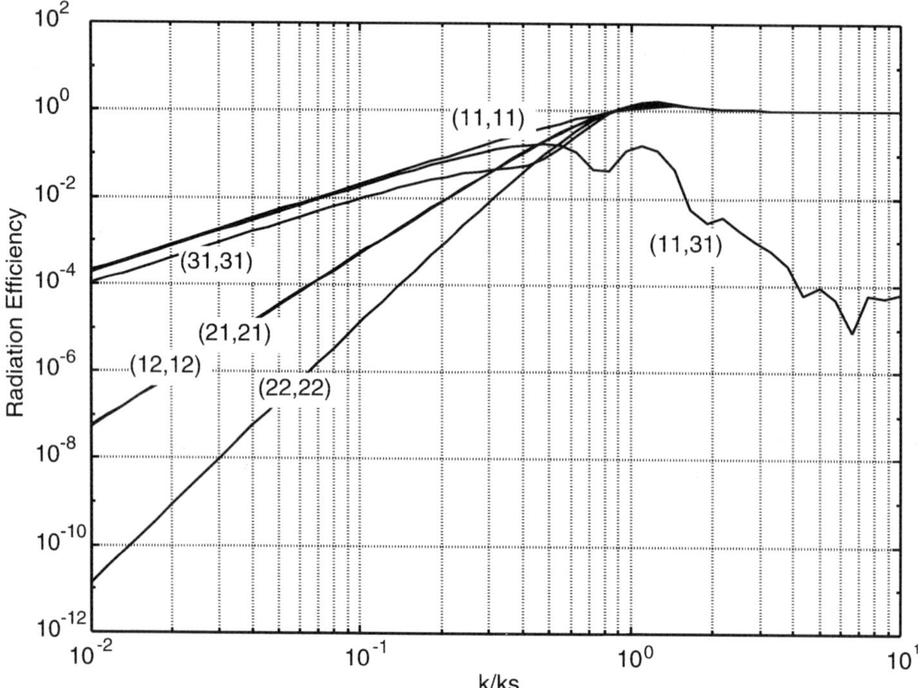


**Figure C.3:** Self and mutual radiation efficiencies for the first five modes of a panel having pinned boundary conditions and an aspect of $L_y/L_x = 0.875$.

executing the script file, the results presented in Figure C.3 are produced. The first five modes of the panel for this example correspond to the (1, 1), (2, 1), (1, 2), (2, 2), and (3, 1) modes. For this range of modes, only the (1, 1) and (3, 1) modes will interact in the acoustic field below coincidence. As indicated in Figure C.3, the mutual radiation efficiency between the (1, 1) mode and the (3, 1) mode is of the same order of magnitude as the self radiation efficiency of the (1, 1) mode below coincidence and has greater efficiency than the self radiation efficiency of the (3, 1) mode. The physical implication for structural acoustic control is that the (3, 1) mode can be used to control radiation associated with the (1, 1) mode and vice versa. The interested reader can generate radiation efficiencies for higher order modes or for panels with differing ratios by appropriately modifying the script file.

### C.3.1 Beam with Pinned Boundary Conditions

Substituting equation (C.11) into equation (C.17), and dividing by the average mean-squared normal velocity associated with each vibrating mode of the beam, one can express the radiation efficiency of the $m$th structural mode as follows (Wallace, 1972a):

$$\sigma_m = \frac{32k^2 L_x L_y}{m_x^2 \pi^4} \int_0^{\pi/2} \int_0^{\pi/2} \left\{ \frac{\mathrm{trig}_b(\alpha, \beta)}{(1 - (\alpha/m_x\pi)^2 \beta)} \right\}^2 \cdot \sin(\theta)\, d\theta\, d\phi \tag{C.25}$$

where

$$\mathrm{trig}_b(\alpha, \beta) = \begin{cases} \cos\left(\dfrac{\alpha}{2}\right) \sin\left(\dfrac{\beta}{2}\right), & m_x \text{ is odd} \\ \sin\left(\dfrac{\alpha}{2}\right) \sin\left(\dfrac{\beta}{2}\right), & m_x \text{ is even} \end{cases} \tag{C.26}$$

### C.3.2 Plate with Pinned Boundary Conditions

Substituting equation (C.12) into equation (C.17), and dividing by the average mean-squared normal velocity assoicated with each vibrating mode of the plate, one can express the radiation efficiency of the $m$th structural mode as follows (Wallace, 1972b):

$$\sigma_m = \frac{64k^2 L_x L_y}{m_x^2 m_y^2 \pi^6} \int_0^{\pi/2} \int_0^{\pi/2} \left\{ \frac{\mathrm{trig}_p(\alpha, \beta)}{(1 - (\alpha/m_x\pi)^2)(1 - (\beta/m_y\pi)^2)} \right\}^2 \cdot \sin(\theta) d\theta\, d\phi \tag{C.27}$$

where

$$\mathrm{trig}_p(\alpha, \beta) = \begin{cases} \cos\left(\dfrac{\alpha}{2}\right) \cos\left(\dfrac{\beta}{2}\right), & m_x \text{ is odd, } m_y \text{ is odd} \\ \cos\left(\dfrac{\alpha}{2}\right) \sin\left(\dfrac{\beta}{2}\right), & m_x \text{ is odd, } m_y \text{ is even} \\ \sin\left(\dfrac{\alpha}{2}\right) \cos\left(\dfrac{\beta}{2}\right), & m_x \text{ is even, } m_y \text{ is odd} \\ \sin\left(\dfrac{\alpha}{2}\right) \sin\left(\dfrac{\beta}{2}\right), & m_x \text{ is even, } m_y \text{ is even} \end{cases} \tag{C.28}$$

## C.4 THE WAVENUMBER TRANSFORM

In Appendix A, the Fourier transform was discussed. For structural acoustics, a Fourier transform of the spatial variables is often used to compute various quantities. This transform is commonly termed the wavenumber transform and can be expressed as follows:

$$P(k_x, k_y, z) = \int_0^{\infty} \int_0^{\infty} p(x, y, z) \exp(jk_x x) \exp(jk_y y) dx\, dy \tag{C.29}$$

As noted by Fuller et al. (1996), the complex pressure within a harmonic sound field can be expressed in terms of the Helmholtz equation:

$$(\nabla^2 + k^2)p(x, y, z) = 0 \tag{C.30}$$

Applying the wavenumber transform, one obtains

$$\left(-k_x^2 - k_y^2 + \frac{\partial}{\partial z} + k^2\right) \cdot P(k_x, k_y, z) = 0 \tag{C.31}$$

The solution of this homogeneous equation takes the following form:

$$P(k_x, k_y, z) = P \exp(-jk_z z) \tag{C.32}$$

where

$$k_z = \sqrt{k^2 - k_x^2 - k_y^2}$$

On the surface of a vibrating structure, the fluid momentum and pressure are related by Euler's equation:

$$j\omega\rho_o v_n(x, y) = -\frac{\partial}{\partial z}\, p(x, y, 0) \tag{C.33}$$

Applying the wavenumber transform, substituting equation (C.32), and performing simple algebra, one obtains

$$P = \frac{\omega\rho_o V_n(k_x, k_y)}{k_z} \tag{C.34}$$

Thus, one can express the wavenumber transform of the pressure field in terms of the wavenumber transform of the velocity of the vibrating surface:

$$P(k_x, k_y, z) = \frac{\omega\rho_o V_n(k_x, k_y)}{k_z} \exp(-jk_z z) \tag{C.35}$$

The pressure can be obtained from an inverse transform; however, for now we proceed with the development of an expression for sound power radiated into a half-space. The time-averaged power can be expressed in terms of a product of the complex pressure and velocity on the surface of the vibrating panel as follows:

$$\overline{P} = \frac{1}{2}\,\mathrm{Re}\left\{\int_{-\infty}^{\infty}\int_{-\infty}^{\infty} p(x,y,0)v_n^*(x,y)dx\,dy\right\} \tag{C.36}$$

Expressing the complex pressure and velocity as a function of inverse wavenumber transforms (also in consideration of Parseval's theorem), one can express the power as follows:

$$\overline{P} = \frac{1}{8\pi^2}\,\mathrm{Re}\left\{\int_{-\infty}^{\infty}\int_{-\infty}^{\infty} P(k_x,k_y,0)V_n^*(k_x,k_y)dk_x\,dk_y\right\} \tag{C.37}$$

However, from Euler's equation, the wavenumber transform of the pressure at the surface is related to the wavenumber transform of the velocity, and thus substituting equation (C.35) into equation (C.37), one obtains

$$\overline{P} = \frac{\omega\rho_o}{8\pi^2}\,\mathrm{Re}\left\{\int_{-\infty}^{\infty}\int_{-\infty}^{\infty} \frac{|V_n^*(k_x,k_y)|^2}{k_z}\,dk_x\,dk_y\right\} \tag{C.38}$$

Thus, the time-averaged power radiated to the half-space is uniquely related to the wavenumber transform of the normal velocity of the vibrating surface. Recall that

$$k_z = \sqrt{k^2 - k_x^2 - k_y^2}$$

and hence if $k^2 < k_x^2 + k_y^2$, then the denominator of equation (C.38) is complex and the associated real portion of the integral is zero. This region is termed the subsonic region and corresponds to nonpropagating acoustic waves with a wave speed less than that of sound in the fluid medium. Thus, one can restrict the limits of integration to that region for which $k^2 \geq k_x^2 + k_y^2$, termed the supersonic region since the wave speed matches or exceeds that of sound in the fluid medium and corresponds to propagating waves. Each spectral line of the wavenumber transform maps to a sound pressure at a unique angular position in space. Greater details of the wavenumber transform are included in the texts by Fahy (1985) and Fuller et al. (1996).

## BIBLIOGRAPHY

Baumann, W. T., W. R. Saunders, and H. H. Robertshaw, 1991. "Active Suppression of Acoustic Radiation from Impulsively Excited Structures," *Journal of the Acoustical Society of America*, **90**(6), 3202–3208.

Bies, D. A., 1976. "Uses of Anechoic and Reverberant Rooms," *Noise Control Engineering Journal*, **7**, 154–163.

Cunefare, K. A., 1991. "The Minimum Multimodal Radiation Efficiency of Baffled Finite Beams," *Journal of the Acoustical Society of America*, **90**(5), 2521–2529.

Fahy, F., 1985. *Sound and Structural Vibration*, Academic Press, New York.

Fuller, C. R., S. J. Elliott, and P. A. Nelson, 1996. *Active Control of Vibration*, Academic Press, New York.

Lomas, N. S. and S. I. Hayek, 1977. "Vibration and Acoustic Radiation of Elastically Supported Rectangular Plates," *Journal of Sound and Vibration*, **52**(1), 1–25.

Pierce, A. D., 1981. *Acoustics*, McGraw-Hill, New York.

Wallace, C. E., 1972a. "Radiation Resistance of a Baffled Beam," *Journal of the Acoustical Society of America*, **51**(3), 936–945.

Wallace, C. E., 1972b. "Radiation Resistance of a Rectangular Panel," *Journal of the Acoustical Society of America*, **51**(3), 946–952.

# ABOUT THE DISK

## INTRODUCTION

In each chapter of the text, examples are presented with corresponding MATLAB script files, which can be used to re-create many of the figures presented. All of the script files are clearly identified in boldface font within each chapter, and the files can be found on the enclosed disk. The files are arranged within subdirectories named by chapter on the disk. For example, scripts discussed in Chapter 3 can be found in the subdirectory labeled chap3 on the disk.

## SYSTEM REQUIREMENTS

The script files will execute on any computing platform (Windows, Macintosh, or UNIX) configured with MATLAB version 4.2c.1 or higher. However, the user should configure the background of the figures to be white to ensure that all line types are clearly displayed. This is the configuration used by the authors and can be accomplished by typing whitebg at the MATLAB prompt.

## SOFTWARE REQUIREMENTS

All script files were written for execution on MATLAB version 4.2c.1. MATLAB version 5.0 was released near the completion of this text, and thus there was insufficient time to check all of the files under version 5.0. If the user experiences difficulties with any of the files under version 5.0, the user should contact The MathWorks, as backwards compatibility for script files should be guaranteed. Simulink version 1.3c was used to generate a few block diagrams in Chapter 9. However, Simulink is not required to execute any of the script files. A number of toolboxes are required to execute the script files, including:

1. Robust Control Toolbox
2. Mu Analysis and Synthesis Toolbox
3. Control System Toolbox
4. Signal Processing Toolbox

## INSTALLATION INSTRUCTIONS

The script files are compressed on the enclosed floppy disk. The installation program (Windows only) will uncompress and copy the MATLAB script files to your hard drive. To run the installation program, do the following:

1. Insert the floppy disk into the floppy disk drive.
2. Windows 3.1 or Windows NT 3.5.1: Go to Program Manager. From the File pull-down menu, choose Run. Window 95 or Windows NT 4.0: From the Start button, choose Run.
3. Type A:\SETUP (where A is the letter of your floppy disk drive). Hit the Enter key.
4. The setup program screen will appear. The default destination directory is C:\ADAPT. If you wish to change the destination directory, you may do so now. Hit the Continue button.
5. The setup program will copy the MATLAB script files to your hard drive in the C:\ADAPT or user-defined directory.

**NOTE:** If you would like to run the script files on a Macintosh or UNIX system, you need to first uncompress the files by running the installation program on a PC system with Windows. You can then take the uncompressed files and use them in Macintosh or UNIX. If you do not have access to a PC computer to uncompress the files, please contact our technical support number below.

## HOW TO USE THE MATLAB FILES

The script files are saved in separate chapter subdirectories. For example, in order to run a script file in Chapter 4, simply change the MATLAB working directory to the chap4 subdirectory on the drive in which you installed the software. Enter the script filename at the command prompt and press Enter. The file will execute and produce any corresponding figures.

## USER ASSISTANCE

If you need basic assistance with installation or if you have a damaged disk, please call the John Wiley & Sons product support number at (212) 850-6753, weekdays between 9 A.M. and 4 P.M. Eastern Standard Time.

For problems associated with execution under MATLAB version 5.0 or higher, contact The MathWorks, Inc.

To place additional orders or to request information about other Wiley products, please call (800) 225-5945.

# INDEX

For information about the disk, refer to "About the Disk" on pages 459–461.

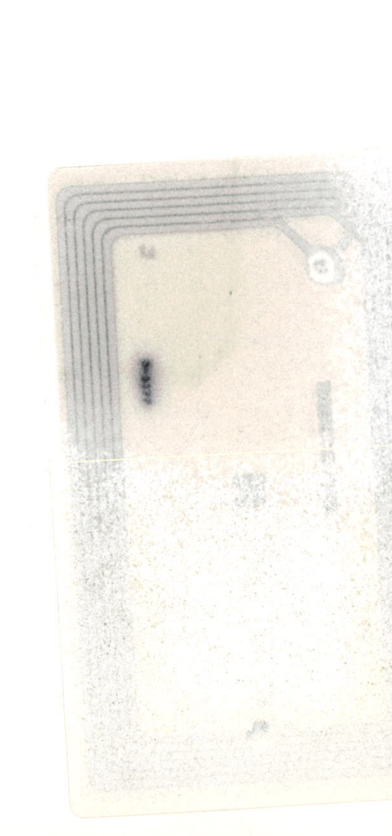